HYDROGELS

DESIGN, SYNTHESIS AND APPLICATION

Kruger Brentt
Publishers

HYDROGELS

DESIGN, SYNTHESIS AND APPLICATION

Richard Sayler
Driss Cherqaoui

Kruger Brentt
Publishers
2025

Kruger Brentt Publishers UK. LTD.
Company Number 9728962
Regd. Office: 68 St Margarets Road, Edgware, Middlesex HA8 9UU

ISBN: 978-1-78715-296-0

For information on all our publications visit our website at http://krugerbrentt.com/

PREFACE

"Hydrogels: Design, Synthesis, and Application" offers an in-depth exploration into the design principles, synthesis methods, and diverse applications of hydrogels, providing insights into the cutting-edge research and practical implications of this versatile class of materials. This book serves as an essential resource for researchers, engineers, scientists, and students interested in understanding and leveraging the unique properties of hydrogels across various disciplines.

Hydrogels are three-dimensional networks of crosslinked hydrophilic polymers capable of absorbing and retaining large amounts of water. Their tunable properties, biocompatibility, and responsiveness to environmental stimuli make them highly attractive for a wide range of applications, including drug delivery, tissue engineering, biosensing, wound healing, and environmental remediation.

As editors of this volume, we have endeavored to compile a diverse array of perspectives, methodologies, and case studies from leading experts and practitioners in the field of hydrogels. Through a blend of theoretical foundations, practical insights, and real-world examples, this book covers a wide range of topics, including hydrogel synthesis techniques, characterization methods, structure-property relationships, and emerging applications.

Our overarching goal in presenting this book is to provide readers with a comprehensive understanding of the design, synthesis, and application of hydrogels. By offering insights into the latest advancements, best practices, and emerging trends in hydrogel research and development, we aim to inspire innovation, collaboration, and interdisciplinary research in this rapidly evolving field.

We extend our sincere gratitude to all the contributors who have generously shared their expertise, experiences, and insights in the field of hydrogels. It is our hope that this book will serve as a valuable reference and guide for researchers, engineers, scientists, and students as they explore the diverse and exciting possibilities of hydrogels in addressing pressing challenges and advancing knowledge across various domains.

Richard Sayler
Driss Cherqaoui

CONTENTS

1 Chapter

ADVANCES IN HYDROGELS

INTRODUCTION: AN OVERVIEW

Hydrogels are flexible materials that are defined by their interconnected networks created by crosslinking. These materials offer a wide range of possible uses since they are highly compatible with living organisms, can be broken down naturally, have a strong affinity for water, and possess mechanical properties similar to those of natural tissue. In recent years, scientists have effectively developed a variety of hydrogels that demonstrate excellent performance and have the ability to fulfill many roles. This has greatly expanded the range and potential applications for their uses. The purpose of this Special Issue is to provide a comprehensive and relevant resource for researchers in related fields by focusing on the most recent breakthroughs in hydrogels. We have collected a total of thirteen primary research publications and three significant reviews from thirteen different countries, specifically Canada, China, Thailand, Mexico, India, Saudi Arabia, Chile, Germany, the Czech Republic, Colombia, Romania, Israel, and the USA.

Hydrogels can be produced through many methods of crosslinking. Photo-crosslinking has attracted considerable attention due to its advantages of mild preparation conditions, as well as convenient and straightforward operation. Liu and colleagues conducted a study where they presented a thorough analysis of various types of photo-crosslinked hydrogel monomers, the methods employed to produce photo-crosslinked hydrogels with different structures, and the prospective applications of these hydrogels in the field of biomedical engineering [1]. Gradilla-Orozco et al. created a hydrogel made of poly(potassium acrylate-co-acrylamide) with a complex fractal-like design using photo-crosslinking [2]. Sabel-Grau and colleagues developed a hydrogel by crosslinking it by a photochemical process, utilizing a molecule known as poly(ethylene glycol)-diacrylate (PEG-DA) [3]. The scientists utilized erythrosin B or eosin Y as a novel photoinitiator, leading to improved multifunctionality of the hydrogels that were formed. The characteristics of hydrogels are affected by different crosslinking mechanisms and the extent of crosslinking. To examine the effect of crosslinking degree on the adhesive characteristics of hydrogels, Li et al. generated hydrogels through various techniques: freeze-thaw cycles for polyvinyl alcohol (PVA) hydrogels, thermal crosslinking for polyacrylamide (PAM) hydrogels, and photo-crosslinking for polyvinyl alcohol-bearing styrylpyridinium group (PVA-SbQ) hydrogels [4]. It was shown that the adhesive properties of these hydrogels decreased as the level of crosslinking increased. However, the adhesion might be improved by maintaining the adhesive functional groups and increasing the flexibility of the polymer chains. Moreover, the development of hydrogels that exhibit both sensitivity and extraordinary qualities in terms of injectability and biodegradability is especially attractive. Steinman and Domb synthesized a copolymer named PEG-PLA-SS-PLA-PEG by chemically bonding poly(ethylene glycol) (PEG) and poly(lactic acid) (PLA) using urethane linkages. This copolymer exhibits thermoresponsive behavior, forming a hydrogel when the temperature exceeds 32 °C. Nevertheless, it exhibits rapid collapse upon reaction with $NaBH_4$ and gradual disintegration through hydrolytic degradation [5]. This hydrogel can be used in drug-delivery systems that have both long-lasting and fast-acting qualities when they come into contact with reducing agents.

CATEGORY OF HYDROGELS

The hydrogels are classified into two categories: natural hydrogels and synthetic hydrogels. Natural hydrogels are frequently obtained from many sources, have a low cost, and demonstrate excellent biocompatibility, biodegradability, and other favorable attributes. As a result, they have attracted considerable attention. Heger et al. investigated the influence of lecithin (L-α-phosphatidylcholine), a naturally occurring phospholipid, on three hydrogels that were crosslinked using different ways (physically crosslinked agarose, ionically crosslinked alginate, and a chemically crosslinked mixture of PVA and chitosan) [6]. The discovery was made that lecithin possesses the capacity to modify the internal composition and physical properties of hydrogels. Múnera-Tangarife and colleagues produced a see-through film by utilizing carboxymethyl cellulose (CMC) obtained from natural sources. The film demonstrated exceptional barrier properties, efficiently preventing the passage of oxygen and lipids. The researchers examined the factors that affect the drying process of CMC utilizing refractance window-conductive hydro-drying (RW-CHD) [7]. Furthermore, Călina et al. utilized e-beam radiation crosslinking to create a variety of superabsorbent hybrid hydrogel compositions by combining natural xanthan gum (XG), carboxymethyl cellulose (CMC), and graphene oxide (GO) [8].With the deepening of research, various multifunctional hydrogels with excellent performance have played an increasingly important role in biomedical fields such as tumor treatment, cell scaffolds, wound repair, biological detection, the food industry and controlled drug release. Wang et al. developed a simple and powerful alginate–Ca^{2+} hydrogel (ACH) to load commercial copper sulfide (CuS) powders for local tumor NIR-II photothermal therapy (PTT) [9]. This hydrogel exhibited a good photothermal capacity and stability. Under 1064 nm NIR-II laser irradiation for 5 min, the temperature enhancement of hydrogels with different concentrations of CuS increased from 17.3 °C to 38.1 °C. These hydrogels had low cytotoxicity toward 4T1 cells, but the toxicity increased in a CuS-concentration- and laser-power-density-dependent manner under radiation. Animal experiments also showed that the large CuS particles in the hydrogel could very efficiently accumulate in tumor tissues and clearly suppress the tumors without causing evident inflammatory lesions or organ damage. Wangsawangrung and co-workers developed a quercetin/hydroxypropyl-β-cyclodextrin (HP-β-CD) inclusion complex-loaded polyvinyl alcohol (PVA) hydrogel that was physically crosslinked through multiple freeze–thaw cycles [10]. The quercetin/HP-β-CD inclusion complex was prepared via the solvent evaporation method, which showed significant antioxidant activity compared with free quercetin. An MTT assay also showed that the viability of mouse fibroblast NCTC 929 clone cells cultured with various concentrations of extracted media from these hydrogels was greater than 70%, indicating that these hydrogels showed low cytotoxicity. All the results illustrate that these hydrogels with the quercetin/HP-β-CD inclusion complex were attractive candidates for wound healing. We also constructed thermosensitive hydrogel scaffolds for cell culture [11]. These hydrogels were formed through the thermally induced gelation of thermosensitive microgels, which were prepared by the radical polymerization of 2-methyl-2-propenoic acid-2-(2-methoxyethoxy) ethyl ester (MEO2MA) and oligoethylene glycol methyl ether methacrylate (OEGMA). The prepared thermosensitive hydrogels could be used for the in situ embedding and three-dimensional (3D) culture of MCF-7 breast cancer cells. The cells grew rapidly in the 3D scaffold and maintained a high proliferative capacity.

Zou and colleagues [12] created fluorescence sensors for commercial applications using molecularly imprinted polymeric hydrogel (MIPG). The MIPG_ZON, which stands for Molecularly Imprinted Polymer

Gel for the detection of zearalenone (ZON), was initially synthesized utilizing 4-vinylpyridine (4-VPY) as the functional monomer and ethylene glycol dimethacrylate (EGDMA) as the crosslinker for the ZON. The MIPG_ZON optical sensor demonstrated exceptional stability and reproducibility. It was capable of detecting ZON in commercial maize juice in a linear manner within the concentration range of 0–10 μM, with a limit of detection (LOD) of 1.6 μM. Additionally, a fluorescent probe based on MIP was created to specifically detect glucuronic acid for the purpose of cell imaging. Casas-Forero et al. created four different types of commercial confectionery hydrogels by incorporating a cryoconcentrated blueberry juice (CBJ) that is abundant in polyphenols, anthocyanins, and flavonoids into gelatin gel (GG), aerated gelatin gel (AGG), gummy (GM), and aerated gummy (AGM), respectively [13]. The hydrogel structures provide protection to the CBJ during in vitro digestion, hence increasing the bioaccessibility of the CBJ. Ahmed and his colleagues have created a nanogel matrix soft confectionery that can be taken orally to supplement vitamin D. This innovation has improved the bioavailability, stability, and patient compliance of the vitamin D supplement [14]. In addition, Sahan et al. emphasized the biomedical uses and potential future developments of biomimetic hydrogels in the examination of cancer mechanobiology [15]. In addition, Almoshari [16] conducted a review on innovative hydrogels that have new formulations for the topical delivery of medicinal substances.

REFERENCES

1. Liu, J.; Su, C.; Chen, Y.; Tian, S.; Lu, C.; Huang, W.; Lv, Q. Current Understanding of the Applications of Photocrosslinked Hydrogels in Biomedical Engineering. Gels 2022, 8, 216.
2. Gradilla-Orozco, J.L.; Hernández-Jiménez, J.; Robles-Vásquez, O.; Cortes-Ortega, J.A.; Renteria-Urquiza, M.; Lomelí-Ramírez, M.G.; Rendón, J.G.T.; Jiménez-Amezcua, R.M.; García-Enriquez, S. Physicomechanical and Morphological Characterization of Multi-Structured Potassium-Acrylate-Based Hydrogels. Gels 2022, 8, 627.
3. Sabel-Grau, T.; Tyushina, A.; Babalik, C.; Lensen, M.C. UV-VIS Curable PEG Hydrogels for Biomedical Applications with Multifunctionality. Gels 2022, 8, 164.
4. Li, Z.; Yu, C.; Kumar, H.; He, X.; Lu, Q.; Bai, H.; Kim, K.; Hu, J. The Effect of Crosslinking Degree of Hydrogels on Hydrogel Adhesion. Gels 2022, 8, 682.
5. Heger, R.; Kadlec, M.; Trudicova, M.; Zinkovska, N.; Hajzler, J.; Pekar, M.; Smilek, J. Novel Hydrogel Material with Tailored Internal Architecture Modified by "Bio" Amphiphilic Components—Design and Analysis by a Physico-Chemical Approach. Gels 2022, 8, 115.
6. Múnera-Tangarife, R.D.; Solarte-Rodríguez, E.; Vélez-Pasos, C.; Ochoa-Martínez, C.I. Factors Affecting the Time and Process of CMC Drying Using Refractance Window or Conductive Hydro-Drying. Gels 2021, 7, 257.
7. Călina, I.; Demeter, M.; Scărișoreanu, A.; Micutz, M. Development of Novel Superabsorbent Hybrid Hydrogels by E-Beam Crosslinking. Gels 2021, 7, 189.
8. Steinman, N.Y.; Domb, A.J. Instantaneous Degelling Thermoresponsive Hydrogel. Gels 2021, 7, 169.
9. Wang, X.; Yang, Z.; Meng, Z.; Sun, S.-K. Transforming Commercial Copper Sulfide into Injectable Hydrogels for Local Photothermal Therapy. Gels 2022, 8, 319.
10. Wangsawangrung, N.; Choipang, C.; Chaiarwut, S.; Ekabutr, P.; Suwantong, O.; Chuysinuan, P.; Techasakul, S.; Supaphol, P. Quercetin/Hydroxypropyl-β-Cyclodextrin Inclusion Complex-Loaded Hydrogels for Accelerated Wound Healing. Gels 2022, 8, 573.

11. Liu, Y.; Luo, Y.-N.; Zhang, P.; Yang, W.-F.; Zhang, C.-Y.; Yin, Y.-L. The Preparation of Novel P(OEGMA-co-MEO2MA) Microgels-Based Thermosensitive Hydrogel and Its Application in Three-Dimensional Cell Scaffold. Gels 2022, 8, 313.
12. Zou, L.; Ding, R.; Li, X.; Miao, H.; Xu, J.; Pan, G. Typical Fluorescent Sensors Exploiting Molecularly Imprinted Hydrogels for Environmentally and Medicinally Important Analytes Detection. Gels 2021, 7, 67.
13. Casas-Forero, N.; Trujillo-Mayol, I.; Zúñiga, R.N.; Petzold, G.; Orellana-Palma, P. Effects of Cryoconcentrated Blueberry Juice as Functional Ingredient for Preparation of Commercial Confectionary Hydrogels. Gels 2022, 8, 217.
14. Ahmed, M.Z.; Gupta, A.; Warsi, M.H.; Ali, A.M.A.; Hasan, N.; Ahmad, F.J.; Zafar, A.; Jain, G.K. Nano Matrix Soft Confectionary for Oral Supplementation of Vitamin D: Stability and Sensory Analysis. Gels 2022, 8, 250.
15. Sahan, A.Z.; Baday, M.; Patel, C.B. Biomimetic Hydrogels in the Study of Cancer Mechanobiology: Overview, Biomedical Applications, and Future Perspectives. Gels 2022, 8, 496.
16. Almoshari, Y. Novel Hydrogels for Topical Applications: An Updated Comprehensive Review Based on Source. Gels 2022, 8, 174.

Chapter 2 CURRENT KNOWLEDGE REGARDING THE USES OF PHOTOCROSSLINKED HYDROGELS IN THE FIELD OF BIOMEDICAL ENGINEERING

1. INTRODUCTION: AN OVERVIEW

Hydrogels are three-dimensional structures formed by linking together polymer chains. They possess exceptional hydrophilicity and have the capacity to absorb substantial quantities of water or tissue fluid. Simultaneously, their volumes have the potential to increase by thousands of times compared to their anhydrous state [1]. The water absorption and water retention properties of hydrogels are strongly influenced by the molecular structure of the crosslinking network, the hydrophilicity of the material, and the degree of crosslinking of the monomers. Insufficient water absorption occurs when the crosslinking network structure is very dense. The greater the hydrophilicity of the monomer structure, the more effective the hydroscopicity. An excessive degree of crosslinking will result in a compact hydrogel structure, hence decreasing water absorption. Hydrogels exhibit a soft and rubbery texture, comparable to live tissue, when their water content falls within a specific range. This characteristic showcases their exceptional ability to interact well with cells and tissues, known as biocompatibility [2]. Thus, hydrogels possess significant utility in the field of biomedical engineering.

Crosslinking is essential in hydrogels to prevent the dissolution of hydrophilic-polymer chains or segments. Hydrogels can be categorized as physical or chemical based on variations in the crosslinking mechanisms of polymers. Physically crosslinked hydrogels typically consist of polymers that create three-dimensional network architectures through various mechanisms such as hydrogen bonding, ionic bonding, hydrophobic bonding, chain entanglement, microcrystal forms, and electrostatic interactions [3]. Although crosslinking is a straightforward and convenient process, it leads to an uneven network structure, weak mechanical strength, and typically reversible crosslinking. Physical gels can be broken down by variations in temperature, pH, or ionic strength in the surrounding conditions. This restricts its utilization in intricate internal settings. On the other hand, chemical crosslinking, which often involves the formation of covalent bonds, is generally irreversible and hence remains stable even when conditions change. Furthermore, hydrogels that are chemically crosslinked typically exhibit superior mechanical stability [4]. Chemically crosslinked hydrogels are frequently used in biomedical applications. Various techniques can be used to chemically crosslink hydrogels, including crosslinking polymerization of complementary functional groups, enzyme-induced polymerization, photo- or heat-induced free-radical polymerization, and high-energy irradiation-crosslinking polymerization [5]. Two recommended techniques of crosslinking are crosslinking polymerization with complementary functional groups and enzyme-induced polymerization. These methods are gentle and do not introduce extra functional molecules [6,7]. However, the range of small molecules and macromolecules that may be used for these two crosslinking methods is somewhat limited in practical applications. As a result, there are only a few hydrogels that can be created using these polymerization processes [8]. On the other hand, light-

or heat-induced radical polymerization is a widely used technique for chemical crosslinking. These methods involve the addition of initiators to a solution containing molecules that are linked together. The initiators cause the molecules to break apart into free radicals when exposed to light or heat. These free radicals then attack the polymer chain together with the crosslinking molecules, starting a chain reaction that leads to the completion of polymerization [9]. Hence, the utilization of photo- or heat-induced free-radical polymerization is particularly prevalent due to its straightforwardness and rapidity. Due to advancements in cross-disciplinary collaboration and scientific study, there has been a rise in the creation and utilization of photocrosslinked hydrogels in the field of biomedical engineering. Hence, it is imperative to provide a concise overview and evaluation of the topic, together with a glance into potential future growth prospects. Initially, we present the constituents of photocrosslinked hydrogels, encompassing both artificial monomers and altered monomers derived from organic substances. Following that, we provide a concise overview of hydrogels with various structures, including films, fibers, microspheres, microneedles, and amorphous (injectable) hydrogels, which are formed using these monomers. Next, we provide a concise overview of the most recent uses of these hydrogels in the field of biomedical engineering. Based on the current state of development, we present our perspectives on the future potential of photocrosslinked hydrogels in the field of biomedical engineering. In conclusion, our intention is for this review to provide researchers with a more comprehensive comprehension of this topic and stimulate additional advancements.

2. CHEMICALLY SYNTHESIZED MOLECULES FOR PREPARATION OF PHOTOCROSSLINKED HYDROGELS

In order to create hydrogels that are photocrosslinked and have excellent biocompatibility, numerous polymerizable small molecules have been produced. Typically, these molecules include unsaturated carbon-carbon bonds, and the polymer network is created through the breaking and adding of bonds using an initiator and light. These molecules can be categorized into four types based on the type of polymerizable group they possess: ethylene, acrylic, acrylamide, and acrylate (Table 1).

Table 1. Four different photocrosslinked molecules and their molecular structures,.

Functional Group Category	Single Structural Formula	Polymer Structural Formula
Vinyl	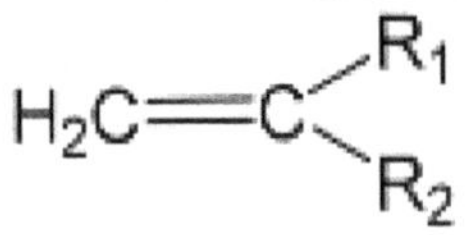	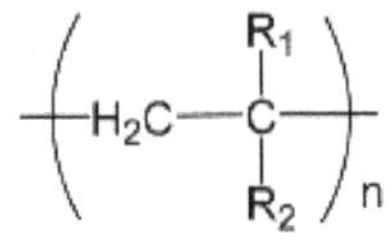
Acrylic class	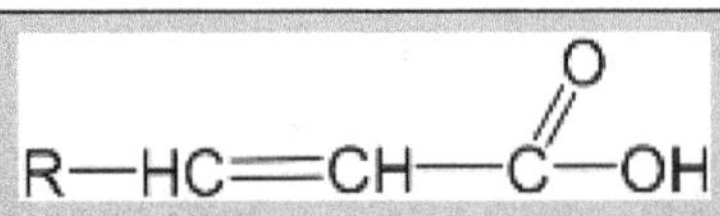	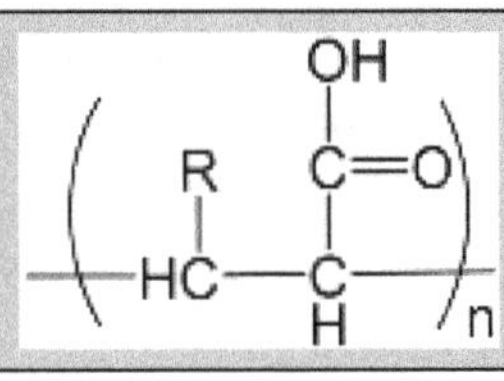
Acrylamide	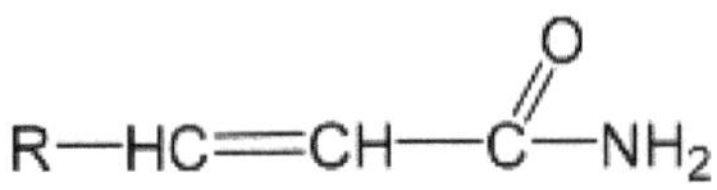	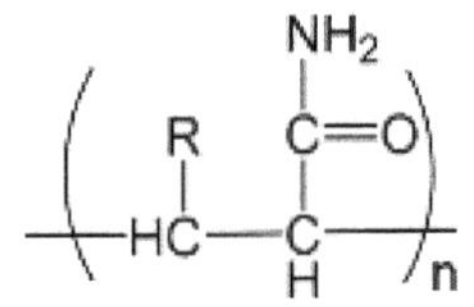
Acrylates	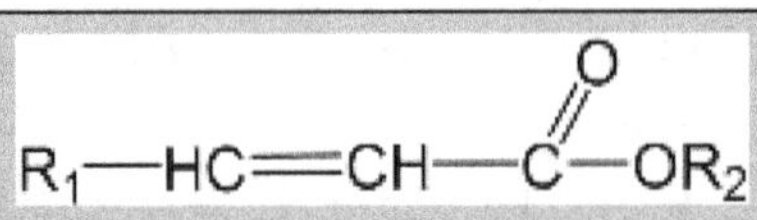	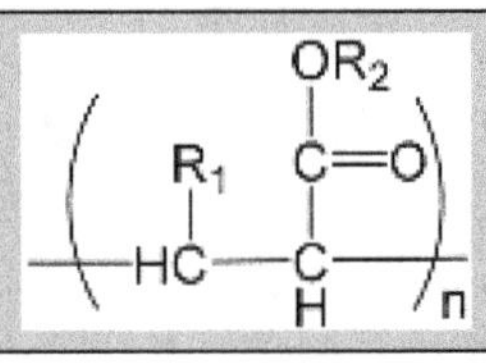

2.1. Ethylenes

Structurally, ethylene is the simplest functional group among polymerizable molecules. Photopolymerization of ethylene molecules was also the earliest method discovered and applied. In early studies, styrene and other monomers were mainly used to synthesize resins due to their poor water solubility, and they were usually polymerized by high-energy radiation or heat [10]. At present, thevinyl molecules used in hydrogel preparations are mainly N-vinyl pyrrolidone (NVP). NVP itself has low viscosity, high reactivity, and limitedskin irritation [11], so it is widely used as a typical ethylene molecule [12]. Kao et al. first reported NVP photocrosslinking with four different comonomers to prepare a series of UV-curable bioadhesives with a high water uptake ranging from 25 to 350 wt% [13]. Subsequently, Fechine et al. studied the effects of crosslinking NVP with other polymerizable molecules such as hydrogen peroxide [14,15]. Lee and Devine et al. provideda detailed discussion on the network structure of copolymerized NVP and polyethylene glycol diacrylate (PEGDA) hydrogels and found that the molecular weight of the main chain was mainly related to the NVP content and was not affected by polymerization time [16,17]. In addition to NVP, a variety of other ethylene molecules have been reported for photocrosslinking polymerizations. For example, Sahiner et al. successfully prepared photocrosslinked bulk polyethylene phosphonic acid (PVPA) by mixing and crosslinking PEGDAs with different molecular weights [18,19]. Ren et al. synthesized a hydrogen-bonded calcium-crosslinked PVDT-PAA hydrogel from 2-vinyl-4,6-diamino-1,3,5-triazine (VDT), acrylic acid (AA), and PEGDA through one-step photopolymerization [20]. However, due to the restriction of water solubility and reactivity, vinyl monomers are still rarely used in the preparation of photocrosslinked hydrogels [21,22].

2.2. Acrylic Acid

Acrylic-acid (AA) molecules have photoactive groups and good water solubility, and due to the free carboxyl structure remaining after polymerization, theycan swell or shrink with environmental pH changes, electric fields, enzyme reactions, and temperature, so that the hydrogel exhibits a tunableresponse. In addition, these free carboxyl groups can interact with other groups under certain conditions, so the hydrogel network can be functionalized by reactions with carboxyl groups or interactions with solutes [23,24,25,26]. Therefore, photocrosslinked hydrogels based on AA molecules are very common and are often used for biological adsorption and environmental purification. For example, Liu et al. successfully synthesized a b-cyclodextrin or polyacrylic-acid nanocomposite (b-CD/PAA/GO) grafted with graphene oxide (GO) based on polyacrylic acid (PAA) through an esterification reaction, and prepared a composite PAA hydrogel, with an adsorption capacity of up to 248 mg/g fordye molecules in wastewater [27]. Hu et al. prepared a new PAA hydrogel by improving and optimizing the crosslinking agent usedin the AA polymerization process, and the adsorption capacity fordye molecules reached a record-breaking 2100 mg/g under neutral conditions [28]. Ma and Kong et al. combined PAA hydrogels with organic montmorillonite or GO, and the resultingcomposite hydrogel showed good adsorption capacitiesfor lead ions (Pb^{2+}) with an adsorption capacity of 223.84 mg/g [29] and cadmium ions (Cd^{2+}) with maximum adsorption capacity up to 316.4 mg/g [30]. In addition, PAA hydrogels can be used in the construction of flexible devices due to their good biocompatibility and flexibility. Based on this, Lu et al. combined PAA with nanocellulose to prepare a flexible hydrogel that can be used for skin sensing [31]. Clearly photocrosslinked hydrogels based on AA derivatives show a wide range of applications.

2.3. Acrylamide

Unlike liquid AA, acrylamide (AAm) molecules are usually solid and exhibit good water solubility. The AAm molecule does not have a free carboxyl group, but instead has a neutral amide bond, so polyacrylamide

(PAAm) hydrogels do not exhibit pH responsiveness but show an equilibrium water content in the range of 94.73–96.26 wt%. Thanks to this, PAAm hydrogels maintain stability in environments with variable solutes [32]. However, the amide bond can also be partially hydrolyzed into a carboxyl group under certain conditions, and the PAAm hydrogel can function as a partial PAA hydrogel. For example, Wang et al. used this principle to prepare enzyme-functionalized microspheres for detection and cleaning of objects [33].

If a hydrophobic isopropyl group is introduced to the other end of the acrylamide molecule, isopropyl acrylamide (NIPAm) is obtained. After photocrosslinking, the resulting polyisopropyl acrylamide (PNIPAm) has a critical phase-transition temperature. When the temperature of the PNIPAm hydrogel is increased above the critical phase-transition temperature, the volume of the hydrogel shrinks significantly, and vice versa. It is worth mentioning that the critical phase-transition temperature of PNIPAm hydrogels is 32 °C and its low critical-dissolution temperature is close to the physiological temperature, it has important application value in biomedical engineering [34]. Zhang et al. prepared PNIPAm-hydrogel microspheres loaded with drugs, which shrank at physiological temperatures and released the loaded drugs for wound repair and disease treatment [35]. When PNIPAm hydrogel is combined with a substance susceptible to photothermal conversions, the new material exhibits photoresponsiveness, which can be used to deter counterfeiting [36,37].

2.4. Acrylates

The reaction of an acrylic derivative (acrylic or methacrylic acid) with the terminal hydroxyl group of another molecule yields an acrylate that can be used for photocrosslinking. For example, polyethylene glycol acrylate (PEGMA), polyethylene glycol diacrylate (PEGDA), polyethylene glycol dimethacrylate (PEGDMA) and other molecules that were obtained after modification at the end of polyethylene glycol (PEG) can be photocrosslinked to form the corresponding polymer network. Compared with acrylic acid and acrylamide hydrogels, acrylate hydrogels are not sensitive to changes in environmental temperature, pH, or other conditions, and are not convenient for functional-group modification; when the surrounding environment changes, this kind of hydrogel maintains good stability. In addition, some acrylate hydrogels, such as PEGDA hydrogels, show good water absorption (about 60 wt%), good biocompatibility, and good adhesion resistance [38]. Therefore, this kind of hydrogel can be used in biomedicine to construct a stable hydrogel skeleton for direct contact with cells or human tissues, biological analyses, and other applications [39]. Hou et al. prepared photocrosslinkedfibers composed of PEGDA with different molecular weights and studied the conversion of acrylate bonds in the hydrogel and the mechanical properties of the fibers in detail. Hou et al. prepared a hydrogel microsphere based on a PEGDA hydrogel and realized the detection of glycoprotein molecules in the solution [40].

3. CHEMICALLY MODIFIED NATURAL MATERIALS

Compared with synthetic materials, natural materials exist widely in nature and have incomparable advantages in sourcing and storage. In addition, natural materials usually have good biocompatibility; they are safer than synthetic materials for use in the biomedical field. At present, hydrogels prepared from natural materials play important roles in cell culture, tissue engineering, and other fields. However, natural materials usually do not have photocrosslinkable groups. In addition to a few complementary functional-group reactions, crosslinking based on natural materials usually relies on physical crosslinking. In practice, the mechanical strength and stability are poor. Therefore, researchers have proposed various strategies for modifying natural materials. The introduction of photocrosslinking groups makes the preparation of natural material hydrogels simpler and more convenient, solves the problems of mechanical strength and poor stability, and further expands their applications. Natural materials used for modification and preparation of crosslinked hydrogels are mainly divided into two categories depending on their composition: polysaccharides and proteins or peptides.

3.1. Polysaccharides

Polysaccharides are carbohydrates with complex and large molecular structures formed by dehydration and condensation of multiple monosaccharide molecules. They are widely distributed in nature and play important roles [41]. For example, peptidoglycan and cellulose are components of the cytoskeletal structure of plants and animals, and starch and glycogen are important energy-storage materials used by animals and plants. The monosaccharides that make up various polysaccharides often have free functional groups; there are many active groups on the polysaccharide chain that can be modified toallow photocrosslinking. Common modified crosslinked hydrogels include alginate, hyaluronic acid, chitosan, heparin, chondroitin sulfate, gellan gum, cyclodextrin, and dextran, among others.

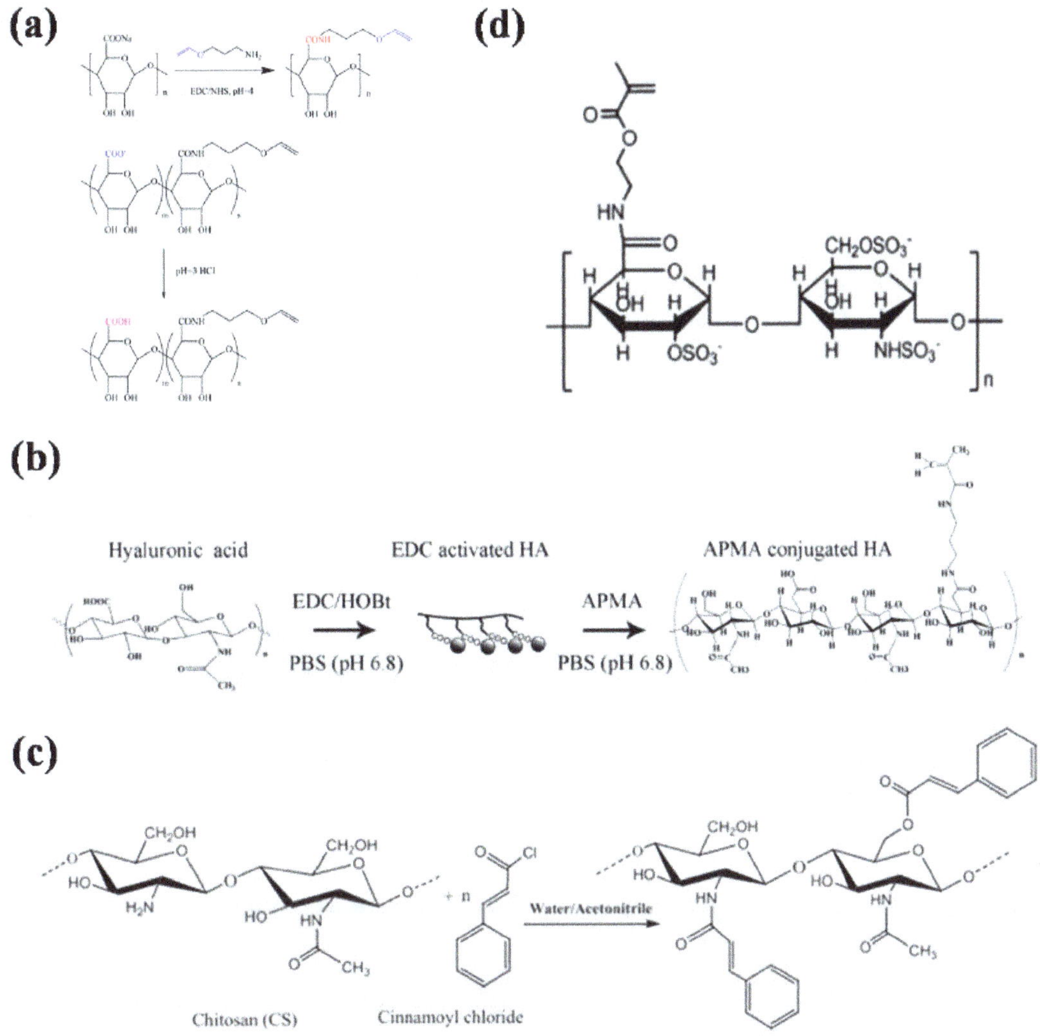

Figure 1. Synthetic route map or structural formula of the photocrosslinked hydrogel. (a) Synthetic route for aminopropyl vinyl ether-modified alginic acid [45]. (b) Synthetic route to N-(3-aminopropyl) methylacrylamide-modified hyaluronic acid [49]. (c) Synthetic roadmap for cinnamoyl chloride chloride-modified [51]. (d) Structural formula of heparin modified by methyl acrylate [52].

Alginate is a natural polysaccharide that can form a physical hydrogel by chelating its independent hydroxyl group with divalent and trivalent metal and heavy-metal ions. It has wide application value for drug delivery, wound repair, and tissue engineering [42,43]. However, physically crosslinked alginate hydrogels have limitations in practical applications due to their uncontrollable gluing speed and instability after gelation. Therefore, many researchers have modified alginate to graft photocrosslinking groups, and realize chemical crosslinking [44]. For example, Xu et al. chemically grafted aminopropyl vinyl ether after activating the carboxyl groups of sodium alginate and obtainedalginic acid functionalized with vinyl ethers (Figure 1a) [45]. Photoinitiators can affect the crosslinking of hydrogels under UV irradiation. Bukhair et al. obtained cinnamoyl-modified photocrosslinkable alginic acid via multistep modificationand demonstrated that these crosslinked alginate hydrogels have better mechanical properties and stabilities, as well as biocompatibility with physically crosslinked alginate hydrogels due to the formation of cyclobutane bridges connecting the alginate polysaccharide chains through the ($2\pi + 2\pi$) cycloaddition reaction of the inserted cinnamoyl moieties [46]. They have great potential in biomedical applications.

Hyaluronic acid is also a linear macromolecular mucopolysaccharide. It has unique viscoelasticity, excellent water retention, biocompatibility, and nonimmunogenicity. It also has important physiological and biological functions and is widely used in clinical procedures [47,48]. However, natural hyaluronic acid has poor stability, is sensitive to hyaluronidase, and lacks mechanical strength. Therefore, chemical modifications are needed to prevent degradation and improve its mechanical strength, and the modified groups mainly include carboxyl, hydroxyl, and the amino group exposed by deacetylation. Lee et al. used Pluronic F127 to modify hyaluronic acid. The resulting polymers exhibited thermosensitive sol–gel transition behaviors over the temperature range of 20–40 °C. After modifying the functional groups of N-(3-dimethylamino propyl)methacrylamide, the HA-F127 polymer was polymerized to form a stable photocrosslinked hydrogel (Figure 1b) [49]. Jenjob et al. prepared microspheres by photocrosslinking bisphosphonates (alendronate) with methyl methacrylate-modified hyaluronic acid. The adsorption efficiency of the microspheres for cationic bone morphogenetic protein 2 (BMP2) reached 91.0% [50].

Chitosan is formed by removing some acetyl groups from the natural polysaccharide chitin. Its chemical structure is similar to that of hyaluronic acid. It is the only cationic polymer in nature [53]. It shows good biodegradability, biocompatibility, bacteriostasis, and other functions. Chitosan is usually insoluble in water and alkali solutions and needs to be dissolved with acid, so chitosan solutions are usually acidic. In addition, the commonly used gelation method for chitosan involves use of aldehyde-containing molecules (such as glutaraldehyde) as crosslinking agents to react with the amino groups of chitosan to form a Schiff base, leading to gelation [34]. In biomedical applications, acidic materials cause irritation to skin or wounds, while residual crosslinking agents have strong biological toxicity. To overcome these shortcomings, Zhou et al. modified chitosan with ethylene groups; the material was dissolved in water and mixed with polyvinyl alcohol modified by methacrylic acid. Photocrosslinked hydrogels were formed after a photoinitiatorwas added [54]. Monier et al. modified chitosan molecules with cinnamyl chloride. The crosslinked chitosan hydrogel showed good stability (Figure 1c) [51].

Heparin is a mucopolysaccharide sulfate composed of glucosamine, L-ieduraldehyde glycoside, N-acetylglucosamine, and D-glucuronic acid. It is strongly acidic. It is a natural anticoagulant in animals. Heparin specifically binds to growth factors, so it has important value in growth-factor-delivery systems [55,56]. Based on this, Yoon et al. modified heparin with N-methylacrylamide hydrochloride and then crosslinked it with cryloyl-modified Pluronic F127 to obtain composite hydrogels for the controlled release of growth factors [57]. Jeong et al. successfully prepared electrospun fiber scaffolds by mixing methyl acrylate-modified alginate and methyl acrylate-modified heparin with polyoxyethylene (PEO) through electrospinning and photocrosslinking, and found that they were effective inregulating cell behavior (Figure 1d) [52].

Chondroitin sulfate is a glycosaminoglycan covalently linked to proteins to form proteoglycans. It is widely distributed in the extracellular matrix and cell surfaces of animal tissues. The sugar chain is polymerized by alternating glucuronic acid and N-acetylgalactosamine disaccharides. It relieves pain and promotes cartilage regeneration [58]. After chemical modification, a chondroitin sulfate hydrogel can be obtained through photocrosslinking. The hydrogel has good biocompatibility and biodegradability, and has broad application prospects in drug delivery and bone repair [59,60]. For example, Kim and others prepared methacrylated PEGDA or chondroitin sulfate hydrogel and used it as a biomineralized three-dimensional scaffold for binding and deposition of charged ions [61]. Ornell et al. used the photocrosslinked methacryl group to covalently modify chondroitin sulfate and form injectable hydrogels, which can be used for sustained release of drugs over a certain period of time (Figure 2a) [62].

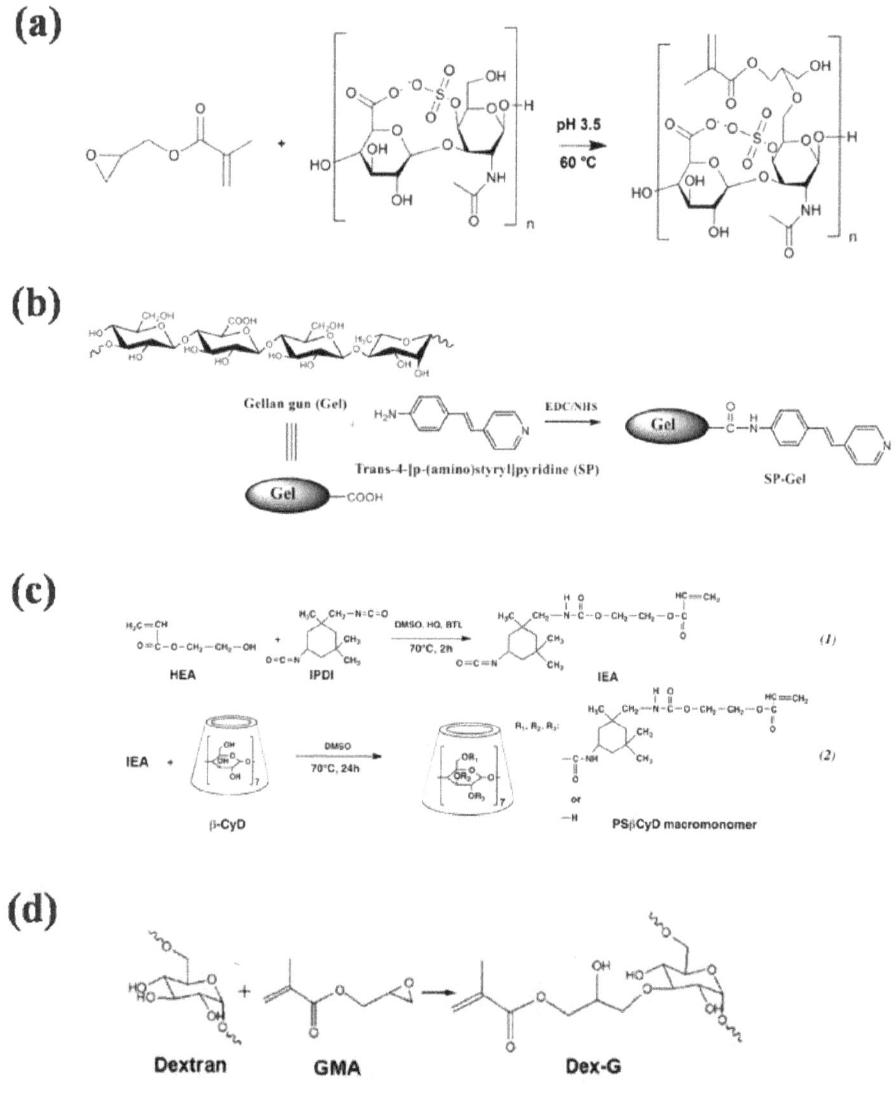

Figure 2. Synthetic routes for some crosslinked hydrogels. (a) Route map for the synthesis of chondroitin methacryloyl sulfate [62]. (b) Synthetic route for gellan gum modified by trans-4-aminophenyl pyridine [66]. (c) Synthetic route to cyclodextrin modified by isophorone diisocyanate and 2-hydroxyethyl acrylate [72]. (d) Synthetic route to glycidyl methacrylate modified dextran [73].

Gellan gum is a linear polysaccharide composed of glucose, glucuronic acid, and rhamnose. It is heat-resistant, acid-resistant, and enzyme-resistant and has good chemical stability. Gellan gum is insoluble in nonpolar organic solvents and cold water, but can be dissolved in hot water to form a transparent solution. After cooling, it becomes a transparent and solid gel [63,64,65]. Because of its good biocompatibility and tunability, researchers have tried to use gellan glue for tissue engineering. However, the gel formed by cations is hard and brittle, which limits its application. For this reason, a variety of photocrosslinking groups were used to modify gellan gum and obtain hydrogels with good biocompatibility and mechanical properties. For example, Oliveriraet al. modified gellan gum with trans-4-aminophenyl pyridine, and the product can be used for catalase immobilization after photocrosslinking (Figure 2b) [66]. Mano et al. modified gellan gum with methacrylic acid to prepare injectable gellan gum, which can be used as a self-generated osteogenic material [67].

Cyclodextrins are a series of cyclic oligosaccharides produced with amylose under the action of cyclodextrin glucosyltransferase. The three common cyclodextrins contain 6, 7, and 8 glucose units and arecalled α-, β-, and γ-cyclodextrins, respectively. These molecules are particularly attractive because they can form inclusion complexes with hydrophobic guests exhibiting appropriate molecular sizes and have high versatility. In addition, they can be chemically modified by hydroxyl substitution [68,69]. For example, Cosola et al. prepared a cyclodextrin modified with polyacrylate, which can prepare photocrosslinked materials with different morphologies via 3D technology [70]. Yamasaki et al. modified cyclodextrin with isophorone diisocyanate and 2-hydroxyethyl acrylate to obtain photocrosslinked cyclodextrin [71]. Microspheres of the prepared photocrosslinked polymer showed good separation efficiency for phenol (Figure 2c) [72].

Glucan is a homopolysaccharide formed with glycosidic linkages between glucose. Based on the types of glycosidic bonds, it can be divided into α-glucan and β-glucan. The common glucan is dextran, a common α-glucan. It has good biocompatibility and is widely used in the biomedical field [74]. Dextran is rich in active hydroxyl functional groups, so it can be chemically modified. Casadei et al. modified it with methacrylate to obtain photocrosslinked dextran, which enablesthe controllable release of polymers [75]. Yin et al. used glycidyl methacrylate to modify dextran and crosslinked it with methacrylic acid ethylene glycol-modified concanavalin A and polyethylene glycol dimethacrylateto obtain hydrogels (Figure 2d) [73]. The hydrogelsshowed good biocompatibility and glucose responsiveness and are expected to be used in glucose biosensors and intelligent insulin delivery.

3.2. Proteins/Peptides

Polypeptides are compounds composed of α-amino acids connected by peptide bonds. They are also intermediate products in protein hydrolysis. Amino acids form polypeptides via dehydration condensation. After winding and folding, polypeptides can form macromolecules with certain spatial structures, namely proteins. Proteins and peptides play important roles in the growth and development of life. Because of their widespread sources, nontoxic degradation products, and even benefits to the human body, proteins and polypeptide products are widely used in the biomedical field. Because the amino acids that make up proteins and peptides have free side-chain groups, they can be endowed with new functions through chemical modification, such as photocrosslinking groups. Common proteins used for modification with photocrosslinking groups include collagen, gelatin, keratin, silk fibroin, and albumin. Their application scope is greatly expanded through chemical modification, which gives them wide influence in the biomedical field.

Collagen is a protein formed by three peptide super-chain helices; it accounts for 25–30% of all proteins in mammals and is the most abundant protein [76,77]. Common collagens include type I, type II, type III, type V, and type XI. Collagen is widely used in foods, medicines, tissue engineering, cosmetics, and other fields because of its good biocompatibility, biodegradability, and biological activity [78]. Photocrosslinking of collagen hydrogel can be realized with chemical modifications and photocrosslinking. The degradation rate after the collapse of collagen hydrogels is slow, and they can be maintained for a long time to achieve specific functions. For example, Yang et al. photocrosslinked methacrylated type II collagen and used it for encapsulation of bone-marrow mesenchymal stem cells (BMSCs) (Figure 3a). The crosslinked collagen maintained its triple-helix structure, which provides a good microenvironment for proliferation and differentiationof BMSCs [79].

Gelatineis formed by partial hydrolysis of collagen, and it can be dissolved in hot water and forms a gel after cooling. This excellent feature provides flexible application scenarios. However, in most cases, a more stable hydrogel structure is desired, and the temperature range for gelatine gels is relatively narrow, which makes it difficult to meet the demand. Therefore, researchers have modified gelatinewith methacryloyl to enable photocrosslinking. Methacryloylated gelatine (GelMA) has good biocompatibility and structural stability after photocrosslinking. Therefore, methacryloylated gelatine shows great application value in the biomedical field [84]. The Khademhosseini team used gelatineas a raw material to synthesize GelMA, and used it in cell culture and microchips to verify its application value in cell-responsive microengineered hydrogels (Figure 3b) [80]. Fu et al. prepared structural colored microspheres with GelMA and used them for cell culture. They found that they had good biocompatibility and are expected to be used to construct liver chips [85].

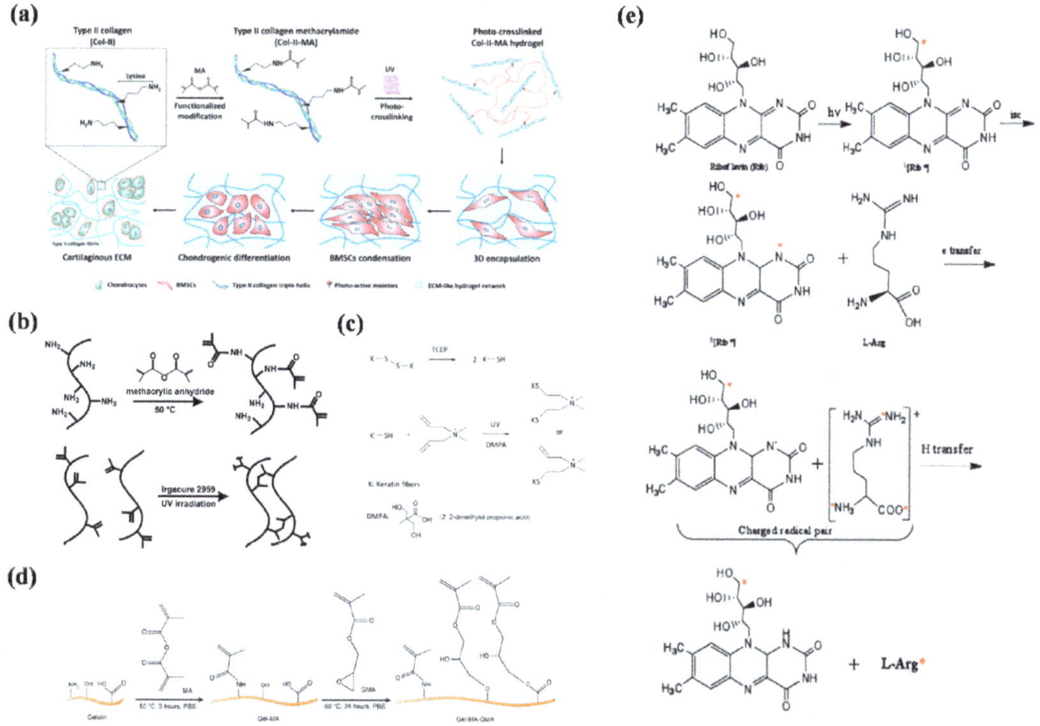

Figure 3. Part of the synthetic route maps or related schematic diagrams for crosslinked hydrogels. (a) Chemical modification of type II collagen and photocrosslinking hydrogel for bone-marrow mesenchymal stem-cell culture [79]. (b) Synthetic route of methacryloylated gelatin and its gelling diagram [80]. (c) Synthesis roadmap of two-step modified silk fibroin [81]. (d) Schematic diagram for crosslinking between keratin and 2,2-dimethoxy-2-phenylacetophenone [82]. (e) Schematic diagram of free-radical formation in the rib/L-Arg system [83].

Keratin is the main protein constituting hair, horn, claws, and the outer layers of animal skin. Because keratin contains cystine, it has a high proportion of disulfide bonds and plays acrosslinking role in protein peptide chains. Therefore, keratin has particularly stable chemical properties and high mechanical strength [86,87]. Studies have shown that keratin extracted from human hair fibers contains a leucine–aspartate–valine (LDV) cell-adhesion motif. Therefore, keratin has great application potential for cell culture and tissue engineering. To provide more application scenarios for keratin, it can be modified. For example, Yu and Hu et al. successfully realized photocrosslinking of keratin by using click chemistry and introducing 2,2-dimethoxy-2-phenylacetophenone (Figure 3c) [81,88].

Silk fibroin is a natural high-molecular fibrin extracted from silk. It has good mechanical and physicochemical properties and a long application history. To improve plasticity, many researchers have tried a variety of methods to modify it chemically (Figure 3d) [82]. Qi and others used silk fibroin modified by methacryl groups to obtain an injectable silk-fibroin hydrogel after photocrosslinking [89]. However, the lower side chains of silk fibroin decreased the content of methacryl groups, so the photocrosslinked silk-fibroin hydrogel had lower mechanical strength. Therefore, it was necessary to modify other active groups on silk fibroin to improve the mechanical strength. For example, Ju et al. obtained silk-fibroin hydrogels with good biocompatibility and mechanical properties by modifying the silk-fibroin hydroxyl with methacrylic acid and then performing crosslinking [90].

Albumin is a protein in plasma that maintains body nutrition and osmotic pressure, and it accounts for approximately 50% of all plasma proteins. It has good biocompatibility and solubility. Therefore, when preparing biomedical materials with albumin as a raw material, the inherent defects of synthetic materials can be avoided. For example, Chiriac and collaborators used riboflavin and arginine as natural initiators to crosslink bovine serum albumin (BSA) and obtain BSA hydrogels (Figure 3e) [83]. In vitro and in vivo experiments showed that the hydrogel had good biocompatibility.

4. PREPARATION OF PHOTOCROSSLINKED HYDROGELS WITH DIFFERENT MORPHOLOGIES

In biomedical applications, different applications have different morphological requirements for common hydrogels. Therefore, it is crucial to prepare a hydrogel with a specific morphology. Common photocrosslinked hydrogel morphologies mainly include fibers, microspheres, thin films, microneedles, amorphous shapes, and so on. At present, many researchers have effectively explored processing methods for hydrogels with different morphologies, which has greatly expanded the application potential of hydrogels in biomedicine.

4.1. Fiber

Fibrous products such as gauze are widely used in the life science and biomedical fields. Inspired by this, many researchers have developed new hydrogel-fiber products. Due to their high surface-to-volume ratios and high porosities, they exhibit good water absorbance and air permeability and can replace some functions, such as those of traditional gauze in the biomedical field [91,92]. Usually, a hydrogel precursor solution is converted into fibers, photocrosslinked, and finally accumulated in woven-fiber products. At present, the main method used for fiber preparation is electrospinning [93,94,95]. When the solution is squeezed out under a high-pressure electric field, the liquid becomes filamentous due to the electric field. After the solvent evaporates, the polymer, polymer mixture, composite materials, and other molecules form fiber shapes. Electrospun fibers have many remarkable properties, such as high surface-to-volume ratios and functional tunability; they can be applicable in many areas including drug delivery, wound dressings, tissue engineering, membranes or filters, electronics, sensors, and energy [92]. Furthermore, when photocrosslinking technology is applied, the mechanical performance and stability of the fibercan be improved further [96,97]. For example, Tang et al. developed a novel high-throughput nanofiber-composite ultrafiltration membrane (Figure 4a) [98]. They first prepared chemically crosslinked polyvinyl-alcohol (PVA) nanofiberscaffolds on a nonwoven substrate, and then prepared a polyvinyl-alcohol (UV-PVA) barrier layer via UV crosslinking. The results showed that the 5 wt% UV-PVA solution coating provided an ultrafiltration membrane with high throughput and high retention rate after UV curing for 20 s. It had good pollution resistance and could be used for the separation of oil and water emulsions.

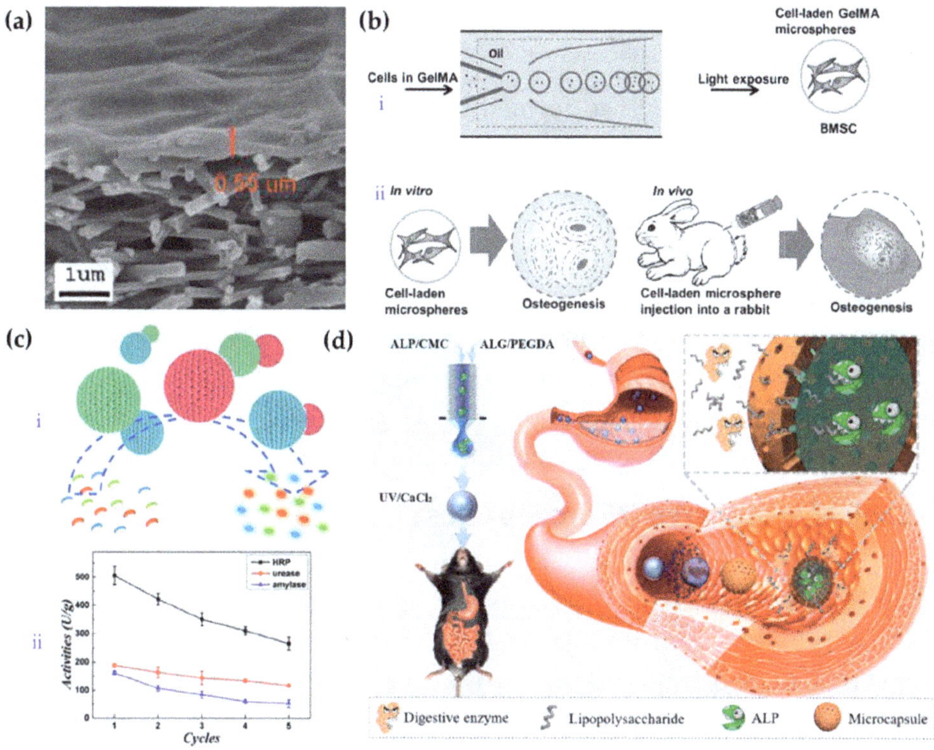

Figure 4. Schematic diagram of the preparation and application of hydrogels with fiber and microspherical morphologies. (a) Electron micrograph of ahigh-flux nanofiber-composite ultrafiltration membrane [98]. (b) Schematic diagram of the preparation and application of photocrosslinked GelMA-hydrogel microsphere [99]. (c) Enzyme-functionalized antiproteolytic-hydrogel microspheres were used for biocatalysis [100]. (d) Schematic diagram for preparation and application of ALP microcapsules [101].

Microspherical hydrogels have excellent mobility and mass delivery, and thus have received considerable attention in fields such as biomedical detection and drug delivery. Specific structural differences can be used to subdivide the micropellets into homogeneous micropellets, antiopal micropellets [32], and nuclear-shell micropellets (microcapsules) [101].

4.2. Microballoons

4.2.1. Homogeneous Micropellets

Homogenous hydrogel microspheres are usually formed by photocrosslinking after the hydrogel precursor solution is dispersed into liquid droplets. There are many ways to disperse solutions into droplets, but the principle is basically consistent; all methods use emulsification to generate droplets, such as with stirring, microfluidics, and electrospraying. Among these techniques, microfluidic technology has obvious advantages in preparing microspheres due to its accurate fluid control [99,102]. For example, Zhao et al. obtained hydrogel microspheres loaded with cells and growth factors by mixing cells and growth factors with GelMA solution, using microfluidics to cut them into droplets, and then photocrosslinking them. BMSCs in microspheres showed significant osteogenic effects both in vitro and in vivo, and significantly increased mineralization, thus promoting bone regeneration (Figure 4b) [99]. Combining microfluidic technology with electrojet technology allowed the application of an electric field as an additional shear force to simplify the microfluidic device. For example, Zhang et al. successfully used microfluidic electroinjection technology to prepare photocrosslinked chondroitin-sulfate microspheres, which showed good value for drug loading and wound repair [103].

4.2.2. Antiopal Micropellets

Opal has a periodic, ordered structure formed by nanoparticles. When the nanoparticles have a specific size, the opal reflects light of a specific wavelength and thus shows a structural color. Negative replication using opal as a template yields an antiopal material. This material has the same structural color characteristics and a continuous porous structure, which provide good prospects for applications in mass transfer (especially in the fields of drug loading and sensing) [100]. For example, Wang and others prepared inverse-opal hydrogel microspheres by using colloidal-crystal microspheres assembled with silicon-dioxide nanoparticles as templates and AAm and N,N′-methylenebis-(acrylamide) (Bis) as photocrosslinked hydrogel skeletons to copy the template microspheres [32]. Afterhydrolysis and enzymatic immobilization of the microspheres, the resulting material had biocatalytic functionality and is expected to be used for treatment of complex water bodies (Figure 4c).

4.2.3. Nuclear-Shell Microspheres (Microcapsules)

Antiopal hydrogel microspheres were obtained by using corrosion for removal of nanoparticle templates. If the corrosion was incomplete, the prepared microspheres had a nuclear-shell structure consisting of a hydrogel or nanoparticle-composite core and an antiopal hydrogel shell. These nuclear-shell microspheres exhibit responsiveness similar to that of antiopal hydrogel microspheres, and they also have high stability and can be used for encoding [104]. Based on this, Xu et al. developed photocrosslinked nuclear-shell hydrogel microspheres that can be used for miRNA detection by modifying the corresponding aptamer [105].

Homogeneous hydrogel microspheres are generally prepared by single-emulsion devices. Core-shell hydrogel microcapsules can be prepared if a double-emulsion device is used and the inner solute is not crosslinked. For example, Zhao et al. developed microcapsules containing an alkaline phosphatase (ALP) solution by applying microfluidic electrojet technology. The shells of the microcapsule were formed by calcium alginate and photocrosslinked PEGDA, which can preserve the activity of ALP in the digestive tract and thus be used for intestinal endotoxin cleaning (Figure 4d) [101].

4.3. Thin Film

Among the many morphologies of photocrosslinked hydrogels, hydrogels with thin-film morphologies are most widely used because of their excellent size ranges and mechanical strength. Thin films are usually made of homogeneous structures, and by adding functional materials, hydrogel functionality can be enhanced in various ways. For example, incorporation of mesoporous active nanoparticles or a vascular endothelial growth factor can allow the hydrogel membrane to play an obvious promoting role in bone and wound healing [106,107]. Using the opal structure as the template, a hydrogel thin film with the antiopal structure was prepared. The nanoporosityand structure color of the antiopal structure enabled enhanced drug loading, sensing, and driving functions for hydrogels, withgood application prospects [37,108]. For example, Zhang et al. constructed photoresponsive antiopal hydrogel films that can be driven to capture and release objects under the action of light (Figure 5a) [109]. In some cases, two sides of the hydrogel film can be made to have different properties and functions by using different functionalization treatments to obtain a Janus structure thin film. Such films show good results when applied in complex environments. Zhang et al. prepared a Janus sponge dressing that was hydrophilic on one side and hydrophobic on the other. When the dressing was applied to the wound, the exudate was discharged from the wound to maintain the microenvironment of the wound, thus accelerating wound repair (Figure 5b) [110].

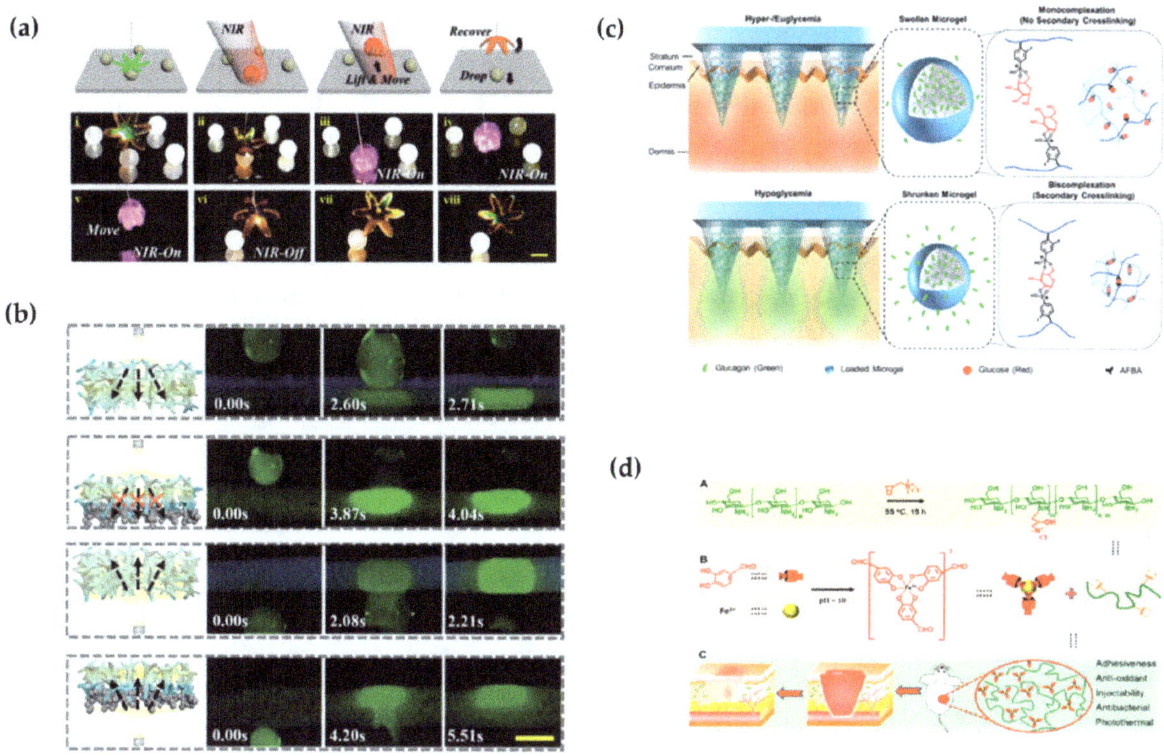

Figure 5. Schematic diagram for application of thin films, microneedles, and injectable morphologies of hydrogels. (a) A light-responsive antiopal hydrogel thin film captures and releases objects under light [109]. (b) One-way liquid discharge capacity of Janus sponge dressings [110]. (c) Schematic representation of the intelligent release of glucagon drugs contained by microneedles in vivo [111]. (d) Preparation principle for an injectable hydrogel with a double dynamic covalent bond and schematic diagram of its application in wound healing [112].

4.4. Microacupuncture Needle

Microneedles are usually arrays of microneedles, typically less than 1mm in length. Due to their size, they only penetrate the surface of the skin without causing bleeding; they cause much less pain than a traditional needle, and the trauma can heal in hours. Therefore, microneedles have obvious advantages in transdermal administration [113,114,115]. Compared with traditional metal- and silicon-based microneedles, hydrogel microneedles have good biocompatibility, large drug loads, and controllable drug release, and they have great development prospects. For example, Ghavaminejad et al. prepared microgel-loaded glucagon using phenylboric acid as a functional group, and then mixed the microgel with methacrylated hyaluronic acid (MeHA) and added it to the template [111]. After photocrosslinking, a photocrosslinked hydrogel microneedle was obtained. This microneedle can release glucagon to raise blood glucose under hypoglycemic conditions, thus preventing hypoglycemia after insulin injections for patients with diabetes (Figure 5c). The mechanical properties of the microneedles can also be controlled by adjusting the morphology. For example, Yu et al. designedan amifostine-loaded armored microneedle AAMN to havestronger mechanical properties than traditional conical microneedles, much higher mechanical strength than conical structuresand high skin permeability, and they can be better used for transdermal administration of amifostine [115].

4.5. Amorphous (Injectable) Hydrogels

Certain hydrogels have inadequate mechanical characteristics as a result of a restricted amount of high-molecular content or insufficient crosslinking between big molecules. Nevertheless, despite the fact that these hydrogels do not retain their precise intricate formations and visual characteristics, they can nevertheless be employed to envelop uneven wounds and can be administered through injection into unevenly shaped areas of tissue loss, hence offering distinct benefits in the field of medical trauma healing. Thus, these hydrogels possess significant utility in the field of biomedical engineering. Liang et al. chemically altered chitosan by introducing quaternary ammonium groups and combined it with trivalent iron to create an injectable hydrogel. The hydrogel also contained protocatechuic aldehyde, which is a compound with both catechol and aldehyde functional groups. It has the potential to facilitate the healing of wounds and infections caused by methicillin-resistant Staphylococcus aureus (Figure 5d) [112]. Furthermore, injectable hydrogels can be supplemented with various functional elements to fulfill specific requirements. As an illustration, Zhao et al. effectively created injectable hydrogels for diabetes treatment by combining drug-loaded hydrogel microspheres with injectable hydrogels [116].

5. BIOMEDICAL APPLICATIONS

Photocrosslinked hydrogels are widely used in biomedicine, such as for biosensors, flexible wearable devices, medicine or tissue engineering, cell microcarriers, organ chips, and so on. It should be noted that when constructing hydrogels, appropriate components should be selected according to the application scenarios. For example, when a hydrogel is used for sensing, it should exhibit antiadhesion properties. When applied directly to humans, the hydrogel components should be biocompatible and preferably adaptable.

5.1. Biomedical Sensor

Hydrogels often consist of diverse functional groups that can interact with molecules or ions in the surrounding environment, leading to alterations in the physical and chemical characteristics of the hydrogels. Through the identification of these alterations, the relevant molecules or ions can be examined and perceived. Qin et al. fabricated hydrogel microspheres with a Janus structure that exhibited pH-responsive volume fluctuations. Consequently, these volume changes resulted in variations in the intensities of surface-enhanced Raman scattering (SERS) signals and fluorescence signal. The pH of the solution might be calculated by collecting and analyzing these two signals, as shown in Figure 6a [117].

When a hydrogel has an ordered micro/nanostructure, it can show structural color, which will change with volume changes of the hydrogel. Therefore, structural color can be used to detect and analyzethe target molecule. The Sun research group has performed a series of studies in this area. They used microspheres or membranes self-assembled with colloidal nanoparticles of silicon wafers as templates, used hydrogels for repeated preparation, and prepared a series of structurally colored hydrogel microspheres and films with different functions. These microspheres showed good sensing ability for detection of tumor markers, glycoproteins, DNA, and heavy-metal ions (Figure 6b) [118].

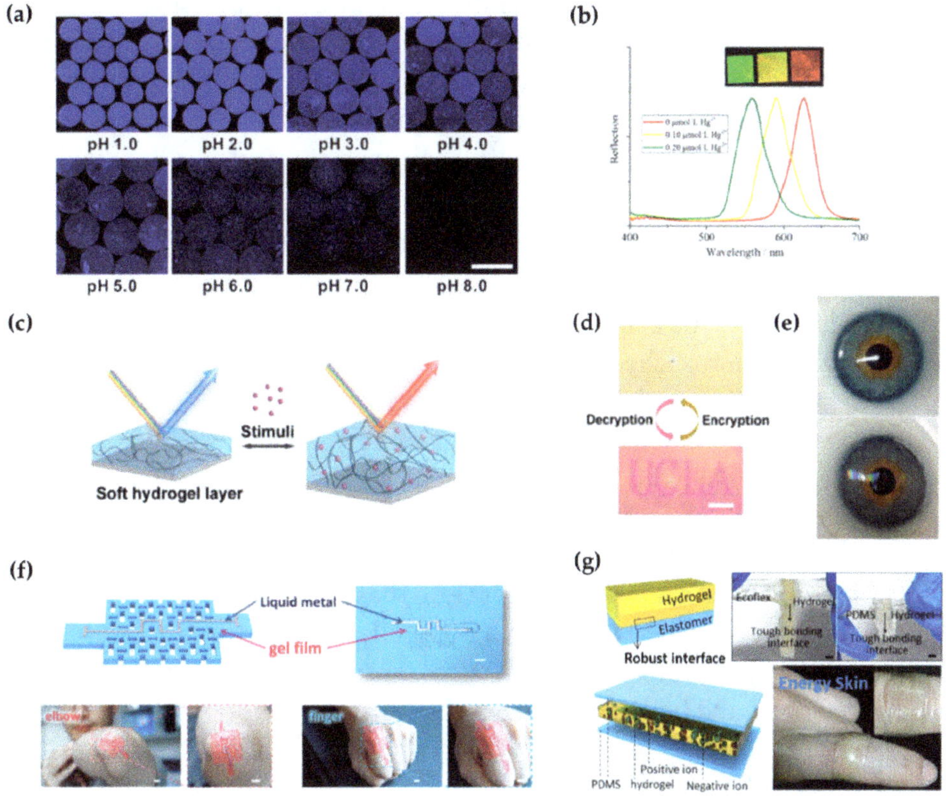

Figure 6. Applications of photocrosslinked hydrogels in biomedical sensors and flexible wearable devices. (a) The volume and fluorescence intensity of pH-responsive hydrogel microspheres changed under different pH conditions [117]. (b) The color of a mercury-ion-responsive hydrogels with structural color in mercury-ion solutions with different concentrations and reflection spectra of responses [118]. (c,d) Schematic diagram of hydrogel interferometer response to the target and demonstration of information encryption [119]. (e) Commercial contact lenses (top) and biosensing contact lenses (bottom) [120]. (f) Origami hydrogel for motion-signal sensing [121]. (g) Self-powered triboelectric nanogenerator that collects motion energy [122].

In addition to construction with micro/nanostructures, optical hydrogels can be directly prepared by using the principle of thin-film interference and the swelling characteristics of hydrogels. For example, Qin's team proposed a hydrogel interferometercolor-change system that was simple to prepare, quick to respond, and easily patterned, and demonstrated various applications in detection and information encryption [119]. This kind of hydrogel does not need a fine micro/nanoconstruction unit, but is used to fabricate a film by rotating the coating on a highly reflective substrate with grafted functional groups. According to the principle of film interference, the hydrogel-film thickness is controlled to cause instant discoloration (Figure 6c). To achieve a portable detection device, researchers have also developed the necessary mobile phone software [119]. When an unknown sample must be analyzed, one takes a picture with the mobile phone, and the software automatically analyzes the subject and provides the test results. Reversible encryption and decryption can be realized according to whether the film is damp or not (Figure 6d) [119].

5.2. Flexible Wearable Devices

The concept of hydrogels was first proposed by Wichterle et al. in 1960. They constructed a three-dimensional hydrogel network comprising hydroxyethyl methacrylate (HEMA) containing a small amount of the crosslinking agent EGDMA to overcome the poor biocompatibility and stability of plastic products at that time. This hydrogel has a soft texture, good light permeability, adjustable mechanical properties and water content, and a certain degree of biological inertia. Based on these excellent properties, Wichterle used it forthe preparation of contact lenses [123]. Subsequently, many researchers made improvements on this material and developed cosmetic pupils with their own structural colors, which can be used for detection of substances in tear drops (Figure 6e) [120,124,125,126].

In addition to contact lenses, many flexible hydrogels are used in combination with sensing elements to produce flexible electronic devices that convert and conduct signals [127,128]. For example, Yu et al. developed a photocrosslinked acrylic hydrogel film with controllable thickness and excellent mechanical properties. During polymerization of the hydrogels, Zr4+ added to the solution coordinated with some of the carboxyl groups in polyacrylic acid, and the resulting hydrogels exhibited high stability. Akirigami structure for the hydrogel was obtained by photolithographic polymerization, so the hydrogel exhibited elevated ductility and flexibility to wrap the curved surface. After combining this origami hydrogel with a liquid metal, the resulting hydrogel sensor was used to sense arm or finger-bending motions (Figure 6f) [121]. To solve the need for a power supply for flexible electronic devices, Wang's research group developed a stretchable triboelectricnanogenerator (TENG) based on elastomer hydrogels. The tensioning, transparent, ultrathin single-electrode TENG with a double-layer structure fit firmly to human skin and deformed as the human body moved. TENGs can also capture energy during deformation processes (pressing, stretching, bending, and twisting) to drive electronic devices (Figure 6g) [122].

5.3. Drug Delivery and Tissue Engineering

The traditional methods of drug administration such as injection and oral administration need to be given frequently, and will cause the problem of high drug concentration in a short time, resulting in side effects. Therefore, sustainable and low-dose drug delivery has important application prospects. The hydrogel material has a three-dimensional network structure that "locks" the drug in a grid, allowing it to be released slowly. As a result, hydrogels could be used for drug delivery, which in turn could play a role in tissue engineering. For example, Lei et al. prepared a photocrosslinked methacrylated hyaluronic acid (HAMA) microsphere loaded with vascular endothelial growth factor (VEGF) by a microfluidic electrospray technique and used in the treatment of thin endometrium. The combination of VEGF and HAMA can promote endometrial regeneration and embryo implantation, while hyaluronic-acid hydrogel scaffolds can work with VEGF to promote endometrial hyperplasia after degradation. Therefore, this drug-loaded hydrogel has a good application prospect (Figure 7a) [129,130,131].

By selecting suitable responsive hydrogels or changing the structure of nanopores, hydrogels can be designed to have tunable drug-delivery effects. The stimuli used to trigger the hydrogel's response are usually light, heat, molecules, or ions. For example, Zhao et al. developed a glucose-responsive injectable hydrogel. Injected into the body, this hydrogel can maintain quasi-homeostasis in the normal range of blood glucose levels for approximately two weeks [116]. Salahuddin et al. prepared a light-controlled hydrogel. When exposed to light, the mesh of the hydrogel changes, altering the diffusion rate of the molecules being convenient for the controlled release of the drug [134]. Yang et al. prepared an antiopal microsphere and used its nanoporous structure to load drugs injected into the joint cavity. Elevated local temperatures during exercise or arthritis can promote the release of drugs, and vice versa. Therefore, this ingenious drug-delivery system could play an important role in the treatment of osteoarthritis (Figure 7b) [132].

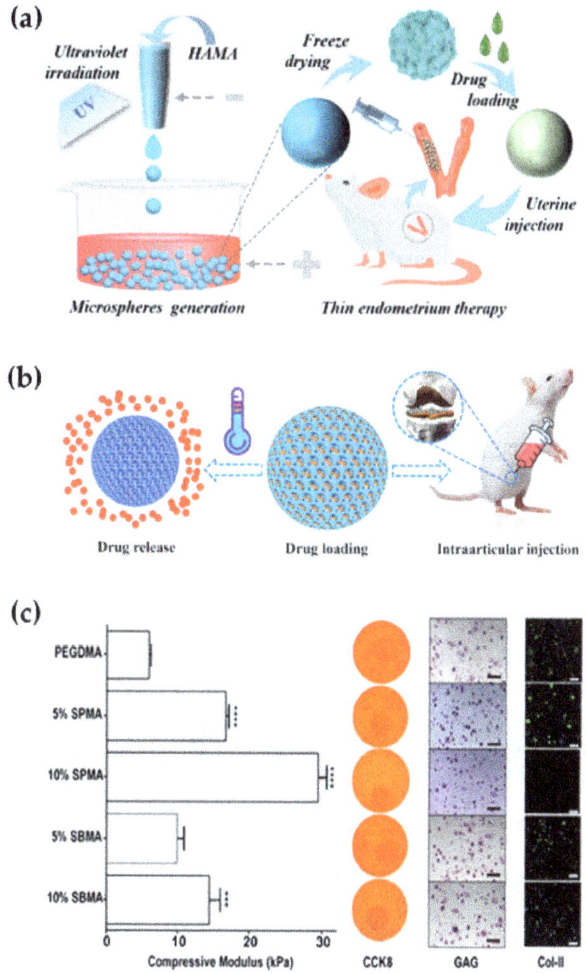

Figure 7. Application of photocrosslinked hydrogels in drug delivery, tissue engineering, and cell microcarriers. (a) Schematic diagram for the preparation of hydrogel microspheres for thin endometrium therapy [129]. (b) Application and principles of structural colored microspheres for treatment of osteoarthritis [132]. (c) Effects of different hydrogels on regulation of cell metabolism [133].

5.4. Cellular Microcarrier

Photocrosslinked hydrogels can be used to load cells as microcarriers for cell culture, and biocompatibility and bioadhesion of hydrogels can be evaluated according to the growth status of the cells. For example, Liu et al. used microcarriers constructed from different hydrogels for cell culture, and found that the growth status of the cells with different microcarriers was inconsistent. This indicated that hydrogel-cell microcarriers can be used to study biological adhesion of different hydrogels [135].

When photocrosslinked hydrogels are used in cell culture, they can also be used as a platform to study cell growth, proliferation, and interaction between cells, which is also the preliminary basis for the application of hydrogels in tissue engineering in vivo. In earlier studies, Burdick's research group prepared hydrogels with gradient-crosslinking density using microfluidic technology, which were used to study the migration and other behaviors of cells in vitro [136]. Dadsetan et al. obtained hydrogels with different crosslinking degrees and mechanical properties by changing the ratio of crosslinking agent and polymerizer in photocrosslinking hydrogels. When these hydrogels were used for chondrocyte culture, cells showed different adhesion effects and morphologies on hydrogels with different crosslinking densities [137]. Huang et al. found that by adding different ionic residues to the hydrogel, it can improve its mechanical properties and regulate the metabolic activity and collagen secretion of the loaded chondrocytes without changing the polymer content and swelling behavior (Figure 7c) [133]. These results indicated that hydrogels, when used as cell microcarriers, can simulate tissues in vitro and preliminarily monitor and regulate cell behavior [138].

5.5. Bionic Organ

Photocrosslinked hydrogels have the capability to be utilized in cell culture, as previously stated. By incorporating microfluidic and other technologies into the cell-culture process, it becomes possible to achieve numerous cell cocultures and imitate the architecture and functioning of certain organs through cell interactions. The system is referred to as an organ chip or organoid. The application value of these biomimetic organs in the biomedical field is significant. Drug assessment is a crucial application. The medication development process today frequently necessitates numerous iterations of animal or human studies to validate the treatment's effectiveness.

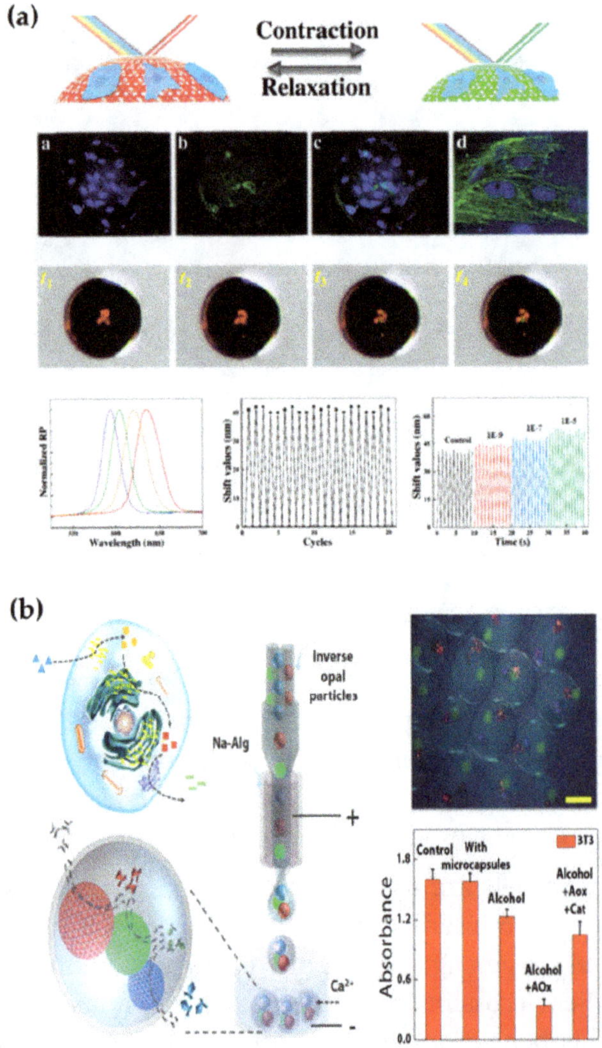

Figure 8. Application of photocrosslinked hydrogels in bionic organoids. (a) Bionic heart chip for cardiac drug evaluation [144]. (b) Bionic liver chip for the study of alcohol metabolism [145].

Animal trials frequently fail to accurately represent the impact of medications on humans due to species variations, and both animal and human research encounter ethical dilemmas. Furthermore, the lengthy duration of both animal trials and human trials significantly amplifies the expenses associated with the development of novel drugs. Organ chips or organoids offer a comparative advantage in addressing ethical concerns and reducing time cycles, making them an effective solution for these issues. Zhang et al. employed a 3D microfluidic cell-culture method to fabricate microchips specifically designed for drug screening of liver, lung, kidney, and other organs affected by organ disorders [139]. Huh's team developed a biomimetic lung-chip microdevice that mimics lung respiration using mechanical force. This device can be utilized to assess the hazardous effects of nanoparticles. In addition, they performed an assessment research on the harmful effects of medications using lung microchips [141]. Toh et al. employed a diverse range of cells to fabricate a liver-organ chip with a three-dimensional structure. Research has demonstrated that this microchip has the ability to replicate the functionality of the liver and can be utilized for the purpose of assessing toxicity [142]. Zhao et al. integrated architecturally colored hydrogels with cardiomyocytes to create an innovative heart chip [143]. They employed optical signals to observe and assess cardiomyocyte activity and treatment efficacy (Figure 8a) [144].

Researchers have developed organ chips not just for medication evaluation but also for the direct therapy of ailments. As an illustration, Fu's team utilized 3D printing technology to create a bionic liver that includes microchannels and nanoparticles. These components enable the liver to perform metabolism and adsorb hazardous compounds [85]. Wang et al. created a bionic enzyme-cascade microcapsule using microfluidic electrojet technology and a structurally e-colored microsphere enzyme carrier. The enzyme-cascade reaction (Figure 8b) can imitate the cascade metabolism of alcohol in the liver within microcapsules [145].

6. CONCISE SUMMARY

Typically, synthetic photocrosslinked hydrogels are created by including molecules that possess double bonds capable of undergoing polymerization upon exposure to light. Examples of such molecules include ethylene, acrylic acid, acrylamide, and acrylic ester. Typically, natural-polymer photocrosslinked hydrogels are altered by the use of acrylates or acrylamide derivatives. These derivatives are created by acylating natural polymers that have reactive groups. Photocrosslinked hydrogels can be synthesized under mild circumstances in the presence of a photoinitiator. Photocrosslinked hydrogels, in comparison to conventional chemical crosslinking methods, have the advantages of in situ polymerization crosslinking, rapid reaction rate, gentle reaction conditions, precise control over geometric shape, and minimal heat release during the reaction.

Photocrosslinked hydrogels are promising biological materials with favorable potential for further development. Nevertheless, they possess some constraints. Typically, the synthetic monomers employed in the production of photocrosslinked hydrogels possess hazardous properties. Eliminating these deleterious chemicals and residual free radicals from photocrosslinked hydrogels poses a formidable challenge. While chemically altered natural materials can release hazardous monomers, their range is still restricted, and the resulting mechanical characteristics are not as easily manageable as those of hydrogels made from synthetic monomers. Furthermore, their exceptional biocompatibility renders them susceptible to infection by numerous germs, hence restricting their use in various domains. Hence, future research should focus on developing natural materials with tailored chemical alterations to effectively suit certain requirements. Regarding the preparation and application, despite the utilization of many methods to create photocrosslinked hydrogels with diverse structures, there remains a significant difficulty in producing photocrosslinked hydrogels that accurately mimic living tissues or organs. The future research will focus on the convergence of cutting-edge processing technologies, novel materials, and the mechanics of human tissue, which will play a crucial role in resolving this issue. We anticipate that by progressively integrating biology, chemistry, engineering, and other fields, the utilization of photocrosslinked hydrogels in biomedical domains will be broadened, leading to much more noteworthy advancements.

AT A GLANCE

Hydrogel materials has significant utility in the field of biomedical engineering. Photocrosslinked hydrogels have gained significant interest because of their diverse range and straightforward preparation methods. In this study, we present a comprehensive analysis of the biomedical-engineering uses of photocrosslinked hydrogels. Initially, we provide the many categories of photocrosslinked hydrogel monomers, followed by a summary of the techniques used to create photocrosslinked hydrogels with diverse morphologies. Afterwards, a range of biomedical uses for photocrosslinked hydrogels are examined. Lastly, this paper discusses the limitations and potential areas for improvement in photocrosslinked hydrogels. This publication aims to provide researchers in related domains with a comprehensive understanding of photocrosslinked hydrogels and to inspire them to explore new avenues for the study of photocrosslinked hydrogels or similar materials.

REFERENCES

1. Naseri, N.; Deepa, B.; Mathew, A.P.; Oksman, K.; Girandon, L. Nanocellulose-based interpenetrating polymer network (IPN) hydrogels for cartilage applications. Biomacromolecules 2011, 3, 1877–1888.
2. Dragan, E.S. Design and applications of interpenetrating polymer network hydrogels. A review. Chem. Eng. J. 2014, 243, 572–590.
3. Norioka, C.; Inamoto, Y.; Hajime, C.; Kawamura, A.; Miyata, T. A universal method to easily design tough and stretchable hydrogels. NPG Asia Mater. 2021, 13, 34.
4. Hennink, W.E.; van Nostrum, C.F. Novel crosslinking methods to design hydrogels. Adv. Drug Deliv. Rev. 2012, 64, 223–236.
5. Zhao, G.; Li, T.; Brochet, D.X.; Rosenberg, P.B.; Lederer, W.J. STIM1 enhances SR Ca2+ content through binding phospholamban in rat ventricular myocytes. Proc. Natl. Acad. Sci. USA 2015, 112, 4792–4801.
6. Sperinde, J.J.; Griffith, L.G. Synthesis and characterization of enzymatically-cross-linked poly(ethylene glycol) hydrogels. Macromolecules 1997, 30, 5255–5264.
7. Sperinde, J.J.; Griffith, L.G. Control and prediction of gelation kinetics in enzymatically cross-linked poly(ethylene glycol) hydrogels. Macromolecules 2000, 33, 5476–5480.
8. Westhaus, E.; Messersmith, P.B. Triggered release of calcium from lipid vesicles: A bioinspired strategy for rapid gelation of polysaccharide and protein hydrogels. Biomaterials 2001, 22, 453–462.
9. Gottlieb, R.; Schmidt, T.; Arndt, K.F. Synthesis of temperature-sensitive hydrogel blends by high-energy irradiation. Nucl. Instrum. Methods Phys. Res. 2005, 236, 371–376.
10. Oezdemir, S.; Oezdemir, E.; Tunca, R.; Haziroglu, R.; Sen, M.; Kantoglu, O.E.; Gueven, O. In vivo biocompatibility studies of poly hydrogels synthesized by γ-rays. Nucl. Instrum. Methods Phys. Res. Sect. B Beam Interact. Mater. Atoms 2003, 208, 395–399.
11. Karakecili, A.G.; Satriano, C.; Gumusderelioglu, M.; Marletta, G. Thermoresponsive and bioactive poly(vinyl ether)-based hydrogels synthesized by radiation copolymerization and photochemical immobilization. Radiat. Phys. Chem. 2008, 77, 154–161.

12. Chen, M.; Shen, Y.; Xu, L.; Xiang, G.; Ni, Z. Highly efficient and rapid adsorption of methylene blue dye onto vinyl hybrid silica nano-cross-linked nanocomposite hydrogel. Colloids Surf. A Physicochem. Eng. Asp. 2021, 613, 126050.
13. Kao, F.J.; Manivannan, G.; Sawan, S.P. UV curable bioadhesives: Copolymers of N-vinyl pyrrolidone. J. Biomed. Mater. Res. 1997, 38, 191–196.
14. Lopergolo, L.C.; Lugao, A.B.; Catalani, L.H. Direct UV photocrosslinking of poly(N-vinyl-2-pyrrolidone) (PVP) to produce hydrogels. Polymer 2003, 44, 6217–6222.
15. Fechine, G.; Barros, J.; Catalani, L.H. Poly(N-vinyl-2-pyrrolidone) hydrogel production by ultraviolet radiation: New methodologies to accelerate crosslinking. Polymer 2004, 45, 4705–4709.
16. Lee, C.Y.; Teymour, F.; Camastral, H.; Tirelli, N.; Hubbell, J.A.; Elbert, D.L.; Papavasiliou, G. Characterization of the network structure of PEG diacrylatehy drogels formed in the presence of N-Vinyl pyrrolidone. Macromol. React. Eng. 2014, 8, 314–328.
17. Devine, D.M.; Higginbotham, C.L. Synthesis and characterisation of chemically crosslinked N-vinyl pyrrolidinone (NVP) based hydrogels. Eur. Polym. J. 2005, 41, 1272–1279.
18. Sahiner, N.; Sagbas, S. The preparation of poly(vinyl phosphonic acid) hydrogels as new functional materials for in situ metal nanoparticle preparation. Colloids Surf. A Physicochem. Eng. Asp. 2013, 418, 76–83.
19. Sagbas, S.; Sahiner, N. A novel p (AAm-co-VPA) hydrogel for the Co and Ni nanoparticle preparation and their use in hydrogel generation from NaBH4. Fuel Process. Technol. 2012, 104, 31–36.
20. Ren, Z.Q.; Zhang, Y.Y.; Li, Y.M.; Xu, B.; Liu, W.G. Hydrogen bonded and ionically crosslinked high strength hydrogels exhibiting Ca2+-triggered shape memory properties and volume shrinkage for cell detachment. J. Mater. Chem. B Mater. Biol. Med. 2015, 3, 6347–6354.
21. Hou, N.; Wang, R.; Wang, F.; Bai, J.; Zhou, J.; Hu, J.; Liu, S.; Jiao, T. Fabrication of hydrogels via host-guest polymers as highly efficient organic dye adsorbents for wastewater treatment. ACS Omega 2020, 5, 5470–5479.
22. Zhang, R.; Peng, H.; Zhou, T.; Li, M.; Guo, X.; Yao, Y. Selective adsorption and separation oforganic dyes by poly(acrylic acid) hydrogels formed withspherical polymer brushes and chitosan. Aust. J. Chem. 2018, 71, 846–854.
23. Jana, S.; Ray, J.; Jana, D.; Mondal, B.; Bhanja, S.K.; Tripathy, T. Removal of vanadium (IV) from water solution by sulfated Katira gum-cl-poly(acrylic acid) hydrogel. Colloids Surf. A Physicochem. Eng. Asp. 2019, 566, 70–83.
24. Anah, L.; Astrini, N. Isotherm adsorption studies of Ni (II) ion removal from aqueous solutions by modified carboxymethyl cellulose hydrogel. Conf. Ser. Earth Environ. Sci. IOP Publ. 2018, 160, 012017.
25. Makhado, E.; Pandey, S.; Nomngongo, P.N.; Ramontja, J. Preparation and characterization of xanthan gum-cl-poly(acrylic acid)/o-MWCNTs hydrogel nanocomposite as highly effective re-usable adsorbent for removal of methylene blue from aqueous solutions. J. Colloid Interface Sci. 2018, 513, 700–714.
26. Gwon, S.; Park, S. Preparation of uniformly sized interpenetrating polymer network polyelectrolyte hydrogel droplets from a solid-state liquid crystal shell. J. Ind. Eng. Chem. 2021, 99, 235–245.
27. Liu, J.; Liu, G.; Liu, W. Preparation of water-soluble β-cyclodextrin/poly(acrylic acid)/graphene oxide nanocomposites as new adsorbents to remove cationic dyes from aqueous solutions. Chem. Eng. J. 2014, 257, 299–308.

28. Hu, X.S.; Liang, R.; Sun, G. Super-adsorbent hydrogel for removal of methylene blue dye from aqueous solution. J. Mater. Chem. A 2018, 6, 17612–17624.

29. Ma, Y.; Lyu, L.; Guo, Y.; Fu, Y.; Wu, T.; Guo, S.; Sun, K.; Wujcik, E.K.; Guo, Z. Porous lignin based poly(acrylic acid)/organo-montmorillonite nanocomposites: Swelling behaviors and rapid removal of Pb (II) ions. Polymer 2017, 128, 12–23.

30. Kong, W.; Yue, Q.; Li, Q.; Gao, B. Adsorption of Cd2+ on GO/PAA hydrogel and preliminary recycle to GO/PAA-CdS as efficient photocatalyst. Sci. Total Environ. 2019, 668, 1165–1174.

31. Lu, F.; Wang, Y.; Wang, C.; Kuga, S.; Huang, Y.; Wu, M. Two-dimensional nanocellulose-enhanced high-strength, self-adhesive, and strain-sensitive poly(acrylic acid) hydrogels fabricated by a radical-induced strategy for a skin sensor. ACS Sustain. Chem. Eng. 2020, 8, 3427–3436.

32. Wang, H.; Gu, H.; Chen, Z.; Shang, L.; Zhao, Z.; Gu, Z.; Zhao, Y. Enzymatic Inverse Opal Hydrogel Particles for Biocatalyst. ACS Appl. Mater. Interfaces 2017, 9, 12914–12918.

33. Wang, H.; Zhang, H.; Zhang, D.; Wang, J.; Tan, H.; Kong, T. Enzyme-functionalized structural color hydrogel particles for urea detection and elimination. J. Clean. Prod. 2021, 315, 128149.

34. Chen, C.; Liu, Y.; Wang, H.; Chen, G.; Ren, J.; Zhang, H. Multifunctional chitosan inverse opal particles for wound healing. ACS Nano 2018, 12, 10493–10500.

35. Zhang, B.; Cheng, Y.; Wang, H.; Ye, B.; Shang, L.; Zhao, Y.; Gu, Z. Multifunctional inverse opal particles for drug delivery and monitoring. Nanoscale 2015, 7, 10590.

36. Zhao, Z.; Wang, H.; Shang, L.; Yu, Y.; Fu, F.; Zhao, Y.; Gu, Z. Bioinspired heterogeneous structural color stripes from capillaries. Adv. Mater. 2017, 29, 1704569.

37. Fu, F.; Chen, Z.; Wang, H.; Liu, C.; Liu, Y.; Zhao, Y. Graphene hybrid colloidal crystal arrays with photo-controllable structural colors. Nanoscale 2019, 11, 10846–10851.

38. Li, L.; Chen, Z.; Shao, C.; Sun, L.; Zhao, Y. Graphene hybrid anisotropic structural color film for cardiomyocytes'monitoring. Adv. Funct. Mater. 2020, 30, 1906353.

39. Manz, A.; Graber, N.; Widmer, H.M. Miniaturized total chemical analysis systems: A novel concept for chemical sensing. Sens. Actuators B Chem. 1990, 1, 244–248.

40. Hou, K.; Hu, Z.; Mugaanire, I.T.; Li, C.; Chen, G.; Zhu, M. Fiber forming mechanism and reaction kinetics of novel dynamic-crosslinking-spinning for poly(ethylene glycol) diacrylate fiber fabrication. Polymer 2019, 183, 121903.

41. Yang, J.; Zhang, Y.S.; Yue, K.; Khademhosseini, A. Cell-laden hydrogels for osteochondral and cartilage tissue engineering. Acta Biomater. 2017, 57, 1–25.

42. Colinet, I.; Dulong, V.; Mocanu, G.; Picton, L.; Cerf, D.L. New amphiphilic and pH-sensitive hydrogel for controlled release of a model poorly water-soluble drug. Eur. J. Pharm. Biopharm. 2009, 73, 345–350.

43. Gupta, A.; Kowalczuk, M.; Heaselgrave, W.; Britland, S.T.; Martin, C.; Radecka, I. The production and application of hydrogels for wound management: A review. Eur. Polym. J. 2019, 111, 134–151.

44. Goh, C.H.; Heng, P.W.S.; Chan, L.W. Alginates as a useful natural polymer for microencapsulation and therapeutic applications. Carbohydr. Polym. 2012, 88, 1–12.

45. Xu, S.; Liang, W.; Xu, G.; Huang, C.; Zhang, J.; Lang, M. A fast and dual crosslinking hydrogel based on vinyl ether sodium alginate. Appl. Surf. Sci. 2020, 515, 145811.

46. Bukhari, A.A.H.; Elsayed, N.H.; Monier, M. Development and characterization of photo-responsive cinnamoly modified alginate. Carbohydr. Polym. 2021, 260, 117771.
47. Huang, H.; Feng, J.; Wismeijer, D.; Wu, G.; Hunziker, E.B. Hyaluronic acid promotes the osteogenesis of BMP-2 in an absorbable collagen sponge. Polymers 2017, 9, 339.
48. Nielsen, J.J.; Low, S.A. Bone-targeting systems to systemically deliver therapeutics to bone fractures for accelerated healing. Curr. Osteoporos. Rep. 2020, 18, 449–459.
49. Lee, H.; Park, T.G. Photocrosslinkable, biomimetic, and thermo-sensitive pluronic grafted hyaluronic acid copolymers for injectable delivery of chondrocytes. J. Biomed. Mater. Res. Part A 2009, 88, 797–806.
50. Jenjob, R.; Nguyen, H.P.; Kim, M.K.; Jiang, Y.; Kim, J.J.; Yang, S.G. Bisphosphonate-conjugated photo-crosslinking polyanionic hyaluronic acid microbeads for controlled BMP2 delivery and enhanced bone formation efficacy. Biomacromolecules 2021, 22, 4138–4145.
51. Monier, M.; Youssef, I.; Abdel-Latif, D.A. Synthesis of photo-responsive chitosan-cinnamate for efficient entrapment of β-galactosidase enzyme. React. Funct. Polym. 2018, 124, 129–138.
52. Jeong, S.I.; Jeon, O.; Krebs, M.D.; Hill, M.C.; Alsberg, E. Biodegradable photo-crosslinked alginate nanofibre scaffolds with tuneable physical properties, cell adhesivity and growth factor release. Eur. Cells Mater. 2012, 24, 331–343.
53. Younes, I.; Rinaudo, M. Chitin and chitosan preparetion from marine sources. Structure, properties and applications. Mar. Drugs 2015, 13, 1133–1174.
54. Zhou, Y.; Dong, Q.; Yang, H.; Liu, X.; Yin, X.; Tao, Y.; Bai, Z. Photocrosslinked maleilated chitosan/methacrylated poly (vinyl alcohol) bicomponent nanofibrous scaffolds for use as potential wound dressings. Carbohydr. Polym. 2017, 168, 220–226.
55. Gospodarowicz, D.; Cheng, J. Heparin protects basic and acidic FGF from inactivation. J. Cell. Physiol. 1986, 128, 475–484.
56. Wissink, M.J.B.; Beernink, R.; Pieper, J.S.; Poot, A.A.; Engbers, G.H.M.; Beugeling, T.; Aken, W.G.V.; Feijen, J. Binding and release of basic fibroblast growth factor from heparinized collagen matrices. Biomaterials 2001, 22, 2291–2299.
57. Yoon, J.J.; Chung, H.J.; Park, T.G. Photocrosslinkable and biodegradable pluronic/heparin hydrogels for local and sustained delivery of angiogenic growth factor. J. Biomed. Mater. Res. Part A 2007, 83, 597–605.
58. Ma, L.; Li, X.; Guo, X.; Jiang, Y.; Li, X.M.; Guo, H.; Zhang, B.; Xu, Y.; Wang, X.; Li, Q. Promotion of endothelial cell adhesion and antithrombogenicity of polytetrafluoroethylene by chemical grafting of chondroitin sulfate. ACS Appl. Bio. Mater. 2019, 3, 891–901.
59. Bai, X.; Lü, S.; Cao, Z.; Ni, B.; Wang, X.; Ning, P.; Ma, D.; Wei, H.; Liu, M. Dual crosslinked chondroitin sulfate injectable hydrogel formed via continuous Diels-Alder (DA) click chemistry for bone repair. Carbohydr. Polym. 2017, 166, 123–130.
60. Li, S.; Ma, F.; Pang, X.; Tang, B.; Lin, L. Synthesis of chondroitin sulfate magnesium for osteoarthritis treatment. Carbohydr. Polym. 2019, 212, 387–394.
61. Kim, H.D.; Lee, E.A.; An, Y.H.; Kim, S.L.; Lee, S.S.; Yu, S.J.; Jang, H.L.; Nam, K.T.; Lm, S.G.; Hwang, N.S.Y. Chondroitin sulfate-based biomineralizing surface hydrogels for bone tissue engineering. ACS Appl. Mater. Interfaces 2017, 9, 21639–21650.

62. Ornell, K.J.; Lozada, D.; Phan, N.V.; Coburn, J.M. Controlling methacryloyl substitution of chondroitin sulfate: Injectable hydrogels with tunable long-term drug release profiles. J. Mater. Chem. B 2019, 7, 2151–2161.

63. Cerqueira, M.T.; Silva, L.P.D.; Santos, T.C.; Pirraco, R.P.; Correlo, V.M.; Reis, R.L.; Marques, A.P. Gellan gum-hyaluronic acid spongy-like hydrogels and cells from adipose tissue synergize promoting neoskin vascularization. ACS Appl. Mater. Interfaces 2014, 6, 19668–19679.

64. Cerqueira, M.T.; Silva, L.P.D.; Santos, T.C.; Pirraco, R.P.; Correlo, V.M.; Marques, A.P.; Reis, R.L. Human skin cell fractions fail to self-organize within a gellan gum/hyaluronic acid matrix but positively influence early wound healing. Tissue Eng. Part A 2014, 20, 1369–1378.

65. Ichibouji, T.; Miyazaki, T.; Ishida, E.; Sugino, A.; Ohtsuki, C. Apatite mineralization abilities and mechanical properties of covalently crosslinked pectin hydrogels. Mater. Sci. Eng. C 2009, 29, 1765–1769.

66. Monier, M.; Shafik, A.L.; El-Mekabaty, A. Designing and investigation of photo-active gellan gum for the efficient immobilization of catalase by entrapment. Int. J. Biol. Macromol. 2020, 161, 539–549.

67. Oliveira, M.B.; Custódio, C.A.; Gasperini, L.; Reis, R.L.; Mano, J.F. Autonomous osteogenic differentiation of hASCs encapsulated in methacrylated gellan-gum hydrogels. Acta Biomater. 2016, 41, 119–132.

68. Serafini, M.R.; Menezes, P.P.; Costa, L.P. Interaction of p-cymene with β-cyclodextrin. J. Therm. Anal. Calorim. 2010, 109, 951–955.

69. Kurkov, S.V.; Loftsson, T. Cyclodextrins. Int. J. Pharm. 2013, 453, 167–180.

70. Cosola, A.; Conti, R.; Grützmacher, H.; Sangermano, M.; Roppolo, L.; Pirri, C.F.; Chiappone, A. Multiacrylated Cyclodextrin: A Bio-Derived Photocurable Macromer for VAT 3D Printing. Macromol. Mater. Eng. 2020, 305, 2000350.

71. Rezanka, M. Synthesis of substituted cyclodextrins. Environ. Chem. Lett. 2019, 17, 49–63.

72. Yamasaki, H.; Odamura, A.; Makihata, Y.; Fukunaga, K. Preparation of new photo-crosslinked β-cyclodextrin polymer beads. Polym. J. 2017, 49, 377–383.

73. Yin, R.; Wang, K.; Han, J.; Nie, J. Photo-crosslinked glucose-sensitive hydrogels based on methacrylate modified dextran–concanavalin A and PEG dimethacrylate. Carbohydr. Polym. 2010, 82, 412–418.

74. Lee, M.H.; Boettiger, D.; Composto, R.J. Biomimetic carbohydrate substrates of tunable properties using immobilized dextran hydrogels. Biomacromolecules 2008, 9, 2315–2321.

75. Pacelli, S.; Paolicelli, P.; Casadei, M.A. New biodegradable dextran-based hydrogels for protein delivery: Synthesis and characterization. Carbohydr. Polym. 2015, 126, 208–214.

76. Yang, K.; Sun, J.; Guo, Z.; Yang, J.; Wei, D.; Tan, Y.; Guo, L.; Luo, H.; Fan, H.; Zhang, X. Methacrylamide-modified collagen hydrogel with improved anti-actin-mediated matrix contraction behavior. J. Mater. Chem. B 2018, 6, 7543–7555.

77. Tytgat, L.; Markovic, M.; Qazi, T.H.; Vagenende, M.; Bray, F.; Martins, J.C.; Rolando, C.; Thienpont, H.; Ottevaere, H.; Ovsianikov, A.; et al. Photo-crosslinkable recombinant collagen mimics for tissue engineering applications. J. Mater. Chem. B 2019, 7, 3100–3108.

78. Song, X.; Dong, P.; Gravesande, J.; Cheng, B.; Xing, J. UV-mediated solid-state crosslinking of electrospinning nanofibers of modified collagen. Int. J. Biol. Macromol. 2018, 120, 2086–2093.

79. Yang, K.; Sun, J.; Wei, D.; Yuan, L.; Yang, J.; Guo, L.; Fan, H.; Zhang, X. Photo-crosslinked mono-component type II collagen hydrogel as a matrix to induce chondrogenic differentiation of bone marrow mesenchymal stem cells. J. Mater. Chem. B 2017, 5, 8707–8718.

80. Nichol, J.W.; Koshy, S.T.; Bae, H.; Hwang, C.M.; Yamanlar, S.; Khademhosseini, A. Cell-laden microengineered gelatin methacrylate hydrogels. Biomaterials 2010, 31, 5536–5544.

81. Hu, Y.; Wang, W.; Yu, D. Preparation of antibacterial keratin fabrics via UV curing and click chemistry. RSC Adv. 2016, 6, 81731–81735.

82. Xuan, M.; Sahoo, J.K.; Cebe, P.; Kaplan, D.L. Photo-crosslinked silk fibroin for 3D printing. Polymers 2020, 12, 2936.

83. Rusu, A.G.; Chiriac, A.P.; Nita, L.E.; Mititelu-Tartau, L.; Tudorachi, N.; Ghilan, A.; Rusu, D. Multifunctional BSA scaffolds prepared with a novel combination of UV-crosslinking systems. Macromol. Chem. Phys. 2019, 220, 1900378.

84. Chiriac, A.P.; Ghilan, A. Advancement in the biomedical applications of the nanogel structures based on particular polysaccharides. Macromol. Biosci. 2019, 15, 1903104.

85. Fu, F.; Shang, L.; Zheng, F.; Chen, Z.; Wang, H.; Wang, J.; Gu, Z. Cells cultured on core-shell photonic crystal barcodes for drug screening. ACS Appl. Mater. Interfaces 2016, 8, 13840–13848.

86. Chevallay, B.; Abdul-Malak, N.; Herbage, D. Mouse fibroblasts in long-term culture within collagen three-dimensional scaffolds: Influence of crosslinking with diphenylphosphorylazide on matrix reorganization, growth, and biosynthetic and proteolytic activities. J. Biomed. Mater. Res. 2000, 49, 448–459.

87. Wu, Y.L.; Lin, C.W.; Cheng, N.C.; Yang, K.C.; Yu, J. Modulation of keratin in adhesion, proliferation, adipogenic, and osteogenic differentiation of porcine adipose-derived stem cells. J. Biomed. Mater. Res. Part B Appl. Biomater. 2017, 105, 180–192.

88. Yu, D.; Cai, J.Y.; Church, J.S.; Wang, L. Modifying surface resistivity and liquid moisture management property of keratin fibers through thiol–Ene click reactions. ACS Appl. Mater. Interfaces 2014, 6, 1236–1242.

89. Qi, C.; Liu, J.; Jin, Y.; Xu, L.; Wang, G.; Wang, Z.; Wang, L. Photo-crosslinkable, injectable sericin hydrogel as 3D biomimetic extracellular matrix for minimally invasive repairing cartilage. Biomaterials 2018, 163, 89–104.

90. Ju, J.; Hu, N.; Cairns, D.M.; Liu, H.; Timko, B.P. Photo-cross-linkable, insulating silk fibroin for bioelectronics with enhanced cell affinity. Proc. Natl. Acad. Sci. USA 2020, 117, 15482–15489.

91. Aduba, D.C.; Hammer, J.A.; Yuan, Q.; Andrew Yeudall, W.; Bowlin, G.L.; Yang, H. Semi-interpenetrating network (sIPN) gelatin nanofiber scaffolds for oral mucosal drug delivery. Acta Biomater. 2013, 9, 6576–6584.

92. Rong, L.; Dou, J.; Jiang, Q.; Jing, L.; Ren, X. Preparation and antimicrobial activity of β-cyclodextrin derivative copolymers/cellulose acetate nanofibers. Chem. Eng. J. 2014, 248, 264–272.

93. Maciejewska, B.M.; Wychowaniec, J.K.; Woniak-Budych, M.; Popenda, L.; Jurga, S. UV cross-linked polyvinylpyrrolidone electrospun fibres as antibacterial surfaces. Sci. Technol. Adv. Mater. 2019, 20, 979–991.

94. Rosa, R.M.; Silva, J.C.; Sanches, I.S.; Henriques, C. Simultaneous photo-induced crosslinking and silver nanoparticle formation in a PVP electrospun wound dressing. Mater. Lett. 2017, 207, 145–148.

95. Son, W.K.; Ji, H.Y.; Park, W.H. Antimicrobial cellulose acetate nanofibers containing silver nanoparticles. Carbohydr. Polym. 2006, 65, 430–434.

96. Wang, Y.S.; Cheng, C.C.; Ye, Y.S.; Yen, Y.C.; Chang, F.C. Bioinspired photo-cross-linked nanofibers from uracil-functionalized polymers. ACS Macro Lett. 2012, 1, 159–162.

97. Singh, U.; Mohan, S.; Davis, F.; Mitchell, G. Modifying the thermomechanical properties of electrospun fibres of poly-vinyl cinnamate by photo-cross-linking. SN Appl. Sci. 2019, 1, 31.

98. Tang, Z.; Jie, W.; Yung, L.; Ji, B.; Ma, H.; Qiu, C.; Yoon, K.; Wan, F.; Fang, D.; Hsiao, B.S. UV-cured poly(vinyl alcohol) ultrafiltration nanofibrous membrane based on electrospun nanofiber scaffolds. J. Membr. Sci. 2009, 328, 1–5.

99. Zhao, X.; Liu, S.; Yildirimer, L.; Zhao, H.; Ding, R.H.; Wang, H.N.; Cui, W.G.; Weitz, D. Injectable stem cell-laden photocrosslinkable microspheres fabricated using microfluidics for rapid generation of osteogenic tissue constructs. Adv. Funct. Mater. 2016, 26, 2809–2819.

100. Lee, G.H.; Jeon, T.Y.; Kim, J.B.; Lee, B.; Lee, C.S.; Lee, S.Y.; Kim, S.H. Multicompartment photonic microcylinders toward structural color inks. Chem. Mater. 2018, 30, 3789–3797.

101. Zhao, C.; Chen, G.; Wang, H.; Zhao, Y.; Chai, R. Bio-inspired intestinal scavenger from microfluidic electrospray for detoxifying lipopolysaccharide. Bioact. Mater. 2021, 6, 1653–1662.

102. Schonberg, J.N.; Zinggeler, M.; Fosso, P.L.; Brandstetter, T.; Ruhe, J. One-step photochemical generation of biofunctionalized hydrogel particles via two-phaseflow. ACS Appl. Mater. Interfaces 2018, 10, 39411–39416.

103. Zhang, H.Z.; Xu, R.H.; Yin, Z.W.; Yu, J.; Liang, N.; Geng, Q. Drug-loaded chondroitin sulfate microspheres generated from microfluidic electrospray for wound healing. Macromol. Res. 2022, 30, 36–42.

104. Ye, B.; Ding, H.; Cheng, Y.; Gu, H.C.; Gu, Z.Z. Photonic crystal microcapsules for label-free multiplex detection. Adv. Mater. 2014, 26, 3270–3274.

105. Xu, Y.S.; Wang, H.; Luan, C.X.; Fu, F.F.; Chen, B.A.; Liu, H.; Zhao, Y.J. Porous hydrogel encapsulated photonic barcodes for multiplex microRNA quantification. Adv. Funct. Mater. 2018, 28, 1704458.

106. Xin, T.; Gu, Y.; Cheng, R.; Tang, J.; Sun, Z.; Cui, W.; Chen, L. Inorganic strengthened hydrogel membrane as regenerative periosteum. ACS Appl. Mater. Interfaces 2017, 9, 41168–41180.

107. Luo, Z.; Che, J.; Sun, L.; Yang, L.; Zu, Y.; Wang, H.; Zhao, Y. Microfluidic electrospray photo-crosslinkable κ-Carrageenan microparticles for wound healing. Eng. Regen. 2021, 2, 257–262.

108. Zhao, Z.; Wang, J.; Lu, J.; Yu, Y.R.; Fu, F.F.; Wang, H.; Liu, Y.X.; Zhao, Y.J.; Gu, Z.Z. Tubular inverse opal scaffolds for biomimetic vessels. Nanoscale 2016, 8, 13574–13580.

109. Zhang, Z.H.; Chen, Z.Y.; Wang, Y.; Chi, J.J.; Wang, Y.T.; Zhao, Y.J. Bioinspired bilayer structural color hydrogel actuator with multienvironment responsiveness and survivability. Small Methods 2019, 3, 1900519.

110. Zhang, H.; Chen, C.W.; Zhang, H.; Chen, G.P.; Wang, Y.T.; Zhao, Y.J. Janus medical sponge dressings with anisotropic wettability for wound healing. Appl. Mater. Today 2021, 23, 101068.

111. GhavamiNejad, A.; Li, J.; Lu, B.; Zhou, L.; Lam, L.; Giacca, A.; Wu, X.Y. Glucose-responsive composite microneedle patch for hypoglycemia-triggered delivery of native glucagon. Adv. Mater. 2019, 31, 1901051.

112. Liang, Y.; Li, Z.; Huang, Y.; Yu, R.; Guo, B. Dual-dynamic-bond cross-linked antibacterial adhesive hydrogel sealants with on-demand removability for post-wound-closure and infected wound healing. ACS Nano 2021, 15, 7078–7093.

113. Liu, S.; Yeo, D.C.; Wiraja, C.; Tey, H.L.; Mrsksich, M.; Xu, C. Peptide delivery with poly(ethylene glycol) diacrylate microneedles through swelling effect. Bioeng. Transl. Med. 2017, 2, 258–267.

114. Zhang, T.; Sun, B.; Guo, J.; Wang, M.; Yan, F. Active pharmaceutical ingredient poly(ionic liquid)-based microneedles for the treatment of skin acne infection. Acta Biomater. 2020, 115, 136–147.

115. Yu, X.; Li, M.; Zhu, L.; Li, J.; Jin, Y. Amifostine-loaded armored dissolving microneedles for long-term prevention of ionizing radiation-induced injury. Acta Biomater. 2020, 112, 87–100.

116. Zhao, F.; Wu, D.; Yao, D.; Guo, R.; Wang, W.; Dong, A.; Kong, D.; Zhang, J. An injectable particle-hydrogel hybrid system for glucose-regulatory insulin delivery. Acta Biomater. 2017, 64, 334–345.

117. Yue, S.; Sun, X.T.; Wang, N.; Wang, Y.N.; Wang, Y.; Xu, Z.R.; Chen, M.L.; Wang, J.H. SERS–fluorescence dual-mode pH-sensing method based on Janus microparticles. ACS Appl. Mater. Interfaces 2017, 9, 39699–39707.

118. Ye, B.F.; Wang, H.; Ding, H.B.; Zhao, Y.J.; Pu, Y.P.; Gu, Z.Z. Colorimetric logic response based on aptamer functionalized colloidal crystal hydrogels. Nanoscale 2015, 7, 7565–7568.

119. Qin, M.; Sun, M.; Bai, R.B.; Mao, Y.Q.; Qin, X.S.; Sikka, D.; Zhao, Y.; Qi, H.J.; Suo, Z.G.; He, X.M. Bioinspired hydrogel interferometer for adaptive coloration and chemical sensing. Adv. Mater. 2018, 30, 1800468.

120. Elsherif, M.; Hassan, M.U.; Yetisen, A.K.; Butt, H. Wearable contact lens biosensors for continuous glucose monitoring using smartphones. ACS Nano 2018, 2, 5452–5462.

121. Yu, H.C.; Hao, X.P.; Zhang, C.W.; Zheng, S.Y.; Du, M.; Liang, S.M.; Wu, Z.L.; Zheng, Q. Engineering tough metallosupramolecular hydrogel films with kirigami structures for compliant soft electronics. Small 2021, 17, 2103836.

122. Liu, T.; Liu, M.M.; Dou, S.; Sun, J.M.; Cong, Z.F.; Jiang, C.Y.; Du, C.H.; Pu, X.; Hu, W.G.; Wang, Z.L. Triboelectric-nanogenerator-based soft energy-harvesting skin enabled by toughly bonded elastomer/hydrogel hybrids. ACS Nano 2018, 12, 2818–2826.

123. Wichterle, O.; Lim, D. Hydrophilic gels for biological use. Nature 1960, 185, 117–118.

124. Alexeev, V.L.; Das, S.; Finegold, D.N.; Asher, S.A. Photonic crystal glucose-sensing material for noninvasive monitoring of glucose in tear fluid. Clin. Chem. 2004, 50, 2353–2360.

125. Xie, Z.Y.; Li, L.L.; Liu, P.M.; Zheng, F.Y.; Guo, L.Y.; Zhao, Y.J.; Jin, L.; Li, T.T.; Gu, Z.Z. Self-assembled coffee-ring colloidal crystals for structurally colored contact lenses. Small 2015, 11, 926–930.

126. Deng, J.Z.; Chen, S.; Chen, J.L.; Ding, H.L.; Deng, D.W.; Xie, Z.Y. Self-reporting colorimetric analysis of drug release by molecular imprinted structural color contact lens. ACS Appl. Mater. Interfaces 2018, 10, 34611–34617.

127. Lin, S.T.; Yuk, H.; Zhang, T.; Parada, G.A.; Koo, H.; Yu, C.J.; Zhao, X.H. Stretchable hydrogel electronics and devices. Adv. Mater. 2016, 28, 4497–4505.

128. Wang, M.X.; Chen, Y.M.; Gao, Y.; Hu, C.; Hu, J.; Tan, L.; Yang, Z.M. Rapid self-recoverable hydrogels with high toughness and excellent conductivity. ACS Appl. Mater. Interfaces 2018, 10, 26610–26617.

129. Lei, L.J.; Lv, Q.Z.; Jin, Y.; An, H.; Shi, Z.; Hu, G.; Yang, Y.Z.; Wang, X.G.; Yang, L. Angiogenic microspheres for the treatment of a thin endometrium. ACS Biomater. Sci. Eng. 2021, 7, 4914–4920.

130. Lei, L.J.; Wang, X.G.; Zhu, Y.L.; Su, W.T.; Lv, Q.Z.; Li, D. Antimicrobial hydrogel microspheres for protein capture and wound healing. Mater. Des. 2022, 215, 110478.

131. Lei, L.J.; Zhu, Y.L.; Qin, X.Y.; Chai, S.L.; Liu, G.X.; Su, W.T.; Lv, Q.Z.; Li, D. Magnetic biohybrid microspheres for protein purification and chronic wound healing in diabetic mice. Chem. Eng. J. 2021, 425, 130671.

132. Yang, L.; Liu, Y.X.; Shou, X.; Ni, D.; Kong, T.T.; Zhao, Y.J. Bio-inspired lubricant drug delivery particles for the treatment of osteoarthritis. Nanoscale 2020, 12, 17093–17102.

133. Huang, H.; Tan, Y.; Ayers, D.C.; Song, J. Anionic and zwitterionic residues modulate stiffness of photocrosslinked hydrogels and cellular behavior of encapsulated chondrocytes. ACS Biomater. Sci. Eng. 2018, 4, 1843–1851.

134. Salahuddin, B.; Wang, S.; Sangian, D.; Aziz, S.; Gu, Q. Hybrid gelatinhydrogels in nanomedicine applications. ACS Appl. Bio Mater. 2021, 4, 2886–2906.

135. Liu, W.; Shang, L.R.; Zheng, F.Y.; Lu, J.; Qian, J.L.; Zhao, Y.J.; Gu, Z.Z. Photonic crystal encoded microcarriers for biomaterial evaluation. Small 2014, 10, 88–93.

136. Burdick, J.A.; Khademhosseini, A.; Langer, R. Fabrication of gradient hydrogels using a microfluidics/photopolymerization process. Langmuir 2004, 20, 5153–5156.

137. Dadsetan, M.; Szatkowski, J.P.; Yasemski, M.J.; Lu, L.C. Characterization of photocrosslinked oligo [poly(ethylene glycol) fumarate] hydrogels for cartilage tissue engineering. Biomacromolecules 2007, 8, 1702–1709.

138. Sontheimer-Phelps, A.; Hassell, B.A.; Ingber, D.E. Modelling cancer in microfluidic human organs-on-chips. Nat. Rev. Cancer 2019, 19, 65–81.

139. Zhang, C.; Zhao, Z.Q.; Rahim, N.A.A.; Noort, D.V.; Yu, H. Towards a human-on-chip: Culturing multiple cell types on a chip with compartmentalized microenvironments. Lab Chip 2009, 9, 3185–3192.

140. Huh, D.; Matthews, B.D.; Mammoto, A.; Montoya-Zavala, M.; Hsin, H.Y.; Ingber, D.E. Reconstituting organ-level lung functions on a chip. Science 2010, 328, 1662–1668.

141. Huh, D.; Leslie, D.C.; Matthews, B.D.; Fraser, J.P.; Jurek, S.; Hamiton, G.A.; Throneloe, K.S.; Mcalexander, M.A.; Ingber, D.E. A human disease model of drug toxicity–induced pulmonary edema in a lung-on-a-chip microdevice. Sci. Transl. Med. 2012, 4, 147.

142. Toh, Y.C.; Lim, T.C.; Tai, D.; Xiao, G.F.; Noort, D.V.; Yu, H. A microfluidic 3D hepatocyte chip for drug toxicity testing. Lab Chip 2009, 9, 2026–2035.

143. Fu, F.F.; Shang, L.R.; Chen, Z.Y.; Yu, Y.R.; Zhao, Y.J. Bioinspired living structural color hydrogels. Sci. Robot. 2018, 3, 8580.

144. Wang, H.; Liu, Y.X.; Chen, Z.Y.; Sun, L.Y.; Zhao, Y.J. Anisotropic structural color particles from colloidal phase separation. Sci. Adv. 2020, 6, 1438.

145. Wang, H.; Zhao, Z.; Liu, Y.X.; Shao, C.M.; Boan, F.K.; Zhao, Y.J. Biomimetic enzyme cascade reaction system in microfluidic electrospray microcapsules. Sci. Adv. 2018, 4, 2816.

3 Chapter

CHARACTERIZATION OF PHYSICOMECHANICAL AND MORPHOLOGICAL PROPERTIES OF HYDROGELS BASED ON MULTI-STRUCTURED POTASSIUM ACRYLATE

1. INTRODUCTION: AN OVERVIEW

Hydrogels are three-dimensional polymer networks capable of absorbing significant quantities of water and other fluids without affecting their structure [1,2,3]. The extent of their swelling is determined by the presence of specific functional groups, the degree of crosslinking, the flexibility of the polymer chains, the arrangement of the polymer molecules, and the crystallinity of the components [4], as well as the thermal history [5]. These distinct materials are being employed in several commercial applications, such as ophthalmic devices, biosensors, biological membranes, and drug carriers [1,2]. The literature has also indicated potential uses of hydrogels as soil conditioners and as agents for removing heavy metal ions [6,7,8]. The characteristics of hydrogels are significantly influenced by the synthesis, concentration, and composition of their components, as well as the process of polymerization. Several investigations have been conducted on the synthesis processes, swelling effects, and the impact of crosslinkers on properties [9,10,11,12,13,14,15,16,17,18,19,20,21,22].

UV curing, often known as photopolymerization, is a different method used to create hydrogels. It enables more precise regulation of the rate at which a reaction occurs and is not influenced by the existence of oxygen in the system [23,24,25]. Furthermore, photopolymerization can be employed with the majority of monomers, requiring just a single additive, namely a photoinitiator [26,27]. Additional benefits of photopolymerization include rapid synthesis times and less heat generation [28]. Furthermore, it is crucial to note that photopolymerization is extensively employed for the synthesis of hydrogels that have significant utility in the realms of medicine and biology. Instances of successful cell encapsulation and high cell viability in photopolymerized hydrogels [23], as well as the creation of sophisticated photoinitiators suitable for various medical diseases [29], have been documented. Furthermore, there have been reports of outstanding antibacterial properties in photopolymerizable hydrogels. Lin et al. (2011) synthesized a biocompatible silicone by combining carboxybetaine and a macromer (bis-α, ω-(methacryloxypropyl) poly-dimethylsiloxane) through photopolymerization, utilizing Darocur® TPO as the photo-initiator. The hydrogels demonstrated exceptional resistance to bacterial adhesion and protein adsorption [30].

Regarding acrylic hydrogels synthesized using photopolymerization, numerous published investigations exist, encompassing copolymer systems [31] as well as reinforced hydrogels [22,32,33], among other variations. There is a limited number of studies available on photopolymerized hydrogels that are based on potassium polyacrylates. Ruan et al. (2004) examined the properties of polyacrylate potassium and polyacrylate sodium hydrogels by employing various photoinitiators during the synthesis process. The hydrogels synthesized with Irgacure 1700 and Irgacure 1800 demonstrated the maximum water absorptions, as discovered in their study [34]. In this study, we employed Irgacure 1700 as a key component.

The superficial instabilities that appear during swelling in gels are a phenomenon known since the XIX century [35]. Tanaka et al. (1987) observed the appearance of patterns in polyacrylamide-based gels during phase transition. This phenomenon affected the understanding of the kinetic process in the gels [36]. In another study by Tanaka et al. (1992), the morphologic evolution and kinetics of superficial patterns in acrylic gels during swelling was reported. A dynamic ordering of patterns was observed [37]. Li et al. (1994) reported the presence of hexagonal-, grain-, and bubble-like shape patterns in ionic N-isopropylacrylamide (N-IPA)-based gels. These patterns were observed below, near, and above the transition phase temperature, and their behavior depended on the temperature, time, external constraint, and thermal history [38]. In a study dealing with photopolymerizable polyhydroxythylmethacrylate (PHEMA) hydrogel films with a crosslinking gradient, Guvendiren et al. (2009) reported a method that allowed them to form various osmotically driven surface patterns without organic solvents for swelling. They observed and captured the shape evolution of such patterns, being first hexagonal structures, then peanut shapes, and then lamellar and finally worm-like patterns [39]. The same research group later reported creasing formation in the gradient PHEMA hydrogels using various solvents. They found that the morphology of patterns depended on the equilibrium linear expansion, which was as a function of the solvent–polymer interaction and the concentration of the crosslinker [40]. Recently, Chuang et al. (2021) demonstrated that the UV irradiation dose and the immersion conditions in DI water determined the characteristics of surface patterns in pHEMA-based hydrogels [41]. The maximum characteristic wavelength of the formed wrinkles depended on the initial immersion time. This dependency had a relationship that followed the power law. Furthermore, it is important to mention that surface structures seem to have an important effect on cells attached on hydrogels. In this regard, Saha et al. (2010) reported that wrinkled patterns on the surface of soft hydrogels made of polyacrylamide greatly influenced cell attachment and cell behavior [42]. Figure 1 displays some of the surface patterns that can be generated in acrylic hydrogels.

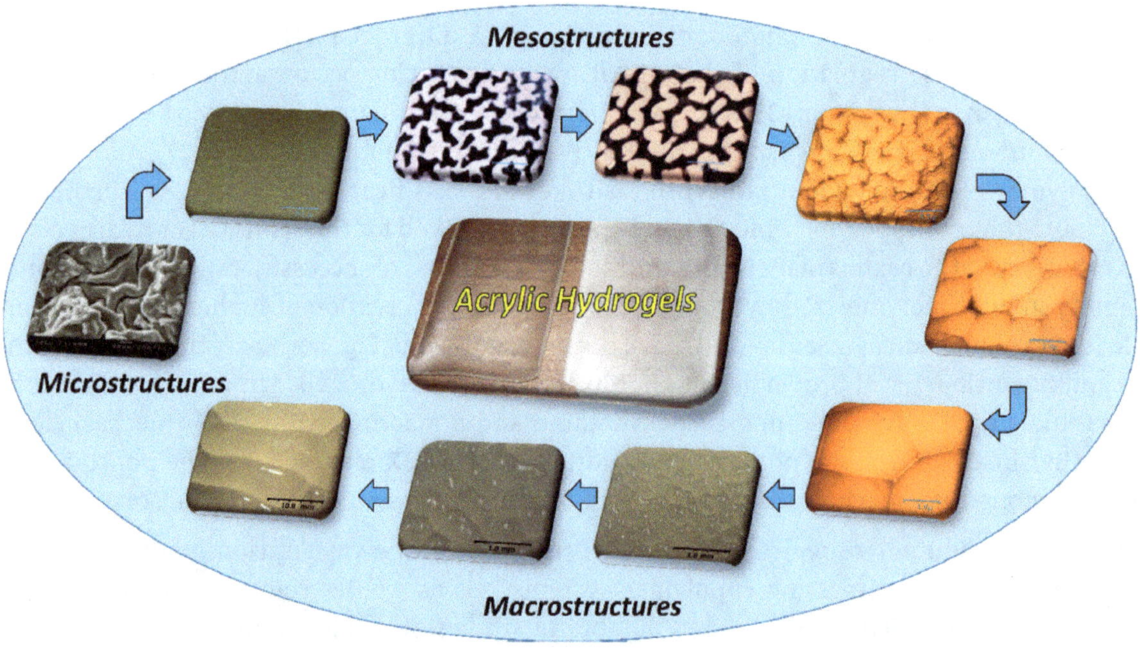

Figure 1. Surface patterns of acrylic hydrogels.

There is a significant interest in polymeric systems that exhibit the spontaneous generation of patterns that can be precisely regulated in terms of their size, order, morphology, and complexity. These materials have a wide range of applications, including coatings, optical filters, batteries, actuators, valves, microfluidic devices, and flexible electronics [43,44,45,46,47]. We synthesized multi-structured hydrogels of poly(potassium acrylate-co-acrylamide) using photopolymerization in this study. By utilizing photo initiation, we were able to regulate the temperature of the synthesis process, resulting in increased structural stability of hydrogels during swelling. The term "multi-structured" was employed to describe the hydrogels' capacity to exhibit patterns or structures at several scales, including micro-, meso-, and macroscales. An evaluation was conducted to determine the impact of the crosslinking agent (glycerol diacrylate or DAG) on the swelling capacity, morphology, and mechanical qualities.

2. RESULTS AND DISCUSSION

Figure 2 illustrates the likely chemical interaction between the acrylic monomers and the crosslinking agent, resulting in the formation of the polymeric network. The cleaning stage [48] was performed to eliminate unwanted residues, including unreacted molecules from the photo-initiator and the monomers, as well as free oligomers.

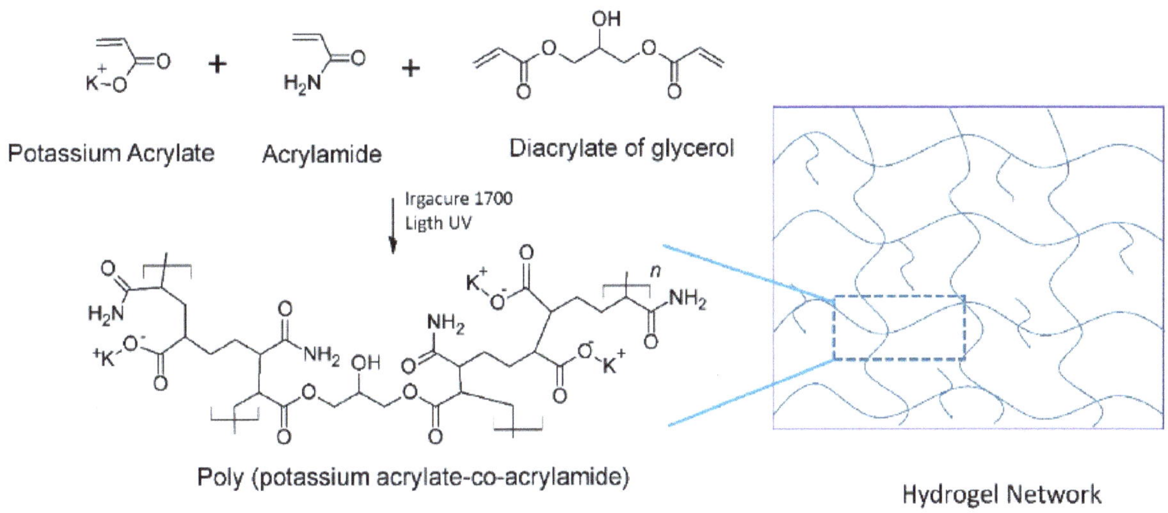

Figure 2. Proposed formation of the polymeric network.

During preparation of hydrogels, it was possible to control the temperature of the reaction by means of photo-polymerization. By controlling the entropy of the reaction solution at low temperatures, an ordered polymeric network was formed, allowing for the preparation of transparent and homogeneous hydrogels. These hydrogels presented a fractal-like structure in three different scales: micro (10-7 to 10-8 m), meso (10-5 to 10-7 m), and macro (10-3 to 10-5 m). Such properties were consistent with those from hydrogels synthetized via redox initiation [20]. The polymer gel fraction (GF) was calculated for all hydrogels (shown in Table 1), as mentioned in the experimental section. All values were higher than 91%, which is consistent with values reported by Rodgers et al., who determined GF's of 89% and 90% for acrylamide and 93% for acrylic acid [29]. Lara-Valencia et al. [10] reported values lower than 90% for acrylic hydrogels obtained by redox initiation and higher than 90% for photoinitiated ones. Wen et al. [31] reported values lower than ours for poly acrylic acid/cellulose nanofibers hydrogels. This could be attributed to the nanofibers, which could have had decreased the UV light transmittance.

Table 1. Maximum swelling and mechanical tests results for hydrogel samples.

Concentration of DAG (wt %)	Polymer Gel Fraction, G_F (%)	Maximum Swelling	K × 10^8	Characteristic Length, λ × 10^6 m	Tensile Strength (Pa)	Young's Modulus (MPa)	Elongation at Break (%)
0.5	91.2 ± 0.8	216.2 ± 7.5	4.28 ± 0.15	1.46 ± 0.038	1979 ± 19	2.64 ± 0.09	1647 ± 52.4
1.0	92.5 ± 1.1	155.1 ± 5.9	4.72 ± 0.18	1.41± 0.036	2087 ± 17	3.66 ± 0.10	1036 ± 43.1
2.0	92.4 ± 0.9	136.8 ± 4.0	5.62 ± 0.16	1.34 ± 0.039	2121 ± 32	3.73 ± 0.18	871 ± 36.0
3.0	93.6 ± 1.4	120.2 ± 4.3	6.03 ± 0.21	1.18 ± 0.025	2206 ± 16	6.48 ± 0.19	660 ± 34.3
4.0	92.1 ± 0.9	99.7 ± 3.9	5.54 ± 0.22	1.08 ± 0.031	2262 ± 34	7.80 ± 0.22	517 ± 32.0
5.0	93.4 ± 1.2	93.7 ± 3.4	5.93 ± 0.21	1.04 ± 0.016	2261 ± 28	8.84 ± 0.43	391 ± 15.2
6.0	93.1 ± 1.3	93.0 ± 3.5	6.86 ± 0.26	0.99 ± 0.007	2343 ± 24	13.43 ± 0.38	349 ± 23.7
7.0	92.9 ± 1.4	79.2 ± 2.3	8.08 ± 0.23	0.97 ± 0.008	2427 ± 20	14.38 ± 0.86	279 ± 21.0
8.0	91.6 ± 0.9	75.0 ± 2.7	7.38 ± 0.26	0.95 ± 0.005	2435 ± 18	15.25 ± 0.44	251 ± 15.5
9.0	92.0 ± 0.9	70.7 ± 2.8	8.34 ± 0.33	0.92 ± 0.008	2482 ± 26	22.34 ± 1.45	210 ± 19.2
10.0	92.7 ± 1.4	69.4 ± 3.3	10.21 ± 0.49	0.91 ± 0.006	2515 ± 33	28.47 ± 0.81	128 ± 12.5

± standard deviation.

2.1. Infrared Spectroscopy FTIR

Figure 3A shows the FTIR spectrum corresponding to acrylic acid. Here, the bands at 3260 cm^{-1} were attributed to the OH groups, while the bands corresponding to the asymmetric and symmetric tension vibrations of CH2 were detected at 2968 cm^{-1} and 1460 cm^{-1}, respectively. The band at 1733 cm^{-1} corresponded to the carboxylic carbonyl group, while the one at 1635 cm-1 was attributed to the vibration of the C=C double bond. The out-of-plane deformation (=C-H) was represented by the band at 812 cm^{-1}. Figure 3B corresponds to the FTIR spectrum of the glycerol diacrylate (DAG), in which the same characteristic bands were observed as in the spectrum shown in Figure 3A. This was because both compounds possess the same functional groups. However, here, we had the presence of a secondary alcohol (-CHOH) that generated bands at 1059 cm^{-1}, 1300 cm^{-1}, and 1413 cm^{-1}. Figure 3C shows the FTIR spectrum of acrylamide. Here, we had the combination of N-H strain and C-N bending in the bands at 1356 cm^{-1} and 675 cm^{-1}, respectively, while the combination of N-H strain and C-N strain occurred at 1610 cm^{-1}. Bands corresponding to the double bond C=O (≈1735 cm^{-1}) and the N-H strain (from 3342 to 3200 cm^{-1}) were also observed. It can be noted that the band for the carbonyl group (C=O) appeared at 1670 cm^{-1}, which could be attributed to the presence of the amino group [10]. In Figure 3D, the spectrum obtained for the acrylamide/potassium acrylate copolymer, with 4 wt % glycerol diacrylate (DAG), is shown. Here, the absence of the C=C double bond signal was evident, and the presence of the characteristic signals for both monomers can be noted. Bands located at 1677 and 1572 cm^{-1} can be attributed to the carbonyl and amino groups, respectively.

2.2. Swelling Kinetics

Figure 4a shows the swelling kinetics of the obtained hydrogels. The swelling of hydrogels with 0.5 wt % DAG was significantly higher than the rest. On the other hand, hydrogels with DAG concentrations from 1 wt % to 10 wt % showed a very similar behavior, having their swelling capacities reduced with increasing crosslinking. It can also be observed that groups of samples followed practically the same behavior (groups of 1, 2, and 3 wt %; groups of 4, 5, and 6 wt %; and groups of 7, 8, and 9 wt %). The maximum swelling ranged from 69 to 217 times the weight of the xerogel. This behavior depended on the amount of DAG (Figure 4). Table 1 shows the values of the maximum swelling for all formulations. Magalhães et al. [49] reported swelling degree values from 144 to 189 for hydrogels where they varied the amount of sodium acrylate in the formulation of poly(sodium acrylate-co-acrylamide) hydrogels. Leitão et al. [50] reported swelling degrees from 620 to 1100 times in acrylamide/potassium acrylate hydrogels crosslinked at 0.05, 0.1, and 0.2% mol with respect to the total mass of monomers.

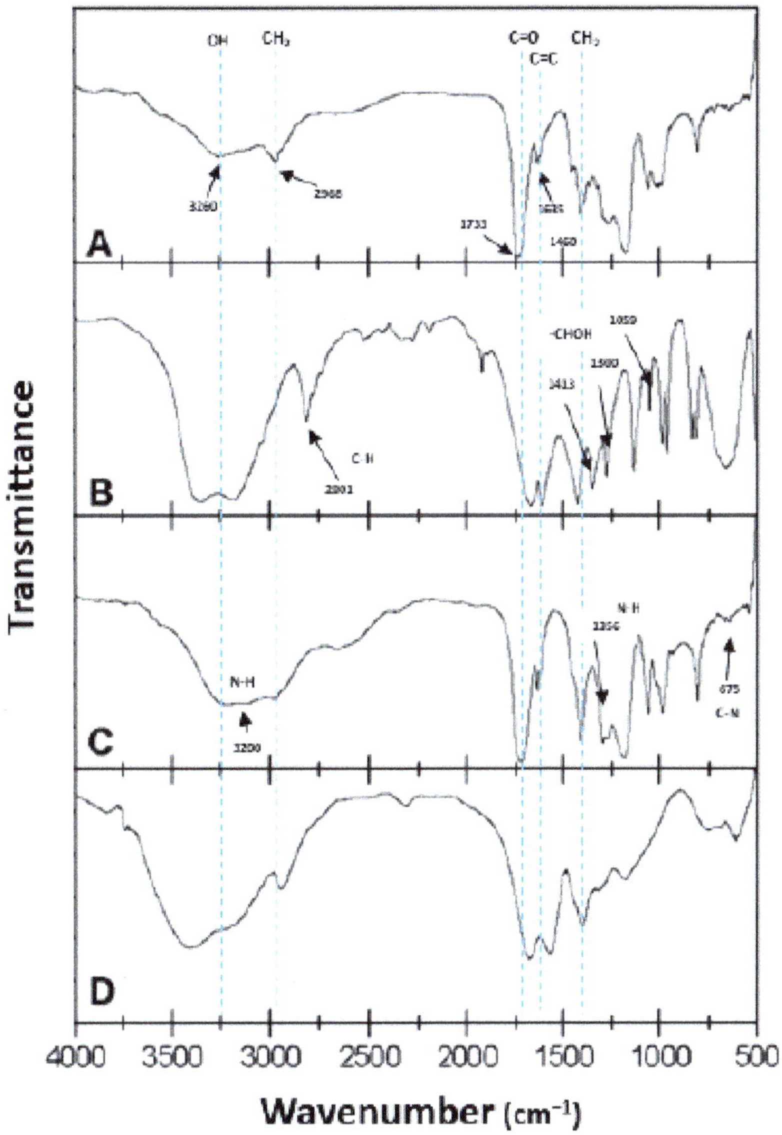

Figure 3. FTIR spectra for (A) acrylic acid, (B) DAG, (C) acrylamide, and (D) hydrogel with 4 wt % DAG.

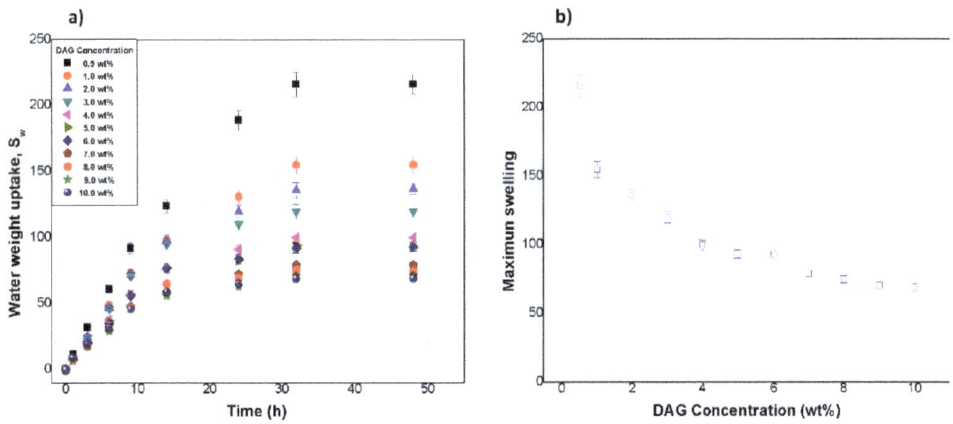

Figure 4. (a) Water weight uptakes of hydrogels as a function of time. (b) Maximum swelling as a function of the crosslinking agent (DAG).

Figure 4b shows the maximum swelling as a function of the crosslinking agent (DAG). A change in slope was observed at 4 wt % of DAG, similar to the characteristic length (λ) of the mesostructures and macrostructures at different times (as can be seen in the figure of the characteristic lengths). This critical value of 4 wt % of DAG is consistent with a critical crosslinking concentration reported in a study from Lopez-Ureta et al. (2008), in which the authors synthetized acrylic acid/acrylamide hydrogels via redox initiation. They also used DAG as a crosslinker [20]. We believe that, starting from this critical concentration (4 wt %), stronger networks were formed, which led to less swelling in the hydrogels. At lower DAG concentrations, the hydrogels physically collapsed near the equilibrium, which can be attributed to weaker networks.

The Schott's model was used in order to analyze the adsorption behavior of hydrogels in water. Figure 5a shows the linear behavior obtained from Equation (5) (see below in the Experimental section), which showed linearity values close to 1, demonstrating that it followed the model adequately.

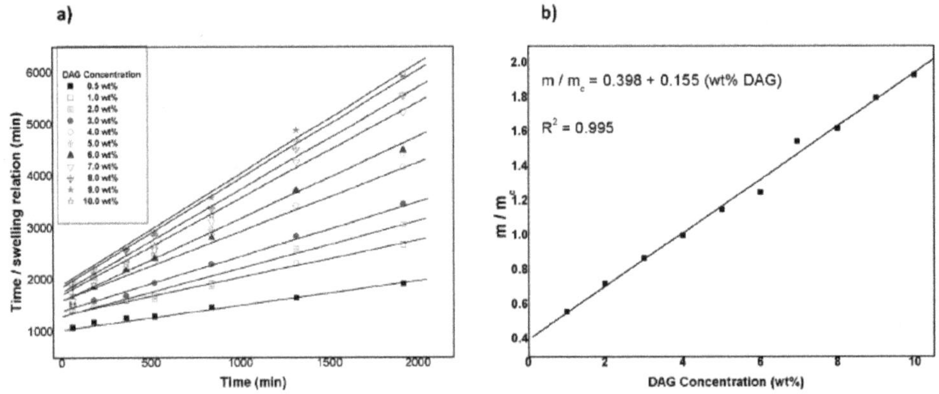

Figure 5. (a) Linearization of the Schott model for all samples. (b) Normalized slopes from the linearization of the Schott model (m/mc) as a function of DAG concentration.

In this context, Figure 5b shows the normalized slopes for all samples (m/mcritical). Here, a linear dependence can be observed. Table 1 shows the values of the constant K for all the samples evaluated.

2.3. Surface Analysis

Figure 6a shows images of the surface of the xeWrogels where it can be observed that with a higher the amount of DAG, the structuring of the hydrogel was more noticeable. A critical crosslinking concentration (CCLC) of around 4 wt % was observed. For DAG concentrations below 4 wWt %, an irregular surface arrangement was noted, while for DAG concentrations higher than 4 wt %, a compact surface, aligned preferentially in one direction, and with a more regular surface texture, was observed. Similar results were reported by López-Ureta et al. [20]. The characteristic length of the structure was in the microscale in xerogels with 10 wt % of DAG but increased when the concentration of DAG decreased. The mesostructure of hydrogels was observed by optical microscopy. When the water penetrated the hydrogels, four distinctive wrinkle patterns were spontaneously formed and transited to random worms, lamellae, peanuts, and ordered hexagonal patterns during swelling, until the maximum swelling was reached. Figure 6b shows images of hydrogels with 1 wt %, 2 wt %, 4 wt %, 6 wt %, and 10 wt % of DAG at short hydration times (10, 40, and 180 s). The order of magnitude of the structures was in the range of 10-6 m. The size of structures increased with hydration time, becoming more defined as the amount of DAG increased. There was evidence in this research that the crosslinking gradient plays a critical role in the evolution of surface patterns and their ordering along with the lateral confinement of the hydrogel, promoting anisotropic osmotic pressure along with the thickness. Similar systems were described [35,36,39,40].

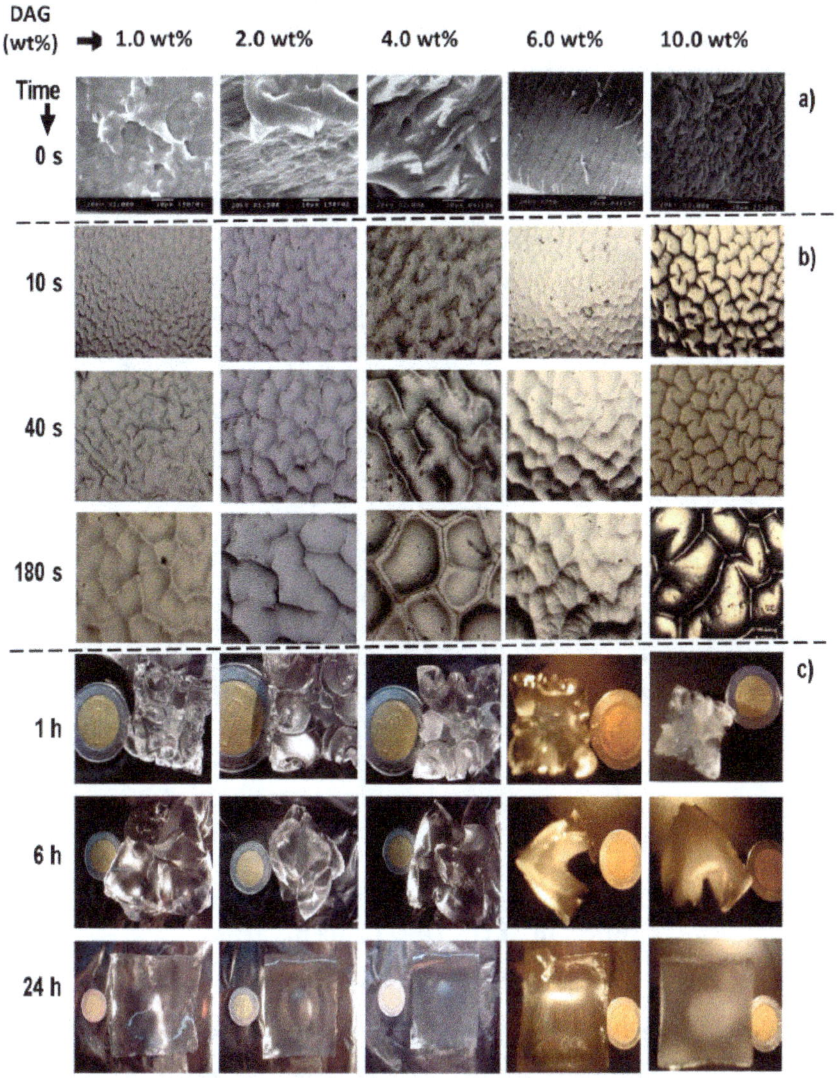

Figure 6. Images for hydrogels with 1 wt %, 2 wt %, 4 wt %, 6 wt %, and 10 wt % of DAG. (a) Micrograph (SEM) of xerogels; (b) polarized light micrographs of swollen hydrogels; (c) macrostructures of swollen hydrogels; all as functions of swelling time.

Figure 5b shows images of hydrogels exhibiting meso- and macrostructures at different swelling times as a function of DGA. The macrostructure was defined from the first minutes of contact of the xerogel with water. After 24 h, all macrostructures disappeared. Equilibrium was reached approximately 32 h after the start of the swelling process. Depending on the degree of crosslinking of the hydrogel, the hydrogel can undergo large volume changes during swelling. During the water absorption process, the resulting compressive stresses can be quite large, even exceeding the elastic modulus of the gel. When this compressive stress becomes large enough (and the material cannot delaminate and buckle macroscopically), an elastic instability arises in which the free surface folds in on itself to locally relieve the compressive stress.

Table 1 shows the average characteristic length values for the mesostructure, measured at 60 s of hydration. Initially, the gel was in an almost stress-free state; however, immersion in a solvent led to swelling of the network until the osmotic stress, due to the mixing of solvent with polymer chains and counterions, was balanced by the elastic strain due to chain stretching. Because of the mechanical constraint provided by the substrate, a gel attached to the surface that was much thinner than its lateral dimensions can only expand in the direction normal to the surface. The result of this uniaxial expansion is that the gel undergoes a state of equibiaxial compressive stress [35].

Figure 7a shows the characteristic length (λ) of the mesostructures at 60 s and 180 s as a function of DAG concentration. The order of magnitude (10-6 m) increased with the water absorption. Similar to Figure 3B and Figure 7b, when the DAG concentration was 4 wt %, there was a clear change in the behavior of the data. This was probable due to the reordering of the polymeric network. Figure 7b displays the relationship between λ and the DAG concentration, but this time at 60 and 360 min. The order of magnitude was 10-3 m. The mesoestructures disappeared when the hydrogels reached their swelling equilibrium.

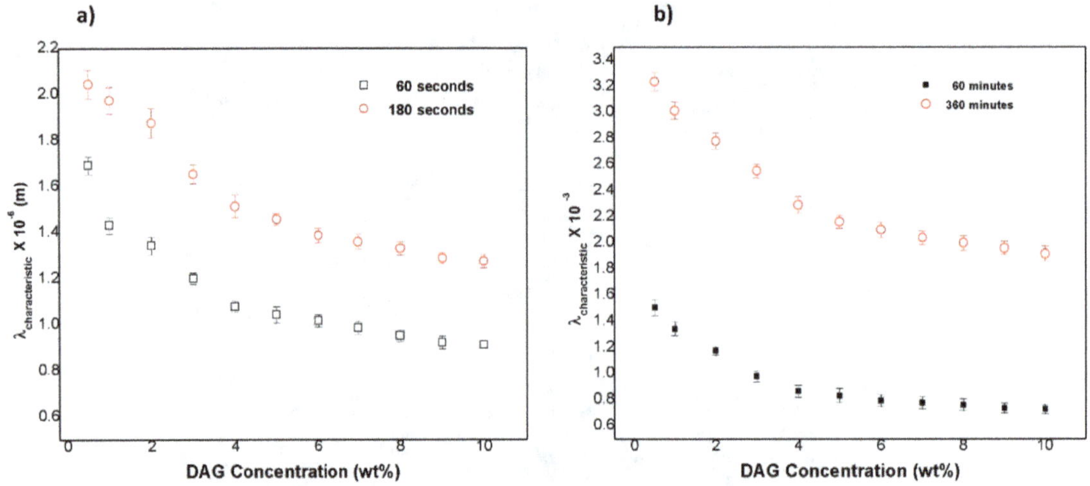

Figure 7. Characteristic length (λ) of structures at (a) 60 s and 180 s and at (b) 60 min and 360 min as a function of the DAG concentration.

The size of the meso- and macrostructures slightly decreased when the DAG concentration also decreased, generating regular "packed" shapes and thus more compacted hydrogels. This type of superficial morphology was observed in all sides of the samples (cubes), demonstrating the tridimensional nature of the patterns. The size of these well-defined structures augmented with the time after immersion. However, the number of these structures remained constant.

2.4. Mechanical Tests

Table 2 shows the values of tensile strength, elongation at break, and Young's modulus. Figure 8a displays tensile test curves for all compositions evaluated, and Figure 8b shows the Young's modulus as a function of the concentration of DAG. It can be observed that the elastic modulus increased when the concentration of DAG increased. In the case of the elongation at break, it was observed that it decreased when DAG concentration increased. For the tensile strength, it generally increased when the concentration of DAG increased as well. These results were expected due to the formation of stronger and stiffer polymeric networks as the DAG concentration increased.

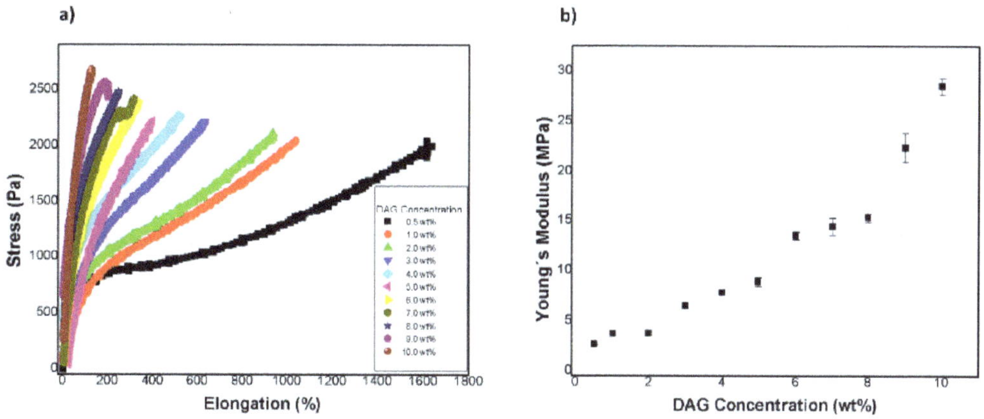

Figure 8. (a) Tensile test curves for all compositions evaluated. (b) Average Young's modulus as a function of the concentration of the crosslinking agent (DAG).

Table 2. Formulations of hydrogels obtained.

Substance	Amount (g)										
Acrylamide	49.0										
Acrylic acid	51.0										
DAG	0.5	1.0	2.0	3.0	4.0	5.0	6.0	7.0	8.0	9.0	10.0
Irgacure 1700	0.03										
Water	100.0										
KOH	neutralizer and generating potassium acrylate										

3. CONCISE SUMMARY

The synthesis temperature was regulated using the photo-polymerization of potassium acrylate and acrylamide. Through photo-polymerization, clear and uniform hydrogels were prepared using potassium acrylate-co-acrylamide. The hydrogels that were formed exhibited a fractal-like structure at several scales, including the micro-, meso-, and macroscales. The swelling of the xerogel reached a maximum value ranging from 69 to 217 times its weight. The extent of swelling was contingent upon the quantity of the crosslinking agent (DAG). The minimum concentration of DAG required for optimal mechanical properties and structural qualities in the hydrogel was determined to be 4 wt %. As the amount of DAG grew, the mechanical resistance also increased, while the elongation at break and swelling capacity values fell. There was a noticeable change in the shape, ability to expand, and stiffness of the material when the weight percentage of DAG reached between 4 and 5%. This work serves as an initial step towards developing improved multi-structured hydrogel materials with diverse uses, including coatings, batteries, flexible electronics, actuators, and optical filters.

4. MATERIALS AND METHODS

4.1. Materials

The monomers used for hydrogel preparation were acrylamide and acrylic acid from Aldrich with purities of 98.5% and 99.3%, respectively. Potassium hydroxide (KOH), 99%, was also purchased from Aldrich. As a photoinitiator, a mixture of 25% bis(2,6-dimethoxybenzoyl)-2,4,4-trimethyl pentylphosphineoxide and 75% 2-hydroxy-2-methyl-1-phenyl-propane-1-one (Irgacure 1700) from Ciba Speciality Chemicals

Inc. was used. Glycerol diacrylate (DAG), 97% purity, was obtained from Industria Azteca Integral.

4.2. Synthesis of Hydrogels

Figure 9 shows a scheme of the complete methodology of this work. The acrylic acid was dissolved in bidistilled water according to the formulation shown in Table 2. The solution was taken to a pH 7 with the addition of a KOH solution at 47 wt %. The solution was kept below 25 °C to avoid thermal polymerization. Subsequently, the acrylamide and DAG were added, the mixture was stirred until the system was homogeneous, and the temperature was lowered to 2 °C. Finally, 1 mL of photoinitiator solution (3 wt % in methanol) was added. The reaction solution was poured into a 0.450 L glass semi-infinite plate reactor, and nitrogen was bubbled for 15 min. It was placed 15 cm away from the lamp (Tecno F15T8-BLB 20 W, 127 v) rich in 366 nm wavelength radiation, inside an isothermal bath at 2 °C for 1 h (to achieve higher conversion). After the reaction time, the hydrogels were removed, and 0.5 × 0.5 × 0.3 mm3 samples were cut and allowed to dry until constant weight, and then were introduced into double-distilled water at 25 °C for three days. The water was changed every 24 h, and they were submerged in the water again for three days at 45 °C. Once the cleaning process was completed, the hydrogels were left to dry at room temperature for 5 days and then put in a vacuum oven at 40 °C until constant weight. The polymer gel fraction (GF) was calculated as follows:

$$G_F(crosslinked\ polymer\ \%) = \left(\frac{W_d}{W_0}\right) * 100 \quad (1)$$

where Wd is the weight of the dry insoluble part of the hydrogel after extraction with water, and W0 is the initial weight of the xerogel [10].

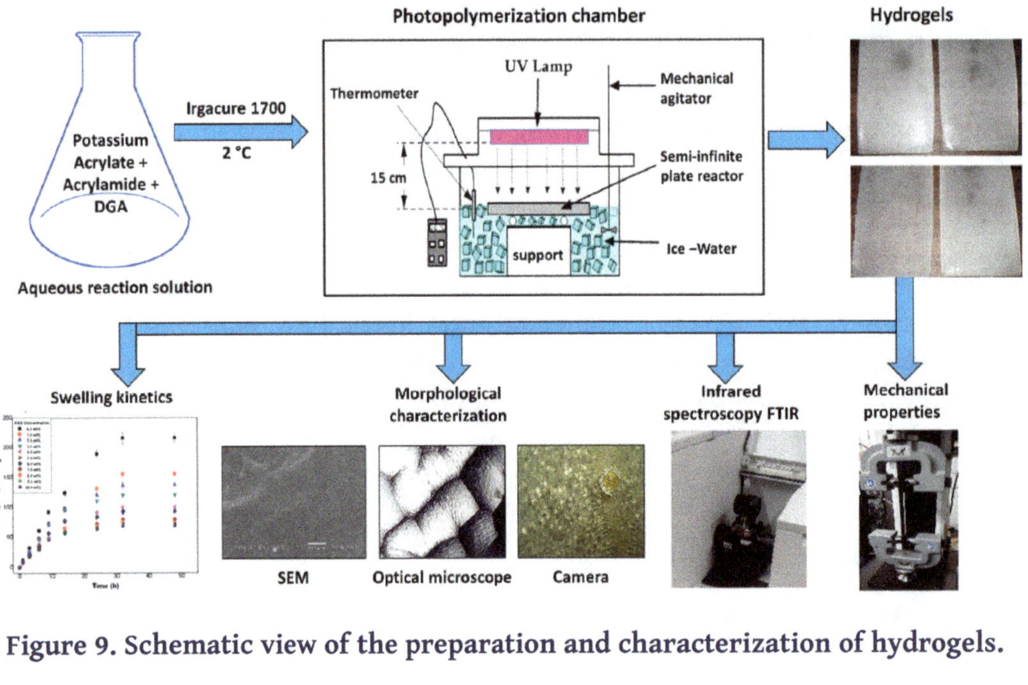

Figure 9. Schematic view of the preparation and characterization of hydrogels.

4.3. Swelling Kinetics

Dried samples were weighed and placed in double-distilled water at 25 °C. The samples were taken out and weighed at different times. Absorbent paper was used to remove the excess of water on the surface of samples. Water absorption was calculated by the difference in weight between the weight of the dry sample and the weight of the swollen sample, using the following equation:

$$S_w \left(\frac{W_t - W_0}{W_0} \right) \tag{2}$$

where Wt and W0 are the weight of the hydrogel at time t and the weight of the xerogel, respectively. To model the swelling kinetics, the second-order model is commonly used, as proposed by Schott [51], which has been used to predict the swelling in acrylic hydrogels [10,20,52]:

$$\frac{dS_w}{dt} = K(S_{w\infty} - S_w)^2 \tag{3}$$

where Sw and Sw∞ are the swellings at time t and at equilibrium swelling, respectively, and K is a constant of the system.

Integrating Equation (3), in the limits of t, Sw, and 0, this gives

$$S_w = \frac{K\,S_w^2\,t}{1 + K\,S_w\,t} \tag{4}$$

Rewriting Equation (4):

$$\frac{t}{S_w} = \frac{1}{K\,S_{w\infty}^2} + \frac{1}{S_{w\infty}} t \tag{5}$$

Equation (5) represents second-order kinetics. In this case, the rate of swelling at any time is directly proportional to the square of the still available swelling capacity, that is, to the solvent uptake that has not yet occurred before reaching the maximum or equilibrium uptake [52].

4.4. Morphological Characterization

The JEOL JSM 5400 LV scanning electron microscope was utilized for morphological characterization. SEM micrographs were acquired for xerogels that were coated with gold. The hydrated samples were seen at various time intervals (10, 20, 30, 40, 40, 60, 120, and 180 s) using an OLIMPUS BX4OF optical microscope equipped with a 40× objective lens to visualize their mesostructure. The images were acquired using an SCC-131A SAMSUNG digital video camera that was modified to be compatible with the microscope. The macrostructure was seen using a Hitachi CCD camera. A coin with a diameter of 21.0 mm was utilized as a reference for size comparison.

4.5. Mechanical Characterization

The hydrogels were mechanically measured using an SFM-10 Universal Testing Machine. The strain rate was adjusted to 50 mm/min at ambient temperature. The experimental data were acquired by conducting a test according to the ASTM D-638-14 Standard Test Method [53], which is a recognized method for determining the tensile properties of plastics. A total of eight specimens were used in the test. The calculation of Young's modulus was derived from the beginning gradient of the tensile curve.

AT A GLANCE

This study utilized a photo-polymerization method to produce potassium acrylate-co-acrylamide hydrogels that exhibit improved mechanical properties. These hydrogels also possess distinct microstructures while dry and distinctive meso- and macrostructures when hydrated. The characteristics of the hydrogels were contingent upon the concentration of the crosslinking agent. An analysis was conducted on the mechanical characteristics, swelling capacity, and morphology, revealing a distinct transition at a critical concentration of the crosslinker. Swelling resulted in the emergence of shape-evolving surface patterns at various levels of size and structure, as observed in morphology. The mechanical properties were significantly affected by these surface features. Hydrogels having organized structures shown superior mechanical capabilities in comparison to hydrogels lacking structure. The study establishes the critical concentration of crosslinking using glycerol diacrylate as a reference for the future production of acrylic hydrogels with many structures and improved characteristics.

REFERENCES

1. Amashta, I.A.K.; Trabanca, O.K.; Trabanca, D.K. Materiales inteligentes: Hidrogeles macromoleculares. Algunas aplicaciones biomédicas. In Anales de la Real Sociedad Española de Química; Real Sociedad Española de Química: Madrid, Spain, 2005; Volume 4, pp. 35–50.
2. Hoffman, A.S. Hydrogels for biomedical applications. Adv. Drug Deliv. Rev. 2012, 64, 18–23.
3. Kiatkamjornwong, S. Superabsorbent polymers and superabsorbent polymer composites. ScienceAsia 2007, 33, 39–43.
4. Khare, A.R.; Peppas, N.A. Swelling/deswelling of anionic copolymer gels. Biomaterials 1995, 16, 559–567.
5. Bautista, F.; Garcia, S.; Jimenez, R.; Lopez, L.; Orozco, E.; Prado, M.; Reyes, I. Influence of the Synthesis Thermal History on the Structure of Acrylate Based Hydrogels; ANTEC-CONFERENCE PROCEEDINGS: Boston, MA, USA, 2005; Volume 5, p. 373.
6. Costa, M.C.G.; Freire, A.G.; Lourenço, D.V.; Sousa, R.R.D.; Feitosa, J.P.D.A.; Mota, J.C.A. Hydrogel composed of potassium acrylate, acrylamide, and mineral as soil conditioner under saline conditions. Sci. Agric. 2021, 79, e20200235.
7. Jasim, L.S.; Aljeboree, A.M. Removal of Heavy Metals by Using Chitosan/Poly (Acryl Amide-Acrylic Acid) Hydrogels: Characterization and Kinetic Study. NeuroQuantology 2021, 19, 31–38.
8. Zdravković, A.; Nikolić, L.; Ilić-Stojanović, S.; Nikolić, V.; Najman, S.; Mitić, Ž.; Ćirić, A.; Petrović, S. The removal of heavy metal ions from aqueous solutions by hydrogels based on N-isopropylacrylamide and acrylic acid. Polym. Bull. 2018, 75, 4797–4821.
9. Washington, R.P.; Steinbock, O. Frontal Polymerization Synthesis of Temperature-Sensitive Hydrogels. J. Am. Chem. Soc. 2001, 123, 7933–7934.
10. Lara-Valencia, V.A.; Dávila-Soto, H.; Moscoso-Sánchez, F.J.; Figueroa-Ochoa, E.B.; Carvajal-Ramos, F.; Fernández-Escamilla, V.V.A.; González-Álvarez, A.; Soltero-Martínez, J.F.A.; Macías-Balleza, E.R.; Enríquez, S.G. The use of polysaccharides extracted from seed of Persea americana var. Hass on the synthesis of acrylic hydrogels. Quim. Nova 2018, 41, 140–150.
11. Isik, B.; Kis, M. Preparation and determination of swelling behavior of poly (acrylamide-co-acrylic acid) hydrogels in water. J. Appl. Polym. Sci. 2004, 94, 1526–1531.
12. Thakur, A.; Wanchoo, R.K.; Singh, P. Structural parameters and swelling behavior of pH sensitive poly (acrylamide-co-acrylic acid) hydrogels. Chem. Biochem. Eng. Q. 2011, 25, 181–194.
13. Cheng, W.M.; Hu, X.M.; Zhao, Y.Y.; Wu, M.Y.; Hu, Z.X.; Yu, X.T. Preparation and swelling properties of poly (acrylic acid-co-acrylamide) composite hydrogels. e-Polymers 2017, 17, 95–106.
14. Sennakesavan, G.; Mostakhdemin, M.; Dkhar, L.K.; Seyfoddin, A.; Fatihhi, S.J. Acrylic acid/acrylamide based hydrogels and its properties—A review. Polym. Degrad. Stab. 2020, 180, 109308.

15. Argade, A.B.; Peppas, N.A. Poly (acrylic acid)-poly (vinyl alcohol) copolymers with superabsorbent properties. J. Appl. Polym. Sci. 1998, 70, 817–829.

16. Kiatkamjornwong, S.; Wongwatthanasatien, R. Superabsorbent polymer of poly [acrylamide-co-(acrylic acid)] by foamed polymerization. I. synthesis and water swelling properties. Macromol. Symp. 2004, 207, 229–240.

17. Pourjavadi, A.; Kurdtabar, M. Collagen-based highly porous hydrogel without any porogen: Synthesis and characteristics. Eur. Polym. J. 2007, 43, 877–889.

18. Tomar, R.S.; Gupta, I.; Singhal, R.; Nagpal, A.K. Synthesis of Poly (Acrylamide-co-Acrylic Acid) based Superabsorbent Hydrogels: Study of Network Parameters and Swelling Behaviour. Polym. -Plast. Technol. Eng. 2007, 46, 481–488.

19. Xie, J.; Liu, X.; Liang, J. Absorbency and adsorption of poly (acrylic acid-co-acrylamide) hydrogel. J. Appl. Polym. Sci. 2007, 106, 1606–1613.

20. Lopez-Ureta, L.C.; Orozco-Guareño, E.; Cruz-Barba, L.E.; Gonzalez-Alvarez, A.; Bautista-Rico, F. Synthesis and characterization of acrylamide/acrylic acid hydrogels crosslinked using a novel diacrylate of glycerol to produce multistructured materials. J. Polym. Sci. Part A Polym. Chem. 2008, 46, 2667–2679.

21. Nesrinne, S.; Djamel, A. Synthesis, characterization and rheological behavior of pH sensitive poly (acrylamide-co-acrylic acid) hydrogels. Arab. J. Chem. 2017, 10, 539–547.

22. Jiménez-Amezcua, R.M.; Villanueva-Silva, R.J.; Muñoz-García, R.O.; Macias-Balleza, E.R.; Flores-Sahagun, T.H.S.; Lomelí-Ramírez, M.G.; Torres-Rendon, J.G.; Garcia-Enriquez, S. Preparation of Agave tequilana Weber Nanocrystalline Cellulose and its Use as Reinforcement for Acrylic Hydrogels. BioResources 2021, 16, 2731–2746.

23. Mironi-Harpaz, I.; Wang, D.Y.; Venkatraman, S.; Seliktar, D. Photopolymerization of cell-encapsulating hydrogels: Crosslinking efficiency versus cytotoxicity. Acta Biomater. 2012, 8, 1838–1848.

24. O'Brien, A.K.; Bowman, C.N. Impact of Oxygen on Photopolymerization Kinetics and Polymer Structure. Macromolecules 2006, 39, 2501–2506.

25. O'Brien, A.K.; Bowman, C.N. Modeling the Effect of Oxygen on Photopolymerization Kinetics. Macromol. Theory Simul. 2006, 15, 176–182.

26. Fouassier, J.P.; Allonas, X.; Burget, D. Photopolymerization reactions under visible lights: Principle, mechanisms and examples of applications. Prog. Org. Coat. 2003, 47, 16–36.

27. Nguyen, K.T.; West, J.L. Photopolymerizable hydrogels for tissue engineering applications. Biomaterials 2002, 23, 4307–4314.

28. Decker, C.; Moussa, K. Photopolymerization of multifunctional monomers in condensed phase. J. Appl. Polym. Sci. 1987, 34, 1603–1618.

29. Rodgers, Z.L.; Hughes, R.M.; Doherty, L.M.; Shell, J.R.; Molesky, B.P.; Brugh, A.M.; Forbes, M.D.E.; Moran, A.M.; Lawrence, D.S. B12-Mediated, Long Wavelength Photopolymerization of Hydrogels. J. Am. Chem. Soc. 2015, 137, 3372–3378.

30. Lin, W.; Zhang, J.; Wang, Z.; Chen, S. Development of robust biocompatible silicone with high resistance to protein adsorption and bacterial adhesion. Acta Biomater. 2011, 7, 2053–2059.

31. Weiqing, R.; Jinliang, Q.; Yuli, H.; Aijie, N. Superabsorbent resin of acrylic acid/ammonium acrylate copolymers synthesized by ultraviolet photopolymerization. J. Appl. Polym. Sci. 2004, 95, 546–555.

32. Wen, Y.; Zhu, X.; Gauthier, D.E.; An, X.; Cheng, D.; Ni, Y.; Yin, L. Development of poly (acrylic acid)/ nanofibrillated cellulose superabsorbent composites by ultraviolet light induced polymerization. Cellulose 2015, 22, 2499–2506.

33. Martínez-Salcedo, S.L.; Torres-Rendón, J.G.; García-Enriquez, S.; Anzaldo-Hernández, J.; Silva-Guzmán, J.A.; de Muniz, G.I.B.; Lomelí-Ramírez, M.G. Physicomechanical Characterization of Poly (acrylic acid-co-acrylamide) Hydrogels Reinforced with TEMPO-oxidized Blue Agave Cellulose Nanofibers. Fibers Polym. 2022, 23, 1161–1170.

34. Ruan, W.; Qiao, J.; Huang, Y.; Niu, A. Synthesis of superabsorbent resin by ultraviolet photopolymerization. J. Appl. Polym. Sci. 2004, 92, 1618–1624.

35. Trujillo, V.; Kim, J.; Hayward, R.C. Creasing instability of surface-attached hydrogels. Soft Matter 2008, 4, 564–569.

36. Tanaka, T.; Sun, S.T.; Hirokawa, Y.; Katayama, S.; Kucera, J.; Hirose, Y.; Amiya, T. Mechanical instability of gels at the phase transition. Nature 1987, 325, 796–798.

37. Tanaka, H.; Tomita, H.; Takasu, A.; Hayashi, T.; Nishi, T. Morphological and kinetic evolution of surface patterns in gels during the swelling process: Evidence of dynamic pattern ordering. Phys. Rev. Lett. 1992, 68, 2794–2797.

38. Li, C.; Hu, Z.; Li, Y. Temperature and time dependencies of surface patterns in constrained ionic N-isopropylacrylamide gels. J. Chem. Phys. 1994, 100, 4645–4652.

39. Guvendiren, M.; Yang, S.; Burdick, J.A. Swelling-induced surface patterns in hydrogels with gradient crosslinking density. Adv. Funct. Mater. 2009, 19, 3038–3045.

40. Guvendiren, M.; Burdick, J.A.; Yang, S. Solvent induced transition from wrinkles to creases in thin film gels with depth-wise crosslinking gradients. Soft Matter 2010, 6, 5795–5801.

41. Chuang, Y.F.; Wei, M.K.; Yang, F.; Lee, S. Water-driven surface wrinkling of poly (2-hydroxyethyl methacrylate) after ultraviolet irradiation. J. Polym. Res. 2021, 28, 1–11.

42. Saha, K.; Kim, J.; Irwin, E.; Yoon, J.; Momin, F.; Trujillo, V.; Schaffeer, D.V.; Healy, K.E.; Hayward, R.C. Surface creasing instability of soft polyacrylamide cell culture substrates. Biophys. J. 2010, 99, L94–L96.

43. Stafford, C.M.; Harrison, C.; Beers, K.L.; Karim, A.; Amis, E.J.; VanLandingham, M.R.; Kim, H.C.; Volksen, W.; Miller, R.D.; Simonyi, E.E. A buckling-based metrology for measuring the elastic moduli of polymeric thin films. Nat. Mater. 2004, 3, 545–550.

44. Sidorenko, A.; Krupenkin, T.; Taylor, A.; Fratzl, P.; Aizenberg, J. Reversible switching of hydrogel-actuated nanostructures into complex micropatterns. Science 2007, 315, 487–490.

45. Holmes, D.P.; Ursiny, M.; Crosby, A.J. Crumpled surface structures. Soft Matter 2008, 4, 82–85.

46. Pandey, A.; Holmes, D.P. Swelling-induced deformations: A materials-defined transition from macroscale to microscale deformations. Soft Matter 2013, 9, 5524–5528.

47. Phadnis, A.; Manning, K.C.; Sanders, I.; Burgin, T.P.; Rykaczewski, K. Droplet-train induced spatiotemporal swelling regimes in elastomers. Soft Matter 2018, 14, 5869–5877.

48. Stockmayer, W.H. Theory of molecular size distribution and gel formation in branched polymers II. General cross linking. J. Chem. Phys. 1944, 12, 125–131.

49. Magalhães, A.S.G.; Almeida Neto, M.P.; Bezerra, M.N.; Ricardo, N.M.; Feitosa, J. Application of FTIR in the determination of acrylate content in poly (sodium acrylate-co-acrylamide) superabsorbent hydrogels. Química Nova 2012, 35, 1464–1467.

50. Leitão, R.C.; Moura, C.P.D.; da Silva, L.R.; Ricardo, N.M.; Feitosa, J.; Muniz, E.C.; Fajardo, R.A.; Rodrigues, F.H. Novel superabsorbent hydrogel composite based on poly (acrylamide-co-acrylate)/nontronite: Characterization and swelling performance. Química Nova 2015, 38, 370–377.

51. Schott, H. Swelling kinetics of polymers. J. Macromol. Sci. Part B 1992, 31, 1–9.

52. Katime, I.; Velada, J.L.; Novoa, R.; Díaz de Apodaca, E.; Puig, J.; Mendizabal, E. Swelling Kinetics of Poly (acrylamide)/Poly (mono-n-alkyl itaconates) Hydrogels. Polym. Int. 1996, 40, 281–286.

53. ASTM International. D638-14 Standard Test Method for Tensile Properties of Plastics; ASTM International: West Conshohocken, PA, USA, 2014.

4 Chapter

PEG HYDROGELS THAT CAN BE CURED USING UV-VIS LIGHT FOR VARIOUS BIOMEDICAL APPLICATIONS, OFFERING SEVERAL FUNCTIONS

1. INTRODUCTION: AN OVERVIEW

Hydrogels are highly appropriate as tissue scaffolds in tissue engineering and medical science because they possess adjustable features such as water content, swellability, diffusivity, and stiffness [1,2]. Explicit control over molecular structure and mechanical properties, such as elasticity, cross-linking degree, or surface morphology, opens up several possibilities [3]. Hydrogels are ideal structures for light-responsive functioning due to their ability to manipulate characteristics using light [4]. Stimuli-responsive hydrogel materials have the ability to alter their mechanical characteristics when exposed to light. However, there has been limited research on the optical characteristics of these materials and their prospective photonic capabilities have received little attention [5]. The ability to manipulate optical properties and incorporate optical functionality creates new possibilities for multifunctional biomedical materials, such as enhanced intraocular lens (IOL) implants.

Cataract, a condition characterized by the permanent cloudiness of the eye's natural lens, is a leading cause of blindness worldwide. The only effective treatment for cataract is the replacement of the clouded lens with an artificial IOL implant. Modern foldable hydrogel lenses are considered state-of-the-art IOLs [6]. Common issues associated with intraocular lenses (IOLs) include the occurrence of calcification after surgery [7] and the development of secondary cataracts [8]. The mechanisms responsible for the occurrence of postoperative clouding, which arises in vivo due to interaction with the biological environment, remain poorly understood.

Here is where the concept of volume holographic structure becomes relevant. Prospective intraocular lenses (IOLs), made from multifunctional biological material that incorporates optical functionality, have the potential to effectively focus light into the retina using an optically structured volume [9]. Consequently, the form and exterior of the IOL remain unoccupied and accessible for other functions. Therefore, any more alterations to the surface are not required in order to achieve specific interactions with the biological environment. To achieve this, we suggest a technique that involves integrating optical structuring within the volume and implementing a targeted surface change. Thus, volume holographic structuring can be utilized to integrate three-dimensional optical structures with precise functionality in terms of diffractive qualities.

Volume holography is a fascinating area of study that involves using photo-responsive polymers to create diffractive structures through a holographic exposure that is spatially manipulated [10].

Holographic elements, such as diffractive structures, has the ability to perform traditional optical functions. Additionally, they have a very flat shape and are lightweight. This creates a significant opportunity to replace traditional refractive optical systems or enhance them with additional capabilities. In order for any scenario to be possible, it is necessary to have appropriate materials that can be patterned with photos and demonstrate utility through structure. Holographic lenses are a specific type of diffractive structures that exhibit classical optical function.

Poly(ethylene glycol) (PEG)-based hydrogels are highly favorable for tissue-engineering scaffolds because of their biocompatibility and inherent ability to resist protein adsorption and cell adhesion [11]. In addition, hydrogels made from poly(ethylene glycol)-diacrylate (PEG-DA) might serve as ideal materials for creating intricate 3D patterns within their structure [12]. PEG macromers ended with acrylate groups undergo fast polymerization in the presence of photoinitiators that produce radicals upon exposure to light [13]. PEG-DA hydrogels are intriguing as scaffolds because they can be customized with specific bioactivity via light-based patterning [12]. Within this particular framework, it was demonstrated that hydrogels can be constructed photolithographically by diffusion processes, which serve as the foundation for volume holography.

In addition, the use of gold nanoparticles in the micropatterning of PEG-based hydrogels enables the creation of functionalized PEG-based hydrogel films [14,15]. The incorporation and arrangement of nanoparticles allows for the alteration of optical characteristics at both the microscale and nanoscale, resulting in the creation of holographic nanoparticle-polymer composite gratings [16].

The selection of the photoinitiator (PI) is crucial in determining the specific photo-response of a given material. The features of the PI exert a significant impact on the production of holographic gratings in the corresponding material [17]. Additionally, it impacts the extent to which specific parameters are fulfilled, such as the ability to withstand humidity [18]. Eosin-Y (EY) and Erythrosin B (EB) are two potential photosensitizers used to create holographic gratings in an AA/PVA photopolymer [19]. EY serves as a photosensitizer because of its exceptional spectroscopic characteristics, rendering it appropriate for utilization with visible light sources and non-hazardous for living organisms [20]. EB is exclusively applicable to free radical polymerization [19].

2. RESULTS AND DISCUSSION

2.1. Gel Formation

PEG-DA was combined with PI, specifically Irgacure, EB, and EY. The films were fabricated using photopolymerization using a 366 nm light source for a duration of 1 hour. The novel gels exhibit favorable optical transparency, mechanical integrity, flexibility, and stability, comparable to existing gels that are based on PEG-DA [21,22,23]. In order to form a gel using PEG-DA with Irgacure, EB, and EY as the photoinitiators, a minimum concentration of photoinitiator was required. No hydrogel was generated when the concentration of PI was reduced. Table 1 displays the minimal concentration required for each of the different PIs. The minimal concentration of PI required to produce the novel hydrogels using PEG-DA and other PIs was determined to be 0.025% for Irgacure, 0.1% for EB, and 0.5% for EY, respectively.

Table 1. Minimum PI concentration to make gel for the different PIs.

Photoinitiator (PI)	Minimum PI Concentration to Make Gel with PEG-DA
Irgacure 2959	0.025%
EB	0.1%
EY	0.5%

2.2. UV-Vis Spectra

Figure 1 shows the UV-Vis spectra before and after crosslinking for the novel PIs. In general, for all new PEG-DA hydrogels, we find spectra shifted somewhat toward higher wavelengths compared to the pure PIs [24,25], while crosslinking tends to cause a small shift toward lower wavelengths, as already substantiated in the literature [26].

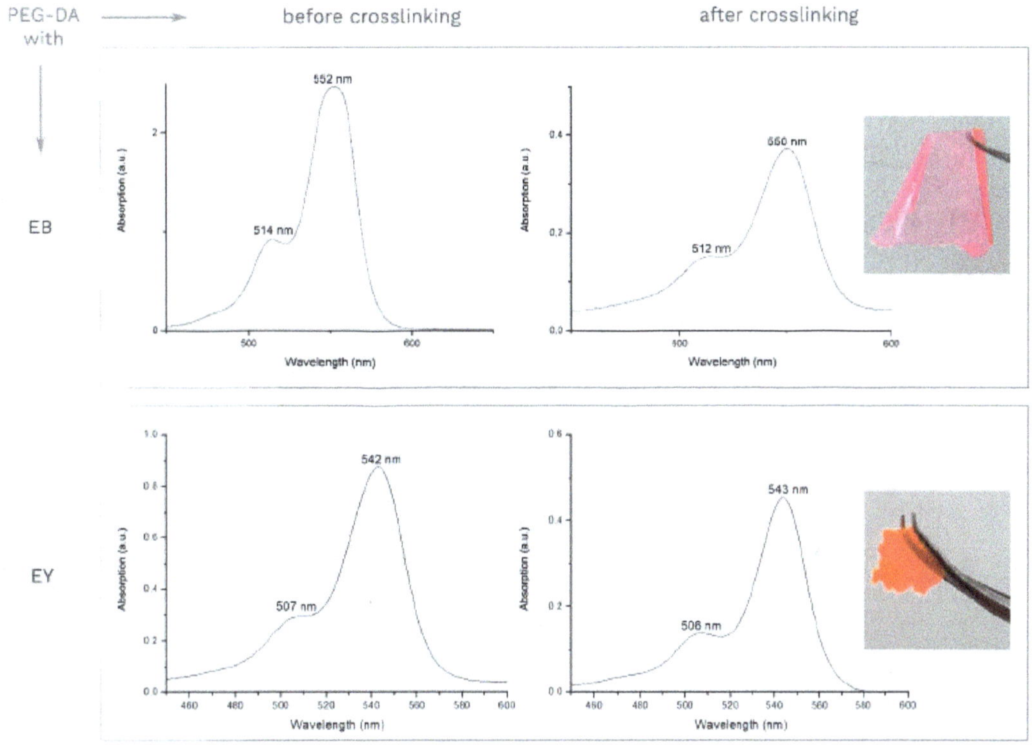

Figure 1. UV-Vis spectra for the novel PIs (EB and EY) with PEG-DA before and after crosslinking, respectively. The hydrogels with new PIs (EB and EY) feature good spectroscopic properties, allowing for use with light sources in the visible range.

While Irgacure, which is an often employed and suitable photoinitiator for biomaterials research, has an absorption maximum in the UV/Vis-spectrum around 300 nm [24], the novel dyes under investigation display a strong absorption of visible light with wavelengths up to 550 nm (see Figure 1).

With EB as PI, the absorption maximum before crosslinking was around 552 nm with a shoulder at 514 nm. After crosslinking, the absorption peak can be observed at 550 nm with a shoulder at 512 nm. The typical color for the hydrogel with EB as PI is pink, as shown in Figure 1.

With EY as PI, the absorption maximum before crosslinking was around 542 nm with a shoulder at 507 nm. After crosslinking, the absorption peak can be observed at 543 nm with a shoulder at 506 nm. The typical color for the hydrogel with EY as PI is orange, as shown in Figure 1.

The major advantage of the new PIs (EB and EY) over the standard Irgacure is their high sensitivity in the optical range of the spectrum, which enables optical patterning—e.g., by volume holography. To counter the disadvantage of strong coloring, the composition could be optimized, e.g., by use of a crosslinker so that the concentrations of EB and EY can be reduced, respectively.

In the next step, the respective optical response must be determined depending on the composition. In some cases (such as with an organic cationic ring-opening polymerization system), competing effects regarding the contribution to the optical grating formation can be observed [27]. It is also known that optical shrinkage can have significant influence on grating formation [28] and that the amount of PEG in a composition affects film shrinkage, as well as its optical properties [29]. Furthermore, we have also observed that photoinitiators may contribute to light-induced modification of optical properties and subsequent pattern formation as well [30].

2.3. Cytotoxicity Tests

A live/dead staining assay has been used to study the cell viability after incubation with PI EB and EY and also EB 0.1% with PEG-DA before and after crosslinking for 24 h. In the live/dead staining assay, dead cells turn up red, while living cells turn up green when observed with a fluorescence microscope. As shown in Figure 2, all cells but one appear to be green, indicating that PIs and hydrogel (PEG-DA with 0.1% PI EB) are cytocompatible and suitable as substrates for studying the behaviour of L929 cells at the biointerface.

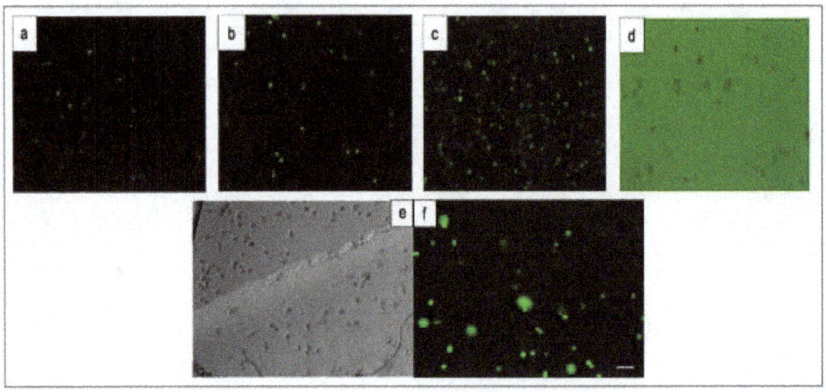

Figure 2. Fluorescent images of cells test by live/dead cell assay with L929 after incubated for 24 h. (a) Control cells; (b) EB; (c) EY; (d) EB 0.1% with PEG-DA; (e) optical micrograph of cross-linked EB/PEG-DA hydrogel; (f) fluorescence image of cross-linked EB/PEG-DA hydrogel. Scale bar depicts 50 μm.

Figure 2 shows fluorescent images of cell tests by live/dead cell assay after being incubated for 24 h with cell line L929. Cell tests shown in Figure 2 confirm cytocompatibility of PEG-DA hydrogels and PIs, hence affirming its aptitude for biomedical applications.

3. CONCISE SUMMARY

Aside from the conventional photoinitiator Irgacure, Erythrosin B (EB) and Eosin Y (EY) were employed as photoinitiators (PI) in the PEG-DA hydrogel. We have identified the minimal concentration of PI required to produce novel hydrogels using PEG-DA for each specific PI. All of the PIs are biocompatible and appropriate for use as substrates to investigate the behavior of L929 cells at the biointerface.

The newly introduced principal investigators (EB and EY) possess favorable spectroscopic characteristics, enabling their use with visible light sources and, consequently, facilitating their employment in volume holography. Cytotoxicity tests were conducted using the L929 cell line to verify the cytocompatibility of hydrogels and PIs. This presents numerous possibilities for PEG-DA hydrogels with PIs as versatile biomedical applications.

The approach of integrating optical structuring of the volume and targeted modification of the surface is especially intriguing for the development of advanced intraocular lens (IOL) implants. By utilizing a multifunctional biomedical material with integrated optical capabilities and operating on the principle of 'function by structure', this novel type of IOL is anticipated to achieve improved functionality [9]. The optical function of an intraocular lens (IOL) that incorporates a holographic lens as a diffractive element is to concentrate the light onto the retina. Holographic elements can be stacked, allowing for overlapping functionality between the distinct elements. The selectivity of a stack leads to a combination of the Bragg selectivity of the individual elements [31].

In addition to improved performance, transferring the optical capabilities from the surface to the interior of the IOL implant offers additional advantages, such as creating a free interface for targeted contact with the biological environment. Free-surface IOLs could provide a more effective solution to the issues associated with conventional IOLs, such as postoperative clouding, which arise from interactions with the biological environment.

To achieve a multifunctional optical material, it is necessary to get a deeper understanding of the processes involved in optical structuring. This includes studying the interaction between polymerization and diffusion, particularly in the context of holographic gratings. The whole strategy involves comprehending the production of holographic gratings as a result of photopolymerization and mass transport processes. Specifically, a light pattern projected into the photosensitive material triggers localized polymerization. The polymerization process is directly proportional to the intensity of light, which leads to the creation of a chemical gradient. This is followed by the diffusion of monomers and their subsequent polymerization. The ultimate grating is created by periodically modifying the optical characteristics in accordance with the pattern of the illuminating light [10].

It is necessary to clarify the specific functions of the separate components in the creation of optical structures in PEG-DA hydrogels including EB and EY. Moreover, it is necessary to investigate whether the inclusion of additional substances, such as crosslinkers or dopants, such as azobenzene-functionalized acrylates or gold nanoparticles, can enhance the creation of optical patterns.

4. MATERIALS AND METHODS

4.1. Preparation of PEG Hydrogels

4.1.1. Chemicals

Poly(ethylene glycol) diacrylate (PEG, Mn 575) and 2-hydroxy-4′-(2-hydroxyethoxy)-2-methylpropiophenone (photoinitiator (PI)—Irgacure 2959), Erythrosin B and Eosin Y were from Sigma-Aldrich Chemie GmbH (Steinheim, Germany). The chemical structures of the hydrogel components are shown in Figure 3.

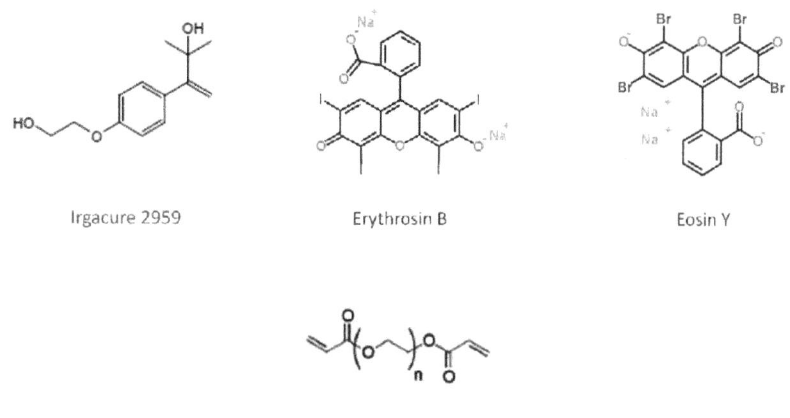

Figure 3. Irgacure 2959, Erythrosin B and Eosin Y are used as photoinitiators (PI); PEG-DA as a precursor for PEG hydrogel.

4.1.2. Synthesis of PEG Based Photopolymer with PIs

PEG-DA was used as precursor. It was mixed with PI (Irgacure, EB and EY, respectively). PI concentration varied between 1% and 6 ppm. The chemical structures of different PIs are shown in Figure 3. For the good mixing of both substances, the mixture was sonicated for around 30 min. At first, the mixture was converted in a cuvette and measured with a UV-Vis spectrometer to obtain spectra before crosslinking. Then, the mixture was dispensed on a glass slide and covered with a thin glass cover slip to achieve a flat and thin hydrogel sample. The glass-sandwich was placed under a UV-light source (366 nm) for 60 min and the glass cover slip was peeled off. A flat, thin standalone hydrogel film was received and also prepared for UV-Vis measurement. Therefore, the hydrogel was placed on a thin glass cover slip and measured with a UV-Vis spectrometer to obtain spectra after crosslinking.

4.2. Cell Culture

4.2.1. Chemicals

Mouse fibroblast L929 cells were provided by Dr. Lehmann, Fraunhofer Institute for Cell Therapy and Immunology, IZI, Leipzig, Germany. RPMI 1640 medium, Trypsin, Fetal Bovine Serum (FBS) and Penicillin/Streptomycin (PS) were provided by PAA Laboratories GmbH, Austria, and cell culture plates are from SPL Live Sciences Inc., Seoul, Korea. The Incubator CB150 Series was from Binder GmbH, Germany. Phosphate Buffered Saline solution (Dulbecco's PBS) was purchased from Sigma-Aldrich Chemie, GmbH, Germany. The counter chamber was from Marienfeld Superior (Paul Marienfeld GmbH & Co., KG, Lauda-Königshofen, Germany).

4.2.2. Cell Culture Experiments

The mouse fibroblasts L929 cells were cultured in the tissue culture plate in RPMI 1640 medium with the addition of 10% Fetal Bovine Serum (FBS) and 1% Penicillin/Streptomycin (PS) in a cell culture plate in an incubator at controlled temperature (37 °C) and CO2 atmosphere (5%). The cells were grown in a cell culture plate and cell culture experiments were performed when a confluency of 75% to 95% was reached.

The hydrogel samples were prior washed with water and kept in a PBS solution for around 30 min before cell culture experiments. As soon as a confluency of at least 75% was reached, the cells were washed with PBS, detached by using trypsin and, after the centrifugation process, a new medium was added on the cells and mixed properly. An amount of 10 µL of this cell medium solution was placed on a cell counter chamber in order to count the cell number by using an optical microscope and to achieve a concentration of 40,000 cells/mL. Depending on the counted cell number, the cell solution was mixed with a defined amount of new medium. The samples were placed in a TCPS plate or on the washed and precut hydrogel. The samples were then cultured within these cells for 24 h, at 37 °C in a 5% CO2 atmosphere in a TCPS.

4.2.3. Live Dead Cytotoxicity Assay

The live dead cytotoxicity assay is a fluorescence-based method for checking the viability of cells. Hereby, the cells are stained with fluoresceindiacetete (FBS) and propidiumiodid (PI) molecules. FBS is dissociated in the cytoplasma of live cells into green fluorescence molecules and, due to the size and charge of the PI molecules, they only can enter into the cell cytoplasm when the cell membranes are damaged and are bound to nucleic acids, which appear then as red fluorescence color. In this manner, the live cells appear as green-stained cells and dead cells appear as red-stained cells in a fluorescent microscope image.

For the live dead assay, a 1:1 v/v solution of PI and FBS in PBS was prepared and added into the cell culture solution in a dark environment. Immediately after the mixtures are prepared, fluorescence images are taken.

4.3. Analytical Instruments

The UV-Vis spectra were acquired using the Cary 4000 UV-Vis Spectrometer manufactured by Agilent Technologies, located in Santa Clara, CA, USA. The spectral range between 300 and 900 nm was measured under normal ambient conditions. The liquid samples, before to crosslinking, were transferred into a cuvette for measurement. The hydrogel films, which were flat and thin, were placed on a thin glass cover slip after crosslinking. These films were then measured using a UV-Vis spectrometer. The results obtained from cell culture experiments were visualized using the optical microscope manufactured by Carl Zeiss, a company based in Germany. The data analysis was conducted using the AxioVision V4.8.2 software, also developed by Carl Zeiss in Oberkochen, Germany.

AT A GLANCE

There is a strong need for biomedical materials that can perform many activities and incorporate optical capabilities. These materials are particularly useful for applications like enhanced intra-ocular lens (IOL) implants. Hence, PEG-DA hydrogels are employed in conjunction with various photoinitiators (PI). Aside from the typical UV PI Irgacure, Erythrosin B and Eosin Y are employed as highly sensitive PI in the optical area of the spectrum. The minimal concentrations of PI required to produce novel hydrogels using PEG-DA and various PIs were determined. Hydrogel films were acquired, suitable for light-based patterning and, consequently, the surface and volume functionalization. Cytotoxicity studies validate the cytocompatibility of hydrogels and mixtures. Utilizing the relationship between structure and function enables the development of biological materials that possess several functions.

REFERENCES

1. Luo, Y.; Shoichet, M.S. A photolabile hydrogel for guided three-dimensional cell growth and migration. Nat. Mater. 2004, 3, 249–253.
2. Higgins, C.; Aisenbrey, E.; Wahlquist, J.; Heveran, C.; Ferguson, V.; Bryant, S.; Mcleod, R. Enhanced mechanical properties of photo-clickable thiol-ene PEG hydrogels through repeated photopolymerization of in-swollen macromer. Soft Matter 2016, 12, 9095–9104.
3. Lensen, M.C.; Schulte, V.; Salber, J.; Dietz, M.; Menges, F.; Möller, M. Cellular responses to novel, micropatterned biomaterials. Pure Appl. Chem. 2008, 80, 2479–2487.
4. Li, L.; Scheiger, J.M.; Levkin, P.A. Design and Applications of Photoresponsive Hydrogels. Adv. Mater. 2019, 31, 1807333.
5. Choi, M.; Choi, J.W.; Kim, S.; Nizamoglu, S.; Hahn, S.K.; Yun, S.H. Light-guiding hydrogels for cell-based sensing and optogenetic synthesis in vivo. Nat. Photonics 2013, 7, 987–994.
6. Izak, A.M.; Werner, L.; Pandey, S.K.; Apple, D.J. Calcification of modern foldable hydrogel intraocular lens designs. Eye 2003, 17, 393–406.
7. Kanclerz, P.; Yildirim, T.; Khoramnia, R. Microscopic Characteristics of Late Intraocular Lens Opacifications. Arch. Pathol. Lab. Med. 2020, 145.
8. Lundgren, B.; Jonsson, E.; Rolfsen, W. Secondary cataract. Acta Ophthalmol. 2009, 70, 25–28.

9. Sabel, T.; Lensen, M.C. Function and structure—Combined optical functionality and specific bio-interaction for multifunctional biomedical materials. J. Med. Mater. Technol. 2017, 1, 10–12.
10. Sabel, T.; Lensen, M.C. Volume Holography: Novel Materials, Methods and Applications. In Holographic Materials and Optical Systems; Naydenova, I., Babeva, T., Nazarova, D., Eds.; InTech: Rijeka, Croatia, 2017.
11. Naahidi, S.; Jafari, M.; Logan, M.; Wang, Y.; Yuan, Y.; Bae, H.; Dixon, B.; Chen, P. Biocompatibility of hydrogel-based scaffolds for tissue engineering applications. Biotechnol. Adv. 2017, 35, 530–544.
12. Hahn, B.M.S.; Miller, J.S.; West, J.L. Three-Dimensional Biochemical and Biomechanical Patterning of Hydrogels for Guiding Cell Behavior. Adv. Mater. 2006, 18, 2679–2684.
13. Nguyen, K.T.; West, J.L. Photopolymerizable hydrogels for tissue engineering applications. Biomaterials 2002, 23, 4307–4314.
14. Yesildag, C.; Ouyang, Z.; Zhang, Z.; Lensen, M.C. Micro-Patterning of PEG-Based Hydrogels with Gold Nanoparticles Using a Reactive Micro-Contact-Printing Approach. Front. Chem. 2019, 6, 1–10.
15. Yesildag, C.; Tyushina, A.; Lensen, M.C. Nano-Contact Transfer with Gold Nanoparticles on PEG Hydrogels and Using Wrinkled PDMS-Stamps. Polymers 2017, 9, 199.
16. Tomita, Y.; Kageyama, A.; Iso, Y.; Umemoto, K.; Kume, A.; Liu, M.; Pruner, C.; Jenke, T.; Roccia, S.; Geltenbort, P.; et al. Fabrication of nanodiamond-dispersed composite holographic gratings and their light and slow-neutron diffraction properties. Phys. Rev. Appl. 2020, 14, 44056.
17. Qi, Y.; Li, H.; Fouassier, J.P.; Lalevée, J.; Sheridan, J.T. Comparison of a new photosensitizer with erythrosine B in an AA / PVA-based photopolymer material. Appl. Opt. 2014, 53, 1052–1062.
18. Mikulchyk, T.; Martin, S.; Naydenova, I. Investigation of the sensitivity to humidity of an acrylamide-based photopolymer containing N-phenylglycine as a photoinitiator. Opt. Mater. 2014, 37, 810–815.
19. Qi, Y.; Gleeson, M.R.; Guo, J.; Gallego, S.; Sheridan, J.T. Quantitative Comparison of Five Different Photosensitizers for Use in a Quantitative Comparison of Five Different Photosensitizers for Use in a Photopolymer. Phys. Res. Int. 2012, 17, 11.
20. Tomal, W.; Ortyl, J. Water-Soluble Photoinitiators in Biomedical Applications. Polymers 2020, 12, 1073.
21. Schulte, V.A.; Diez, M.; Hu, Y.; Möller, M.; Lensen, M.C. Combined Influence of Substrate Stiffness and Surface Topography on the Antiadhesive Properties of Acr-sP(EO-stat-PO) Hydrogels. Biomacromolecules 2010, 11, 3375–3383.
22. Lensen, M.C.; Schulte, V.A.; Diez, M. Cell Adhesion and Spreading on an Intrinsically Anti-Adhesive PEG Biomaterial. In Biomaterials; Pignatello, R., Ed.; IntechOpen: Rijeka, Croatia, 2011.
23. Kelleher, S.; Jongerius, A.; Löbus, A.; Strehmel, C.; Zhang, Z.; Lensen, M.C. AFM Characterisation of elastically micropatterned surfaces fabricated by Fill-Molding In Capillaries (FIMIC) and investigation of the topographical influence on cell adhesion to the patterns. Adv. Eng. Mater. 2012, 14, B56–B65.
24. Kaastrup, K.; Sikes, H.D. Using photo-initiated polymerization reactions to detect molecular recognition. Chem. Soc. Rev. 2016, 45, 532–545.
25. Yang, R.; Soper, S.A.; Wang, W. A new UV lithography photoresist based on composite of EPON resins 165 and 154 for fabrication of high-aspect-ratio microstructures. Sensors Actuators A Phys. 2007, 135, 625–636.

26. Sung, J.; Lee, D.G.; Lee, S.; Park, J.; Jung, H.W. Crosslinking Dynamics and Gelation Characteristics of Photo- and Thermally Polymerized Poly(Ethylene Glycol) Hydrogels. Materials 2020, 13, 3277.

27. Sabel, T.; Zschocher, M. Transition of refractive index contrast in course of grating growth. Sci. Rep. 2013, 3, 1–7.

28. Sabel, T. Spatially resolved analysis of Bragg selectivity. Appl. Sci. 2015, 5, 1064–1075.

29. Dimitrov, O.; Stambolova, I.; Vassilev, S.; Lazarova, K.; Babeva, T.; Mladenova, R. Surface and Morphological Features of ZrO 2 Sol-Gel Coatings Obtained by Polymer Modified Solution. Mater. Proc. 2020, 2, 1–8.

30. Sabel, T. Volume hologram formation in SU-8 photoresist. Polymers 2017, 9, 198.

31. Akbari, H.; Naydenova, I.; Persechini, L.; Garner, S.M.; Cimo, P.; Martin, S. Diffractive Optical Elements with a Large Angle of Operation Recorded in Acrylamide Based Photopolymer on Flexible Substrates. Int. J. Polym. Sci. 2014, 2014, 918285.

5 Chapter

THE IMPACT OF THE DEGREE OF CROSSLINKING IN HYDROGELS ON HYDROGEL ADHESION

1. INTRODUCTION: AN OVERVIEW

The process of bonding materials together sometimes necessitates the utilization of an adhesive substance [1]. These two materials are referred to as adherends, whereas the glue is known as adhesive [2]. Hydrogels are extensively porous polymer networks with a three-dimensional structure. These networks are formed by crosslinking hydrophilic polymers, which allows hydrogels to effectively retain significant quantities of water [2]. Nevertheless, employing adhesive substances is impractical when seeking to establish a connection between human tissues and hydrogels. Hydrogels are employed as both the adherend and adhesive in hydrogel adhesion [1]. Thus, it is necessary to create a sticky hydrogel surface when utilizing hydrogels in the field of biomedicine, specifically for applications like wound treatment and wearable devices [3,4]. The adhesion of hydrogels is crucial in numerous applications in the fields of medicine and engineering [5,6].

Consequently, there is significant interest in developing a perfect hydrogel that exhibits exceptional adhesive properties [7]. Hydrogel adhesion often occurs due to surface energy, intermolecular interactions, and effects that are close to the surface [1,2]. Traditionally, academic researchers have mostly concentrated on studying the outer layer of hydrogels or the boundary between hydrogels and substrates in order to develop robust hydrogel adhesion [3,8]. Research investigations have investigated the correlation between the internal structure of hydrogels and their surface adherence, although only a limited number of studies have been conducted on this topic [9,10].

A polymer precursor solution generally comprises several polymer chains. After the polymer chains undergo crosslinking, the precursor solution will undergo a transformation and become solid hydrogels [11,12]. The degree of crosslinking in hydrogels can impact their internal structure, leading to alterations in the mechanical and swelling properties of the hydrogels. Typically, hydrogels with a greater degree of crosslinking exhibit increased compressive and tensile strength, as well as a reduced equilibrium swelling rate [13,14]. Hydrogel adhesion is a crucial property, alongside mechanical and swelling properties, for hydrogels. In recent years [15,16], it has been extensively utilized in the fields of medicine and engineering. While the adherence of hydrogels is mostly dependent on manipulating the contact surface of soft and moist surfaces, our research has revealed that the degree of crosslinking in hydrogels also has an impact on adhesion. Hydrogels are often created using several crosslinking techniques, such as the freezing-thawing cycle, photo-crosslinking, and thermal-crosslinking [17,18]. In order to demonstrate the universality of this discovery, we selected polyvinyl alcohol (PVA), polyacrylamide (PAM), and polyvinyl alcohol-bearing styrylpyridinium group (PVA-SbQ) hydrogels. These hydrogels were prepared using different methods: freezing-thawing (F-T) cycle, thermal-crosslinking, and photo-crosslinking. The purpose of this study was to investigate the impact of crosslinking on the adhesion of hydrogels, which is achieved through hydrogen bonding. Additionally, this study also examines the relationship between the degree of crosslinking and the adhesion properties of the hydrogel. This study aims to offer valuable insights for enhancing the adhesion of hydrogels by hydrogen bonding.

2. RESULTS AND DISCUSSION

2.1. The Effect of The Crosslinking Degree of PVA Hydrogels on Hydrogel Adhesion, and Mechanical and Swelling Properties

PVA is a conventional raw material for preparing hydrogels because of its satisfactory biocompatibility, biodegradation, and nontoxicity [19,20]. PVA hydrogels can be formed through the well-known F-T method via the physical crosslinking by hydrogen bonds (Figure 1a,b) [21,22]. During the freezing process, water freezes and reduces interaction with hydroxyl groups on PVA chains [23]. As a result, the hydroxyl groups on PVA chains bind to form hydrogen bonding (Figure 1a). Consequently, the PVA solution becomes a hydrogel due to the development of a 3D structure (Figure 1b). With the increased freezing time, there is an increase in the hydrogen bond formation. Hydrogen bonds act as physical crosslinking points in PVA hydrogels. Therefore, the crosslinking degree depends on the freezing time.

There are several methods of testing hydrogel adhesion [1]. Figure 1c shows the modified method for testing the shear adhesive strength of PVA hydrogels used in this study unlike the general method (Figure 1d) [1,24]. The adherends were stuck to the same side of hydrogels to avoid significant errors due to the low adhesive strength of PVA hydrogels. Figure 1e shows that with the increase in the freezing time of PVA hydrogels, the adhesive strength decreased from 1600 Pa to 400 Pa. In addition, the swelling ratio of PVA hydrogels decreased with the increase in crosslinking degree that results from the increased freezing time (Figure 1f). Conversely, the tensile strength, elongation at break and compressive strength of PVA hydrogels improved with longer freezing duration because of the increased crosslinking degree (Figure 1g–i).

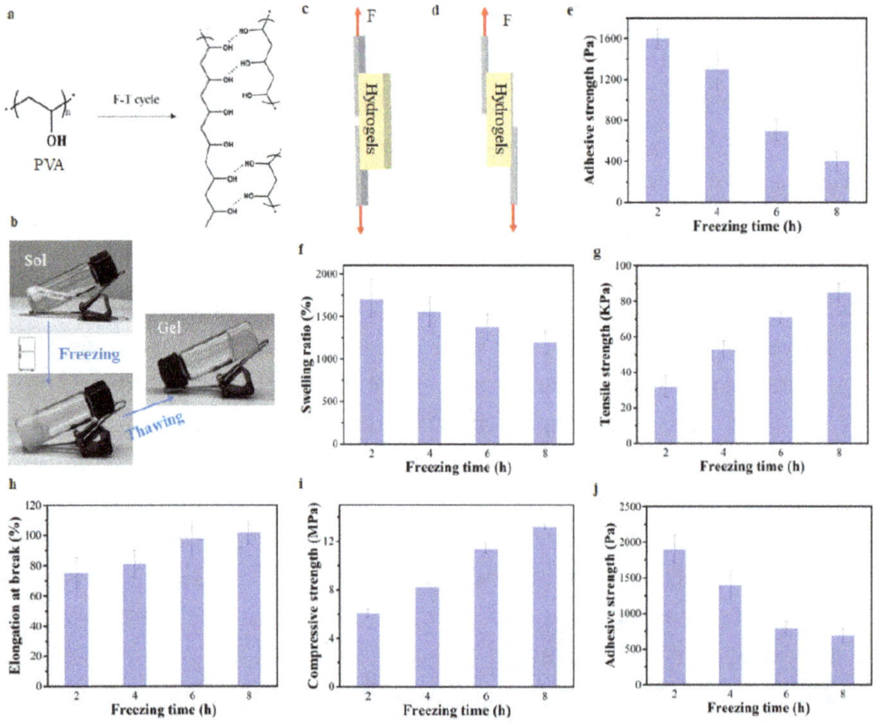

Figure 1. (a) The formation mechanism of PVA hydrogels. (b) The preparation process of PVA hydrogels. (c) The modified method of testing hydrogel adhesion used in this study. (d) The general method of

testing hydrogel adhesion. (e) The adhesive strength of PVA hydrogels prepared by different freezing times. (f) The swelling ratio of PVA hydrogels prepared by different freezing times. (g) The tensile strength of PVA hydrogels prepared by different freezing times. (h) The elongation at break of PVA hydrogels prepared by different freezing times. (i) The compressive strength of PVA hydrogels prepared by different freezing times. (j) The adhesive strength of PVA/CNC hydrogels prepared by different freezing times.

It is well-known that the crosslinking degree benefits mechanical strength because a higher crosslinking degree results in a denser hydrogel structure [12,25,26]. In addition to mechanical strength, the crosslinking degree affects hydrogel adhesion simultaneously. For PVA hydrogels prepared by the F-T cycle, on the one hand, most of the hydroxyl groups on PVA formed hydrogen bonds used for crosslinking during the F-T process, which resulted in the lack of free hydroxyl groups used for adhesion [7,27]. On the other hand, the increase in the crosslinking degree of the PVA hydrogels limited the accessibility of hydroxyl groups due to the decrease in segment mobility. Consequently, the adhesive property between hydrogels and substrates reduced significantly with increased freezing time. This phenomenon was also applied to composite hydrogels. As shown in Figure 1j, the adhesion of polyvinyl alcohol/cellulose nanocrystal (PVA/CNC) composite hydrogels decreased with the increase in freezing duration.

2.2. The Effect of The Crosslinking Degree of PAM Hydrogels on Hydrogel Adhesion, and Mechanical and Swelling Properties

Unlike PVA hydrogels prepared by the F-T cycle, PAM hydrogels are usually synthesized by thermal-crosslinking [28,29]. The covalent bonding formed by the polymerization acts as crosslinking points in these types of hydrogels rather than hydrogen bonding (Figure 2a), which results in the sol–gel transition (Figure 2b). The crosslinking degree depends on the crosslinker content and the thermal-crosslinking time. Therefore, PAM hydrogels with different crosslinking degrees were obtained by adjusting crosslinking time and crosslinker content to investigate the relationship between crosslinking degree and hydrogel adhesion.

As shown in Figure 2c,d, with the increase in the crosslinking time and crosslinker content, the hydrogel adhesion decreased. This observation implied that a higher crosslinking degree was not helpful for hydrogel adhesion. Unlike the PVA hydrogels, the free functional groups used for adhesion did not decrease in PAM hydrogels [1,2]. The decrease in hydrogel adhesion mainly resulted from the limitation of segment mobility. Although there were a lot of free amino groups on the PAM hydrogels, the amino group could not move to the hydrogel surface to interact with some functional groups on substrates as the crosslinking degree of PAM hydrogels increased. In addition, with the increase in crosslinking time and crosslinker content, the swelling ratio of PAM hydrogels simultaneously decreased (Figure 2e,f). The decreased swelling ratio further corroborated that the crosslinking degree of the hydrogel increased and a denser hydrogel structure was formed [30,31,32]. With the increase in crosslinking degree caused by increased crosslinking time, the compressive strength of PAM hydrogels gradually increased (Figure 2g). Conversely, while the increase in the crosslinker content improved the crosslinking degree, the compressive strength of PAM hydrogels did not increase significantly (Figure 2h). This is because the higher crosslinker content resulted in the crosslinking of PAM without complete polymerization. Accordingly, the length of PAM polymer chains that make up the hydrogel network were short, which resulted in the slight decrease in compressive strength. Therefore, it is challenging to enhance the mechanical and adhesive properties of hydrogels simultaneously.

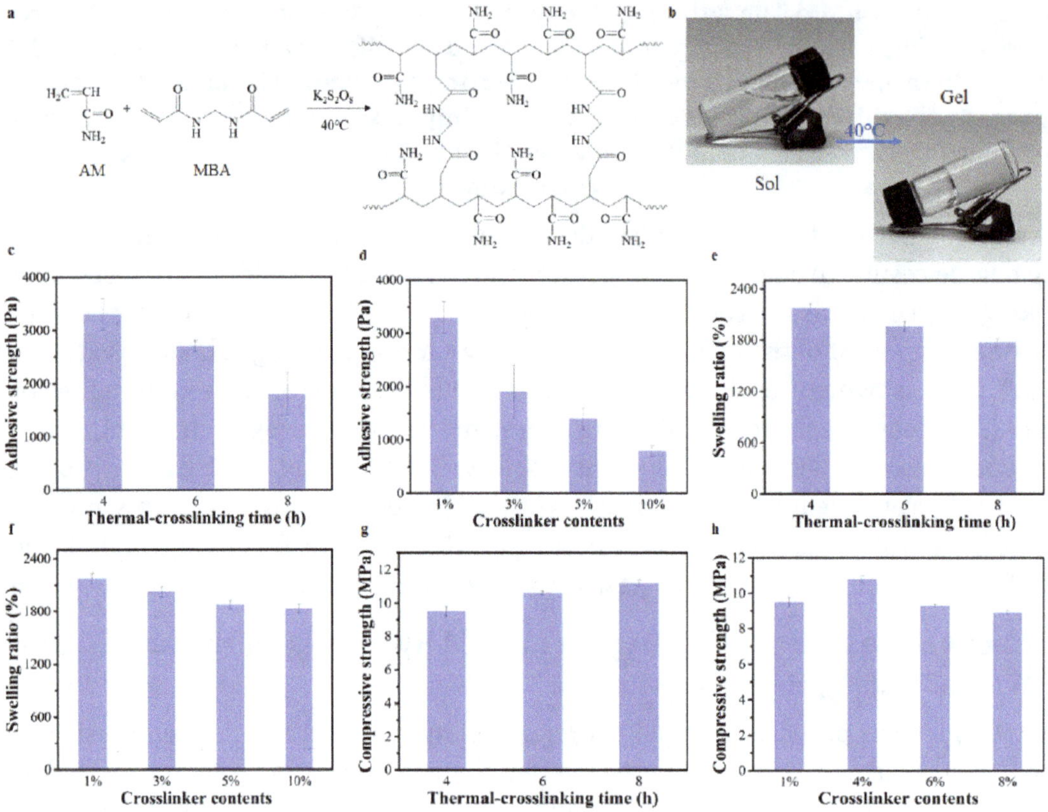

Figure 2. (a) The formation mechanism of PAM hydrogels. (b) The preparation process of PAM hydrogels. (c) The adhesive strength of PAM hydrogels prepared by different thermal-crosslinking times. (d) The adhesive strength of PAM hydrogels prepared with different crosslinker contents. (e) The swelling ratio of PAM hydrogels prepared by different thermal-crosslinking times. (f) The swelling ratio of PAM hydrogels prepared with different crosslinker contents. (g) The compressive strength of PAM hydrogels prepared by different thermal-crosslinking times. (h) The compressive strength of PAM hydrogels prepared with different crosslinker contents.

2.3. The Effect of The Crosslinking Degree of PVA-SbQ Hydrogels on Hydrogel Adhesion, and Mechanical and Swelling Properties

In addition to the F-T cycle and thermal-crosslinking, photo-crosslinking is another common method for preparing hydrogels. To prove the generality of the effect of crosslinking degree on hydrogel adhesion, the PVA-SbQ hydrogels with different degrees were prepared by adjusting the photo-crosslinking duration. Instead of the conventional F-T cycle or using chemical crosslinkers, PVA-SbQ hydrogels could be formed by a fast and facile photo-crosslinking method via the photodimerization of carbon–carbon double bonds (C=C bonds) on SbQ functional groups (Figure 3a,b) [33,34]. As shown in Figure 3c, the adhesion of a PVA-SbQ hydrogel sharply decreased with the increase in photo-crosslinking time. The increased photo-crosslinking time also reduced the swelling ratio of PVA-SbQ hydrogels, which suggested an increase in the crosslinking degree (Figure 3d) [30,31,32]. Conversely, the compressive strength and tensile strength steadily increased as the photo-crosslinking duration was increased (Figure 3e,f). As with the PAM hydrogels, the decrease in hydrogel adhesion primarily resulted from the limitation of segment mobility, and the enhanced mechanical properties were caused by the increase in the crosslinking degree. This proved that photo-crosslinked hydrogels also followed the aforementioned relationship between the degree of crosslinking and hydrogel adhesiveness.

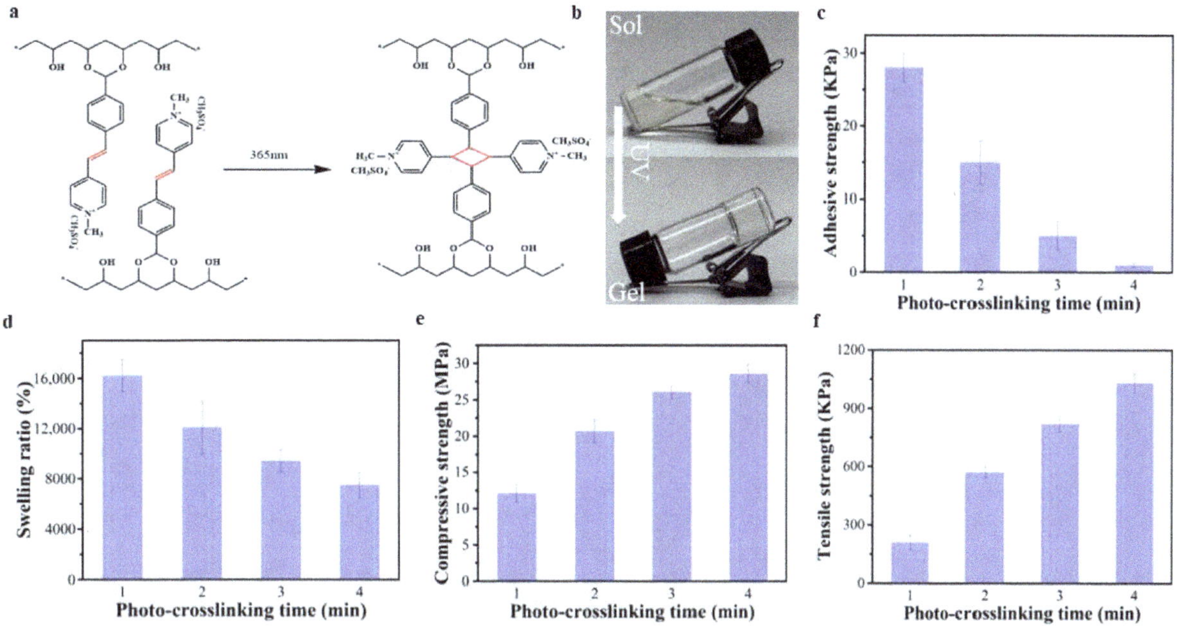

Figure 3. (a) The formation mechanism of PVA-SbQ hydrogels. (b) The preparation process of PVA-SbQ hydrogels. (c) The adhesive strength of PVA-SbQ hydrogels prepared by different photo-crosslinking times. (d) The swelling ratio of PVA-SbQ hydrogels prepared by different photo-crosslinking times. (e) The compressive strength of PVA-SbQ hydrogels prepared by different photo-crosslinking times. (f) The tensile strength of PVA-SbQ hydrogels prepared by different photo-crosslinking times.

2.4. The Strategies for Constructing Strong Hydrogel Adhesion

In the study, three kinds of hydrogels were prepared by the freezing-thawing cycle, thermal-crosslinking and photo-crosslinking, respectively. The hydrogel adhesion was primarily generated by hydrogen bonding because of the amino and hydroxyl groups. In addition to the amino and hydroxyl groups shown in these hydrogels, carboxyl groups can also contribute towards hydrogel adhesion. Therefore, amino, hydroxyl or carboxyl groups are required to construct hydrogel adhesion generated by hydrogen bonding (Figure 4a). However, these functional groups should be free. For instance, the maximum adhesive strength of PVA-SbQ hydrogels is much higher than that of PVA hydrogels (Figure 4b). Although there were many hydroxyl groups in PVA hydrogels, they formed hydrogen bonding used for crosslinking with themselves rather than interacting with the functional groups on the substrate. As a result, these hydroxyl groups are not free and will contribute towards hydrogel adhesion. On the contrary, the fabricated PVA-SbQ hydrogel holds massive free hydroxyl groups as the crosslinking was mainly driven by photodimerization of SbQ groups instead of hydrogen bonds forming between hydroxyl groups. The advantage of hydrogen bond-triggered adhesion is that it is recyclable since the hydrogen bonding formation utilizes dynamic and noncovalent interactions. The cyclic stripping test demonstrated that the adhesive strength of the PVA-SbQ hydrogel did not reduce significantly with the increase in the number of stripping cycles (Figure 4c). In addition to the free functional groups, it was found that the crosslinking degree also affected the hydrogel adhesion. The increase in the crosslinking degree of the hydrogel restricted the mobility of the polymer chains (Figure 4d). Hence, the polymer chains with functional groups were not quickly diffused towards the hydrogel surface to form intimate contact with the substrate. As a result, excessive crosslinking of hydrogels led to a reduced adhesive strength (Figure 4e).

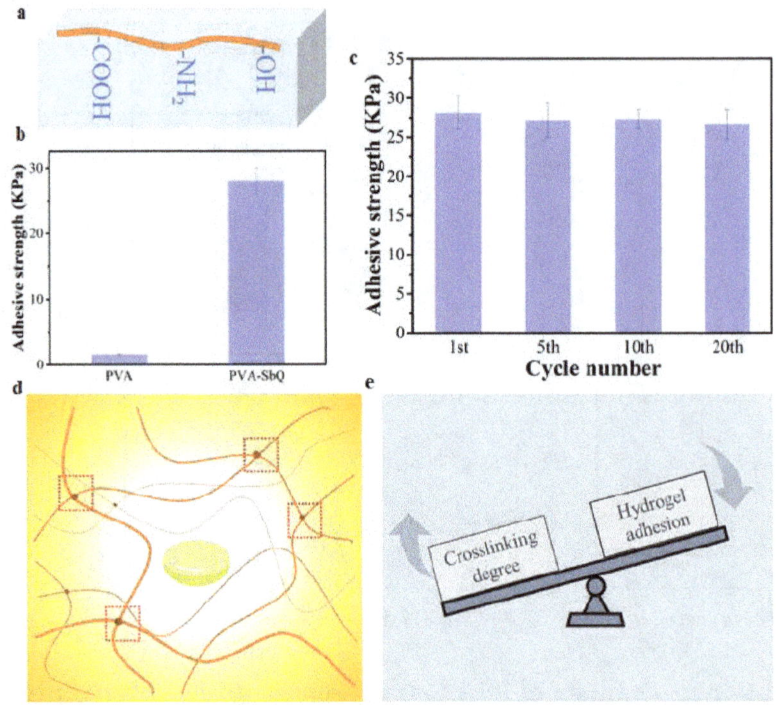

Figure 4. (a) The functional groups required for hydrogel adhesion generated by hydrogen bonds. (b) The maximum adhesive strength of PVA and PVA-SbQ hydrogels. (c) The adhesive strength of PVA-SbQ hydrogels under different stripping cycles. (d) The effect of crosslinking degree of hydrogels on chain movement. (e) The effect of crosslinking degree on hydrogel adhesion.

3. CONCLUSIONS

In this study, we investigated PVA, PAM and PVA-SbQ hydrogels fabricated by the freezing–thawing cycle, thermal-crosslinking and photo-crosslinking, respectively. These three types of hydrogels present representative models of hydrogels formed by different crosslinking mechanisms and extend the observations and derived conclusion to a wide class of materials. The hydrogel adhesion capability decreased with the increase in the crosslinking degree of these hydrogels. This is because a higher crosslinking degree would limit segment mobility. As a result, functional groups on polymer chains could not move to the hydrogel surface to interact with the substrate to generate hydrogel adhesion. Therefore, in addition to free functional groups used to interact with substrates, the flexibility of polymer chains that make up hydrogels is vital for hydrogel adhesion. However, the decrease in crosslinking degree in order to increase the availability of free functional groups and higher segment mobility would result in poor mechanical strength of the hydrogels and limit their functionality. Therefore, it is essential as well as challenging to maintain the balance between hydrogel adhesion and mechanical strength. However, a rational design of a hydrogel crosslinked network and free functional groups can allow better control over these hydrogel characteristics.

4. MATERIALS AND METHODS

4.1. Materials

Polyvinyl alcohol (PVA, Mn ≈ 130,000, 99% hydrolysis), N,N'-methylenebisacrylamide (MBA), potassium persulfate (KPS) and acrylamide (AM) were obtained from Sigma-Aldrich (St. Louis, MO, USA) and used without further purification. Polyvinyl alcohol-bearing styrylpyridinium group (PVA-SbQ) solution was purchased from Shanghai KCKI Printing Technology Co., Ltd., Shanghai, China (polymerization degree was 1700, and the concentration of SbQ was 0.03–0.05 mol kg-1), and cellulose nanocrystal (CNC) aqueous suspension was prepared from cotton linters.

4.2. The Preparation of PVA Hydrogel

First, 8.0 g PVA powder was added to 92 g deionized water to obtain 8 wt% PVA solution. Secondly, the mixtures were put into in a three-mouth flask and then stirred at 95 °C until the PVA powder was completely dissolved. Subsequently, the PVA solution was cooled down to room temperature, until the air bubbles completely disappeared. Lastly, 8 wt% PVA solution was poured into a watch glass and placed in a -20 °C refrigerator for 2, 4, 6 and 8 h respectively to obtain PVA hydrogels with different freezing durations.

4.3. The Preparation of PVA/CNC Hydrogels

First, the CNC aqueous suspension was prepared by acid hydrolysis method. Specific steps are as follows: 3 g cotton fiber was added to 62 mL 65% H2SO4 and treated at 50 °C for 45 min under constant stirring. The mixture was centrifuged and dialyzed against deionized water to remove the residual acid. The aqueous suspension was then ultrasonicated to evenly disperse the CNC and break agglomerates. The dispersed CNC was diluted to obtain a homogenous aqueous dispersion.

Next, 8 wt% PVA solution was added to the CNC aqueous suspension at given CNC/PVA mass ratio of 0.01:1. Subsequently, the PVA/CNC solution was stirred at 90 °C for 2 h and then ultrasonicated for additional 20 min. Finally, the PVA/CNC solution was poured into a watch glass and placed in a -20 °C refrigerator for 2, 4, 6 and 8 h, respectively, to obtain PVA/CNC hydrogels with different freezing durations.

4.4. The Preparation of PAM Hydrogels

An 8 wt% AM solution was obtained by dissolving 8 g AM in 92 g deionized water. Subsequently, different weights of MBA powders were added into the 8 wt% AM solution to obtain samples with different compositions, where the mass ratio between MBA and AM were 1:100, 3:100, 5:100 and 10:100, respectively. These mixtures were then stirred continuously for 30 min. After the addition of the MBA and stirring for 30 min, 0.0192 g KPS was added into the solution and stirred for another 10 min to initiate polymerization. Finally, the prepared hydrogel solutions were placed at a temperature of 45 °C for 4 h to obtain PAM hydrogels with different MBA compositions. Among the various hydrogels, the PAM hydrogels with 1% MBA were selected to be placed in 45 °C for 4, 6 and 8 h, respectively, to obtain and study PAM hydrogels with different thermal-crosslinking times.

4.5. The Preparation of PVA-SbQ Hydrogels

PVA-SbQ hydrogels were directly formed from 8 wt% PVA-SbQ solution with different photo-crosslinking times by being exposed to an F300 UV lamp for 1, 2, 3, 4 min, respectively.

4.6. Characterization

Tensile strength of the hydrogels was evaluated by performing tensile tests on an Instron 5967 electronic universal testing machine. The PVA and PVA/SbQ hydrogels were cut into dumbbell-shaped samples with a diameter of 7–9 mm and a thickness of 3–5 mm. Additionally, the compressive tests were carried out on cylindrical hydrogel samples with a diameter of 7–9 mm and a thickness of 3–5 mm by an Instron 5967 electronic universal testing machine at room temperature. The compression rate was fixed at 2 mm/min. Measurements of each sample were repeated at least three times and the acquired data were reported as average value and standard deviation.

To estimate the adhesive strength of hydrogels, the shear adhesive strength tests were performed. The PVA, PVA/CNC, PAM and PVA/SbQ hydrogels were cut into a rectangular spline with a size of 40 mm × 10 mm × 5 mm and then adhered to metal sheets, where the adhesion area on one side was 10 mm × 10 mm and the other side was 10 mm × 20 mm, as shown in Figure 1c. The force response was recorded while keeping one metal sheet fixed and moving another metal sheet at a speed of 20 mm/min using a universal testing machine. In the experiment, 5 samples were tested in each group to obtain the average value and standard deviation. The adhesive strength was calculated by the maximum load divided by the adhesion area.

To test the swelling ratio of hydrogels, the hydrogel samples were immersed in distilled water for 3 days at least at room temperature, until the hydrogel weight did not increase. In the next step, these swollen hydrogels were placed in a 40 °C oven to evaporate all of the water in hydrogels. The weight of swollen hydrogel and dried hydrogels were recorded, respectively. The equilibrium swelling ratio was calculated by the following equation

$$ESR\% = W_s/W_d \times 100 \quad (1)$$

where Ws and Wd are the weight of the swollen hydrogel at room temperature and dried hydrogels, respectively.

AT A GLANCE

Adhesive hydrogel polymers have significantly contributed to the progress of biomedical engineering. The topic of hydrogel adhesion has garnered significant interest in both research and practical applications. This paper expands the investigation of hydrogel adherence beyond the surface of hydrogels. In this study, we investigated the impact of the internal crosslinking degree of hydrogels, manufactured using various ways, on hydrogel adhesion in order to determine its universality. The results indicate that when the degree of crosslinking increases, the adhesion of the hydrogel decreases dramatically due to the restriction of segment mobility. Furthermore, two uncomplicated approaches to enhance hydrogel adhesion through the utilization of hydrogen bonding were suggested. One method was employed to preserve the functional groups responsible for hydrogel adhesion, while another method was utilized to augment the flexibility of the polymer chains comprising the hydrogels.

REFERENCES

1. Yang, J.; Bai, R.; Chen, B.; Suo, Z. Hydrogel adhesion: A supramolecular synergy of chemistry, topology, and mechanics. Adv. Funct. Mater. 2020, 30, 1901693.
2. Zhang, W.; Wang, R.; Sun, Z.M.; Zhu, X.; Zhao, Q.; Zhang, T.; Cholewinski, A.; Yang, F.; Zhao, B.; Pinnaratip, R.; et al. Catechol-functionalized hydrogels: Biomimetic design, adhesion mechanism, and biomedical applications. Chem. Soc. Rev. 2020, 49, 433–464.
3. Kamoun, E.A.; Kenawy, E.-R.S.; Chen, X. A review on polymeric hydrogel membranes for wound dressing applications: PVA-based hydrogel dressings. J. Adv. Res. 2017, 8, 217–233.
4. Liang, Y.; He, J.; Guo, B. Functional hydrogels as wound dressing to enhance wound healing. ACS Nano 2021, 15, 12687–12722.
5. Bovone, G.; Dudaryeva, O.Y.; Marco-Dufort, B.; Tibbitt, M.W. Engineering hydrogel adhesion for biomedical applications via chemical design of the junction. ACS Biomater. Sci. Eng. 2021, 7, 4048–4076.

6. Caló, E.; Khutoryanskiy, V.V. Biomedical applications of hydrogels: A review of patents and commercial products. Eur. Polym. J. 2015, 65, 252–267.

7. Li, Z.; Wang, D.; Bai, H.; Zhang, S.; Ma, P.; Dong, W. Photo-crosslinking strategy constructs adhesive, superabsorbent, and tough PVA-based hydrogel through controlling the balance of cohesion and adhesion. Macromol. Mater. Eng. 2020, 305, 1900623.

8. Wang, L.; Zhou, M.; Xu, T.; Zhang, X. Multifunctional hydrogel as wound dressing for intelligent wound monitoring. Chem. Eng. J. 2022, 433, 134625.

9. Yang, J.; Steck, J.; Bai, R.; Suo, Z. Topological adhesion II. Stretchable adhesion. Extreme Mech. Lett. 2020, 40, 100891.

10. Steck, J.; Kim, J.; Yang, J.; Hassan, S.; Suo, Z. Topological adhesion. I. Rapid and strong topohesives. Extreme Mech. Lett. 2020, 39, 100803.

11. Bai, H.; Li, Z.; Zhang, S.; Wang, W.; Dong, W. Interpenetrating polymer networks in polyvinyl alcohol/cellulose nanocrystals hydrogels to develop absorbent materials. Carbohydr. Polym. 2018, 200, 468–476.

12. Li, Z.; Bai, H.; Zhang, S.; Wang, W.; Ma, P.; Dong, W. DN strategy constructed photo-crosslinked PVA/CNC/P(NIPPAm-co-AA) hydrogels with temperature-sensitive and pH-sensitive properties. New J. Chem. 2018, 42, 13453–13460.

13. Liu, X.; He, X.; Yang, B.; Lai, L.; Chen, N.; Hu, J.; Lu, Q. Dual physically cross-linked hydrogels incorporating hydrophobic interactions with promising repairability and ultrahigh elongation. Adv. Funct. Mater. 2021, 31, 2008187.

14. Dodero, A.; Pianella, L.; Vicini, S.; Alloisio, M.; Ottonelli, M.; Castellano, M. Alginate-based hydrogels prepared via ionic gelation: An experimental design approach to predict the crosslinking degree. Eur. Polym. J. 2019, 118, 586–594.

15. Liu, X.; Zhang, Q.; Duan, L.; Gao, G. Bioinspired nucleobase-driven nonswellable adhesive and tough gel with excellent underwater adhesion. ACS Appl. Mater. Interfaces 2019, 11, 6644–6651.

16. Liu, X.; Zhang, Q.; Gao, Z.; Hou, R.; Gao, G. Bioinspired adhesive hydrogel driven by adenine and thymine. ACS Appl. Mater. Interfaces 2017, 9, 17645–17652.

17. Chen, Y.; Li, J.; Lu, J.; Ding, M.; Chen, Y. Synthesis and properties of poly (vinyl alcohol) hydrogels with high strength and toughness. Polym. Test. 2022, 108, 107516.

18. Liu, J.; Su, C.; Chen, Y.; Tian, S.; Lu, C.; Huang, W.; Lv, Q. Current understanding of the applications of photocrosslinked hydrogels in biomedical engineering. Gels 2022, 8, 216.

19. Jiang, S.; Liu, S.; Feng, W. PVA hydrogel properties for biomedical application. J. Mech. Behav. Biomed. Mater. 2011, 4, 1228–1233.

20. Wu, S.; Hua, M.; Alsaid, Y.; Du, Y.; Ma, Y.; Zhao, Y.; Lo, C.; Wang, C.; Wu, D.; Yao, B.; et al. Poly (vinyl alcohol) hydrogels with broad-range tunable mechanical properties via the hofmeister effect. Adv. Mater. 2021, 33, 2007829.

21. Hassan, C.M.; Peppas, N.A. Cellular PVA hydrogels produced by freeze/thawing. J. Appl. Polym. Sci. 2000, 76, 2075–2079.

22. Zhang, H.; Xia, H.; Zhao, Y. Poly (vinyl alcohol) hydrogel can autonomously self-heal. ACS Macro Lett. 2012, 1, 1233–1236.

23. Hassan, C.M.; Peppas, N.A. Structure and morphology of freeze/thawed PVA hydrogels. Macromolecules 2000, 33, 2472–2479.

24. Bai, H.; Yu, C.; Zhu, H.; Zhang, S.; Ma, P.; Dong, W. Mussel-inspired cellulose-based adhesive with underwater adhesion ability. Cellulose 2021, 29, 893–906.

25. Cui, T.; Sun, Y.; Wu, Y.; Wang, J.; Ding, Y.; Cheng, J.; Guo, M. Mechanical, microstructural, and rheological characterization of gelatin-dialdehyde starch hydrogels constructed by dual dynamic crosslinking. LWT 2022, 161, 113374.

26. Lin, X.; Zhao, X.; Xu, C.; Wang, L.; Xia, Y. Progress in the mechanical enhancement of hydrogels: Fabrication strategies and underlying mechanisms. J. Appl. Polym. Sci. 2022, 60, 2525–2542.

27. Wang, M.; Bai, J.; Shao, K.; Tang, W.; Zhao, X.; Lin, D.; Huang, S.; Chen, C.; Ding, Z.; Ye, J. Poly (vinyl alcohol) hydrogels: The old and new functional materials. Int. J. Polym. Sci. 2021, 2021, 2225426.

28. Sennakesavan, G.; Mostakhdemin, M.; Dkhar, L.K.; Seyfoddin, A.; Fatihhi, S. Acrylic acid/acrylamide based hydrogels and its properties—A review. Polym. Degrad. Stab. 2020, 180, 109308.

29. Gombert, Y.; Roncoroni, F.; Sánchez-Ferrer, A.; Spencer, N.D. The hierarchical bulk molecular structure of poly (acrylamide) hydrogels: Beyond the fishing net. Soft Matter 2020, 16, 9789–9798.

30. García-García, J.M.; Liras, M.; Garrido, I.Q.; Gallardo, A.; París, R. Swelling control in thermo-responsive hydrogels based on 2-(2-methoxyethoxy) ethyl methacrylate by crosslinking and copolymerization with N-isopropylacrylamide. Polym. J. 2011, 43, 887–892.

31. Zhu, Q.; Barney, C.W.; Erk, K.A. Effect of ionic crosslinking on the swelling and mechanical response of model superabsorbent polymer hydrogels for internally cured concrete. Mater. Struct. Constr. 2015, 48, 2261–2276.

32. Mahkam, M.; Doostie, L. The relation between swelling properties and cross-linking of hydrogels designed for colon-specific drug delivery. Drug Deliv. J. Deliv. Target. Ther. Agents 2005, 12, 343–347.

33. Avramescu, A.; Noguer, T.; Avramescu, M.; Marty, J.-L. Screen-printed biosensors for the control of wine quality based on lactate and acetaldehyde determination. Anal. Chim. Acta 2002, 458, 203–213.

34. Soldatkin, A.P.; Montoriol, J.; Sant, W.; Martelet, C.; Jaffrezic-Renault, N. Development of potentiometric creatinine-sensitive biosensor based on ISFET and creatinine deiminase immobilised in PVA/SbQ photopolymeric membrane. Mater. Sci. Eng. C 2002, 21, 75–79.

6 Chapter

DESIGN AND ANALYSIS OF A NOVEL HYDROGEL MATERIAL WITH CUSTOMIZED INTERNAL STRUCTURE MODIFIED BY "BIO" AMPHIPHILIC COMPONENTS USING A PHYSICO-CHEMICAL APPROACH

1. INTRODUCTION: AN OVERVIEW

Hydrogels are polymers that are attracted to water and have a structure that forms a three-dimensional network. They may absorb a significant amount of water because of the existence of water-attracting parts, which makes them ideal for biomedical uses such as scaffolds [1]. Choosing the relevant elements for creating the ultimate hydrogel enables the production of a practical and effective material with distinct characteristics such as porosity, biocompatibility, and biodegradability. The precise and customizable features of these materials make them suitable and highly sought-after in several fields such as tissue engineering, pharmaceuticals, water treatment, and material engineering.

Hydrogels possess a significant characteristic of emulating and imitating biological systems, specifically the extracellular matrix (ECM). The ECM is a complex framework consisting of various proteins, polysaccharides, and other constituents that provide structural support. The extracellular matrix (ECM) controls various biological activities such as cell survival, growth, proliferation, migration, and differentiation [2]. Designing a customized in vitro environment that accurately replicates the ordered organization of the extracellular matrix (ECM) is a significant challenge and an important objective. Hydrogels have an advantage as prospective materials for scaffolds because they may be adjusted to have specific physico-chemical properties (such as electrical charge and pore size) and mechanical qualities (such as stiffness and tensile strength) that are necessary for cellular activity [3]. Most hydrogels are biocompatible, including naturally derived polymers like agarose, alginate, chitosan, collagen, fibrin, gelatin, hyaluronic acid, and dextran, as well as synthetic gels based on poly(ethylene glycol) (PEG), poly(vinyl alcohol) (PVA), and poly(hydroxyethyl methacrylate) (PHEMA) [4].

One way to change, enhance, or customize the properties of hydrogels is by including hydrophobic or micellar domains into the gel structure. This is because the mechanical properties and porosity microstructure of hydrogels are primarily determined by their 3D network structure [5].

Pure hydrophobic association (HA) hydrogels are hydrogels that are generated through the physical crosslinking of polymers using hydrophobic interactions. These hydrophobic interactions make up 5-20% of the total polymer content in the hydrogels. Most hydrophobic association hydrogels are created through micellar copolymerization [6]. Tuncaboylu et al. made an effort to enhance the weak mechanical strength of self-healing hydrogels by developing hybrid hydrogels. These hybrid hydrogels were created by introducing strong hydrophobic contacts between hydrophilic polymers, facilitated by a physical crosslinker called stearyl methacrylate [7]. Adding NaCl to the reaction solution during the copolymerization of large hydrophobes (specifically stearyl methacrylate (C18)) with the hydrophilic monomer acrylamide (AAm) in an aqueous solution of sodium dodecyl sulfate (SDS) resulted in the enlargement of micelles and the incorporation of the large hydrophobes into the SDS micelles. The rheological tests demonstrated that the hydrophobic connections, which were encased by surfactant micelles, functioned as reversible and breakable crosslinks that were accountable for the fast self-repairing of the hydrogels [7].

One alternate method to increase the strength of the hydrogel network is to incorporate particles as extra crosslinking points, such as latex particles or nanoparticles [6]. Latex particles (LPs) are often produced using emulsion polymerization, which allows for efficient energy dissipation and enhances the mechanical properties of hydrogels. Gu et al. [8] suggested a technique that involved the adsorption of hydrophobic alkyl chains from hydrophobic monomers onto the surface of latex microspheres. This process was followed by the stabilization of the microspheres in the presence of surfactants, resulting in the formation of hydrophobic association centers as the initial physical crosslinking points. Furthermore, the anionic sulfate radicals, which result from the dissociation of the persulfate, were drawn towards the cationic chains of the latex microspheres. These microspheres were produced through a surfactant-free emulsion copolymerization process involving styrene and a vinylidene comonomer that contains a cationic side group. As a result, secondary physical crosslinking centers were formed. The addition of cationic latex microspheres resulted in an enhancement of the tensile and compression strength of the modified hydrogel in comparison to the pure hydrophobic association hydrogel.

Due to their large specific surface area, the addition of inorganic nanoparticles to the hydrogel network can enhance its mechanical properties, particularly in terms of surface structure and charging [6]. Simultaneously, the incorporation of calcium carbonate nanoparticles [9], hydroxyapatite [10], kaolin [11], and laponite particles [12] can also promote hydrogel adherence.

However, the use of polymeric nanoparticles allows for the encapsulation of both hydrophobic and hydrophilic compounds [6]. Furthermore, Arno et al. conducted a study to examine the impact of particle morphology, such as particle shape, size, and surface, on the adhesion and mechanical properties of the resulting calcium-alginate hydrogels [13]. The researchers showed that including 2D platelets into the polymeric network significantly enhanced both the adhesion across hydrogel surfaces and the mechanical strength of the material, in comparison to the 0D spherical or 1D cylindrical equivalents.

Hydrogels can have their properties adjusted by selecting suitable ingredients and crosslinking procedures, as well as by altering the internal structure of the gel using a structural modifier like lecithin during the preparation process. Lecithin is a phospholipid mixture that is amphiphilic, meaning it has both hydrophilic and hydrophobic properties. It mostly consists of distearoylphosphatidylcholine. Lecithin is known for its capacity to improve the bioavailability of medications that are taken together with it. It also has strong biocompatibility. This information is supported by reference [14]. Lecithin has the ability to spontaneously form various liquid-crystalline structures in water systems, which is influenced by the quantity of water and temperature. The structures that are most likely to form under typical laboratory settings are lamellar liquid-crystalline formations [15]. Furthermore, adjusting the proportion of lecithin in the multi-component hydrogel system has the potential to enhance the usability and effectiveness of the developed gels. The internal structure of materials determines their transport and mechanical properties, which can be significantly influenced by any rearrangement.

The most prevalent platforms in this field are liposomes and microemulsions, which are among the various types of lecithin-based systems [16]. Liposomes are a type of soft nanoparticles made of phospholipids that usually have dimensions of approximately 100 nm [17]. Hydrophilic active compounds can be enclosed within the internal water compartments of liposomes, thanks to their closed vesicular structure. On the other hand, hydrophobic compounds can be placed within the bilayer of the liposome. Typically, lecithin-based liposomal hydrogels are employed as carriers. However, these systems still suffer from drawbacks, such as a sluggish and unregulated drug release process [18]. On the other hand, lecithin microemulsion-based gels or organogels provide several benefits compared to liposomal hydrogels. These include a simpler manufacturing process, the lack of organic solvents, and improved storage stability due to the thermodynamic stability of microemulsions [19]. The structure of lecithin microemulsion-based gels consists of lecithin, which functions as both a surfactant and a gelling agent when combined with a nonpolar organic solvent (external phase) or a polar agent, typically water.

A significant amount of research has been conducted on altering the internal structures of hydrogels. However, as far as we know, there has not been a comprehensive study that specifically examines the preparation and targeted modification of the internal structures of biocompatible hydrogels using natural amphiphilic substances and their important mechanical properties.

This study specifically examines the impact of the structure modifier lecithin, which has the ability to self-organize into liquid-crystalline structures, and its concentration on the mechanical properties of hydrogels with varying degrees of crosslinking. The outcomes of this study have the potential to enhance our comprehension of the interplay between lecithin and the hydrogel network, as well as the relationship between lecithin and model medications. The presence of lecithin aggregates in hydrogels can be seen as a representation of phospholipid structures, such as cell membranes, found in actual tissues. Consequently, it can serve as a model to understand how these structures may affect the rheological or transport aspects of the extracellular matrix.

2. RESULTS AND DISCUSSION

On the basis of the prior experience of our team and in an attempt to investigate the effect of different crosslinking strategies on the final properties of hydrogels, the following materials were selected: agarose as a physically crosslinked hydrogel, alginate crosslinked by polyvalent ions as an ionically crosslinked hydrogel, and PVA-chitosan as a chemically crosslinked hydrogel.

As stated in Section 4, for each type of crosslinking, four different samples were investigated. Three samples with lecithin additions at different concentrations (0.5, 1, and 2 wt.%) were labeled according to their lecithin concentration (i.e., "0.5", "1" and "2"). The fourth sample was a reference sample without lecithin, simply marked as "R". The lecithin concentrations were selected on the basis of preliminary experiments focused mainly on estimating the maximum amount of lecithin that could be incorporated into the hydrogel matrix.

2.1. Physical Crosslinking

Agarose was a representative of the physically crosslinked hydrogel matrix, whose properties were affected by lecithin content. Hydrogel samples after preparation as well as samples after the drying and rehydration procedure were studied (schematic figure of the preparation procedure can be seen in the Supplementary Materials Figure S1).

2.1.1. Rheology

Amplitude sweep results for physically crosslinked hydrogels obtained under an applied oscillatory strain of 1 Hz suggest that differences in lecithin concentration have, from a viscoelastic property point of view, a minimal influence on the hydrogel structure after preparation, especially with respect to the width of the linear viscoelastic region (as can be seen in Figure 1a). The storage as well as the loss modulus gradually increased with increasing lecithin concentration, which might be due to the overall higher dry content of the hydrogels. The effect of lecithin concentration on the viscoelastic properties of agarose hydrogels was also minimal in the linear viscoelastic region (LVR), which is the range of the values of storage modulus where the hydrogel is able to resist the applied oscillatory strain and can thus indicate the strength of non-covalent hydrogel nodes. Probably, the strength of the physically crosslinked hydrogel is provided mainly by non-covalent weak interactions (H-bonding) between the chains of agarose. Lecithin only had a small effect on the viscoelastic properties of 1 wt.% aq. agarose. The obtained values marking the end of the LVR were very similar for all samples physically crosslinked (Table 1). The values reported in the tables were either obtained by rheology software (TRIOS TA Instruments) analyses (cross-over point, average moduli values in LVR) or calculated. The end of LVR was obtained by comparing the average value of storage modulus in LVR with each point, where the deviation greater than 5% marked the end of the LVR. The mesh size calculations are described in Section 4.2. The cross-over point ($G' = G''$), the point at which the hydrogel was irreversibly damaged, was very similar for all samples.

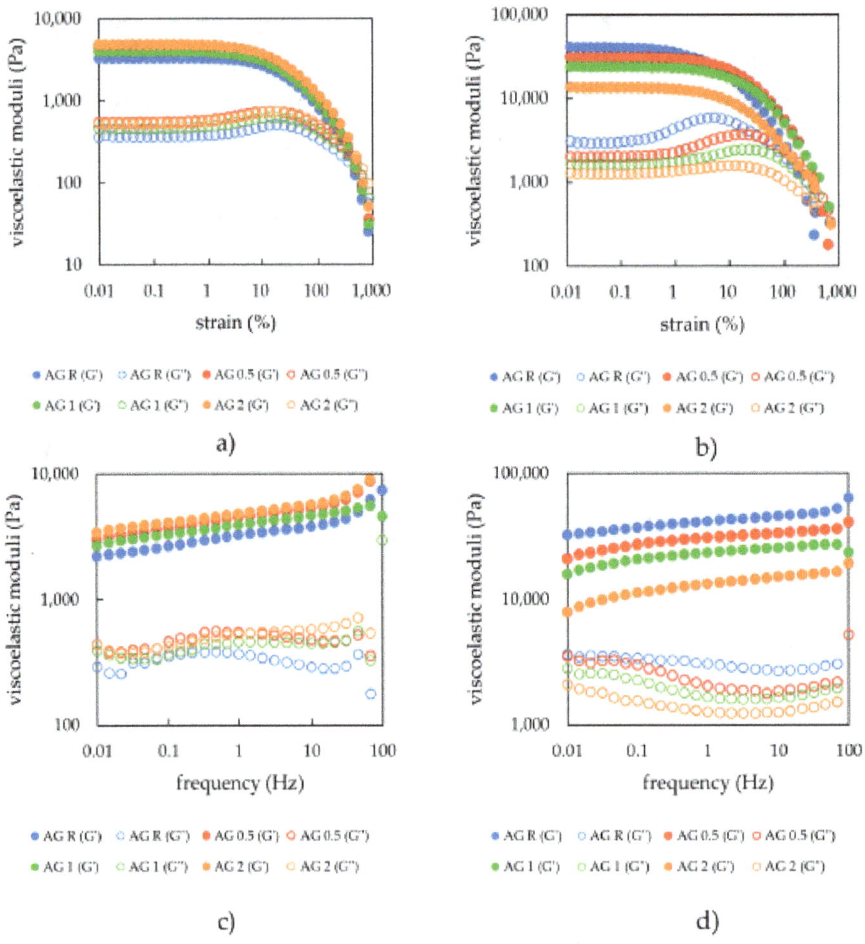

Figure 1. (a) Strain sweep of agarose hydrogels with different lecithin concentrations (0, 0.5, 1, and 2 wt.%) after preparation; (b) strain sweep of agarose hydrogels with different lecithin concentrations (0, 0.5, 1, and 2 wt.%) after drying and rehydration of the xerogels; (c) frequency sweep of agarose hydrogels with different lecithin concentrations (0, 0.5, 1, and 2 wt.%) after preparation; (d) frequency sweep of agarose hydrogels with different lecithin concentrations (0, 0.5, 1, and 2 wt.%) after drying and rehydration of the xerogels.

Table 1. Values for physically crosslinked agarose hydrogels after preparation obtained from strain and frequency sweep tests before drying.

	Cross-Over Point		Average Moduli Values in LVR		End of LVR	Mesh Size
Lecithin Concentration	G'	Strain	G'	G"	Strain	Mesh
(wt.%)	(Pa)	(%)	(Pa)	(Pa)	(%)	(nm)
0 (R)	157.5 ± 4.1	425.8 ± 2.2	3299 ± 277	366 ± 28	2.5 ± 1.0	13.3 ± 0.1
0.5	207.9 ± 2.1	414.1 ± 4.5	4576 ± 12	551 ± 15	1.8 ± 0.0	13.4 ± 0.4
1	194.9 ± 10.7	433.2 ± 10.2	4002 ± 81	461 ± 4	1.8 ± 0.0	12.7 ± 0.1
2	224.5 ± 0.0	468.0 ± 2.9	4880 ± 27	529 ± 8	1.8 ± 0.0	12.9 ± 0.3

The same amplitude sweep tests were performed on samples dried to the xerogel form and again rehydrated. The amount of absorbed water had a significant effect on these samples. As can be seen from Figure 1b as well as from the dry matter content experiments (Section 2.1.2), the samples with the highest lecithin content were able to reabsorb the largest amount of water (twice as much water as the sample without lecithin). This was also reflected in the amplitude sweep results because the moduli values for these hydrogels decreased proportionately. The reference sample had the highest moduli values, whereas the lowest values were observed for the samples with the greatest lecithin concentrations. The moduli values were somewhat larger than those for the samples studied after preparation (Table 2), mainly due to the elevated values of the swelling degrees of the systems after drying and rehydration in comparison with those of the just prepared hydrogels. Lecithin, therefore, favored water absorption. For the physically crosslinked hydrogels, even the cross-over point was affected, and samples with higher lecithin concentrations shifted the cross-over point to higher strain values. This could be the effect of the attractive interactions between lecithin and the polysaccharide chains, leading to the reinforcement of the hydrogels obtained after their drying and rehydration. In the initially prepared hydrogels, lecithin was dispersed to a greater extent in a liquid medium without this (strong) effect. This could be explained by the H-bonding between polysaccharide chains and lecithin, which are more significant for the rehydrated hydrogels because of the absence of water (in xerogel), which could not interfere. The same could be observed for the cross-over point, which again gradually increased with lecithin concentration.

Table 2. Values obtained from strain and frequency sweep tests for physically crosslinked agarose hydrogels after drying and rehydration.

	Cross-Over Point		Average Moduli Values in LVR		End of LVR	Mesh Size
Lecithin Concentration	G'	Strain	G'	G"	Strain	Mesh
(wt.%)	(Pa)	(%)	(Pa)	(Pa)	(%)	(nm)
0 (R)	1814 ± 340.6	250.4 ± 131.7	41,386 ± 10,517	2977 ± 707	0.3 ± 0.1	7.6 ± 1.1
0.5	1005.7 ± 142.9	718.5 ± 129.3	31,216 ± 980	2010 ± 4	1.2 ± 0.5	7.6 ± 0.2
1	542.7 ± 0.0	1148.7 ± 0.0	23,829 ± 3	1642 ± 118	1.4 ± 0.3	8.2 ± 0.1
2	350.8 ± 35.5	1257.6 ± 12.2	13,506 ± 1217	1256 ± 122	0.9 ± 0.0	9.0 ± 0.3

Frequency sweep test results are presented in Figure 1 and show that the shape of the rheograms for all hydrogel samples was very similar. The storage modulus was dominant, which means that the samples act as a fully crosslinked gel material with a fully crosslinked internal structure. The trend of the moduli values was the same as that observed for the amplitude sweep tests and therefore indicates that the lecithin addition increased the values of the storage moduli as well as of the loss moduli, which was well correlated with the higher dry matter content, as previously stated. With increasing oscillation frequency values, the moduli values increased, which means that the hydrogel samples were not completely relaxed, and the degree of relaxation was influenced by the type of crosslinking. Practically, the average relaxation time of the hydrogel network exceeds the period associated with the progressively increasing frequency of the applied oscillatory deformations. The values of the mesh size of the internal structure of the hydrogels calculated from the frequency sweep tests using Equations (3) and (4) are recorded in Table 1 and Table 2. The results for the freshly prepared hydrogels showed the same trend as other rheological data (i.e., that the mesh size does not differ substantially between the concentrations), whereas for the rehydrated xerogels, a slight increase could be observed at higher lecithin concentrations, which can be explained by lecithin fitting itself into the pores and thus increasing its size. This could be explained by lecithin forming lamellar liquid-crystalline structures in absorbed water along with the already mentioned H-bonding between the polysaccharide chains and lecithin. When comparing the absolute values of mesh sizes for freshly prepared and rehydrated hydrogels, we can see that the pores decreased in size after rehydration.

Based on the results of the strain and frequency sweeps performed onto the freshly prepared agarose hydrogels, it can be seen that lecithin, as an amphiphilic natural component, does not lead to a substantially modified viscoelastic behavior of these physically crosslinked hydrogels in the range of lecithin concentrations used (see Figure 1a,c). Agarose, which forms a thermoreversible physical hydrogel in an aqueous medium in the form of a natural linear polysaccharide, was not expected to interact significantly with amphiphilic lecithin. Thus, it was not expected that agarose could significantly interact with amphiphilic lecithin. Lecithin thus serves only as a filler, and does not interfere significantly with the internal structure of the hydrogel. Therefore, lecithin plays an important role in the rehydration of dried samples. Thus, an increasingly higher content of lecithin in the structure of such type of hydrogels causes the viscoelastic moduli storage and loss moduli to gradually decrease. Practically, the presence of lecithin affects the ability of agarose xerogels (hydrogel after drying) to reabsorb water (i.e., to swell) (see Figure 1b,d and Figure 2). The final viscoelastic properties of hydrogels are definitely affected by the amount of dispersion medium (water) after the swelling of xerogels. If the addition of lecithin, as the modifier of the internal architecture of hydrogels, is able to change the swelling properties, it will also definitely change the viscoelastic properties due to the different amount of water. From the applicative point of view, this finding is absolutely essential, given that by choosing a suitable concentration of additive (lecithin), we were able to prepare hydrogels with the required properties (especially viscoelastic) tailored to a specific purpose.

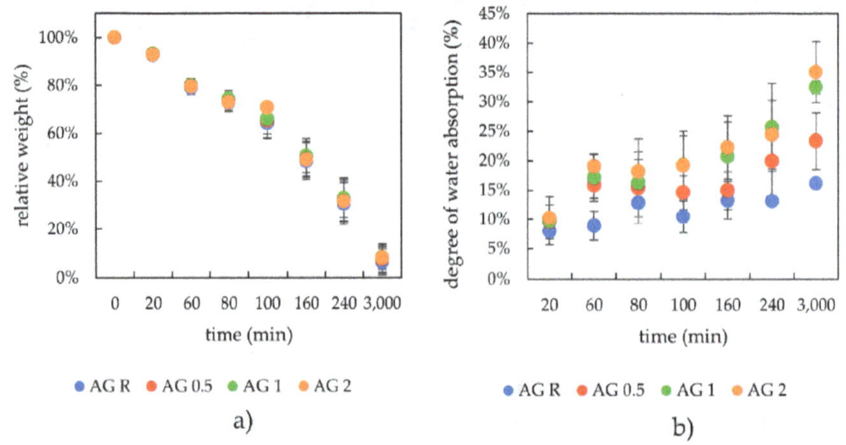

Figure 2. Drying (a) and rehydration (b) of the physically crosslinked agarose hydrogels with different contents of lecithin.

2.1.2. Drying and Rehydration Measurements

The amounts of water and dry matter associated with the studied gels are two of the most important parameters for hydrogel characterization and future applicability. Dry matter affects the behavior of the final material. The same is true for the water inside the hydrogel, which significantly affects, for example, the transport properties. As stated in Section 1, these parameters predetermine the applicative nature of the final system.

The results of the drying kinetics of physically crosslinked hydrogels can be seen in Figure 2a. At the start of these experiments, all weights of the hydrogels (2 ± 0.2 g) and xerogels were comparably the same. It can be seen that the lecithin addition had no influence on the drying kinetics. The most likely explanation is that water retained by lecithin is not bound as tightly as water hydrating agarose. Conversely, during the swelling process, hydrogel with lecithin easily draws water (more easily than the agarose hydrogel solely) and this resulted in the lecithin-agarose samples showing a higher swelling ability with corresponding lower moduli (Figure 1). The swelling experiments demonstrated the influence of lecithin on the swelling capacity. Therefore, the lecithin structures insert themselves into the hydrogel pores and support the water intake. The kinetics of the swelling process was very similar for all samples, with a peculiarity noted at the onset of the experiment, where the samples richer in lecithin (1 and 2 wt.%) revealed a greater rate of water absorption. Additionally, the same systems (agarose with 1 and 2 wt.% of lecithin) were able to absorb the largest amount of water.

2.1.3. Morphological Characterization of Xerogels

Morphological characterization was performed on dried samples; therefore, the results may not correspond to the results obtained from methods where hydrogels are studied in native form (specifically, rheology). From the results obtained by scanning electron microscopy (SEM), the effect of lecithin addition could be observed in sectional view. The surfaces of these xerogels were smooth and with no visible pores on the micrometer scale. In sectional view, the lecithin-free xerogel exhibited a layered structure of polymer fibers with no visible interferences (see Figure 3). The same layered morphology was also observed for xerogels of agarose with different contents of lecithin even though there were regions of fusion of adjacent layers. Overall, the general morphology, practically devoid of pores as revealed by SEM, is most likely due to a compact structure resulting via the air drying procedure applied to hydrogels to finally obtain xerogels.

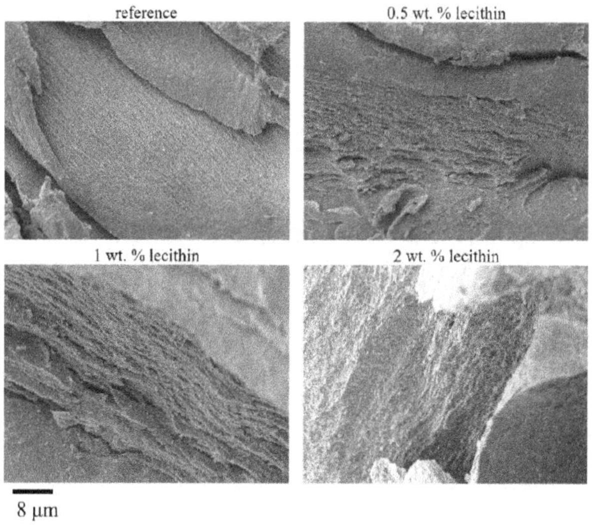

Figure 3. Physically crosslinked agarose xerogels with different lecithin contents observed in sectional view by SEM. Magnification 5000×.

For these xerogels, gas sorption measurements were also performed (Table 3). The low values of the specific surface suggest a lack of the pore structure of xerogels, with a slight dependence on the compactness of layered morphology of these systems in dry state. Even if the results of gas sorption are in line with those of SEM investigation, the gas sorption method is not quite a suitable technique for determining the structure of these xerogels.

Table 3. Specific surface area for physically crosslinked agarose xerogels with the addition of lecithin determined by gas sorption.

Concentration of Lecithin (wt.%)	Specific Surface Area (m^2/g)
0 (R)	3.4
0.5	1.0
1	1.9
2	2.1

2.2. Ionic Crosslinking

Sodium alginate crosslinked by the calcium chloride in the two to one weight ratio was a representative of the ionically crosslinked hydrogel matrix, where the negatively charged poly(guluronic) acid units of alginate (-COO-) interact with the polyvalent ions (Ca2+) to form a bond (schematic figure of the preparation procedure can be seen in Supplementary Materials, Figure S2). The final properties were also affected by lecithin addition. Hydrogel samples, both after preparation and dried and rehydrated, were studied by rheology, drying, and rehydration as well as morphological characterization.

2.2.1. Rheology

Ionically crosslinked hydrogels also underwent amplitude sweep tests. What is immediately observable is the decreasing trend of moduli for the freshly prepared samples as lecithin content increase (see Figure 4a). One of the reasons for this is the water intake during gelling, which increases for samples with ascending lecithin concentration (Section 2.2.2), the amphiphilic component playing a major role in the preparation of ionically crosslinked hydrogels. Larger lecithin addition also modified some characteristics of the hydrogels (see Table 4). The average moduli values in LVR steadily decreased after lecithin addition, thus making the gel softer. The most likely explanation is that after the crosslinking of alginate by calcium ions, free calcium chloride is still present in the system and is able to interact with the added lecithin micelles due to its dissociated form. Higher lecithin content causes a competitive interaction and as a result, lecithin displaces the calcium ions in the crosslinked alginate. Further lecithin could interact with the alginate via quaternary ammonia or with the calcium ions via negatively charged phosphate residues. For the moduli decrease, we could suggest that newly formed nodes are weaker and, in a lesser amount compared with the original alginate gel. Such competitive interactions were observable even during sample preparation, where the precipitate was visible on the surface of the solution. They were also confirmed by viscosity measurements, where the solution of calcium chloride and lecithin had higher viscosity values than expected, based on the viscosity of lecithin in water and of calcium chloride in water (figure is available in Supplementary Materials Figure S3). Other rheological data were very similar for the samples and, as stated earlier, the biggest differences were in the moduli values, thus in the hydrogel strength.

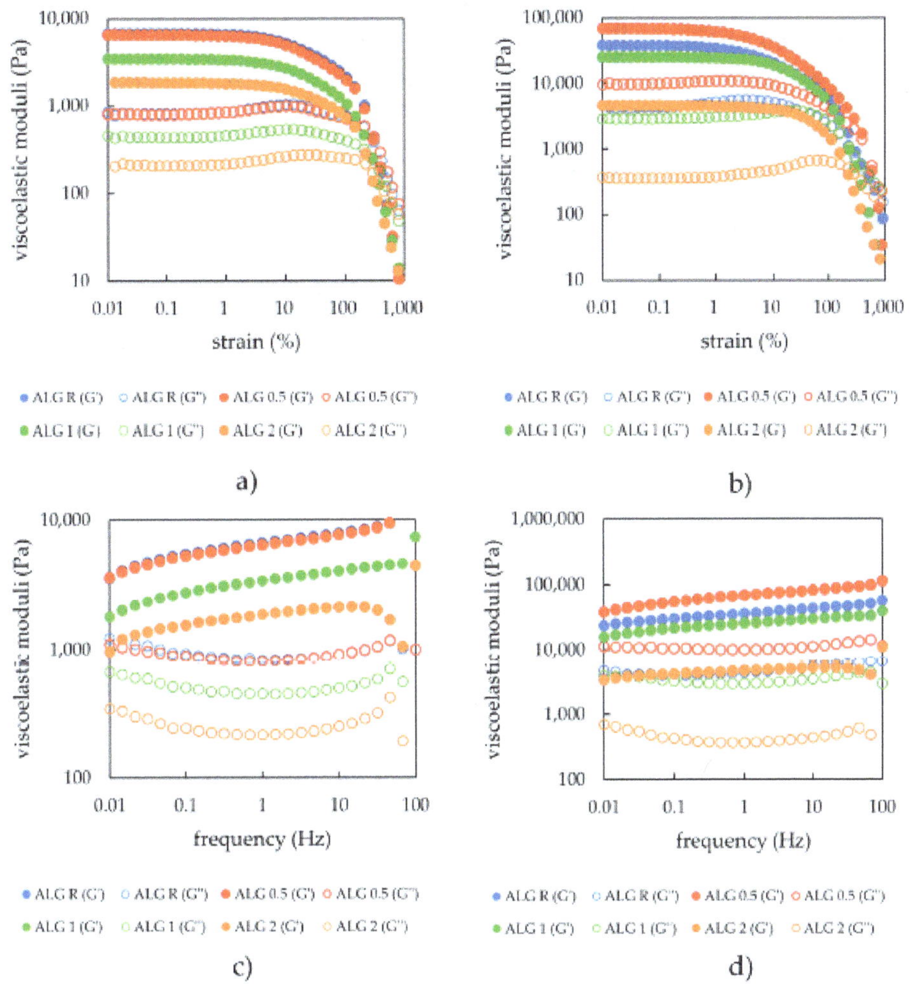

Figure 4. (a) Strain sweep of alginate hydrogels with the addition of different lecithin concentrations (0, 0.5, 1, and 2 wt.%) after preparation; (b) strain sweep of alginate hydrogels with different lecithin concentrations (0, 0.5, 1, and 2 wt.%) after drying and rehydration of the xerogels (frequency applied 1 Hz); (c) frequency sweep of alginate hydrogels with different lecithin concentrations (0, 0.5, 1, and 2 wt.%) after preparation; (d) frequency sweep of alginate hydrogels with different lecithin concentrations (0, 0.5, 1, and 2 wt.%) after drying and rehydration of the xerogels.

Table 4. Values for ionically crosslinked alginate hydrogels after preparation obtained from strain and frequency sweep tests before drying.

	Cross-Over Point		Average Moduli Values in LVR		End of LVR	Mesh Size
Lecithin Concentration	**G′**	**Strain**	**G′**	**G″**	**Strain**	**Mesh**
(wt.%)	**(Pa)**	**(%)**	**(Pa)**	**(Pa)**	**(%)**	**(nm)**
0 (R)	150.4 ± 9.1	260.4 ± 18.6	1667 ± 192	165 ± 23	1.7 ± 0.0	10.9 ± 0.4
0.5	158.6 ± 12.5	275.3 ± 24.4	2138 ± 480	245 ± 66	1.3 ± 0.0	11.0 ± 0.7
1	110.8 ± 1.2	260.8 ± 5.4	1052 ± 1	104 ± 0	1.6 ± 0.3	13.8 ± 1.9
2	65.3 ± 17.9	278.2 ± 9.0	468 ± 15	41 ± 0	2.1 ± 0.4	17.3 ± 1.5

The rehydrated samples followed a similar trend with respect to the moduli values, where these values decreased with increasing lecithin concentration. Average moduli values in LVR reported in the table below (Table 5) were higher than those presented in Table 4 because the rehydrated samples were not able to reabsorb the same amount of water as the freshly prepared hydrogels. Such behavior could be due to a compact arrangement favored by non-covalent interactions (mainly ionic interactions induced by Ca2+ ions onto both alginate and lecithin components) during the drying process.

Table 5. Values for ionically crosslinked alginate hydrogels after drying and rehydration obtained from the strain and frequency sweep tests.

	Cross-Over Point		Average Moduli Values in LVR		End of LVR	Mesh Size
Lecithin Concentration (wt.%)	**G' (Pa)**	**Strain (%)**	**G' (Pa)**	**G" (Pa)**	**Strain (%)**	**Mesh (nm)**
0 (R)	479.2 ± 129.7	210.8 ± 119.7	26,342 ± 13,355	3191 ± 1346	1.6 ± 0.4	4.6 ± 1.4
0.5	894.9 ± 612.4	522.6 ± 51.8	68,513 ± 17,434	9861 ± 1533	0.6 ± 0.6	12.3 ± 2.1
1	1179.5 ± 106.7	209.1 ± 37.3	25,386 ± 741	2912 ± 45	1.2 ± 0.2	8.3 ± 1.8
2	553.5 ± 24.3	189.5 ± 17.4	4599 ± 500	1842 ± 1447	2.4 ± 0.0	7.6 ± 0.5

The rheograms obtained during the frequency sweep tests (expressed as viscoelastic moduli on applied frequency) (Figure 4c) obeyed the same order as those that resulted from the amplitude sweep tests (storage and loss moduli as a function of oscillatory applied strain of 1 Hz) (Figure 4a) for all the studied alginate and alginate-lecithin hydrogels. The calculated mesh size from the rheological (frequency sweep) measurement for freshly prepared ionically crosslinked alginate hydrogels indicated the effect of lecithin on the structural properties of these hydrogels. The higher addition of lecithin causes a higher mesh size (more than 50% if the hydrogels without/with 2 wt.% of lecithin is compared). The effect of lecithin concentration was also not observed for dried and rehydrated hydrogels. Although ionically crosslinked hydrogels have the ability to reabsorb the dispersion medium and again create a network internal structure by water intake, the internal structure of these hydrogels is probably damaged by the air-drying process. Moreover, swelled hydrogels differ in mesh size values in comparison with freshly prepared (e.g., hydrogels with 2 wt.% of lecithin had a mesh size of 17.3 nm while the mesh size of the hydrogels with the same concentration of lecithin after swelling was 7.6 nm). Therefore, the effect of lecithin on the mesh size of hydrogels repeatedly prepared by drying and swelling in water medium was negligible.

2.2.2. Drying and Rehydration Measurements

The drying curves for the alginate-lecithin systems were very similar almost irrespective of the lecithin content, in contrast to the drying dependence obtained for the freshly prepared hydrogels of alginate solely (Figure 5a). The different kinetics regarding the rate of water loss during the drying step could be due to the way lecithin fills the hydrogel pores and holds water within, and also due to the favorable electrostatic Ca2+-lecithin interactions, which influence the hydrogel structure and thus enable it to better hold water. As for the swelling after drying, it can be observed that the samples with higher lecithin concentrations were able to absorb water more rapidly and to a higher capacity, which is again due to the modified hydrogel network due to the presence of lecithin.

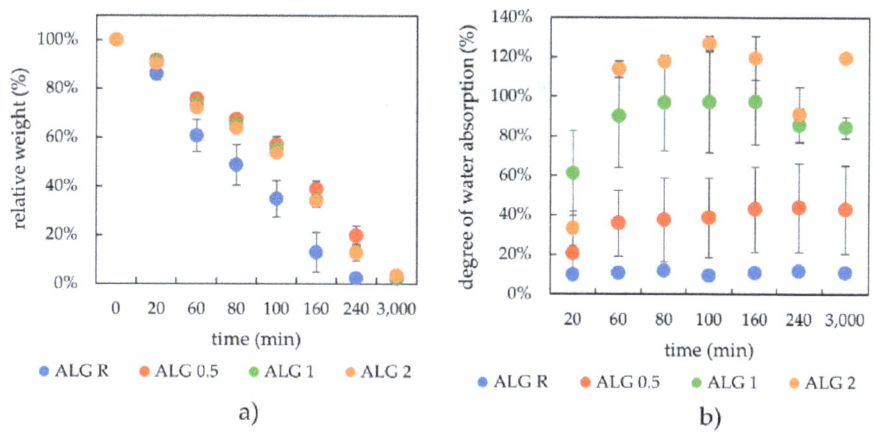

Figure 5. Drying (a) and rehydration (b) of ionically crosslinked alginate hydrogels with different lecithin content.

2.2.3. Morphological Characterization of Xerogels

SEM images taken for xerogels prepared by ionic crosslinking show the effect of lecithin on the surface morphology of the samples (see Figure 6). Surface morphology of lecithin-free samples and of those with 0.5 wt.% lecithin exhibited a roughness due to the many micrometer-sized crystals of CaCl2 resulted after air-drying. Instead, the surface of xerogels with 1 and 2 wt.% lecithin is practically devoid of crystalline aggregates, with some degree of roughness, which led to a more compact structure of these mixed systems in their dry state. The morphological characteristics microscopically revealed are in accordance with the decreasing tendency of the specific surface values (from gas sorption measurements, Table 6) as the lecithin content rose. On the other hand, the lack of CaCl2 crystalline aggregates for the systems with a higher lecithin content (1 and 2 wt.%) could be related to Ca2+ consumption in favorable electrostatic interactions with lecithin anions, which means that the crystalline structures observed in the case of alginate xerogels without lecithin and for those with 0.5 wt.% lecithin could be due to the excess of CaCl2 contained in these explored samples.

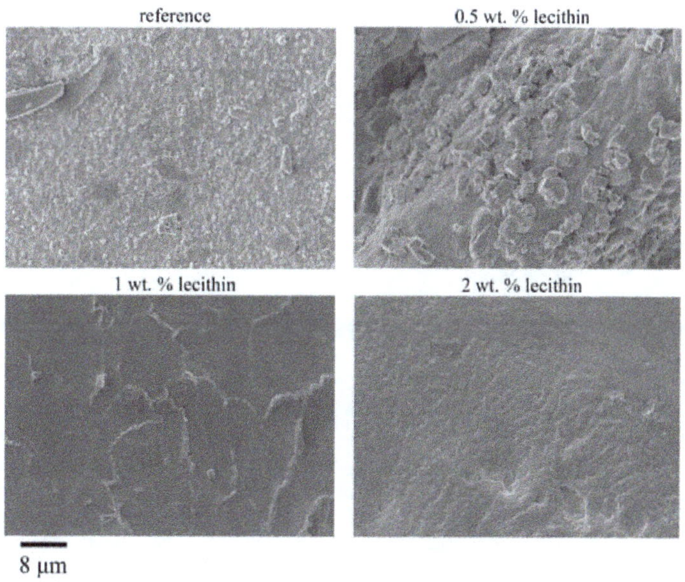

Figure 6. Surface morphologies of ionically crosslinked alginate xerogels with the addition of lecithin revealed by SEM. Magnification 5000×.

Table 6. Specific surface area for ionically crosslinked alginate xerogels with the addition of lecithin determined by gas sorption.

Concentration of Lecithin (wt.%)	Specific Surface Area (m^2/g)
0 (R)	9.1
0.5	6.3
1	5.9
2	4.7

2.3. Chemical Crosslinking

Poly(vinyl alcohol) and chitosan crosslinked by the epichlorohydrin was a representative of the chemically crosslinked hydrogel matrix. Epichlorohydrin reacts with either the hydroxyl group of PVA or amino group of chitosan to form a highly reactive intermediate. This intermediate product reacts with another hydroxyl (PVA) or amino group (chitosan) to form the crosslinked structure. Study of these hydrogels, both in their freshly prepared state, after air-drying at 40 °C and their subsequent rehydration and as xerogels, showed some physico-mechanical properties altered by the lecithin content

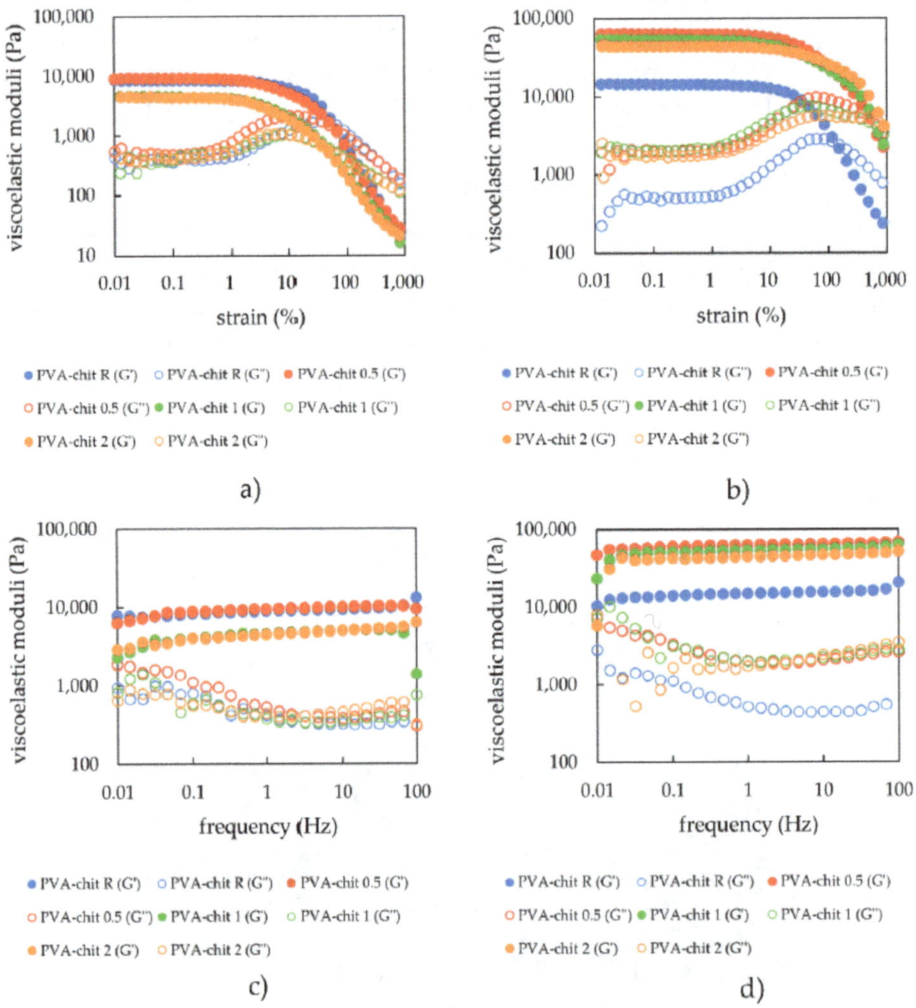

Figure 7. (a) Strain sweep of PVA-chitosan hydrogels with the different lecithin concentrations (0, 0.5, 1, and 2 wt.%) after preparation; (b) strain sweep of PVA-chitosan hydrogels with different lecithin

concentrations (0, 0.5, 1, and 2 wt.%) after drying and rehydration of the xerogels (frequency applied–1 Hz); (c) frequency sweep of PVA-chitosan hydrogels with different lecithin concentrations (0, 0.5, 1, and 2 wt.%) after preparation; (d) frequency sweep of PVA-chitosan hydrogels with different lecithin concentrations (0, 0.5, 1, and 2 wt.%) after drying and rehydration of the xerogels.

2.3.1. Rheology

For chemically crosslinked hydrogels, the amplitude sweep results showed that the addition of lecithin modified the rheological properties of hydrogels (see Figure 7a). However, the highest lecithin concentration did not lead to further changes in the mechanical properties. The same can be said after comparing the data points (see Table 7). At the same time, a higher content of lecithin decreased the values marking the end of the LVR as well as the strength of the hydrogels and the cross- over point values. The results are acceptable after taking into account the preparation and final state of the hydrogel. An important step of the preparation procedure is drying of the liquid mixture, which leads to crosslinking of the nodes and its subsequent rehydration. If lecithin is present, the rehydration is improved.

Table 7. Values for chemically crosslinked PVA-chitosan hydrogels obtained from strain and frequency sweep tests before drying.

	Cross-Over Point		Average Moduli Values in LVR		End of LVR	Mesh Size
Lecithin Concentration	G'	Strain	G'	G"	Strain	Mesh
(wt.%)	(Pa)	(%)	(Pa)	(Pa)	(%)	(nm)
0 (R)	1665.3 ± 43.2	53.8 ± 8.2	8629 ± 304	398 ± 4	1.6 ± 0.3	13.6 ± 0.7
0.5	1005.5 ± 32.4	49.4 ± 18.4	6644 ± 1503	307 ± 44	1.2 ± 0.9	13.8 ± 0.6
1	666.6 ± 5.4	40.2 ± 3.2	4545 ± 129	377 ± 68	0.6 ± 0.1	12.7 ± 0.1
2	631.6 ± 24.7	39.1 ± 4.7	4398 ± 195	421 ± 5	0.7 ± 0.1	12.9 ± 0.1

The same experiments were performed for hydrogel samples dried and rehydrated. The dried and rehydrated hydrogels with lecithin assembled into the pores ended up with modified properties (see Figure 7b), specifically, an increase in moduli values and a decrease in the values marking the cross-over point, in contrast to the reference sample. As can be seen in Figure 7b and Table 8, the presence of lecithin makes the hydrogels obtained after the drying–rehydration step much more deformation resistant, characterized by much higher values of strain at the cross-over point. At the same time, for these mixed rehydrated hydrogels, lecithin, irrespective of its content, exerted a larger influence in the enhancement of the hydrogels' strength (average moduli values in LVR) when compared to the rehydrated systems physically and ionically crosslinked.

Table 8. Values for chemically crosslinked PVA-chitosan hydrogels after drying and rehydration obtained from strain and frequency sweep tests.

	Cross-Over Point		Average Moduli Values in LVR		End of LVR	Mesh Size
Lecithin Concentration	G'	Strain	G'	G"	Strain	Mesh
(wt.%)	(Pa)	(%)	(Pa)	(Pa)	(%)	(nm)
0 (R)	2470.0 ± 494.7	138.5 ± 13.5	14,514 ± 1413	532 ± 33	3.2 ± 0.0	11.6 ± 0.3
0.5	7122.4 ± 633.3	379.1 ± 233.0	62,099 ± 6505	1928 ± 65	5.0 ± 1.0	7.1 ± 0.1
1	4964.6 ± 275.8	502.2 ± 277.5	52,833 ± 10,153	2089 ± 246	3.0 ± 1.7	6.2 ± 1.3
2	4074.2 ± 182.3	900.1 ± 97.5	43,685 ± 3177	1761 ± 211	5.9 ± 2.3	8.1 ± 0.0

The frequency and amplitude sweep results indicated the same tendency discussed above (see comparatively Figure 7). Thus, a critical lecithin concentration is necessary to modify the properties of this type of chemically crosslinked hydrogels (according to the results lying between 0.5 and 1 wt.%); also, there is a maximum concentration above which further modifications do not occur (differences between 1 and 2 wt.% are negligible). The significant difference in the chemically crosslinked hydrogels (comparing to the physically and ionically crosslinked) is the relaxation phenomenon characterized by much longer relaxation times in contrast to covalently crosslinked systems. Covalently crosslinked hydrogels exhibit almost constant values of storage moduli over the whole range of the applied frequencies. The same trend was also observed for the dried and rehydrated samples. Again, for all samples, the storage modulus prevailed in comparison to the loss modulus. The mesh sizes of these samples (Table 7 and Table 8) were not affected by the content of lecithin, a result that can be explained by the character of covalent crosslinking, which is stronger than physical and ionic crosslinking. On the other hand, the same trend of decreasing mesh sizes after rehydration could be observed.

2.3.2. Drying and Rehydration Measurements

As can be seen from Figure 8, the drying and swelling kinetics were not significantly altered by the addition of lecithin. Only a marginal influence was observed for samples with the highest lecithin concentrations, which were able to absorb the most water. This generally smaller influence of lecithin can be explained by the structure of chemically crosslinked hydrogels, which are characterized by a high enough crosslinking density and, consequently, by a smaller pore size morphology. The structure is more organized due to the stronger covalent bonds. The water absorption for this kind of hydrogel possessing stronger covalent cross linkages was very fast and occurred almost immediately during the first minutes of the swelling experiments.

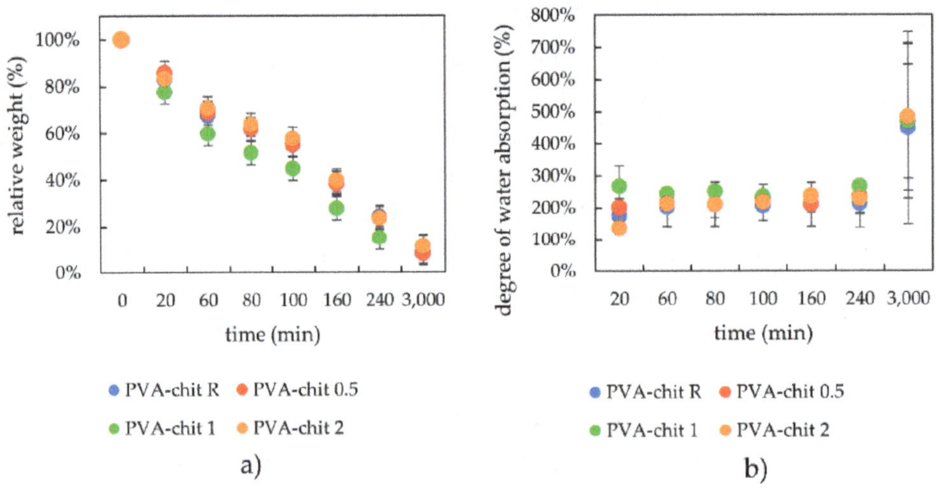

Figure 8. Drying (a) and rehydration (b) of chemically crosslinked PVA-chitosan hydrogels with different lecithin content.

2.3.3. Morphological Characterization of Xerogels

Results on the structural characterization of chemically crosslinked xerogels were similar to those for physically crosslinked hydrogels. The surface morphology of these xerogels looked smooth with no visible pores. In sectional view, SEM images revealed clear layered structures, with an interlayer roughness increasing with lecithin content (Figure 9), which in turn led to a gradual ascension of the value of specific surface (Table 9). Despite this fact, an apparently less corrugated surface observed for lecithin-free hydrogels had a higher specific surface area (Table 9), which might be explained by a greater compactness associated with the layered structure of the mixed xerogels.

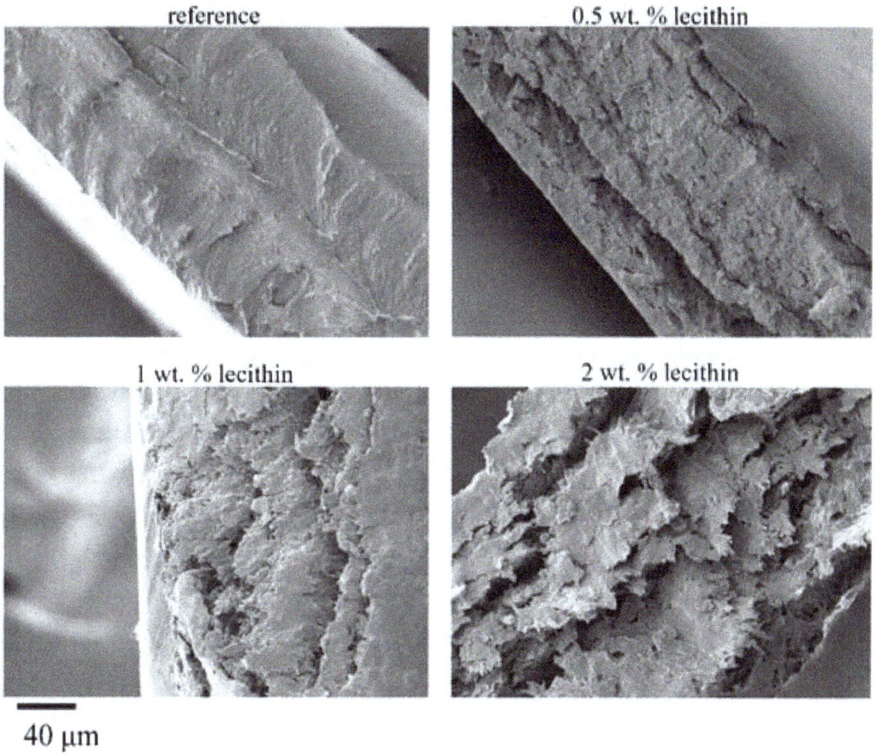

Figure 9. SEM images of chemically crosslinked PVA-chitosan xerogels with the addition of lecithin taken from the sectional view. Magnification 1000×.

Table 9. Specific surface area for chemically crosslinked PVA-chitosan xerogels with different lecithin content determined by gas sorption.

Concentration of Lecithin (wt.%)	Specific Surface Area (m^2/g)
0 (R)	2.9
0.5	0.8
1	1.2
2	1.6

3. CONCISE SUMMARY

This study investigated the impact of lecithin (L-α-phosphatidylcholine) on three hydrogels with distinct crosslinking methods (physically crosslinked agarose, alginate crosslinked by calcium ions, and a mixture of PVA and chitosan crosslinked by epichlorohydrin). The primary objective of this study was to examine the disparities between the gels analyzed immediately after their manufacture and the comparable rehydrated xerogels, which were prepared via swelling. By selecting the lecithin content, we successfully altered some mechanical characteristics of the hydrogels, particularly in the case of the rehydrated ones, by modifying their internal structure. The incorporation of lecithin had the most significant impact on improving the strength of chemically crosslinked PVA-chitosan gels. This finding is somewhat aligned with the mesh size and the amount of water absorbed into the gels' structure after being air-dried. In addition to the rheological data and the kinetics of water loss during hydrogel dehydration, the conclusions were further substantiated by conducting scanning electron microscopy and gas sorption studies on the xerogels. Despite the apparent unsuitability of gas sorption for this material, it effectively confirms the non-porous nature of the xerogels.

In this study, we established that incorporating phospholipid lecithin into the hydrogel matrix can modify their mechanical characteristics. This finding could have significant implications for the utilization of these hydrogels in specific applications. Nevertheless, it is imperative to examine the transport qualities as well. Hence, additional transport experiments are necessary to gain a comprehensive understanding of hydrogel materials and their potential uses.

4. MATERIALS AND METHODS

Hydrogels with distinct gelation mechanisms (physical, ionic, chemical crosslinking) [20] were studied. As an example of a physically crosslinked matrix, the linear thermoreversible polysaccharide agarose (Agarose E, Condalab, Madrid, Spain) at 1 wt.%, was used [21]. As an example of an ionically crosslinked matrix, sodium alginate (Sigma-Aldrich, Prague, Czech Republic) at 2 wt.% crosslinked by calcium chloride (Lach-Ner, Neratovice, Czech Republic) at a two to one weight ratio was chosen [22]. For chemically crosslinked hydrogels, poly(vinyl alcohol) (Sigma-Aldrich, Prague, Czech Republic) mixed with chitosan (low molecular weight, Sigma-Aldrich, Prague, Czech Republic) and crosslinked by epichlorohydrin (Sigma-Aldrich, Prague, Czech Republic) was employed [23]. L-α-Phosphatidylcholine (lecithin) was incorporated into all hydrogel samples before gelation at three different weight percentage concentrations (Sigma-Aldrich, Czech Republic, Prague).

The materials and their concentrations and ratios were selected on the basis of data previously reported [20,21,22,23,24] and can be seen in the table below (Table 10).

Table 10. Concentrations of each individual component in the final hydrogel form (agarose, sodium alginate, calcium chloride, PVA, chitosan, and lecithin).

Physically Crosslinked Hydrogels			
Sample	**Agarose (wt.%)**		**Lecithin (wt.%)**
AG R	1		0
AG 0.5	1		0.5
AG 1	1		1
AG 2	1		2
Ionically Crosslinked Hydrogels			
Sample	**Sodium Alginate (wt.%)**	**Calcium Chloride ($mol \cdot dm^{3}$)**	**Lecithin (wt.%)**
ALG R	2	0.1	0
ALG 0.5	2	0.1	0.5
ALG 1	2	0.1	1
ALG 2	2	0.1	2
Chemically Crosslinked Hydrogels			
Sample	**PVA (wt.%)**	**Chitosan (wt.%)**	**Lecithin (wt.%)**
PVA R	7.8	2.5	0
PVA 0.5	7.8	2.5	0.5
PVA 1	7.8	2.5	1
PVA 2	7.8	2.5	2

4.1. Water Loss during Drying and Rehydration Measurements

The ability to hold, release, and absorb water was tested by different approaches. Water loss was monitored by means of simple drying tests. All samples were dried either in the laboratory dryer at 40 °C and regularly weighed, or in a semi-automatic moisture analyzer (IR-35, Denver Instrument, Denver, CO, USA), where the weight was recorded automatically. The relative weight of the hydrogel (x) during drying was calculated using the following formula:

$$x = \frac{m_t}{m_0} \cdot 100 \quad (1)$$

where mt is the weight of the gel at time t, and m0 is the weight of the hydrogel in the swollen state.

Often very small weight losses of water from the hydrogel samples made using drying scales more difficult. For this reason, drying kinetics were mostly studied using the combination of laboratory driers and analytical scales, upon which samples were weighed every twenty minutes. After the samples were dried to the xerogel form, they were inserted into a water bath, where they were kept until they reached their maximum water absorption capacity. The degree of water absorption (ma) was calculated by:

$$m_a = \frac{m_t}{m_x} \cdot 100 \quad (2)$$

where mt is the weight of the hydrogel at time t, and mx is the weight of the xerogel. The hydrogel samples were regularly weighed on analytical scales to study their swelling kinetics.

4.2. Rheology

Hydrogels are semi-solid materials that exhibit distinctive mechanical characteristics lying between those of solids and liquids. Therefore, rheology is indeed an appropriate technique for studying their behavior [25,26,27,28,29]. The mechanical properties of the prepared hydrogels were determined by rheological characterization using a rotational rheometer (Discovery HR-2, TA Instruments) employing cross-hatched 20 mm plate–plate geometry to avoid potential sensor wall-slippage during measurement. The complex rheological procedure consisted of strain sweep and frequency sweep tests. The strain sweep test is a useful tool for obtaining information about samples if fluid-like or gel-like behavior under different values of applied strain prevails. In addition, it is possible to determine the region where the deformation is non-destructive (the linear viscoelastic region-LVR) as well as the behavior of the sample when the LVR strain limit is exceeded. The other mentioned test, the frequency sweep test, serves the purpose of describing hydrogel behavior in the non-deformation range (LVR) and provides information about different crosslinking sites (if applicable) in the internal structure of the hydrogel. Both tests were carried out on freshly prepared samples and rehydrated ones. The rehydrated samples were first dried to constant mass in the laboratory dryer for two days at a constant temperature of 40 °C and further rehydrated for three days in distilled water. Freshly prepared agarose and alginate samples were measured within a gap of 1000 μm. The gap for rehydrated samples varied according to the thickness of the gel, which depended on its swelling capacity, 500 μm for agarose gels and 1000 μm for alginate gels. PVA-chitosan hydrogels (both fresh and rehydrated) were measured within a gap of 200 μm due to the limited thickness of the prepared hydrogel foils. Prior to each applied test, samples were allowed to temper and rest for 180 s after loading into the measuring gap.

To obtain a suitable value of constant amplitude strain for the linear viscoelastic region (LVR), which was an essential parameter for ongoing frequency sweep tests, strain sweep tests were conducted first within the amplitude strain range of 0.01–1000% under a constant frequency of oscillation of 1 Hz in at least two repetitions, using a freshly loaded sample for each test. From these measurements, a strain of 0.1% was chosen as a suitable value of deformation for ongoing frequency tests, because this strain value lays within the LVR for all fresh and rehydrated samples. The range of oscillating frequencies for the frequency sweep tests was set to 0.01–100 Hz. Like the former strain sweep tests, the frequency sweep tests were also conducted in at least two repetitions. A summary of settings for both rheology tests is presented in Table 11.

Table 11. Summary of settings for rheology measurements (conditioning step, amplitude sweep, and frequency sweep).

Conditioning Step			
Temperature		25 °C	
Time		180 s	
Amplitude Sweep		**Frequency Sweep**	
temperature	25 °C	temperature	25 °C
strain	0.01–1000%	strain	0.1%
points per decade	8	points per decade	6
frequency	1 Hz	frequency	0.01–100 Hz

Routine techniques that are usable for the characterization of the internal structures of many materials (e.g., scanning electron microscopy) have some limitations in the study of hydrogels. One of the most limiting factors is that the structures of hydrogels are mostly studied in a dried state. The internal structures of a hydrogel in the presence of water and in the absence of water must certainly differ. Moreover, the preparation of the hydrogel in its dried state is also critical because the dispersion medium (water) must be removed (mostly by evaporation or by sublimation if lyophilization is used). Unfortunately, both of these processes (evaporation as well as sublimation) have a significant impact on the final xerogel morphology. Simply, the fragile internal structure of the hydrogel may be critically damaged by the removal of the dispersion medium. Thus, such a resulting structure (specifically, the porous structure) revealed by scanning electron microscopy often has low informative value with respect to the internal structure of the hydrogel in its swollen state. Therefore, an alternative way to determine the pore size (and then obtain information about the internal structure of the hydrogel) must be found. An interesting solution to this problem is offered by the rheological characterization of the hydrogel, which involves the calculation of the mesh size.

Mesh size, as one of the most critical parameters in hydrogel characterization, was calculated by means of relaxation spectra (relaxation moduli G and relaxation time λ) from the frequency sweep oscillation measurements in accordance with the Maxwell model [30]. The frequency sweep (viscoelastic moduli as a function of oscillation frequency) was interpolated by continuous relaxation spectra in TRIOS software (TA Instruments, New Castle, DE, USA).

Typical relaxation spectra can be found in the Supplementary Materials (Figure S5). On the basis of previous rheological investigation [25], it was concluded that the optimal number of Maxwell elements was 4, in order to fit the frequency sweep measurements of the hydrogels. Four relaxation moduli were obtained from continuous relaxation spectra analyses. The sum of relaxation moduli was calculated in order to determine the crosslinking density [31] (see Equation (3), where ρx represents the crosslinking density (mol·m-3)) and provides information on the density of the junction in the swollen hydrogel form. G (Pa) is the sum of 4 relaxation moduli, R (J·mol-1·K-1) represents the universal gas constant, and T is the thermodynamic temperature in Kelvins.

$$\rho_X = \frac{G}{RT} \tag{3}$$

If all criteria are met (in particular, frequency sweep measurements are realized in the linear viscoelastic region and the mechanical properties of hydrogels with different crosslinking are consistent with rubber elasticity theory [32]), finally the mesh size can be calculated using Equation (4), where ξ is the mesh size (unit: m) and NA represents Avogadro's number.

$$\xi = \sqrt[3]{\frac{6}{\pi \rho_X N_A}} \tag{4}$$

4.3. Morphological Characterization of Xerogels

Since the structure affects properties that are crucial for hydrogel applications, determining the hydrogel morphology is one of the most important characterizations. There are many direct (microscopy) and indirect (scattering-based) methods to characterize hydrogel morphology [33]. Several direct visualization techniques (light microscopy, laser scanning confocal microscopy, and micro-computed tomography) that can handle swollen hydrogels have considerable disadvantages (e.g., limited resolution) [34]. On the other hand, commonly used scanning electron microscopy includes a critical step (i.e., the inevitable solidification of the sample using drying or freezing, during which the collapse of the structure or the creation of artifacts can occur) [35,36]. Kaberova et al. [37] tested the usability of scanning electron microscopy and concluded that the results from this method should always be confirmed by microscopy techniques applicable for gels in their swollen state.

For the characterization of dry samples, the specific surface area (the Brunauer–Emmett–Teller (BET) approach) is typically determined. The specific surface area is not suitable for characterizing hydrogels because of the already mentioned artifacts that appear during the preparation of dried samples. However, it can be used, for example, for the characterization of materials used in a dried state and that can form hydrogels (adsorbent) [38], or for the confirmation of reversible porosity [39].

The structure of the xerogels was studied in this work. Specifically, scanning electron microscopy and gas sorption were chosen as suitable techniques for determining the internal architecture of xerogels. Since the mechanical properties were studied for hydrogels right after preparation and also for swollen hydrogels after dehydration, it seemed convenient to investigate the structural properties of the hydrogels in these forms. Since this form is a dry form, it was possible to avoid deformation of the structure caused by the preparation of hydrogels for scanning electron microscopy.

4.3.1. Scanning Electron Microscopy

To determine changes in hydrogel structure, xerogels of all prepared samples were subjected to direct visualization using scanning electron microscopy. The samples were dried in a laboratory dryer at 40 °C. A few small specimens were taken from each studied sample to maintain objective observation. These specimens were subsequently gold-coated in a sputtering device (POLARON) and investigated using a ZEISS EVO LS 10 scanning electron microscope.

Both the surface morphologies and sectional images of samples were recorded. Observations were realized in secondary electron (SE) mode and the accelerating voltage was set to 5 kV to avoid charging of the samples.

4.3.2. Gas Sorption

A NOVA 2200e high-speed gas sorption analyzer (Quantachrome Instruments) was used to determine the specific surface area. The samples were weighed into a measuring cell (0.05–0.1 g). The measuring cell was placed in a degassing station, where the degassing process was carried out at 75 °C for 20 h. After cooling, the degassed sample was weighed to four decimal places. The samples were placed in a measuring station. The adsorption and desorption isotherms were measured under liquid nitrogen (77 K) from 0.05–0.95 of the relative pressure P/P0. The obtained data were processed by NovaWin software and specific surface area was calculated by the multi-point BET method.

AT A GLANCE

Currently, hydrogels are utilized in a wide range of applications, including both industrial and biological fields such as tissue engineering, medication delivery systems, cosmetics, water treatment, and more. In order to meet the specific requirements of different applications, it is important to have the ability to alter the characteristics of hydrogel materials, especially their transport and mechanical properties that are directly linked to their structure. These properties play a critical role in determining the suitability of hydrogels for use in contemporary material engineering. Hence, the feasibility of fabricating hydrogel materials with adjustable characteristics is a tangible subject that continues to be investigated. One straightforward method to modify these qualities is to make changes to the internal structure by incorporating an additional component. The inclusion of natural compounds is advantageous because of their compatibility with living organisms and their ability to break down naturally over time. Thus, this study centered on hydrogels that were altered using lecithin, a naturally occurring component in our body's tissues. Hydrogels were synthesized using several methods of crosslinking, including physical, ionic, and chemical techniques. The mechanical properties of the materials were observed and these studies were enhanced by conducting drying and rehydration measurements, and were further supported by analyzing the physical characteristics of the xerogels. By including natural lecithin, it becomes feasible to alter essential characteristics of hydrogels, including porosity and mechanical properties, hence influencing their ultimate applications.

REFERENCES

1. Aswathy, S.; Narendrakumar, U.; Manjubala, I. Commercial hydrogels for biomedical applications. Heliyon 2020, 6, e03719.
2. Kular, J.K.; Basu, S.; Sharma, R.I. The extracellular matrix: Structure, composition, age-related differences, tools for analysis and applications for tissue engineering. J. Tissue Eng. 2014, 5, 2041731414557112.
3. Geckil, H.; Xu, F.; Zhang, X.; Moon, S.; Demirci, U. Engineering hydrogels as extracellular matrix mimics. Nanomedicine 2010, 5, 469–484.
4. Gadjanski, I. Recent advances on gradient hydrogels in biomimetic cartilage tissue engineering. F1000Research 2017, 6, 2158.
5. Pekař, M. Hydrogels with Micellar Hydrophobic (Nano)Domains. Front. Mater. 2015, 1, 35.
6. Zhang, Y.; Chen, Q.; Dai, Z.; Dai, Y.; Xia, F.; Zhang, X. Nanocomposite adhesive hydrogels: From design to application. J. Mater. Chem. B 2021, 9, 585–593.
7. Tuncaboylu, D.C.; Argun, A.; Algi, M.P.; Okay, O. Autonomic self-healing in covalently crosslinked hydrogels containing hydrophobic domains. Polymer 2013, 54, 6381–6388.

8. Gu, S.; Duan, L.; Ren, X.; Gao, G.H. Robust, tough and anti-fatigue cationic latex composite hydrogels based on dual physically cross-linked networks. J. Colloid Interface Sci. 2017, 492, 119–126.

9. Li, A.; Jia, Y.; Sun, S.; Xu, Y.; Minsky, B.B.; Stuart, M.A.C.; Cölfen, H.; von Klitzing, R.; Guo, X. Mineral-Enhanced Polyacrylic Acid Hydrogel as an Oyster-Inspired Organic–Inorganic Hybrid Adhesive. ACS Appl. Mater. Interfaces 2018, 10, 10471–10479.

10. Cui, C.; Wu, T.; Gao, F.; Fan, C.; Xu, Z.; Wang, H.; Liu, B.; Liu, W. An Autolytic High Strength Instant Adhesive Hydrogel for Emergency Self-Rescue. Adv. Funct. Mater. 2018, 28, 1804925.

11. Fan, X.; Wang, S.; Fang, Y.; Li, P.; Zhou, W.; Wang, Z.; Chen, M.; Liu, H. Tough polyacrylamide-tannic acid-kaolin adhesive hydrogels for quick hemostatic application. Mater. Sci. Eng. C 2020, 109, 110649.

12. Rajabi, N.; Kharaziha, M.; Emadi, R.; Zarrabi, A.; Mokhtari, H.; Salehi, S. An adhesive and injectable nanocomposite hydrogel of thiolated gelatin/gelatin methacrylate/Laponite® as a potential surgical sealant. J. Colloid Interface Sci. 2020, 564, 155–169.

13. Arno, M.C.; Inam, M.; Weems, A.C.; Li, Z.; Binch, A.L.A.; Platt, C.I.; Richardson, S.M.; Hoyland, J.A.; Dove, A.P.; O'Reilly, R.K. Exploiting the role of nanoparticle shape in enhancing hydrogel adhesive and mechanical properties. Nat. Commun. 2020, 11, 1420.

14. Zhang, H.; Hao, R.; Ren, X.; Yu, L.; Yang, H.; Yu, H. PEG/lecithin–liquid-crystalline composite hydrogels for quasi-zero-order combined release of hydrophilic and lipophilic drugs. RSC Adv. 2013, 3, 22927–22930.

15. Shchipunov, Y.A. Lecithin. In Encyclopedia of Surface and Colloid Science, 3rd ed.; Somasundaran, P., Ed.; CRC Press: Boca Raton, FL, USA, 2015; Volume 3, pp. 3674–3693.

16. Elnaggar, Y.S.; El-Refaie, W.M.; El-Massik, M.A.; Abdallah, O.Y. Lecithin-based nanostructured gels for skin delivery: An update on state of art and recent applications. J. Control. Release 2014, 180, 10–24.

17. Thompson, B.R.; Zarket, B.C.; Lauten, E.H.; Amin, S.; Muthukrishnan, S.; Raghavan, S.R. Liposomes Entrapped in Biopolymer Hydrogels Can Spontaneously Release into the External Solution. Langmuir 2020, 36, 7268–7276.

18. Li, D.; An, X.; Mu, Y. A liposomal hydrogel with enzyme triggered release for infected wound. Chem. Phys. Lipids 2019, 223, 104783.

19. Talaat, S.M.; Elnaggar, Y.S.R.; Abdalla, O.Y. Lecithin Microemulsion Lipogels Versus Conventional Gels for Skin Targeting of Terconazole: In Vitro, Ex Vivo, and In Vivo Investigation. AAPS PharmSciTech 2019, 20, 161.

20. Maitra, J.; Shukla, V.K. Cross-linking in Hydrogels—A Review. Am. J. Polym. Sci. 2014, 4, 25–31.

21. Trudicova, M.; Smilek, J.; Kalina, M.; Smilkova, M.; Adamkova, K.; Hrubanova, K.; Krzyzanek, V.; Sedlacek, P. Multiscale Experimental Evaluation of Agarose-Based Semi-Interpenetrating Polymer Network Hydrogels as Materials with Tunable Rheological and Transport Performance. Polymers 2020, 12, 2561.

22. Kuo, C.K.; Ma, P.X. Ionically crosslinked alginate hydrogels as scaffolds for tissue engineering: Part 1. Structure, gelation rate and mechanical properties. Biomaterials 2001, 22, 511–521.

23. Garnica-Palafox, I.M.; Sánchez-Arévalo, F.M.; Velasquillo, C.; García-Carvajal, Z.; Garcia-Lopez, J.; Ortega-Sánchez, C.; Ibarra, C.; Luna-Barcenas, G.; Solís-Arrieta, L. Mechanical and structural response of a hybrid hydrogel based on chitosan and poly(vinyl alcohol) cross-linked with epichlorohydrin for potential use in tissue engineering. J. Biomater. Sci. Polym. Ed. 2014, 25, 32–50.

24. Mendes, A.C.L.; Shekarforoush, E.; Engwer, C.; Beeren, S.; Gorzelanny, C.; Goycoolea, F.M.; Chronakis, I.S. Co-assembly of chitosan and phospholipids into hybrid hydrogels. Pure Appl. Chem. 2016, 88, 905–916.
25. Smilek, J.; Jarábková, S.; Velcer, T.; Pekař, M. Compositional and Temperature Effects on the Rheological Properties of Polyelectrolyte–Surfactant Hydrogels. Polymers 2019, 11, 927.
26. Mourycová, J.; Datta, K.K.R.; Procházková, A.; Plotěná, M.; Enev, V.; Smilek, J.; Másilko, J.; Pekař, M. Facile synthesis and rheological characterization of nanocomposite hyaluronan-organoclay hydrogels. Int. J. Biol. Macromol. 2018, 111, 680–684.
27. Derkach, S.R.; Ilyin, S.O.; Maklakova, A.A.; Kulichikhin, V.G.; Malkin, A.Y. The rheology of gelatin hydrogels modified by κ-carrageenan. LWT-Food Sci. Technol. 2015, 63, 612–619.
28. López-Marcial, G.R.; Zeng, A.Y.; Osuna, C.; Dennis, J.; García, J.M.; O'Connell, G.D. Agarose-Based Hydrogels as Suitable Bioprinting Materials for Tissue Engineering. ACS Biomater. Sci. Eng. 2018, 4, 3610–3616.
29. Gila-Vilchez, C.; Bonhome-Espinosa, A.B.; Kuzhir, P.; Zubarev, A.; Duran, J.D.G.; Lopez-Lopez, M.T. Rheology of magnetic alginate hydrogels. J. Rheol. 2018, 62, 1083–1096.
30. Gradzielski, M.; Hoffmann, I. Polyelectrolyte-surfactant complexes (PESCs) composed of oppositely charged components. Curr. Opin. Colloid Interface Sci. 2018, 35, 124–141.
31. Pescosolido, L.; Feruglio, L.; Farra, R.; Fiorentino, S.; Colombo, I.; Coviello, T.; Matricardi, P.; Hennink, W.E.; Vermonden, T.; Grassi, M. Mesh size distribution determination of interpenetrating polymer network hydrogels. Soft Matter 2012, 8, 7708–7715.
32. Flory, P.J. Principles of Polymer Chemistry; Cornell University Press: New York, NY, USA, 1953.
33. Raghuwanshi, V.S.; Garnier, G. Characterisation of hydrogels: Linking the nano to the microscale. Adv. Colloid Interface Sci. 2019, 274, 102044.
34. Suchý, T.; Šupová, M.; Bartoš, M.; Sedláček, R.; Piola, M.; Soncini, M.; Fiore, G.B.; Sauerová, P.; Kalbáčová, M.H. Dry versus hydrated collagen scaffolds: Are dry states representative of hydrated states? J. Mater. Sci. Mater. Med. 2018, 29, 20.
35. Muthulakshmi, L.; Pavithra, U.; Sivaranjani, V.; Balasubramanian, N.; Sakthivel, K.M.; Pruncu, C.I. A novel Ag/carrageenan–gelatin hybrid hydrogel nanocomposite and its biological applications: Preparation and characterization. J. Mech. Behav. Biomed. Mater. 2021, 115, 104257.
36. Marmorat, C.; Arinstein, A.; Koifman, N.; Talmon, Y.; Zussman, E.; Rafailovich, M. Cryo-Imaging of Hydrogels Supermolecular Structure. Sci. Rep. 2016, 6, 25495.
37. Kaberova, Z.; Karpushkin, E.; Nevoralová, M.; Vetrík, M.; Šlouf, M.; Dušková-Smrčková, M. Microscopic Structure of Swollen Hydrogels by Scanning Electron and Light Microscopies: Artifacts and Reality. Polymers 2020, 12, 578.
38. Bhagat, S.D.; Kim, Y.-H.; Yi, G.; Ahn, Y.-S.; Yeo, J.-G.; Choi, Y.-T. Mesoporous SiO2 powders with high specific surface area by microwave drying of hydrogels: A facile synthesis. Microporous Mesoporous Mater. 2008, 108, 333–339.
39. Weber, J.; Bergström, L. Mesoporous Hydrogels: Revealing Reversible Porosity by Cryoporometry, X-ray Scattering, and Gas Adsorption. Langmuir 2010, 26, 10158–10164.

Chapter 7 VARIABLES INFLUENCING THE DURATION AND METHOD OF CMC DRYING UTILIZING REFRACTANCE WINDOW OR CONDUCTIVE HYDRO-DRYING

1. INTRODUCTION

The scarcity of non-renewable resources for packaging materials has shifted attention towards biopolymers. Over the past ten years, there has been extensive study on biodegradable films made from natural raw materials such carboxymethyl cellulose (CMC). These films are non-toxic and non-allergenic, and they create a transparent coating that effectively blocks oxygen and lipids [1,2]. These films have applications in food packaging, adhesives, biocomposites, and hydrogel films.

Antosik et al. [1] fabricated films using CMC as carriers for a pressure-sensitive adhesive. The solution containing polysaccharides was placed into a mold made of polystyrene and subjected to a drying process at a temperature of 60 °C for a duration of 24 hours. Hasheminya et al. [2] fabricated films using Kefiran, essential oil, and CMC. The films were produced by pouring the solution onto glass plates and allowing it to dry at a temperature of 25 °C for a duration of 72 hours. Furthermore, CMC was employed for its ability to bind in tea waste bioplastics. The drying procedure occurred in an oven for a duration of 3 hours at a temperature of 70 °C [3]. Cheng et al. [4] created a composite film by combining CMC, konjac glucomannan, and palm olein in order to enhance the film's ability to prevent moisture from penetrating. The solutions were desiccated at ambient temperature for approximately 20 hours. Li et al. [5] created a film using starch and CMC. The samples were dried for 12 hours at room temperature. Hydrogels are another instance of CMC films. When made using CMC and polyvinylpyrrolidone [6], they can provide novel prospects for designing effective packaging materials. These works employed the casting technique to desiccate the films on a small scale in the laboratory. Given the significance of producing this genre of film on an industrial scale, it is necessary to have a drying technique that can be easily expanded, operates without interruption, and is energy efficient.

Refractance Window® (RW) drying is a method created by MCD Technologies, Inc. (Tacoma, WA, USA) [7,8]. The process used for high thicknesses is referred to as conductive hydro-drying (CHD) [9]. The substance to be dehydrated is positioned on a membrane that is in direct contact with water (at a temperature of 95–97 °C) at normal atmospheric pressure. The water is circulated again at a temperature below its boiling point in order to supply the required thermal energy for the drying process. Several writers have stated that this film is highly transparent to infrared radiation [7,8,10]. The transfer of thermal energy from the water to the food occurs through conduction and radiation via the plastic film, while the moisture is transported by the airflow passing over the layer of food [9]. The rise of this technology has been significant in the past five years, with 75% of the articles on the subject being published during this period. This increase can be attributed to the technology's advantages in terms of quality and cost.

An advantage of RW-CHD drying is its comparatively short drying time in comparison to other drying procedures, such as solar drying, tray drying, or freeze drying. In tray drying, the items must undergo drying over a duration of 3 to 5 hours at elevated temperatures, while freeze drying requires a processing period ranging from 18 to 24 hours [11]. Ochoa-Martínez et al. [12] discovered that, in the case of the RW method, the moisture content diminishes swiftly to a level below 5% wb within around 30 minutes for 1-mm samples and 60 minutes for 2-mm samples. By comparison, it required approximately 240 minutes to get comparable outcomes using the tray drying method. Baeghbali et al. [13] reported a drying time of 20-24 hours for freeze drying, while the observed time for diverse items using RW was 5-7 minutes. In their study, Nindo et al. [14] provided the residence times (in hours) for drying asparagus puree using several technologies: freeze drying (18–24), tray drying (2.5–5.5), spouted bed (1.2–2.3), microwave (0.5–1.6), and RW (0.074). Extended drying durations have a negative impact on the quality of the product. The increased rate of drying in RW-CHD can be due to the transmission of heat by radiation between the heated water and the food [14,15]. Conduction heat transfer also has a notable impact in CHD [16].

In contrast, RW dryers exhibit superior thermal efficiency as compared to traditional dryers, with a thermal efficiency that is twice as high as freeze dryers [8]. Baeghbali et al. [13] conducted a comparative analysis of various drying procedures to assess the overall energy efficiency in the drying process of pomegranate juice. The RW dryer exhibited the highest energy efficiency at 31.56%, followed by the spray dryer at 12.92% and the freeze dryer at 1.12%. In addition, Baeghbali et al. [17] found that the total energy consumption for drying carrot puree using RW dryers ranged from 0.375 to 0.525 kW, which was significantly lower than the 70 to 84 kW required for freeze drying. Bernaert et al. [18] also found similar results, with the energy efficiency of RW being three times greater than spray drying and 40 times higher than freeze drying.

Furthermore, the drying equipment used in this process exhibits lower energy consumption and operation expenses compared to tray, freeze, drum, and spray drying techniques. In addition, it has minimal carbon dioxide emissions [19]. For instance, the price of the WR drier might range from one-third to one-half of the cost of freeze drying [8,15,20]. When compared to conventional drying methods like drum drying or convection drying, production costs can be reduced by as much as 70% [21,22].

The impact of process variables, such as sample thickness and heating water temperature, on the physicochemical qualities of various foods, particularly fruits and vegetables, and to a lesser degree, meats, dairy products, and other types of food, has been extensively investigated. Mahanti et al. [23] performed a comprehensive examination on this topic. Optimal food temperatures and brief drying periods minimize the degradation of food qualities. Additional crucial process variables include the composition and gauge of the film employed, which impact the proportion of radiation transmission. These variables have not undergone any research or investigation. The majority of the pertinent studies have employed polyethylene terephthalate (PET) films. Previous researchers [9,16,24] conducted computations to ascertain the proportionate impact of different heat transfer processes in RW-CHD drying. Under no circumstances was the infiltration of infrared radiation detected, which is a crucial factor that impacts heat transmission in this particular technology. There is a lack of literature data on this variable specifically for infrared wavelengths, with the exception of Ginzburg's study [25] on different food materials and Almeida et al.'s research [26] on potatoes in near-infrared drying. Furthermore, the heat fluxes emitted by water, water-film, and water-film-CMC systems have not been quantified in order to define RW-CHD drying.

Understanding the variables that impact drying time and energy usage in RW-CHD, including sample thickness, heating water temperature, and film type and thickness, is crucial. Only the first two categories, namely fruits and vegetables, have been extensively researched for various items. Meats, dairy products, and other categories have received comparatively less attention. This method has not been utilized for the purpose of drying gels, such as CMC. Due to the elevated moisture content and extended processing duration, it is imperative to employ exceptionally efficient drying techniques for the dehydration of CMC films, which can serve as eco-friendly packaging materials.

The primary aim of the study was to assess the variables that influence the duration of drying and the amount of energy used in the RW-CHD drying process of CMC. The study aimed to quantify the depth of radiation penetration into CMC and assess the impact of various operational variables (such as CMC thickness, heating water temperature, and film type and thickness) on drying time and energy consumption.

The study revealed that higher temperatures and thinner layers resulted in shorter drying times. Likewise, the amount of energy used reduced as the temperature and thickness decreased. Nevertheless, the time required for drying and the amount of energy needed per unit weight of the product remained the same when drying at either 1.5 or 3.5 mm. The determination of the sample size to be dried (product thickness) is crucial in considering the cost of operation and the ultimate quality of the product when subjected to prolonged exposure to high temperatures. When comparing temperature and CMC thickness, it was found that the type of film had a negligible impact on time and energy usage. This was attributed to the films' exceptional radiation transparency, which indicates a minimal resistance to the flow of heat, regardless of their type or thickness. However, the depth to which radiation can penetrate in CMC was found to be 1.20 ± 0.19 mm. This feature is not affected by the kind and thickness of the film or the temperature of the heating water. It is solely determined by the material itself. The temperature has no impact on this property since the range of wavelengths evaluated at the temperatures examined is quite narrow. When the thickness is around this threshold, a greater amount of radiation energy is utilized, resulting in a higher rate of heating during the initial drying period (3 minutes).

2. RESULTS AND DISCUSSION

2.1. Drying Kinetics

Figure 1 shows the drying kinetics of CMC with thicknesses of 1.5, 2.5, and 3.5 mm for the three films (PP, LDPE, and PET) at 90 and 70 °C. As expected, as the sample thickness decreased and the drying temperature increased, the water loss was higher for every film.

Considering that the air conditions in contact with the sample during drying were 60.4 ± 6.5% relative humidity and 26.9 ± 1.7 °C, the equilibrium moisture content for CMC is 0.17 kg water/kg dry solid (15% wb) according to the sorption isotherm reported by Torres et al. [27]. In the kinetics shown, values close to this equilibrium value can be observed.

An initial part with a constant slope was observed in these curves, corresponding to the constant drying rate period, up to an approximate moisture value of 5 kg water/kg ss (83% wb). The high drying rate in this period was notorious, unlike tray drying, which is very short with a low drying rate [12]. During this period, there was water saturation on the surface and easy replacement of evaporated water due to the low resistance to mass transfer in the interior. In RW-CHD, the sample exhibited rapid heating (in the first seconds of drying), maintaining the continuous transport of water inside the sample to the surface, in contrast to the tray drying, where the sample heating was slow, and heat transfer was the limiting mechanism. Moisture removal from the surface to the air occurred by convective diffusion, the driving force of which is the water concentration difference. Parrouffe et al. [28] found that infrared radiation had no significant influence on reducing drying time beyond the constant rate period due to the low moisture content at the surface, which generally has a low penetration depth or a low absorption. The drying rate in the constant rate period (kg water/kg ss-min) corresponding to the slope of the curves in the linear region is presented in Figure 2.

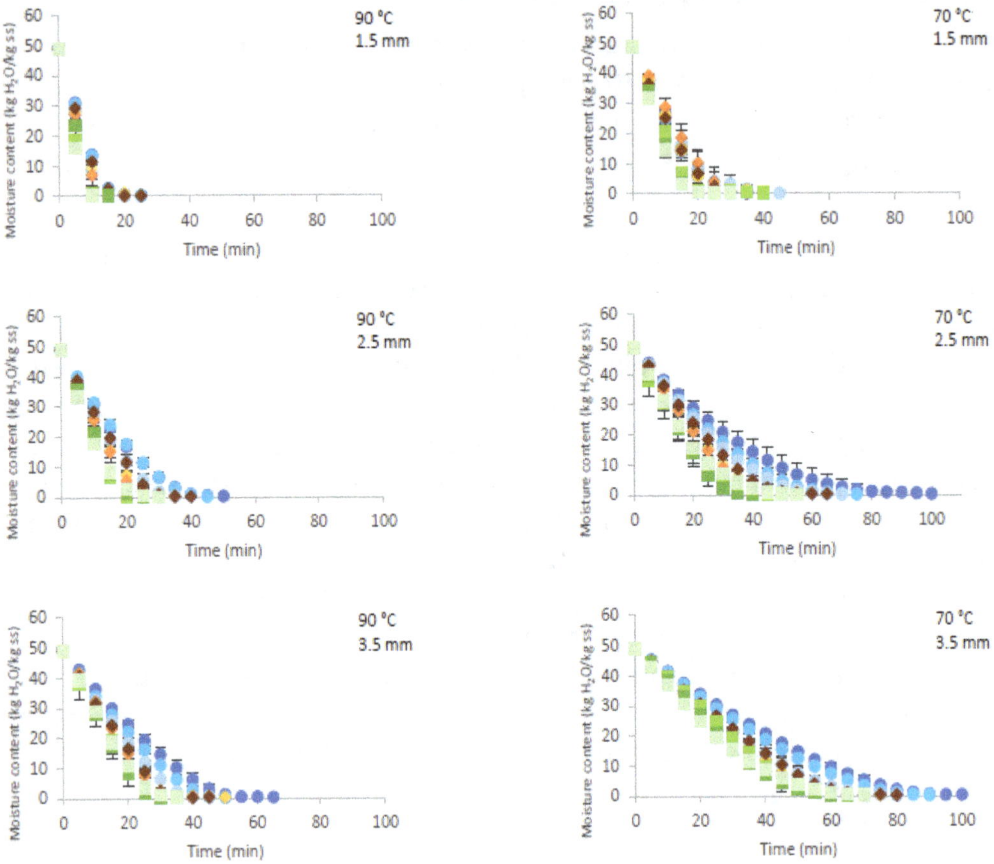

Figure 1. Effect of drying time, water temperature, and CMC thickness on moisture content (•PP 0.83, •PP 0.55, •PP 0.38, ♦ LDPE 0.15, ♦ LDPE 0.10, ♦ LDPE 0.03, ■ PET 0.25, ■ PET 0.18, ■ PET 0.08).

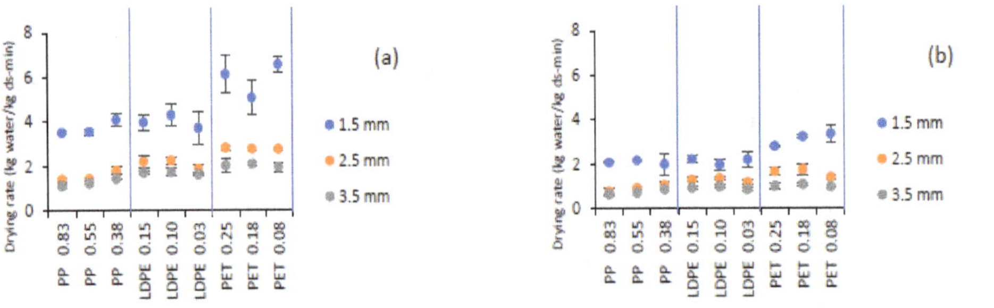

Figure 2. Effect of type and thickness of plastic films and CMC thickness on the drying rate: (a) 90 °C and (b) 70 °C.

In the analysis of variance, it was observed that the main factors (type-film thickness, CMC thickness, and heating water temperature) and their interactions had a significant effect on the drying rate ($p < 0.05$). According to the F-Ratio value, the relative importance of CMC thickness (1548.63) related to temperature (429.83) and film type-thickness (73.59) was highlighted. Tukey tests for temperature, CMC thickness, and film thickness type are presented in Table 1, Table 2 and Table 3.

Table 1. Effect of water temperature on CMC drying kinetics.

Temperature (°C)	Drying Rate (kg Water/kg ds-min)	Drying Flux (kg Water/h-m²)	Time (min)
70	1.53 [a]	3.38 [a]	52.37 [a]
80	2.14 [b]	4.73 [b]	37.52 [b]
90	2.78 [c]	6.05 [c]	29.33 [c]
Values followed by the same letter are not significantly different ($p < 0.05$, Tukey's test).			

Table 2. Effect of CMC thickness on drying kinetics.

CMC Thickness (mm)	Drying Rate (kg Water/kg ds-min)	Drying Flux (kg Water/h-m²)	Time (min)
1.5	3.49 [a]	4.96 [a]	22.67 [a]
2.5	1.70 [b]	4.66 [b]	42.58 [b]
3.5	1.26 [c]	4.55 [b]	53.97 [c]
Values followed by the same letter are not significantly different ($p < 0.05$, Tukey's test).			

Table 3. Effect of type and thickness of plastic film on CMC drying kinetics.

Film Type-Thickness (mm)	Drying Rate (kg Water/kg ds-min)	Drying Flux (kg Water/h-m²)	Time (min)
PP 0.83	1.55 [d]	3.46 [e]	55.86 [a]
PP 0.55	1.70 [cd]	3.75 [de]	48.85 [b]
PP 0.38	1.90 [bcd]	4.13 [cd]	42.80 [c]
LDPE 0.15	2.06 [bc]	4.68 [b]	38.31 [cd]
LDPE 0.10	2.10 [b]	4.85 [b]	37.66 [cd]
LDPE 0.03	1.93 [bcd]	4.59 [bc]	39.42 [cd]
PET 0.25	2.68 [a]	5.77 [a]	30.36 [e]
PET 0.18	2.66 [a]	5.62 [a]	30.56 [e]
PET 0.08	2.76 [a]	5.64 [a]	33.86 [de]
Values followed by the same letter are not significantly different ($p < 0.05$, Tukey's test).			

According to the Tukey analysis presented in Table 1, the mean value of the drying rate was significantly different for the three temperatures and the three CMC thicknesses studied. The rate increased with increasing heating water temperature and decreasing CMC thickness (higher values were obtained for a CMC thickness of 1.5 mm) and higher differences for PET films. The mean value of the drying rate increased by 81.7% when increasing the temperature from 70 to 90 °C and was 177.0% when decreasing the CMC thickness from 3.5 to 1.5 mm. Similar behavior was observed by Zotarelli et al. [24] in mango drying, with an increase of 40% when decreasing the thickness from 2 to 3 mm.

Table 2 shows that the three thicknesses of each film evaluated had a similar mean drying rate value. The highest rate was obtained for the PET films, while the other two materials presented lower values that were statistically equal.

Although the tests with PET for a sample thickness of 1.5 mm showed a higher drying rate in the constant period (moisture content dropped from 98% to 80% wb), statistical analysis of the results showed that the film type had little effect on time compared to the temperature effect and CMC thickness. According to Tsilingiris [29], a slight change in chemical composition, thickness, and measurement processes should produce substantial changes in the optical properties of the film.

Figure 3 shows the drying flux, which corresponds to the water evaporation capacity of the equipment per unit of drying area (kg water/h-m2). These values were between 2 and 9 kg/h-m2, similar to those obtained by Zotarelli et al. [24] (between 2.67 and 10.75 kg/h-m2) when drying mango pulp (thickness of 2, 3, and 5 mm) and heating water of 75, 85 and 95 °C. In addition, Nindo et al. [14] obtained a value of 10 kg/h-m2 when drying pumpkin puree, with thicknesses between 0.4 and 0.6 mm in the pilot plant and of 3.1, 3.9, and 4.6 kg/h-m2 at the industrial level.

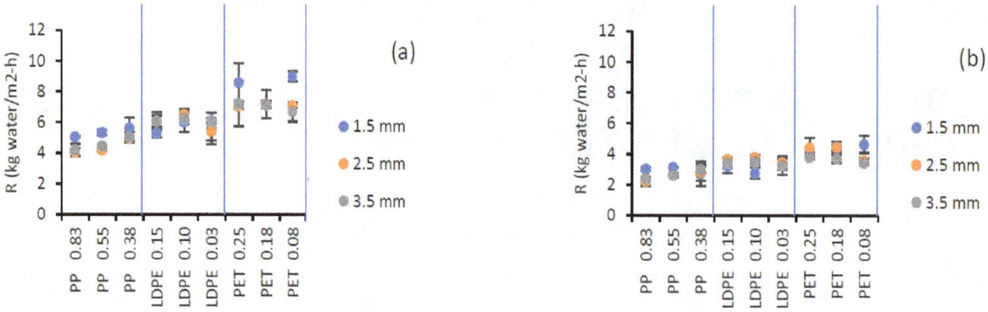

Figure 3. Effect of type and thickness of plastic films and CMC thickness on the drying flux (R): (a) 90 °C and (b) 70 °C.

All factors and their interactions affected the drying flux ($p < 0.5$). According to the F-ratio, the water temperature (700.62) had the most significant effect. Film thickness (92.80) and CMC thickness (17.84) had far less significant effects. According to Tukey's test (Table 1), the three temperatures had effects that differed in terms of significance. The mean value of the drying flux increased with increasing temperature. A similar effect was observed in the drying of mango pulp [24]. By increasing the temperature from 70 to 90 °C, the drying flux increased by 79%.

On the other hand, Tukey's test for CMC thickness (Table 2) showed that the mean value of the drying flux was higher and statistically different for CMC thickness of 1.5 mm. The mean drying flux obtained with 2.5 and 3.5 mm CMC thicknesses was not significantly different. The mean drying flux value increased by 9.0% when reducing the thickness from 3.5 to 1.5 mm. Zotarelli et al. [24] observed similar behavior when decreasing the thickness of mango pulp.

The Tukey test (Table 3) for film type and thickness showed that the mean value of the drying flux was statistically similar for the three thicknesses of each of the films and different for each type of film. The highest drying flux was obtained with PET, followed by LDPE and PP.

Figure 4 shows the drying time required to reach a final moisture content of 0.312 kg water/kg ss (this corresponds to the highest final moisture content of all treatments). Statistically, it was observed that all factors had a significant effect ($p < 0.05$) on drying time (min). According to the F-ratio, the most significant effect was CMC thickness (993.90), followed by temperature (540.26). Shende and Datta [30] showed the same behavior for mango drying. Film type-thickness (93.34) showed a minimal effect. Azizi et al. [31] also found that increasing the thickness of PET film from 0.1 to 0.3 mm does not have any significant effect on the drying time of kiwifruit irrespective of water temperature and sample thickness.

In studies of fruit drying using PET, drying times that were comparable to those presented in Figure 4 were obtained. For cornelian cherry pulp with a thickness of 1 mm at 90, 95, and 98 °C, the drying time was between 15 and 20 min [32]; a time of 20 min was obtained for mango pulp with a thickness of 2 mm at 95 °C [30]; and a time of 40 min was recorded for tomato pulp with a thickness between 1 and 1.5 mm at 90 °C and 60 min at 75 °C [22]. The increase in drying time with increasing food thickness was also observed in the drying of papaya puree [33] and mango slices [12].

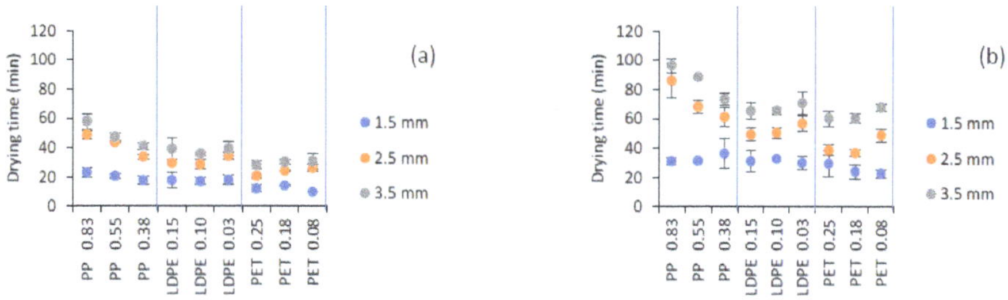

Figure 4. Effect of type and thickness of plastic films and CMC thickness on the drying time: (a) 90 °C and (b) 70 °C.

According to Tukey's test, significantly different means were obtained for the temperatures and CMC thicknesses evaluated (Table 1 and Table 2). As expected, the mean value of drying time decreased with increasing temperature and decreasing thickness. A decrease from 90 to 70 °C increased the drying time by 78.6%; an increase in thickness from 1.5 to 3.5 mm increased the drying time by 138.1%. However, there was a direct relationship between time and the amount of final product obtained, which in turn was related to CMC thickness (R2 = 0.9992) (Table 4), meaning that it was equivalent to drying with 1.5 or 3.5 mm in terms of the final amount obtained. This is important for deciding on the amount of sample to dry in terms of final product quality and operating costs. According to Table 3, for PP, the mean value of drying time was significantly different for each of its thicknesses and higher than for the other films evaluated (the evaluated thicknesses of PP were very high compared to those of LDPE and PET). The lowest average value of the drying time was obtained with the PET films (regardless of their thickness), which agrees with the higher drying rate obtained.

Table 4. Relation between the dry product and the sample thickness.

Thickness (mm)	Drying Time (min)	Wet Sample Weight (g)	Dry Sample Weight (g)	Drying Index (g Dry Sample/h)
1.5	22.7	33.3 ± 1.2	0.9 ± 0.0	2.6 ± 0.9
2.5	42.6	64.8 ± 3.8	1.7 ± 0.1	2.6 ± 0.8
3.5	54.0	84.9 ± 3.0	2.2 ± 0.1	2.7 ± 0.8

When the yield analysis (g/h-m2) was conducted for the conditions studied (Table 5), it was observed that the highest values were obtained for 3.5 mm food thickness and 90 °C. The highest yield was obtained for PET 0.25. The best performances for LDPE and PP were observed with LDPE 0.1 and PP 0.38, respectively. In addition to the yield, it was necessary to consider the temperatures to which the product was subjected during drying, which affected the quality of the final product.

Table 5. Effects of process factors on the drying yields (g/h-m2).

Temperature (°C)	70			80			90		
Thickness (mm)	1.5	2.5	3.5	1.5	2.5	3.5	1.5	2.5	3.5
PP 0.83	61.5	45.2	51.0	62.2	66.3	70.4	81.8	78.3	86.5
PP 0.55	60.4	54.7	55.9	89.3	72.4	73.1	95.0	89.1	101.5
PP 0.38	49.1	58.0	64.1	83.3	80.2	88.1	101.2	107.0	113.1
LDPE 0.15	60.7	76.0	74.1	72.1	96.4	99.8	97.8	123.0	120.8
LDPE 0.10	56.5	73.6	71.5	80.1	97.2	106.0	106.4	132.9	132.6
LDPE 0.03	64.8	68.5	70.3	94.8	105.4	106.2	108.8	110.8	126.6
PET 0.25	62.1	90.8	83.7	122.0	120.6	120.2	150.7	157.2	165.7
PET 0.18	73.9	91.0	75.8	118.2	123.8	105.2	129.5	139.8	147.2
PET 0.08	79.7	69.3	67.9	125.3	93.1	95.3	180.5	128.6	148.6

2.2. Temperature Profiles

Figure 5 shows the temperature profiles of the main components of the system (water, water–film interface, film–CMC interface, and CMC–air interface) for some of the conditions studied.

The interfaces of PP-CMC and PET-CMC films were rapidly heated by radiation and conduction (within the first three minutes of drying) to approximately 10–20 °C below the source temperature. During drying, their temperature decreased by 5–10 °C as the CMC absorbed energy from the film to equilibrate. In general, they may not heat up to the source temperature due to their high transmissivity (70–90 %) and low conductivity (PP: 0.23–0.26 W/mK; PET: 0.16–0.20 W/mK) [34]. On the other hand, when using LDPE, the interface temperature with CMC increased rapidly. Due to its high conductivity (0.25–0.33 W/mK) [34], it reached a value close to the heating water (about 5 °C below the water temperature) and remained constant during drying. The film interface temperature that did not depend on the food was always constant, while that which was in contact with the food was a function of the food temperature.

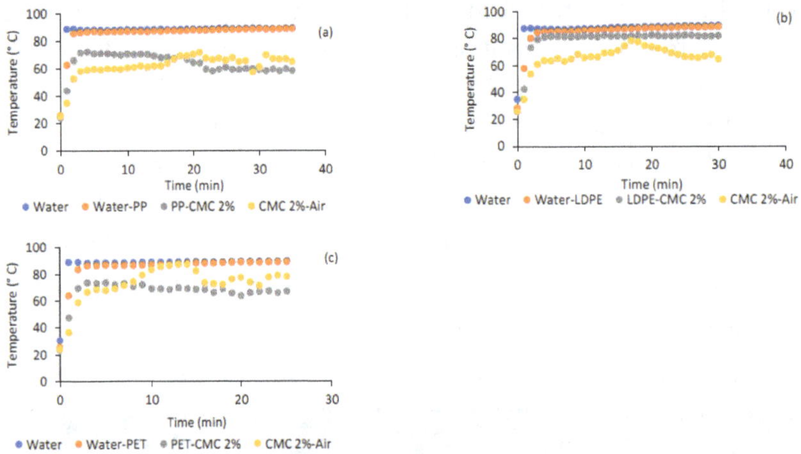

Figure 5. Temperature profiles of CMC-2.5 mm drying at water temperature of 90 °C: (a) PP 0.38 mm, (b) LDPE 0.15 mm, and (c) PET 0.18 mm.

The energy absorbed by the sample was used to heat (sensible heat) and to evaporate the water (latent heat). In general, the CMC temperature remained below the water temperature due to evaporative cooling and the effect exerted by the air temperature (26.9 ± 1.7 °C). On the other hand, the CMC initially absorbed energy from the film and warmed up to equilibrium with the temperature of the film–CMC interface (the film lost energy and the food gained energy); subsequently, heating of the sample was observed, and finally, they equilibrated at the same temperature. An increase in the sample temperature up to water temperature was only observed in the drying with LDPE-0.03 due to the low film thickness and high conductivity. The sample temperature rose to that of the water in tests carried out at 70 °C with 1.5 mm thickness, regardless of the film used.

2.3. Heat Flux Emitted by Water, Water–Film, and Water–Film–CMC Systems

The radiative heat flux emitted by the water at 70, 80, and 90 °C and the radiative heat flux transmitted through the film (water–film system) determined during the first 50 s were measured (Figure 6). It was observed that the heat flux was significantly higher for times longer than 50 s (these data are not presented). The heating of the film after this time (Figure 5) increased its emissivity, causing an increase in heat flux. Statistically, a high-temperature effect was observed.

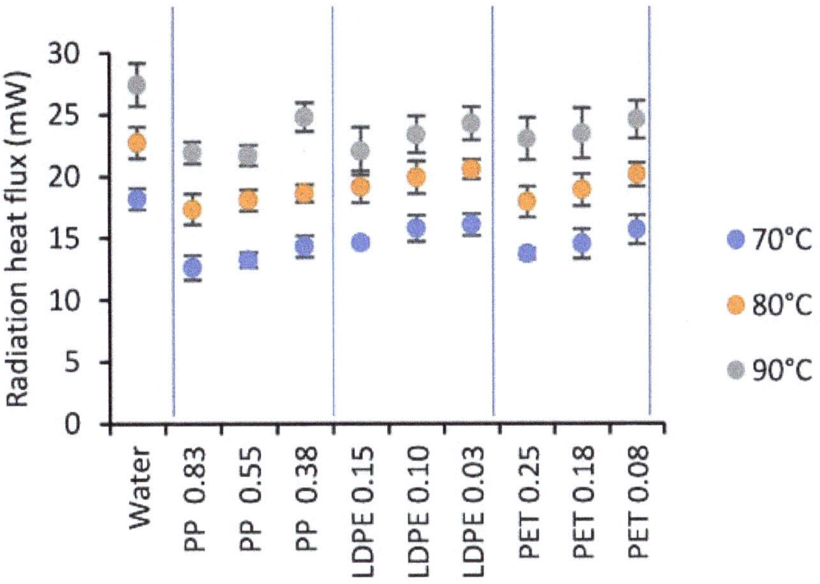

Figure 6. Radiation heat transfer for water and water-drying film systems.

According to Figure 6, all films showed high transmittance to radiation, with the heat flux being similar to that of the water. These results agree with those presented by Tsilingiris [29] for PP and LDPE films. That confirms that the films were transparent to radiation to a high degree, which allowed the heating of the sample, as discussed in the previous section.

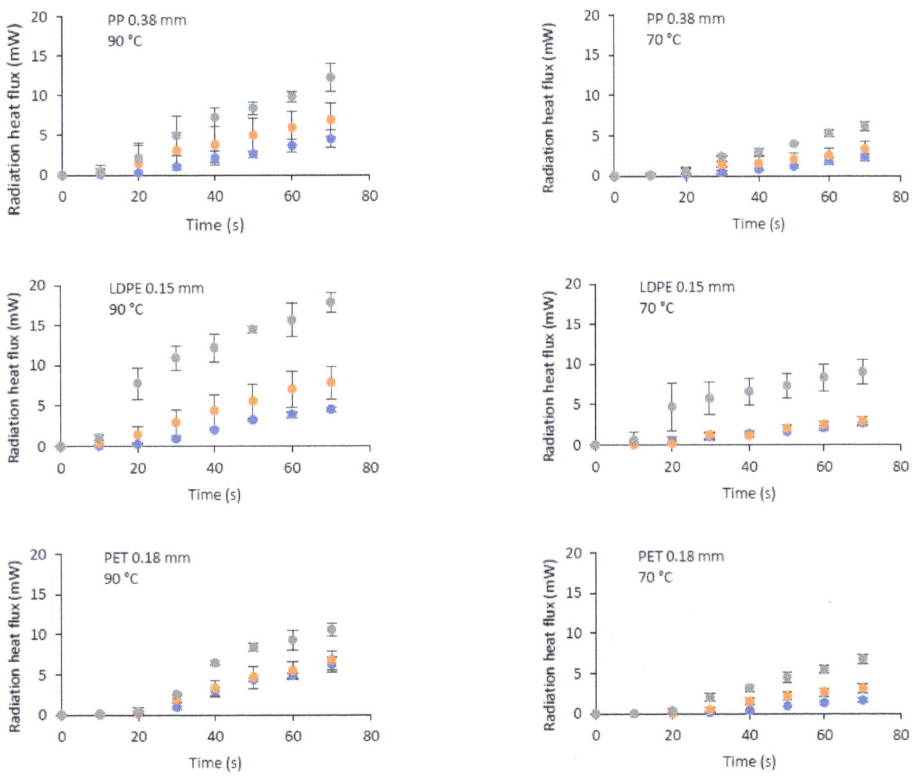

Figure 7. Effect of the type of plastic film, water temperature, and CMC thickness (•1.5 mm, •2.5 mm y •3.5 mm) on the radiation heat flux leaving the CMC.

On the other hand, the radiation heat flux leaving the CMC during the first 70 s of drying is presented in Figure 7 for some of the conditions studied. It was observed that the radiative heat flux was higher for heating water at 90 °C when compared to 80 or 70 °C. It was also observed that it increased with time; initially, this heat flux was low due to the high radiation absorption of the food; subsequently, the flux increased due to the combined effect of the radiation emission of the food upon heating and the lower absorption.

2.4. Penetration Depth of the Infrared Radiation in CMC

The penetration depth is defined as the depth at which the radiation intensity decays by 37% (1/e) of its initial value [35,36]. In other words, 63% of the incident radiation is used [11,30]. Depending on the material, electromagnetic waves can travel inside the food or be absorbed on its surface [37]. Using (Equation (1)), the parameters $I0$ and **μ-1** were obtained (Table 6).

$$I = I_0 e^{-\mu x} \tag{1}$$

where I is the radiation passing through the solid (water–film–food system) of thickness x, $I0$ is the initial radiation reaching the solid, and μ is the attenuation coefficient.

Table 6. Beer–Lambert Law parameters for radiant heat flow.

Water Temperature (°C)	70			80			90		
Film Type-Thickness (mm)	μ^{-1}	*Ln (Io)*	R^2	*μ-1*	*Ln (Io)*	R^2	*μ-1*	*Ln (Io)*	R^2
PP 0.83	1.33	2.35	0.94	1.35	2.69	0.96	0.73	1.37	1.00
PP 0.55	1.10	2.54	1.00	1.34	2.78	0.96	0.83	1.20	0.99
PP 0.38	1.05	2.57	0.98	0.73	3.04	0.99	0.89	1.13	0.98
LDPE 0.15	1.21	2.71	0.93	1.31	2.91	1.00	0.93	1.08	0.95
LDPE 0.10	1.22	2.79	0.99	1.23	2.97	1.00	0.74	1.36	0.98
LDPE 0.03	1.26	2.69	0.96	1.27	2.96	0.99	0.66	1.52	0.94
PET 0.25	1.33	2.33	0.91	1.41	2.73	0.96	0.88	1.13	0.97
PET 0.18	0.70	2.73	1.00	0.95	2.71	0.97	0.84	1.19	0.86
PET 0.08	1.16	2.67	0.99	1.33	2.94	0.99	0.80	1.25	0.98

The ANOVA and Tukey's test showed no significant effect of temperature or film type-thickness, so the values obtained for all penetration depths were averaged (Table 6). The average value of the radiation penetration depth in 2% CMC was 1.20 ± 0.19 mm. Although the penetration depth is known to depend on the nature of the material and temperature of the emitting source (wavelength) [37], the value obtained depended only on the material (CMC 2%) in the range of temperatures evaluated (70, 80, and 90 °C). It was, as expected, independent of the type and thickness of the films evaluated.

The total emissive power includes the energy of all wavelengths in the radiation spectrum [37]. The wavelength at which the maximum emissive power of radiation occurs, λmax, depends on the emitter temperature (T, K) and is given by Wien's displacement law, presented in Equation (2) [38].

$$\lambda_{max} = \frac{2898}{T} \tag{2}$$

According to Equation (2), for heating water temperatures between 70 and 90 °C, the wavelengths at which the maximum emissive power is obtained are 8.4 μm and 8.0 μm, respectively. This slight difference in wavelengths for the studied temperature range explains the lack of an effect of this factor on the penetration depth.

There is little information available in the literature on the penetration depth of infrared radiation [26,37]. For foods with high humidity (>80), penetration depth values between 1 and 7 mm were found, depending on the spectral peaks of the radiation source [25]. For potatoes with humidity between 67 and 82%, radiation penetration depth values between 0.45 and 2.85 mm were obtained [26], and for carrots, 1.5 mm was reported [25].

The amount of infrared radiation that a vegetable or animal sample can absorb depends on the wavelength of the emitter and the sample composition [37]. Foods are complex mixtures of biochemical molecules, biological polymers, inorganic salts, and water. Water shows high absorption at wavelengths of 3.0, 4.7, 6.0, and between 12 and 15.3 μm [39]. These values do not coincide in the wavelength range where the highest emissive power of water is obtained (8.4 μm and 8.0 μm at the temperatures used). On the other hand, carbohydrates (i.e., CMC) show two strong absorption bands at 3 μm and between 7 and 10 μm [39], coinciding with the emission bands of water. Accordingly, resonant absorption occurs in carbohydrates, which heat the water in the CMC. Subsequently, a thermal heating effect of the water occurs due to the dipole orientation in the electromagnetic field of radiation.

2.5. Energy Consumption

Figure 8 shows the results of energy consumption. According to the analysis of variance, film thickness, CMC thickness, and the heating water temperature had a significant effect on energy consumption ($p < 0.05$). According to the F-ratio, the most significant factor was CMC thickness (907.56), while the heating water temperature (101.8) and film thickness (72.56) had a much smaller effect.

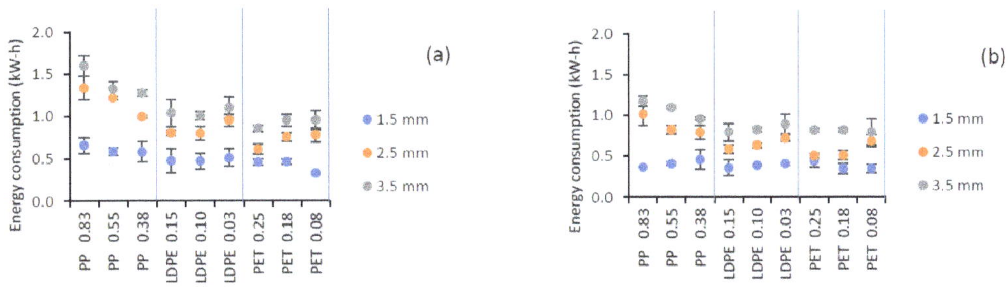

Figure 8. Effects of type and thickness of plastic films, CMC thickness, and water temperature ((a) 90 °C and (b) 70 °C) on energy consumption.

Tukey's test showed significant differences for both thickness and temperature. Energy consumption increased by 125.0%, thereby increasing the CMC thickness ($p < 0.05$) from 1.5 to 3.5 and 26.9% with the increase in temperature from 70 to 90 °C. This behavior was expected considering the greater amount of water to evaporate at the higher thickness and the more significant amount of energy required with increasing temperature.

The Tukey test for film thickness and type showed that the mean energy consumption value was higher for PP 0.83 mm, followed by PP 0.55 and 0.38 mm. There was no significant difference in the mean value of energy consumption ($p < 0.05$) among the other films, probably due to a combined effect of transmissivity and conductivity.

In the present work, the energy consumption varied between 0.44 and 0.99 kWh depending on the thickness of the sample. On the other hand, Baeghbali et al. [13] reported a consumption of 4.31 ± 0.82 kWh (using a pilot-scale continuous dryer in the drying of pomegranate juice). When analyzing the specific energy consumption ratio (SER) [19], which in turn was related to the CMC thickness (Table 7), a direct relationship was found (R2 = 0.9989). This means that the energy consumption, in terms of the final amount of product obtained for the three thicknesses studied, was similar. As mentioned in the time calculation, this information is essential for deciding on the amount of sample to dry depending on the final product quality and operating costs. Menon et al. [19], in their review of energy efficiency in drying technologies, reported SER values between 0.13 and 1.9 kWh/kg for super-heated steam drying, between 0.7 and 37.1 kWh/kg for microwave drying, and between 1.41 and 3.11 kWh/kg for impinging heat drying.

Table 7. Specific energy consumption (SER).

CMC Thickness (mm)	Energy Consumption (kWh)	SER (kWh/kg Water)
1.5	0.44 ± 0.09	13.6 ± 2.8
2.5	0.79 ± 0.21	12.5 ± 2.8
3.5	0.99 ± 0.20	12.0 ± 2.3

3. MATERIALS AND METHODS

3.1. Carboxymethylcellulose (CMC)

CMC powder (Gelycel F1 3500, specification 10031) with 4.14% humidity (Agenquímicos, Santiago de Cali, Colombia) was used. It was diluted in distilled water to obtain a 2% solution (98% moisture) using a blender (Samurai, Innova, Colombia) until a uniform mixture was obtained. The solution was left at rest for 24 h to release all air bubbles. Metal frames of 9.5 cm × 29.5 cm were used to obtain CMC sheets of the desired thickness (1.5, 2.5, and 3.5 mm).

3.2. Equipment

A stationary RW-CHD dryer (HS-50, Ceirobots, Santiago de Cali, Colombia) (Figure 9) was used. Two drying trays (10.3 cm × 30.3 cm), the bottoms of which were composed of plastic film, were in contact with the hot water. CMC was placed on the plastic films. The water was heated by two resistors (14.7 A each) in a 20 L tank, and the temperature was maintained through a recirculation system. A fan was used to remove the water vapor produced during drying. Three films of different thicknesses were evaluated, each according to commercial availability: polypropylene (PP) of 0.83, 0.55, and 0.38 mm; low density polyethylene (LDPE) of 0.15, 0.10, and 0.03 mm; and polyethylene terephthalate (PET) of 0.25, 0.18, and 0.08 mm.

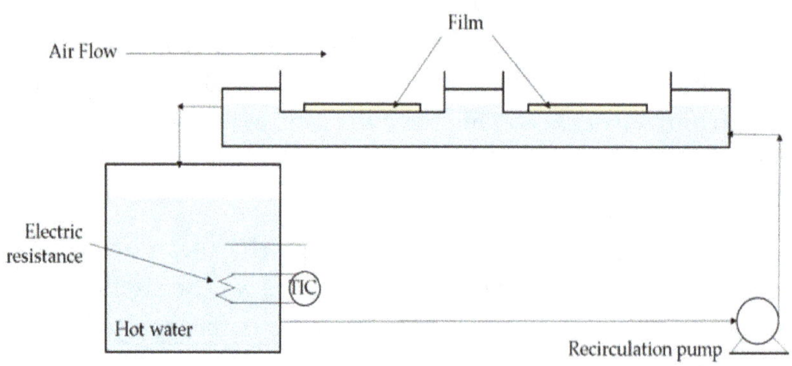

Figure 9. Schematic diagram of an RW-CHD dryer.

3.3. Experimental Design

The factors evaluated were the type of film with different thicknesses (PP-0.83, PP-0.55, PP-0.38, LDPE-0.15, LDPE-0.10, LDPE-0.03, PET-0.25, PET-0.18, PET-0.08), the thickness of the CMC (1.5, 2.5, and

3.5 mm), and the temperature of the heating water (70, 80, and 90 °C). A 9 × 3 × 3 complete randomized factorial design was used. All tests were performed in triplicate for a total of 243 treatments. The response variables evaluated were moisture content and drying time, temperature profiles, radiation heat flux, radiation penetration depth, and energy consumption. In addition, the radiant heat flux for the water and water–film system was determined at 70, 80, and 90 °C.

3.4. Response Variables

3.4.1. Moisture Content and Drying Time

Moisture content (db) was determined every 5 min. The drying tray was re-removed from the dryer, and the bottom dried and weighed. A balance (Ohaus®, Ad-venturerTM, Shanghai, China, accurate to 0.01 g) was used. The initial and final moisture content of the sample was determined by oven drying (Thermo Scientific, Heratherm OGS60, Langenselbold Germany) at 70 °C for 24 h [32]. The drying time required to reach a moisture content of 0.312 kg water/kg ss of CMC was determined to compare treatments (this value was selected considering that it was the minimum value obtained in some treatments).

3.4.2. Temperature Profiles

The temperatures of the hot water, the water–film interface, and the film-water-food interface were measured every 5 s using J-type thermocouples (Omega, 5TC-TT-J-30-72, diameter 0.25 mm, Norwalk, CT, USA) and a data acquisition system (Comark Instruments, model Diligence Evg N3014, Norwich, UK), with an accuracy of 0.2 °C. A thermographic camera (Flir, Model E4, Wilsonville, OR, USA) was used to measure the temperature of the food–air interface.

3.4.3. Radiative Heat Flow

The radiative heat flux emitted by the water, the water–film system, and the water–film–food system was measured for all experimental design conditions. A potentiometer (Molectron Detector Inc., PowerMax 5200, Portland, OR, USA) with a radiation measurement scale between 0 and 10 W (12 scale ranges; the scale from 0 to 100 mW was used) and an accuracy of 0.1 mW was used. This equipment used a pyroelectric sensor to measure the radiation heat flux in a spectral range of 0.25 to 11 µm (from ultraviolet to infrared). The diameter of the measuring sensor was 19 mm. The sensor was placed 4 cm from the radiant-heat-flux-emitting surface. The measurements were performed before the heating of the film or CMC, and the radiative heat flux was calculated by dividing the heat flux by the sensor area (0.000284 m2). Measurements were taken between 30 and 70 s. Before 30 s, the system had not stabilized, and after 80 s, conduction heat transfer was likely to occur due to heating of the film. All these measurements were taken on the drying film without food.

3.4.4. Radiation Penetration Depth

The radiation penetration depth (μ-1 μ-1 or x0.37) is defined as the depth at which the radiation intensity decays by 37% (1/e) of its initial value [35,36]. Using the radiation attenuation law, also known as the Beer–Lambert Law [29,40] (Equation (1)), the values of the intercept ($ln(I0)$ ($I0$)) and the slope (-µ) were determined. For this, the experimental values of I vs. x (1.5-, 2.5-, and 3.5-mm thickness CMC, for heating water at 70, 80, and 90 °C) were used.

The radiation penetration depth was calculated for a radiation heat flux at 30 s, thus considering only the radiation emitted by the hot water passing through the film and the food. Data corresponding to 10 and 20 s were not considered due to the system's instability at the beginning of the measurement.

3.4.5. Energy Consumption

Energy consumption was determined using a Peacefair meter, model PZEM-022 (China), for alternating current in the voltage range of 80 to 260 V, 50/60 Hz, up to 100 A. The sensing element of this equipment was placed at the power input of the drying equipment.

3.5. Statistical Analysis

Statgraphics Centurion 19 software (Statgraphics Technologies, Inc., version 19.2.01, 2021, The Plains, VA, USA) was used to perform analysis of variance and determine the dependence between process variables and response variables, with a 95% confidence level ($p < 0.05$). It was also used to determine the independence between means using Tukey's test.

AT A GLANCE

Extensive investigation has been conducted on biodegradable films made from natural raw materials like carboxymethyl cellulose (CMC). This research has been carried out due to the fact that these films have the ability to produce transparent films that possess effective barrier characteristics against oxygen and lipids. Given the significance of producing this genre of film on an industrial scale, a drying procedure that can be expanded and maintained consistently is necessary. The Refractance window-conductive hydro drying (RW-CHD) technique is a sustainable and energy-efficient technology that shows great promise for drying this particular chemical. The aim of this study was to assess the variables (CMC thickness, heating water temperature, and film type) and the depth of radiation penetration that impact the duration of drying and energy usage. The study revealed that when the temperature increased and the thickness reduced, the drying time decreased. Likewise, the amount of energy used reduced as the temperature and thickness decreased. Nevertheless, the amount of time required for drying and the energy expended per unit weight of the resulting product were the same regardless of the thicknesses tested. The impact of film type on time and energy usage was minimal in comparison to the influence of temperature and CMC thickness. The depth of radiation penetration into the CMC was measured to be 1.20 ± 0.19 mm. When the thickness approached this number, the radiation energy was more effectively utilized, resulting in a higher heating rate at the start of the drying process.

REFERENCES

1. Antosik, A.K.; Wilpiszewska, K.; Czech, Z. Carboxymethylated Polysaccharide-Based Films as Carriers for Acrylic Pressure-Sensitive Adhesives. Int. J. Adhes. Adhes. 2017, 73, 75–79.
2. Hasheminya, S.-M.; Mokarram, R.R.; Ghanbarzadeh, B.; Hamishekar, H.; Kafil, H.S.; Dehghannya, J. Influence of Simultaneous Application of Copper Oxide Nanoparticles and Satureja Khuzestanica Essential Oil on Properties of Kefiran–Carboxymethyl Cellulose Films. Polym. Test. 2019, 73, 377–388.
3. Liu, M.; Arshadi, M.; Javi, F.; Lawrence, P.; Davachi, S.M.; Abbaspourrad, A. Green and Facile Preparation of Hydrophobic Bioplastics from Tea Waste. J. Clean. Prod. 2020, 276, 123353.
4. Cheng, L.H.; Abd Karim, A.; Seow, C.C. Characterisation of Composite Films Made of Konjac Glucomannan (KGM), Carboxymethyl Cellulose (CMC) and Lipid. Food Chem. 2008, 107, 411–418.
5. Li, Y.; Shoemaker, C.F.; Ma, J.; Shen, X.; Zhong, F. Paste Viscosity of Rice Starches of Different Amylose Content and Carboxymethylcellulose Formed by Dry Heating and the Physical Properties of Their Films. Food Chem. 2008, 109, 616–623.

6. Gregorova, A.; Saha, N.; Kitano, T.; Saha, P. Hydrothermal Effect and Mechanical Stress Properties of Carboxymethylcellulose Based Hydrogel Food Packaging. Carbohydr. Polym. 2015, 117, 559–568.
7. Magoon, R.E. Method and Apparatus for Drying Fruit Pulp and the Like. U.S. Patent 4,631,837A, 30 December 1986.
8. Nindo, C.I.; Tang, J. Refractance Window Dehydration Technology: A Novel Contact Drying Method. Dry. Technol. 2007, 25, 37–48.
9. Ortiz-Jerez, M.J.; Ochoa-Martínez, C.I. Heat Transfer Mechanisms in Conductive Hydro-Drying of Pumpkin (Cucurbita Maxima) Pieces. Dry. Technol. 2015, 33, 965–972.
10. Kudra, T.; Mujumdar, A.S. Advanced Drying Technologies, 2nd ed.; Taylor & Francis: Boca Raton, FL, USA, 2009; ISBN 978-1-4200-7387-4.
11. Abonyi, B.I.; Tang, J.; Edwards, C.G. Evaluation of Energy Efficiency and Quality Retention for the Refractance WindowTM Drying System; Washington State University: Tacoma, WA, USA, 1999; p. 38.
12. Ochoa-Martínez, C.I.; Quintero, P.T.; Ayala, A.A.; Ortiz, M.J. Drying Characteristics of Mango Slices Using the Refractance WindowTM Technique. J. Food Eng. 2012, 109, 69–75.
13. Baeghbali, V.; Niakousari, M.; Farahnaky, A. Refractance Window Drying of Pomegranate Juice: Quality Retention and Energy Efficiency. LWT-Food Sci. Technol. 2016, 66, 34–40.
14. Nindo, C.I.; Feng, H.; Shen, G.Q.; Tang, J.; Kang, D.H. Energy Utilization and Microbial Reduction in a New Film Drying System. J. Food Process. Preserv. 2003, 27, 117–136.
15. Clarke., P.T. Refractance WindowTM—"Down Under". In Proceedings of the Proceedings on the 14th International Drying Symposium, São Paulo, Brasil, 22–25 August 2004; pp. 813–820.
16. Ortiz-Jerez, M.J.; Gulati, T.; Datta, A.K.; Ochoa-Martínez, C.I. Quantitative Understanding of Refractance WindowTM Drying. Food Bioprod. Process. 2015, 95, 237–253.
17. Baeghbali, V.; Niakosari, M.; Kiani, M. Design, Manufacture and Investigating Functionality of a New Batch Refractance Window System. In Proceedings of the 5th International Conference on Innovations in Food and Bioprocess Technology, Pathumthani, Thailand, 7–9 December 2010.
18. Bernaert, N.; Van Droogenbroeck, B.; Van Pamel, E.; De Ruyck, H. Innovative Refractance Window Drying Technology to Keep Nutrient Value during Processing. Trends Food Sci. Technol. 2019, 84, 22–24.
19. Menon, A.; Stojceska, V.; Tassou, S.A. A Systematic Review on the Recent Advances of the Energy Efficiency Improvements in Non-Conventional Food Drying Technologies. Trends Food Sci. Technol. 2020, 100, 67–76.
20. Nindo, C.I.; Tang, J.; Cakir, E.; Powers, J.R. Potential of Refractance Window Technology for Value Added Processing of Fruits and Vegetables in Developing Countries; American Society of Agricultural and Biological Engineers: Portland, OR, USA, 2006.
21. Kaspar, K.L.; Park, J.S.; Mathison, B.D.; Brown, C.R.; Massimino, S.; Chew, B.P. Processing of Pigmented-Flesh Potatoes (Solanum Tuberosum L.) on the Retention of Bioactive Compounds: Potato Processing on Nutrient Retention. Int. J. Food Sci. Technol. 2012, 47, 376–382.
22. Abul-Fadl, M.M.; Ghanem, T.H. Effect of Refractance-Window (RW) Drying Method on Quality Criteria of Produced Tomato Powder as Compared to the Convection Drying Method. World Appl. Sci. J. 2011, 15, 953–965.

23. Mahanti, N.K.; Chakraborty, S.K.; Sudhakar, A.; Verma, D.K.; Shankar, S.; Thakur, M.; Singh, S.; Tripathy, S.; Gupta, A.K.; Srivastav, P.P. Refractance WindowTM-Drying vs.Other Drying Methods and Effect of Different Process Parameters on Quality of Foods: A Comprehensive Review of Trends and Technological Developments. Future Foods 2021, 3, 100024.

24. Zotarelli, M.F.; Carciofi, B.A.M.; Laurindo, J.B. Effect of Process Variables on the Drying Rate of Mango Pulp by Refractance Window. Food Res. Int. 2015, 69, 410–417.

25. Ginzburg, A.S. Application of Infrared Radiation in Food Processing. Chemical and Process Engineering Series; Leonard Hill: London, UK, 1969.

26. Almeida, M.; Torrance, K.E.; Datta, A.K. Measurement of Optical Properties of Foods in Near- and Mid-Infrared Radiation. Int. J. Food Prop. 2006, 9, 651–664.

27. Torres, M.D.; Moreira, R.; Chenlo, F.; Vázquez, M.J. Water Adsorption Isotherms of Carboxymethyl Cellulose, Guar, Locust Bean, Tragacanth and Xanthan Gums. Carbohydr. Polym. 2012, 89, 592–598.

28. Parrouffe, J.M.; Dostie, M.; Navarri, P.; Andrieu, J.; Mujumdar, A.S. Heat and Mass Transfer Relationship in Combined Infrared and Convective Drying. Dry. Technol. 1997, 15, 399–425.

29. Tsilingiris, P.T. Comparative Evaluation of the Infrared Transmission of Polymer Films. Energy Convers. Manag. 2003, 44, 2839–2856.

30. Shende, D.; Datta, A.K. Optimization Study for Refractance Window Drying Process of Langra Variety Mango. J. Food Sci. Technol. 2020, 57, 683–692.

31. Azizi, D.; Jafari, S.M.; Mirzaei, H.; Dehnad, D. The Influence of Refractance Window Drying on Qualitative Properties of Kiwifruit Slices. Int. J. Food Eng. 2017, 13, 20160201;

32. Tontul, I.; Eroğlu, E.; Topuz, A. Convective and Refractance Window Drying of Cornelian Cherry Pulp: Effect on Physicochemical Properties. J. Food Process. Eng. 2018, 41, e12917.

33. Ocoró-Zamora, M.; Ayala-Aponte, A. Influence of Thickness on the Drying of Papaya Puree (Carica papaya L.) through Refractance WindowTM Technology. DYNA 2013, 80, 147–154.

34. Múnera-Tangarife, R.D. Evaluación de Los Factores Que Afectan La Transferencia de Calor En El Secado Por Ventana de Refractancia-Hidrosecado; Universidad del Valle: Santiago de Cali, Colombia, 2021.

35. Richardson, P. (Ed.) Thermal Technologies in Food Processing; Woodhead Publishing in Food Science and Technology; CRC Press: Boca Raton, FL, USA; Woodhead: Cambridge, UK, 2001; ISBN 978-1-85573-558-3.

36. Sun, D.-W. Thermal Food Processing: New Technologies and Quality Issues; CRC Press: Boca Raton, FL, USA, 2012; ISBN 978-1-4398-7679-4.

37. Pawar, S.B.; Pratape, V.M. Fundamentals of Infrared Heating and Its Application in Drying of Food Materials: A Review: Mapping of Infrared Drying of Foods. J. Food Process. Eng. 2017, 40, e12308.

38. Krishnamurthy, K.; Khurana, H.K.; Soojin, J.; Irudayaraj, J.; Demirci, A. Infrared Heating in Food Processing: An Overview. Comp. Rev. Food Sci Food Saf. 2008, 7, 2–13.

39. Sandu, C. Infrared Radiative Drying in Food Engineering: A Process Analysis. Biotechnol. Prog. 1986, 2, 109–119.

40. Siegel, R.; Howell, J.R.; Mengüc, M.P. Thermal Radiation Heat Transfer; CRC Press: Boca Raton, FL, USA, 2016; ISBN 978-1-4987-5774-4.

8 Chapter

CREATION OF INNOVATIVE SUPERABSORBENT HYBRID HYDROGELS BY THE PROCESS OF E-BEAM CROSSLINKING

1. INTRODUCTION: AN OVERVIEW

The specialized scientific literature contains reports on a range of composite hydrogels that possess distinct properties and can be utilized in various applications. These include the delivery or controlled release of drugs [1,2,3], water purification adsorbents [4,5,6], dyestuff adsorbents in different fields [7], and energy-storage devices [8,9]. The majority of hydrogels that incorporate graphene oxide (GO) in their composition were synthesized by the use of chemical crosslinking agents.

Huang et al. synthesized superabsorbent hydrogels using various concentrations of graphene oxide (GO) in a polymerization process called in situ radical solution polymerization. The resulting hydrogels demonstrated a maximum swelling capacity of 1100% [10,11].

The process of hydrogel crosslinking using ionizing radiation is a widely accepted technique that is efficient and fast, resulting in the production of sterile goods with a durable and uniform network structure.

The degree of crosslinking, which substantially influences the swelling degree, can be adjusted by adjusting the absorbed dose. Due to its numerous hydrophilic groups on its surface, GO is a highly effective material that can be combined with various polymers to create new hydrogels with exceptional superabsorbent capabilities [12,13,14]. CMC and XG, which are natural polymers, experience increased susceptibility to deterioration when exposed to irradiation in aqueous solutions, primarily due to the indirect influence of water. Li et al. [15] conducted a study to examine the impact of e-beam irradiation on XG, a component commonly used in the food sector to assess its resistance to irradiation. The study demonstrated that the viscosity of irradiated aqueous XG-based solutions decreased as the irradiation dose increased (5–50 kGy), indicating the degradation or depolymerization of polysaccharide macromolecules.

Şen and Hayrabolulu et al. conducted comprehensive studies on the γ-radiation of XG in both solid and aqueous states. The researchers demonstrated that γ-radiation in air causes chain scission in XG. This impact is more pronounced when the dose rate is reduced, since it enhances oxidative degradation during irradiation [16,17]. The study group utilized γ-radiation crosslinking synthesis to generate superabsorbent hydrogels based on XG. This synthesis method involved exposing the hydrogels to a very low absorbed dosage of 1-2 kGy [18].

A separate study examined the impact of different levels of irradiation dose (ranging from 0 to 30 kGy) on the viscosity of CMC solutions. Additionally, this study devised techniques to regulate the degradation process [19].

Frequently, when natural polymers are exposed to irradiation, the process of crosslinking occurs simultaneously with the destruction of the polymer chain. Notwithstanding this significant drawback, natural polymers are favored over petroleum-derived synthetic polymers for ecological reasons, as well as because natural polymers include inherent characteristics of biocompatibility and biodegradability. Therefore, biomedical materials are not only more convenient to manipulate, but also highly compatible with cells and tissues.

The primary motivation for including graphene oxide (GO) into hybrid hydrogels with natural polymers like xanthan gum (XG) and carboxymethylcellulose (CMC) is to enhance the mechanical properties of the resulting hydrogel beyond those of the original polymers. Natural polymers are widely recognized for their inferior mechanical qualities, despite their exceptional absorption capabilities resulting from their very hydrophilic characteristics.

GO enhances the flexibility of polymer chains and improves their resistance to changes in temperature and chemical reactions. Graphene oxide (GO) is a substance that can enhance the mechanical strength of natural polymers. It is easily disseminated in polymer mixtures, which improves the compatibility of these compounds [20,21].

Graphene oxide (GO) can play a crucial role in the formation of some hydrogels, namely during the lyophilization process, which involves the removal of water from the hydrogels. GO hydrogels retain their chemical composition while exhibiting characteristics such as high porosity, a large specific surface area, and a strong attraction to different molecules. Recent research has demonstrated that hydrogel systems based on graphene oxide (GO) have a higher ability to carry drugs and release them over a longer period of time. This makes them well-suited for advanced applications in drug administration. This finding is supported by reference [22]. In our prior research, we created superabsorbent hydrogels by exposing them to e-beam irradiation without air, using modest radiation doses (0.5-3 kGy). These hydrogels were made from a concentrated solution of polymers in water, which had a paste-like consistency [23]. The hydrogels mentioned before were characterized by a reduced ability to swell and a greater density of crosslinking. The objective of this study was to create and analyze a novel hybrid hydrogel made entirely of biodegradable polymers (XG and CMC) combined with GO. The aim was to produce a hydrogel with optimal structural, rheological, and swelling properties, suitable for applications in biomedical engineering [24] and hygienic products.

2. RESULTS AND DISCUSSION

2.1. Sol-Gel Analysis

Figure 1 shows the variation of the gel fraction (GF) depending on the absorbed dose of the XGCMCGO hydrogels with compositions of 50:50, 80:20, and 70:30. For these hydrogel compositions, it was observed that the highest percentage of the GF was obtained at the lowest absorbed dose: 78% for XGCMCGO (50:50) and 73% for XGCMCGO (80:20); while for XGCMCGO (70:30), at the lowest dose, it obtained the lowest percentage of the GF, namely 60%. The XGCMCGO (70:30) hydrogel showed an almost constant value of GF over the entire range of applied radiation doses.

For the hydrogel samples having the compositions of 50:50 and 70:30, the percentage of GF decreased with an increasing absorbed dose. The exception to this rule was the hydrogel with equal concentrations of polymers (XGCMCGO (50:50)), for which the determination of the GF was no longer possible at doses above 7.5 kGy, due to the advanced degradation of the samples after irradiation.

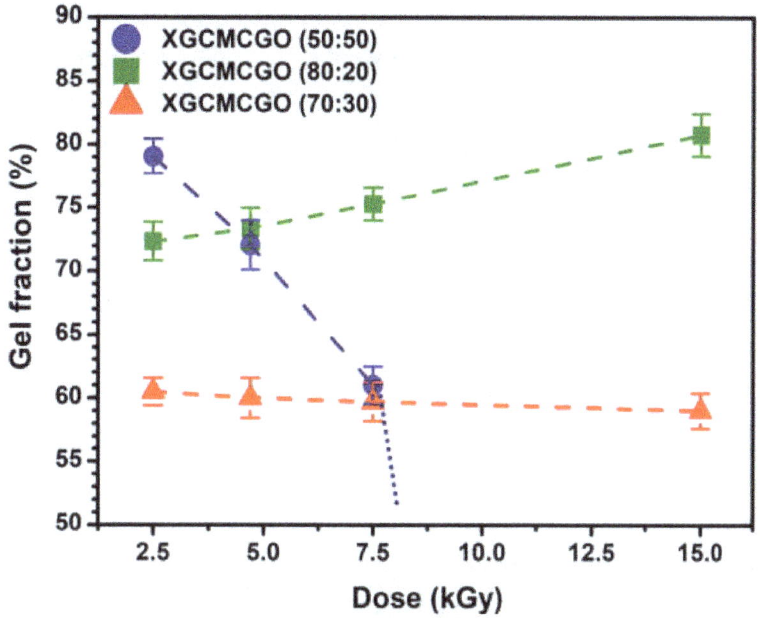

Figure 1. Gel fraction as a function of the absorbed dose during e-beam irradiation for the XGCMCGO hydrogels.

For the XGCMCGO (80:20) hydrogel, the percentage of the GF increased with an increasing absorbed dose; the highest percentage of the GF was 78% at an irradiation dose of 15 kGy.

The obtained values were comparable to those of other studies regarding the e-beam crosslinking of similar hydrogel compositions. For example, Sung et al. prepared CMC/GO hydrogels by e-beam irradiation at 30 kGy and obtained a gel fraction of 67.3% [20].

Another study, performed by Said et al., which investigated the formation of CMC/AA hydrogels in aqueous solutions under the effect of e-beam irradiation, showed that the gel fraction increased quickly with a dose of 50 kGy, reaching a value of 70%, then in the dose range of 50–100 kGy, showed a slight increase [25].

The rate of hydrogel formation depends on the p0/q0 ratio of the polymer. If the crosslinking process predominates over the degradation process, an insoluble gel is formed as a detriment to the degradation process [26]. When polymers are subjected to ionizing radiation, crosslinking and main chain scission occur simultaneously. The quantitative estimation of the extent of crosslinking and degradation can be made using the Charlesby–Rosiak equation. The p0/q0 ratios calculated using the Charlesby–Rosiak equation for the XGCMCGO hydrogels are presented in Table 1.

Table 1. Sol-gel parameters calculated according the Charlesby–Rosiak Equation (3).

Hydrogel	p_0/q_0	Dg (kGy)
XGCMCGO (50:50)	1.04	0.23
XGCMCGO (80:20)	0.64	0.24
XGCMCGO (70:30)	0.66	0.79

The lowest value of the p0/q0 of 0.64 was obtained for the hydrogel composition of 80:20, which contained the highest XG concentration. The XGCMCGO (50:50) and XGCMCGO (70:30) hydrogels showed p0/q0 ratios of 1.04 and 0.66, respectively. It is well known in the radiation-chemistry field that when various polymeric composition subjected to a treatment with ionizing radiation show a p0/q0 less than

2, the crosslinking processes predominated [27]. In the present study, these values were slightly higher compared to those obtained for hydrogels prepared with high concentrations of acrylic acid (70% AA) in data presented in our previous study, where the p0/q0 ratio was equal to 0.28 [23].

The gelation dose (Dg) is another parameter that shows the minimum dose required to obtain an insoluble gel during irradiation. Dg values decreased as the concentration of polymers and crosslinking agent (N'N—methylenebis(acrylamide), NMBA) increased. We observed that the gelation occurred at a very low dose (0.23 kGy) for the XGCMCGO (50:50) hydrogel.

The radiation-chemical yield of crosslinking (G(X)) decreased with an increase in the irradiation dose for all the XGCMCGO hydrogel compositions. The maximum value of this parameter was 132.2 μmol/J for the XGCMCGO (70:30) hydrogel at a dose of 2.5 kGy. For the radiation-chemical yield of scission (G(S)), the maximum value was 252.5 μmol/J for the XGCMCGO (50:50) hydrogel at a dose of 2.5 kGy (Table 2).

Table 2. Radiation-chemical yield of crosslinking G(X) and chain scission G(S) calculated for the XGCMCGO hydrogels.

Dose (kGy)	Radiation-Chemical Yields (μmol/J)	XGCMCGO (50:50)	XGCMCGO (80:20)	XGCMCGO (70:30)
2.5	G(X)	121.4	113.7	132.2
	G(S)	252.5	100.1	121.6
4.7	G(X)	54.4	65.8	73.5
	G(S)	113.2	57.9	67.6
7.5	G(X)	33.6	46.8	53.4
	G(S)	69.8	41.2	49.1
15	G(X)	-	27.6	27.9
	G(S)	-	24.3	25.6

As shown in the results presented in Table 2, G(S) > G(X) for the XGCMCGO (50:50) hydrogel, with approximately double the value. Therefore, it was obvious that in this case, the degradation processes predominated.

On the other hand, we observed that G(X) > G(S) for the XGCMCGO (80:20) and XGCMCGO (70:30) hydrogels. In these cases, the crosslinking predominated.

2.2. Swelling Degree

To characterize the network structure and determine the effective crosslinking density of the XGCMCGO hydrogels, the swelling properties in deionized water (DI) were first investigated.

Figure 2 shows the variation of the swelling degree (SD) at equilibrium, as a function of time and the absorbed dose, for the XGCMCGO hydrogels. All hydrogels showed good swelling ability. The maximum SD obtained at a dose of 15 kGy was ~6000%, which classified these XGCMCGO (70:30) hydrogels as "superabsorbent hydrogels". In the case of hydrogels with low concentrations of CMC, XG, AA, and NMBA, the SD had a value of 2250% for the XGCMCGO (50:50) hydrogel obtained at 4.7 kGy. At irradiation doses greater than 5 kGy, this hydrogel was degraded in deionized water (DI) in less than 2 h. For the XGCMCGO (80:20) hydrogel, the SD showed a maximum value of 2300% at a dose of 7.5 kGy.

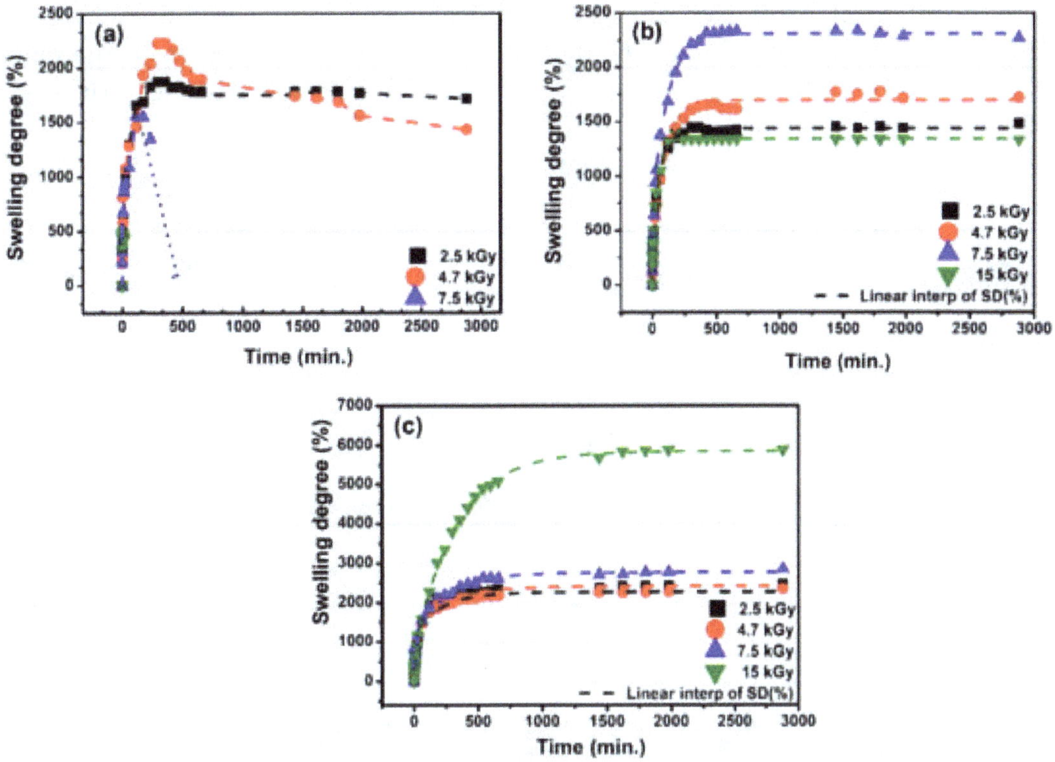

Figure 2. The swelling degree of the hydrogels in DI water: (a) XGCMCGO (50:50); (b) XGCMCGO (80:20); (c) XGCMCGO (70:30).

Moreover, we can specify that these hydrogels had the ability to reach equilibrium in less than 6 h and were stable in deionized water; of these, the most suitable proved to be the XGCMCGO (70:30) hydrogel. Previous studies of superabsorbent hydrogels based on CMC/AA/montmorillonite obtained by γ-irradiation showed a swelling degree of 16,000% [28]. Compared to our study, the above study used a high concentration of polymers; i.e., 10% CMC and 30% AA. Sultana et al. synthesized copolymer hydrogels from acrylamide/CMC by γ-irradiation, and they obtained an SD value in DI water of 795% [29]. We have shown that by using low concentrations of polymers and GO and appropriate mass ratios between them, with a moderate radiation dose rate, stable hydrogels with different degrees of swelling were obtained.

Superabsorbent hydrogels based on xanthan gum/polyacrylic acid/graphene oxide were prepared as absorbers for removing methylene blue dye from the water. For these hydrogels, the maximum swelling value in the neutral medium was found to be 2100% for a GO concentration of 1% [30]. Superabsorbent resin based on acrylic acid/CMC/GO showed the highest swelling capacity with 0.6% of GO in hydrogel composition, about 750 g g-1 in distilled water. The above study demonstrated that the incorporation of a moderate amount of GO in a CMC-based hydrogel can improve the water capacity of the material [31]. The CMC/GO composite superabsorbents prepared by e-beam irradiation at 10 kGy showed a swelling capacity of 140 g g-1 in distilled water [32]. After analyzing the data, we concluded that the XGCMCGO hydrogels presented in this study, in addition to having the lowest concentration of 0.1% GO, had the best swelling properties in a neutral environment.

2.3. Diffusion of Water

The analysis of water diffusion mechanisms in swollen polymeric systems has received considerable attention in recent years due to the important applications of swollen hydrogels in biomedical and pharmaceutical engineering. When a hydrogel is in contact with water, water enters the hydrogel by diffusion and the hydrogel network expands, resulting in the swelling of the hydrogel. Diffusion involves the migration of water into pre-existing or dynamically formed spaces between the macromolecular chains of the hydrogel [33]. To follow this process, Equation (7) was applied to the initial stage of swelling process, up to 60% of maximum swelling (equilibrium state) [34]. Table 3 shows the values of water diffusion mechanisms characteristic to the XGCMCGO hydrogels.

Table 3. Parameters k, n, and D for the XGCMCGO hydrogels.

XGCMCGO (50:50)						
Dose (kGy)	k	*n*	R^2	D (cm^2/s)	R^2	Mechanism
2.5	-3.42	0.29	0.91	0.02	0.98	Fickian
4.7	-3.41	0.42	0.96	0.03	0.98	Fickian
7.5	-3.47	0.47	0.99	0.03	0.98	Fickian
15	-3.08	0.28	0.94	0.03	0.98	Fickian
XGCMCGO (80:20)						
Dose (kGy)	k	*n*	R^2	D (cm^2/s)	R^2	Mechanism
2.5	-3.59	0.52	0.96	0.02	0.92	Non—Fickian
4.7	-3.44	0.45	0.97	0.03	0.95	Fickian
7.5	-3.53	0.59	0.99	0.03	0.93	Non—Fickian
15	-2.98	0.38	0.97	0.04	0.99	Fickian
XGCMCGO (70:30)						
Dose (kGy)	k	*n*	R^2	D (cm^2/s)	R^2	Mechanism
2.5	-3.15	0.44	0.98	0.03	0.98	Fickian
4.7	-2.94	0.52	0.99	0.04	0.97	Non—Fickian
7.5	-2.93	0.45	0.98	0.04	0.99	Fickian
15	-2.94	0.46	0.99	0.03	0.99	Fickian

A value of $n \leq 0.5$ indicated a Fickian diffusion mechanism. In this case, the rate of solvent diffusion was much lower compared to the macromolecular chain relaxation of the polymer. A value of n in the range $0.5 < n < 1$ indicated a non-Fickian transport, in which the diffusion of the solvent into the hydrogel structure was rapid compared to the relaxation rate of the macromolecular chain of the polymer [35].

Figure 3a–c show graphically, in logarithmic scale, the values of the parameter F = f(t). Figure 3d–f show graphically F = f (t 0.5). The values of the coefficients n, k, and D were calculated from the slope and intersection of the lines resulting from the swelling kinetics. The calculated values for the diffusion exponents n were less than 0.5 for the XGCMCGO (50:50) hydrogel at all irradiation doses, thus indicating that the solvent transport mechanism was of a Fickian type.

The XGCMCGO (80:20) hydrogel also presented a Fickian diffusion mechanism, but only in the cases of irradiation doses of 4.7 and 15 kGy. For the other irradiation doses (2.5 kGy and 7.5 kGy) the diffusion mechanism was non-Fickian. When the diffusion was non-Fickian, the relaxation time of the macromolecular chain and the diffusion had the same order of magnitude. As the solvent diffused into the hydrogel, the rearrangement of the macromolecular chains did not occur immediately.

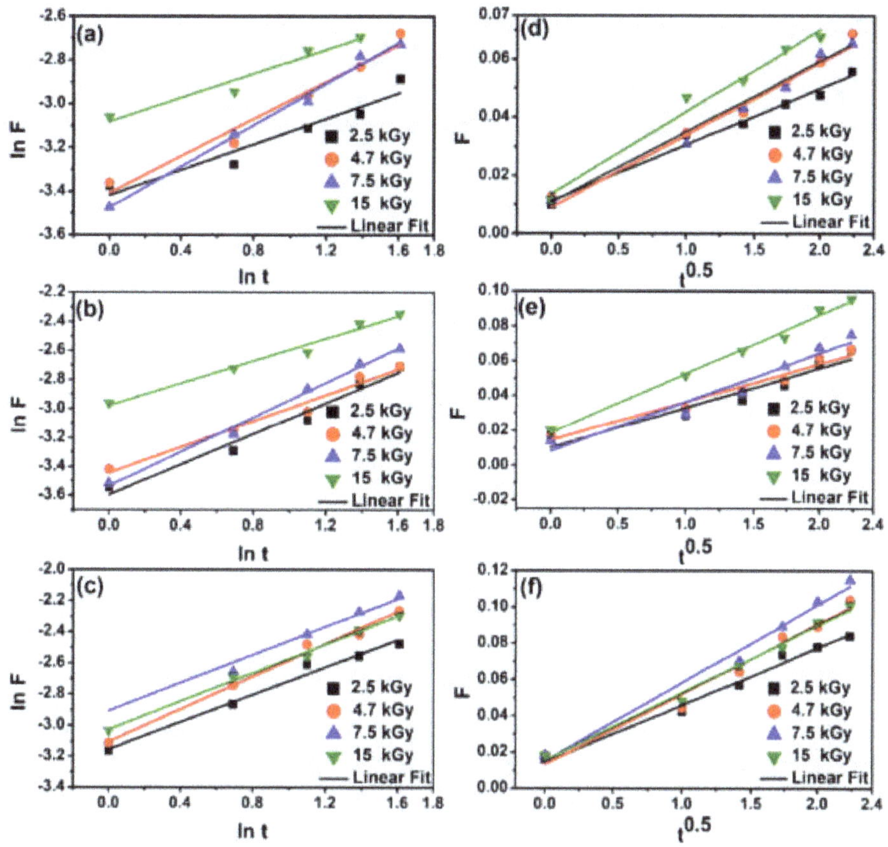

Figure 3. Swelling kinetics curves for the hydrogels: (a) XGCMCGO (50:50); (b) XGCMCGO (80:20); (c) XGCMCGO (70:30). Plots of F versus t0.5 for the hydrogels: (d) XGCMCGO (50:50); (e) XGCMCGO (80:20); (f) XGCMCGO (70:30).

The mechanism of water diffusion of the XGCMCGO (70:30) hydrogel was of a Fickian type; the exception was the hydrogel crosslinked with 4.7 kGy, in which case a value of n greater than 0.5 was obtained, thus exhibiting a non-Fickian diffusion behavior. In another similar study of a hybrid hydrogel obtained from acrylamide and XG, a value of n equal to 0.48 was reported [36]. Another study showed the obtaining of a GO-based hydrogel in which the inorganic component was incorporated in different concentrations. For this, values of n were obtained in the range of 0.5–1, which indicated a non-Fickian diffusion, which is specific to crosslinked hydrogels [37].

The values of the diffusion coefficient (D) were found in the range of 0.02–0.04 cm2/s, depending on the irradiation dose and the composition of the hydrogel. Compared to other studies, these values were much lower, suggesting the formation of high-molecular-weight hydrogels [38], as can be seen in the rheological measurements. For example, following a study on hydrogels obtained from CMC/acrylamide/GO by radical polymerization, the values obtained for the diffusion coefficient were in the range of 0.66–1.26 cm2/s [39].

2.4. ATR-FTIR Spectroscopy

The characteristic FTIR spectra for the natural polymers (XG and CMC) and GO are shown in Figure 4. The FTIR spectrum characteristic of XG showed: 3308 cm-1 (O–H groups); 2917 cm-1 (CH3 and CH2 functional groups); 1700–1730 cm-1 (C = O); 1607 cm-1 (C = O); 1405 cm-1 (C–O group); and 1025 cm-1 (C–O group). The FTIR spectra of CMC showed: 3303 cm-1 (O–H group); 2923 cm-1 (C–H stretching vibrations);

1586 cm−1 (COO− group); 1420 cm−1 (CH2); 1320 cm−1 (O-H groups); and 1030 cm−1 (C–O group). Regarding the FTIR spectrum of GO, it showed: 3231 cm−1 (O–H); 1732 cm−1 (C = O and C–H groups); 1620 cm−1 (C = C); 1275 cm−1 (C–OH); and 1053 cm−1 (C–O).

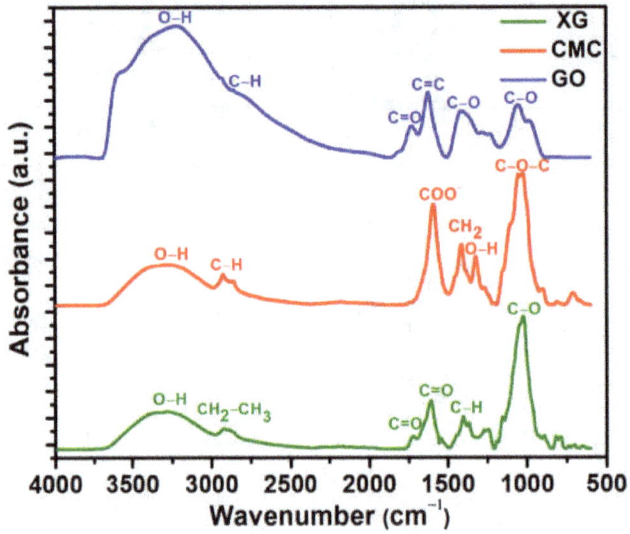

Figure 4. The FTIR spectra of pure XG, CMC, and pure GO.

Figure 5a,b shows the characteristic FTIR spectra for nonirradiated polymeric blends with various ratios (XG:CMC) without GO (left side) and containing GO (right side). According to the acquired FTIR spectra, the displacement of the characteristic bands located in the range of 3600–600 cm-1 toward smaller wavenumbers corresponded to the formation of hydrogen bonds due to the miscibility of the polymers. In addition, increases in the band intensities in this area were proportional to the concentrations of natural polymers (XG and CMC) included in the polymeric blend.

For the XGCMCGO unirradiated polymeric blend, the hydrophilic functional groups (-COOH, -C = O, -OH, and -C-O-C) of GO played a key role in improving the compatibility between the polymer matrix and GO.

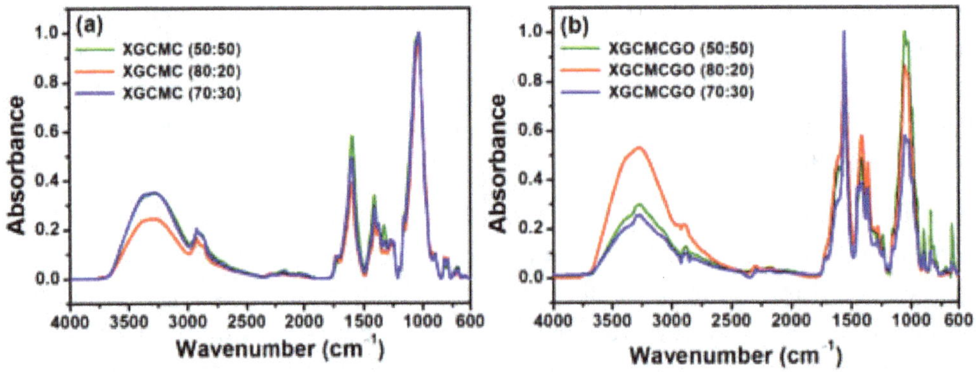

Figure 5. (a) The FTIR spectra of XGCMC without GO dried hydrogel; (b) the FTIR spectra of XGCMCGO dried hydrogel.

The FTIR-ATR spectra, corresponding to XGCMCGO hydrogels irradiated at doses in the range of 2.5–15 kGy, are shown in Figure 6. It can be seen that the intensity of the absorption bands varied with the irradiation dose.

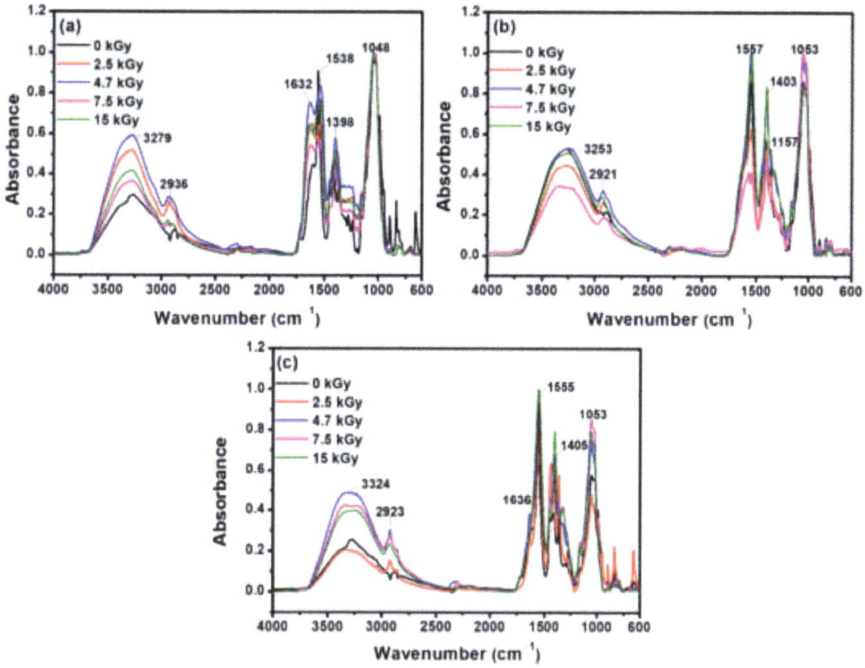

Figure 6. The FTIR spectra of unirradiated and irradiated dried hydrogels: (a) XGCMCGO (50:50); (b) XGCMCGO (80:20); (c) XGCMCGO (70:30).

The intensity of specific absorption bands in the range of 3300–3000) cm-1 for XGCMCGO (50:50) increased with the absorbed dose. The position of the bands assigned to the groups: C–H varied from 2929 cm-1 to 2940 cm-1; C = O varied from 1614 cm-1 to 1637 cm-1; and COO– varied from 1538 cm-1 to 1542 cm-1, after e-beam irradiation at a dose of 15 kGy.

Table 4. The elastic modulus (G′), the molecular weight between two crosslinks of the hydrogel chains (MC), the crosslink density (νe), and network mesh size (ξ) of the XGCMCGO hydrogels.

XGCMCGO (50:50)				
Dose (kGy)	G′ (Pa)	M_c (kg/mol)	v_e (mol/m^3)	ξ (nm)
2.5	869	132.7	0.76	140.1
4.7	712	157.3	0.64	158.4
7.5	646	160.3	0.63	170.6
15	200	-	-	-
XGCMCGO (80:20)				
Dose (kGy)	G′ (Pa)	M_c (kg/mol)	v_e (mol/m3)	ξ (nm)
2.5	912	210.7	0.49	138.5
4.7	933	195.9	0.52	132.8
7.5	1020	172.1	0.59	128.8
15	1052	146.5	0.70	123.6
XGCMCGO (70:30)				
Dose (kGy)	G′ (Pa)	M_c (kg/mol)	v_e (mol/m3)	ξ (nm)
2.5	1016	250.4	0.41	145.9
4.7	840	241.3	0.43	171.8
7.5	795	207.8	0.49	191.7
15	766	200.1	0.52	210.6

In the case of the XGCMCGO (80:20) hydrogel, the value of the band assigned to the O-H group decreased with the irradiation dose from 3268 cm−1 to 3253 cm−1. The intensity of the bands assigned to CH groups increased significantly for the irradiated polymeric blends. The decrease of the wavenumber from 1558/1412/1193/1054 cm−1, due to the application of a dose of 15 kGy, to 1557/1403/1157/1053 cm−1, respectively, may have been caused by the different degrees of crosslinking of the hydrogel. As the absorbed dose increased, the crosslinking density also increased (see Table 4).

For the XGCMCGO (70:30) hydrogels, the intensity of the absorption bands in the range of 3400–2800 cm−1 increased with a dose of 7.5 kGy. At this dose, the position corresponding to the OH group shifted to smaller wavenumbers, from 3333 to 3324 cm−1. The band assigned to the COO–group was shifted from 1556 cm−1 to 1550 cm−1 when the hydrogel was irradiated with a dose of 15 kGy. A similar trend was observed for the other bands in the range of 1500–800 cm−1. The intensity of the characteristic bands for the XGCMCGO (70:30) hydrogel increased when it was irradiated at doses up to 7 kGy due to the fact that the crosslinking density decreased, a direct consequence of increased swelling. These results were supported by the data obtained from swelling experiments (~6000% at a dose of 15 kGy).

2.5. Characterization of the Network Structure

The characterization of the network structure of a hydrogel is a complex procedure due to the many types of networks encountered in the practice; namely: regular, irregular, and weakly or strongly crosslinked networks.

Basic structural parameters for the XGCMCGO hydrogels such as the elastic modulus (G′), average molecular weight between two successive crosslinks (MC), crosslink density (νe), and mesh size (ξ) were calculated according to the equations presented in Section 4.7. Cn was taken as a weighted average of the characteristic ratios of the polymers: XG = 271 [40], CMC = 10 [41] and AA = 6.7 [42]. Mr is the monomeric unit of XG, CMC, and AA, taken as an average (XG = 933 g/mol, CMC = 234 g/mol, and AA = 72 g/mol) [43].

Using the values of the elastic modulus (G′) determined by rheological analysis for each hydrogel composition and based on the theory of rubber elasticity, we determined MC. The network parameters calculated for the XGCMCGO hydrogels are presented in Table 4.

The values of the MC parameter for all hydrogel compositions were in the range of 132.7–250.4 kg/mol. For the XGCMCGO (80:20) and (70:30) hydrogel compositions, the corresponding Mc values decreased with an increase in the absorbed dose; while for the XGCMCGO (50:50) hydrogel compositions, the corresponding MC values increased. This increase demonstrated once again the degradation of this polymeric mixture under the action of ionizing radiation.

For the XGCMCGO (50:50) and (70:30) hydrogel compositions, the corresponding ξ values increased with an increase in the irradiation dose; while for the XGCMCGO (80:20) hydrogel compositions, the corresponding ξ values decreased with an increase in the irradiation dose. The decrease of the parameter ξ may have been due to the increase in the crosslinking density. Therefore, a large number of active or free radical centers were formed on the polymer chains, and a large number of small chains were formed, which led to crosslinking and the formation of a more compact network structure with a smaller mesh size. Figure 7 shows the appearance of the polymer mixture before and after irradiation with e-beams.

For this reason, νe is one of the most important structural parameters for characterizing a class of hydrogels that have superabsorbent properties. According to the values obtained, it was shown that this parameter increased as the absorbed dose increased in the cases of the XGCMCGO (80:20) and (70:30) hydrogels. The behavior was completely different in the case of XGCMCGO (50:50), where the values for νe decreased with an increase in the absorbed dose. All the values obtained for νe were in the range of 0.41–0.76 mol/m3, and showed the obtaining of hydrogels with a moderately crosslinked network structure that allowed the absorption of an increased amount of fluids.

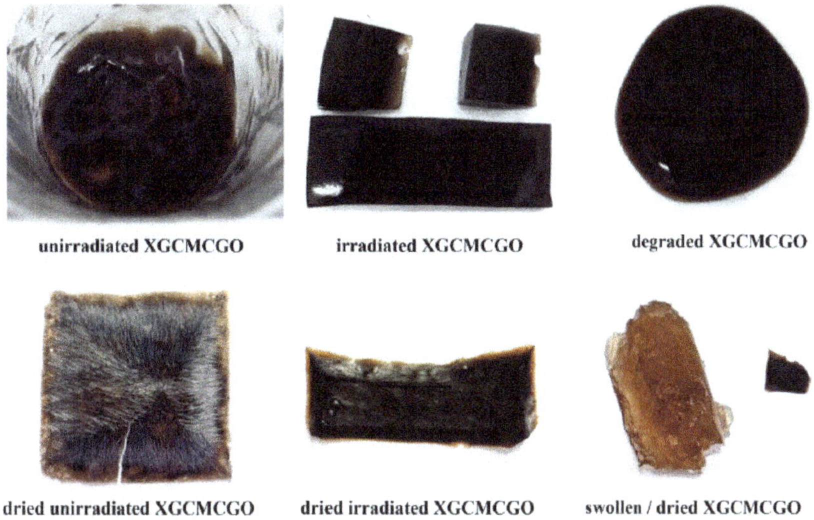

Figure 7. XGCMCGO hydrogels in different hypostases.

Knowing the hydrogel pore size is important to understanding the mechanism for transport of bioactive substances in the macromolecular network of such materials. For example, controlling the rate of a drug diffusion is essential, because it reflects the space available for a drug molecule to diffuse inside or outside the swollen hydrogel network [44].

The values of ξ obtained for the XGCMCGO hydrogels were in the range of 123.6–210.6 nm. These values were similar to those obtained by other authors who developed hydrogels with applicability in tissue engineering, with ξ values previous reported in the range of 139–258 nm [45].

2.6. Rheological Analysis

The hydrogels were evaluated to determine the viscoelastic parameters G' and G''. The elastic modulus (G') shows the elasticity of the crosslinked bonds and the total elasticity of the material, while the loss or viscous modulus (G'') provides a perspective on the viscosity of the material. The ratio between G' and G'' indicates the total resistance to deformation of the material [46].

The mechanical properties of hydrogels are important when selecting material for biomedical and other applications. They depend on the composition of the hydrogel and its water content. Hydrogels with a higher water content generally have a better permeability and biocompatibility [47]. However, there are some disadvantages, as a high degree of swelling is accompanied by a decrease in mechanical strength. For many applications, the combination of a high swelling degree and good mechanical properties is very important. Many approaches have been used to improve the mechanical properties of hydrogels, including the copolymerization of hydrophilic and hydrophobic monomers, increasing the crosslinking density, and varying the polymerization conditions [48].

Figure 8 shows the variation of the elastic and viscous (G' and G'') moduli as a function of angular frequency (ω) and the absorbed dose for the XGCMCGO hydrogels. We observed that G' was greater than G'' for all the hydrogel compositions. This is a critical requirement for hydrogels, as a G' greater than the G'' suggests that the hydrogel has a higher elastic behavior [49].

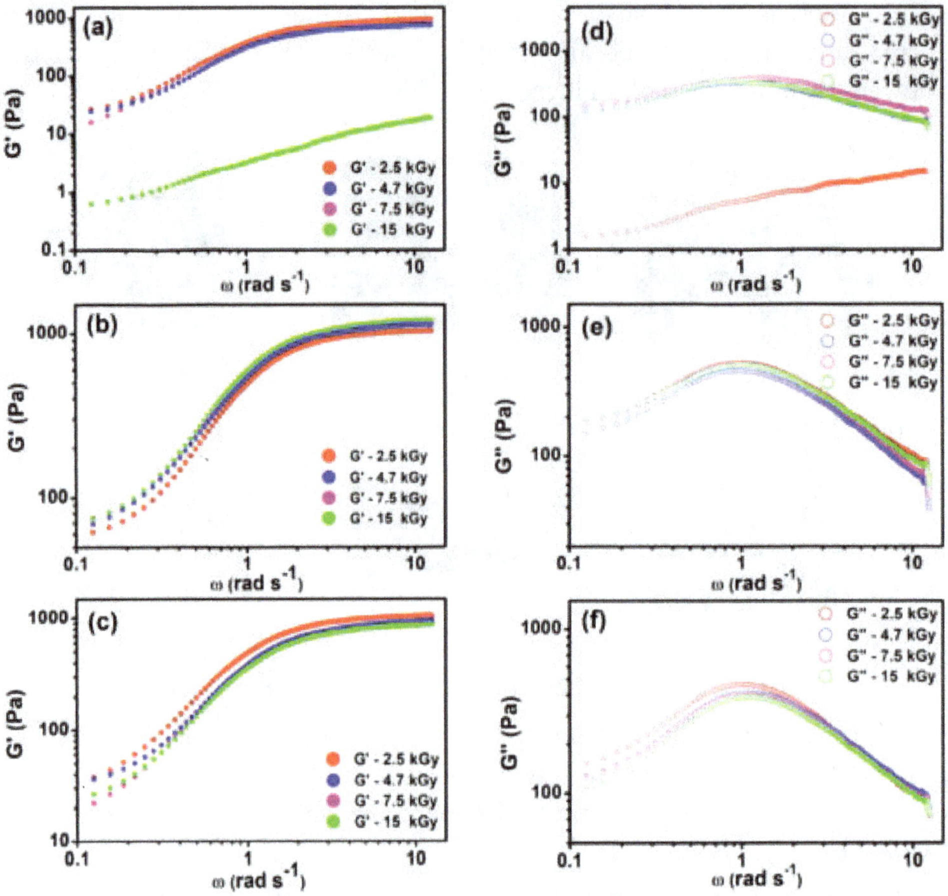

Figure 8. Elastic modulus (G′), as a function of the angular frequency (ω), of the (a) XGCMCGO (50:50), (b) XGCMCGO (80:20), and (c) XGCMCGO (70:30) hydrogels and their evolutions as a function of the absorbed dose. Viscous modulus (G″), as a function of the angular frequency (ω), of the (d) XGCMCGO (50:50), (e) XGCMCGO (80:20), and (f) XGCMCGO (70:30) hydrogels, and their evolutions as a function of the absorbed dose.

Rheological analysis confirms the crosslinking or degradation processes that occur depending on the composition of the polymer mixture. Moreover, the effect of irradiation at various irradiation doses is highlighted, and the G′ and G″ is dependent on the composition of the hydrogel and the absorbed dose. A rheological analysis was performed on swollen hydrogel samples. The determined values of G′ and G″ are shown in Table 4.

For the XGCMCGO (80:20) hydrogel, the G′ increased with an increase in the absorbed dose, with a maximum value for G′ = 1052 Pa at 15 kGy. The XGCMCGO (50:50) hydrogel had a G′ = 869 Pa at 2.5 kGy. When this polymeric blend was irradiated with 15 kGy, the G′ decreased very drastically down to 12 Pa. In the case of the XGCMCGO (70:30) hydrogel, G′ = 913 Pa was obtained at 2.5 kGy, and at 15 kGy, G′ = 766 Pa.

The G′ decreased with the absorbed dose in the cases of the XGCMCGO (50:50) and (70:30) hydrogels. We concluded that the decrease in G′ reflected the reduction of the crosslinking density (νe), while the decrease in G″ was due to the inhibition of the viscous behavior, showing a complete degradation of the polymer blend.

The XGCMCGO (80:20) hydrogel had a very well defined elastic behavior depending on the absorbed dose. In a recent study, a superabsorbent hydrogel based on CMC/starch/GO presented a value of G′ = 8050 Pa [50]. In another study, hydrogels obtained from CMC by redox polymerization for soft-tissue healing applications had a value of G′ = 814 Pa [45]. As shown by the rheological analysis and the sol-gel analysis, the XGCMCGO (80:20) hydrogel had the best properties.

3. CONCISE SUMMARY

We synthesized innovative and intricate XGCMCGO hydrogels by e-beam crosslinking for possible application as a substrate in biomedical engineering. The use of e-beams for irradiating XGCMC polymeric blends has been shown to be an effective technique, particularly when incorporating GO into their composition. By subjecting polymeric mixtures to varying amounts of irradiation, we were able to produce hydrogels with diverse characteristics.

The mechanical strength of XGCMCGO hybrid hydrogels surpassed that of the pure CMC hydrogel, while also exhibiting superior flexibility and enhanced swelling behavior compared to the basic XG hydrogel.

The gel fraction and swelling characteristics of the hydrogels produced were contingent upon the polymer composition and absorbed dosage.

The hydrogels showed superabsorbent capacity and reached equilibrium in less than 6 h, especially the XGCMCGO (70:30) hydrogel, which reached the highest swelling degree of about 6000%, at 15 kGy. The crosslinking process predominated compared to the degradation process.

By characterizing the network structure, we observed that the hybrid hydrogels with high XG concentration showed the best structural properties. The crosslinking density increased with the absorbed dose.

The mesh size obtained from the experimental data of rheological analysis was in the range of 123–210 nm; these values were comparable to those of several hydrogels.

The interaction between the hydrogel components was highlighted by FT-IR analysis, and the increase in absorption band intensities for the characteristic functional group, as well as their shifting toward lower wavenumbers, were correlated with the crosslinking degree, which decreased at a dose of 15 kGy.

For all hydrogel compositions, G′ > G″, thus suggesting that the hydrogels had a higher elastic behavior. The G′ had values in a wide range of 12–1052 Pa, with the maximum value obtained at a dose of 15 kGy for the XGCMCGO (80:20) hydrogel composition.

Therefore, the prepared XGCMCGO superabsorbent hydrogels belong to a class of environmentally friendly materials, and might have potential practical applications in many areas, such as in biomedical engineering and hygienic products.

4. MATERIALS AND METHODS

4.1. Materials

Sodium carboxymethylcellulose (CMC, Mw = 2.5 × 105 g/mol), N′N-methylene-bis-acrylamide (NMBA 99%, Mw = 154.17 g/mol), acrylic acid anhydrous 99% containing MEHQ as inhibitor (AA, Mw = 72.06 g/mol), and NaOH were purchased from Merck KGaA, Darmstadt, Germany. A commercial xanthan gum (XG—food grade, produced by Jungbunzlauer, Wien, Austria) in powder form with a molecular weight of Mw = 1.6 × 106 g/mol was used. Ultra-highly concentrated single-layer graphene oxide (6.2 g/L) was purchased from Graphene Laboratory Inc., New York, NY, USA.

4.2. Synthesis of Hydrogels and E-Beam Irradiation

In this experiment, three different hydrogels based on XGCMCGO with different content of XG and CMC in the presence of AA, NaOH, and NMBA were prepared. XG (5 wt %) and CMC (2 wt %) were dissolved in DI water at room temperature. After complete solubilization of the XG and CMC, they were mixed with each other in different compositions. For each XG:CMC ratio, 75 mL of the mixture was prepared. The ratios of XG:CMC in the mixtures were 50:50; 80:20, and 70:30. For a better understanding of the experiments, sample compositions are presented in Table 5.

Table 5. Sample composition details.

Hydrogel	Chemical Composition (%, w/v)					
	XG	CMC	GO	AA	NaOH	NMBA
XGCMCGO (50:50)	2.1	0.8	0.1	0.8	0.4	2
XGCMCGO (80:20)	3.3	0.3	0.1	2.8	1.4	1.5
XGCMCGO (70:30)	2.9	0.5	0.1	5.9	2.9	0.5

Then, 15 mL of each homogeneous solution of XGCMCGO was packed in a hermetically sealed polyethylene zip bag and subjected to e-beam irradiation at predetermined doses (2.5–15 kGy). The e-beam sample irradiation was performed in air at room temperature (25 °C) using a linear electron accelerator (National Institute for Laser, Plasma and Radiation Physics, Măgurele, Romania) at a fixed beam energy of 6 MeV (average beam current of 10 µA, pulse length of 3.75 µs, pulse repetition rate of 50 Hz, and average dose rate of 1 kGy/min) [51]. The dosimetry was performed using graphite calorimeters.

4.3. Sol-Gel Analysis

The hydrogel samples were dried in a vacuum oven to a constant weight and then immersed in DI water for 48 h at room temperature (25 °C). After 48 h, the swollen hydrogels were removed from the water and dried at 30 °C to a constant weight. The gel fraction (GF) and soluble fraction (s) were calculated as follows:

$$G(\%) = \left(\frac{W_d}{W_i}\right) \tag{1}$$

$$s = 1 - G \tag{2}$$

where Wi is the initial weight of dried sample after irradiation, Wd is the weight of the dried insoluble part of sample after immersion for 48 h, and s is soluble fraction of the polymer. All measurements were carried out in triplicate for each sample, and all values were expressed as mean value and standard deviation of three independent samples.

The gelation doses (the doses necessary to produce the first insoluble gel fraction) and the degradation vs. crosslinking ratios for the XGCMCGO hydrogels were calculated using a customized computer program for sol-gel analysis (Gelsol95), which was based on the Charlesby–Rosiak formula [26]:

$$s + \sqrt{s} = \frac{p_0}{q_0} + \left(2 - \frac{p_0}{q_0}\right)\left(\frac{D_v + D_g}{D_v + D}\right) \tag{3}$$

where p0 is the degradation density (i.e., average number of main chain scissions per monomer unit and per unit dose), q0 is the crosslinking density (i.e., fraction of monomer units crosslinked per unit dose), D is the absorbed dose (kGy), Dg is the gelation dose (kGy), and DV is the virtual dose (kGy) (the dose necessary to transform the real sample into a sample with the molecular weight distribution of Mw/Mn = 2).

The radiation yields of crosslinking and degradation (scission) were calculated using the following equations:

$$G(X) = \frac{4.9 \cdot 10^2 \cdot c}{M_C \cdot D \cdot \rho} \tag{4}$$

$$G(S) = G(X) \cdot 2\frac{p_0}{q_0} \tag{5}$$

where G(X) is the radiation yield of crosslinking (expressed as number of moles of crosslinking bonds per Joule), G(S) is the radiation yield of chain scission (mol/J), MC (kg/mol) is the average molecular weight between two successive crosslinks, c (g/L) is the polymer concentration in irradiated solution, D is absorbed dose (J/kg), and ρ (kg/m3) is the polymer density [52].

4.4. Swelling Degree

The swelling properties of hydrogels were explored by placing the dried hydrogels in DI water at room temperature for 48 h to reach swelling equilibrium. At specified times, the swelled hydrogels were taken out of the distilled water, blotted with paper, weighed, and immersed again.

The swelling degree (SD(%)) was calculated as a function of the dry (Wd) and swollen (Ws) hydrogel weights using Equation (6) [53]:

$$SD(\%) = \frac{(W_s - W_d)}{W_d} \cdot 100 \tag{6}$$

4.5. Diffusion of Water

The most basic law of Fick's was used for the explanation of swelling kinetics and diffusion of the polymeric structures. The following equation was used to determine the nature of diffusion of water into hydrogels [54]:

$$F = \frac{M_t}{M_\infty} = kt^n \tag{7}$$

where F is the fraction of swelling due to the water uptake, Mt is the adsorbed water at time t, M∞ is the adsorbed water at equilibrium, k is a proportionality constant, and n is the diffusional exponent. The first 60% of the water uptake data were fitted to Equation (7), and the corresponding values of k and n were obtained.

4.6. ATR-FTIR Spectroscopy

The changes in chemical structure of the crosslinked hydrogels were investigated. ATR-FTIR spectra of unirradiated and irradiated samples were taken with a PerkinElmer Spectrum 100 FTIR Spectrometer. The samples for FTIR analysis were first dried in a vacuum oven at 30 °C for 72 h. The samples were subjected to wavenumbers ranging from 4000 to 600 cm−1 at ambient temperature and a resolution of 4 cm−1, averaged from 50 scans/sample.

4.7. Characterization of Network Structure

Network parameters of the XGCMCGO hydrogels, such as the average molecular weight between two crosslinks (Mc), crosslinking density (νe), and mesh size (ξ), were determined by using the swelling and rheological measurements. Using elastic modulus (G′) values determined from rheological measurements and based on the rubber elasticity theory, Mc could be determined using the following equation [55]:

$$M_C = \frac{A\rho RT(\nu_{2r})^{2/3}(\nu_{2s})^{1/3}}{G'} \tag{8}$$

where R is the universal gas constant (8.314 m3 Pa/molK), T is the absolute experimental temperature (298.15 °K), ν2r is the polymer volume fraction after e-beam crosslinking, ν2s is the polymer volume fraction of the crosslinked hydrogel in swollen state, ρ (kg/m3) is the polymer density, and the factor A equals 1 for an affine network and 1–2/ϕ for a phantom network.

The effective crosslink density (νe) of the hydrogels was calculated using Equation (9):

$$\nu_e = \frac{\rho}{M_C} \tag{9}$$

The polymer volume fractions (ν2r and ν2s) were determined using Equation (10):

$$\nu_{2r(s)} = \frac{\left[1 + \left(w_{2r(s)} - 1\right)\cdot\rho_{hydrogel}\right]^{-1}}{\rho_{solvent}} \tag{10}$$

where ρhydrogel and ρsolvent are the densities of the hydrogel and solvent (kg/m3), respectively; and w2r(s) is the weight of the hydrogel after e-beam crosslinking after swelling (g). The weight swelling ratio of hydrogels after crosslinking (w2r) was calculated as: w2r = hydrogel mass after irradiation/hydrogel dry mass. The weight swelling ratio of hydrogels after swelling (w2s) was calculated as: w2s = hydrogel mass after swelling/hydrogel dry mass.

The mesh size of the polymer network (ξ) was determined using Equation (11) [56]:

$$\xi = \nu_{2s}^{-1/3}\cdot\left[C_n\left(\frac{2M_C}{M_r}\right)\right]^{-1/2}\cdot l \tag{11}$$

where Cn is the Flory characteristic ratio, Mr is the average molecular weight of the repeating unit, and l is the carbon–carbon bond length (0.154 nm).

4.8. Dynamic Rheological Measurements

Dynamic rheological measurements of the hydrogels were performed by employing an MFR 2100 Micro Fourier Rheometer (GBC, Australia) equipped with a home-made temperature control jacket connected to a Lauda E100 circulating water bath.

The operating parameters of the instrument during rheological investigation were as follows: gap between plates—400 μm; displacement amplitude—0.03 μm (to fall into the linear viscoelasticity domain); frequency domain—0.005–2.000 Hz (with a step of 0.005 Hz, which led to angular frequencies, in rad/s, of 2π times higher than the corresponding frequencies taken in Hz); equilibration time for each of the isothermal measurements—20 min; and 30 scans per rheogram.

The dynamic rheological parameters of storage modulus (G′) and loss modulus (G″) were determined to evaluate the stability of the hydrogel network. All rheological measurements were performed in triplicate at the same constant temperature of 23 °C, and all values were expressed as mean value and standard deviation.

AT A GLANCE

This study involved the synthesis of several superabsorbent hybrid hydrogel compositions using xanthan gum (XG), sodium carboxymethylcellulose (CMC), and graphene oxide (GO). The synthesis was carried out by e-beam radiation crosslinking. We conducted a comprehensive research of the impact of graphene oxide (GO) concentration on the chemical structure of hydrogels. This analysis included sol-gel analysis, swelling degree measurement, water diffusion assessment, ATR-FTIR spectroscopy, evaluation of network structure, and dynamic mechanical analysis. The gel fraction and swelling properties of the hydrogels that were generated were influenced by the compositions of the polymers used and the amount of radiation absorbed. The hybrid XGCMCGO hydrogels exhibited a high degree of water absorption and achieved equilibrium in under 6 hours. The XGCMCGO (70:30) hydrogel exhibited the most swelling capacity, reaching approximately 6000%, when exposed to an irradiation dose of 15 kGy. The elastic (G') and viscous (G'') moduli exhibited a significant dependence on the absorbed dose. At higher levels of crosslinking, the G' parameter was shown to surpass 1000 Pa. For the XGCMCGO (80:20) hydrogel compositions, the Mc and ξ parameters decreased as the absorbed dose rose, indicating a rise in crosslinking density. This result confirms the formation of a superabsorbent hydrogel with a stable structure.

REFERENCES

1. Qiu, Y.; Park, K. Environment-sensitive hydrogels for drug delivery. Adv. Drug Deliv. Rev. 2001, 53, 321–339.
2. Miyata, T.; Uragami, T.; Nakamae, K. Biomolecule-sensitive hydrogels. Adv. Drug Deliv. Rev. 2002, 54, 79–98.
3. Gupta, P.; Vermani, K.; Garg, S. Hydrogels: From controlled release to pH-responsive drug delivery. Drug Discov. Today 2002, 7, 569–579.
4. Guo, S.; Dong, S. Graphene nanosheet: Synthesis, molecular engineering, thin film, hybrids, and energy and analytical applications. Chem. Soc. Rev. 2011, 40, 2644–2672.
5. Chen, Y.; Chen, L.; Bai, H.; Li, L. Graphene oxide–chitosan composite hydrogels as broad-spectrum adsorbents for water purification. J. Mater. Chem. A 2013, 1, 1992–2001.
6. Gao, H.; Sun, Y.; Zhou, J.; Xu, R.; Duan, H. Mussel-Inspired Synthesis of Polydopamine-Functionalized Graphene Hydrogel as Reusable Adsorbents for Water Purification. ACS Appl. Mater. Interfaces 2013, 5, 425–432.
7. Wang, J.; Su, S.; Qiu, J. Biocompatible swelling graphene oxide reinforced double network hydrogels with high toughness and stiffness. New J. Chem. 2017, 41, 3781–3789.
8. Peng, S.; Han, X.; Li, L.; Zhu, Z.; Cheng, F.; Srinivansan, M.; Adams, S.; Ramakrishna, S. Unique Cobalt Sulfide/Reduced Graphene Oxide Composite as an Anode for Sodium-Ion Batteries with Superior Rate Capability and Long Cycling Stability. Small 2016, 12, 1359–1368.
9. Xu, H.; Adolfsson, K.; Xie, L.; Hassanzadeh, S.; Pettersson, T.; Hakkarainen, M. Zero-Dimensional and Highly Oxygenated Graphene Oxide for Multifunctional Poly(lactic acid) Bionanocomposites. ACS Sustain. Chem. Eng. 2016, 4.
10. Huang, Y.; Zeng, M.; Ren, J.; Wang, J.; Fan, L.; Xu, Q. Preparation and swelling properties of graphene oxide/poly(acrylic acid-co-acrylamide) super-absorbent hydrogel nanocomposites. Colloids Surf. A Physicochem. Eng. Asp. 2012, 401, 97–106.

11. Lei, H.; Xie, M.; Zhao, Y.; Zhang, F.; Xu, Y.; Xie, J. Chitosan/sodium alginate modificated graphene oxide-based nanocomposite as a carrier for drug delivery. Ceram. Int. 2016, 42.
12. Geim, A.K.; Novoselov, K.S. The rise of graphene. Nat. Mater. 2007, 6, 183–191.
13. Balandin, A.A.; Ghosh, S.; Bao, W.; Calizo, I.; Teweldebrhan, D.; Miao, F.; Lau, C.N. Superior Thermal Conductivity of Single-Layer Graphene. Nano Lett. 2008, 8, 902–907.
14. Lee, C.; Wei, X.; Kysar, J.; Hone, J. Measurement of the Elastic Properties and Intrinsic Strength of Monolayer Graphene. Science 2008, 321, 385–388.
15. Li, Y.; Ha, Y.; Wang, F.; Li, Y. Effect of Irradiation on the Molecular Weight, Structure and Apparent Viscosity of Xanthan Gum in Aqueous Solution. Adv. Mater. Res. 2011, 239–242, 2632–2637.
16. Sen, M.; Hayrabolulu, H.; Taskin, P.; Torun, M.; Cutrubinis, M.; Güven, O. Radiation induced degradation of xanthan gum in the solid state. Radiat. Phys. Chem. 2015, 124.
17. Hayrabolulu, H.; Cutrubinis, M.; Güven, O.; Sen, M. Radiation induced degradation of xanthan gum in aqueous solution. Radiat. Phys. Chem. 2017, 144.
18. Hayrabolulu, H.; Demeter, M.; Cutrubinis, M.; Şen, M. Radiation synthesis and characterization of xanthan gum hydrogels. Radiat. Phys. Chem. 2021, 188, 109613.
19. Choi, J.; Lee, H.-S.; Kim, J.-H.; Lee, K.W.; Lee, J.; Seo, S.J.; Kang, K.W.; Byun, M.W. Controlling the radiation degradation of carboxymethylcellulose solution. Polym. Degrad. Stab. 2008, 93, 310–315.
20. Sung, Y.; Kim, T.-H.; Lee, B. Syntheses of carboxymethylcellulose/graphene nanocomposite superabsorbent hydrogels with improved gel properties using electron beam radiation. Macromol. Res. 2016, 24, 143–151.
21. Lee, S.; Lee, H.; Sim, J.H.; Sohn, D. Graphene oxide/poly(acrylic acid) hydrogel by γ-ray pre-irradiation on graphene oxide surface. Macromol. Res. 2014, 22, 165–172.
22. Pinelli, F.; Nespoli, T.; Rossi, F. Graphene Oxide-Chitosan Aerogels: Synthesis, Characterization, and Use as Adsorbent Material for Water Contaminants. Gels 2021, 7, 149.
23. Calina, I.; Demeter, M.; Vancea, C.; Scarisoreanu, A.; Meltzer, V. E-beam radiation synthesis of xanthan-gum/carboxymethylcellulose superabsorbent hydrogels with incorporated graphene oxide. J. Macromol. Sci. Part A 2018, 55, 260–268.
24. Das, B.; Prasad, K.E.; Ramamurty, U.; Rao, C. Nano-indentation studies on polymer matrix composites reinforced by few-layer graphene. Nanotechnology 2009, 20, 125705.
25. Said, H.M.; Alla, S.G.A.; El-Naggar, A.W.M. Synthesis and characterization of novel gels based on carboxymethyl cellulose/acrylic acid prepared by electron beam irradiation. React. Funct. Polym. 2004, 61, 397–404.
26. Olejniczak, J.; Rosiak, J.; Charlesby, A. Gel/dose curves for polymers undergoing simultaneous crosslinking and scission. Int. J. Radiat. Appl. Instrum. Part C. Radiat. Phys. Chem. 1991, 37, 499–504.
27. Mozalewska, W.; Czechowska-Biskup, R.; Olejnik, A.K.; Wach, R.A.; Ulański, P.; Rosiak, J.M. Chitosan-containing hydrogel wound dressings prepared by radiation technique. Radiat. Phys. Chem. 2017.
28. Salmawi, K.; El-Naggar, A.; Ibrahim, S. Gamma Irradiation Synthesis of Carboxymethyl Cellulose/ Acrylic Acid/Clay Superabsorbent Hydrogel. Adv. Polym. Technol. 2016, 37, 515–521.

29. Sultana, S.; Islam, M.R.; Dafader, N.C.; Haque, M.E. Preparation of carboxymethyl cellulose/acrylamide copoly-mer hydrogel using gamma radiation and investigation of its swelling behavior. J. Bangladesh Chem. Soc. 2012, 25, 132–138.

30. Hosseini, S.M.; Shahrousvand, M.; Shojaei, S.; Khonakdar, H.A.; Asefnejad, A.; Goodarzi, V. Preparation of superabsorbent eco-friendly semi-interpenetrating network based on cross-linked poly acrylic acid/xanthan gum/graphene oxide (PAA/XG/GO): Characterization and dye removal ability. Int. J. Biol. Macromol. 2020, 152, 884–893.

31. Wang, Z.; Ning, A.; Xie, P.; Gao, G.; Xie, L.; Li, X.; Song, A. Synthesis and swelling behaviors of carboxymethyl cellulose-based superabsorbent resin hybridized with graphene oxide. Carbohydr. Polym. 2017, 157, 48–56.

32. Kim, B.; Kim, T.-H.; Lee, B. Optimal synthesis of carboxymethylcellulose-based composite superabsorbents. Korean J. Chem. Eng. 2021, 38, 215–225.

33. Saraydın, D.; Isikver, Y.; Sahiner, N.; Güven, O. The Influence of Preparation Methods on the Swelling and Network Properties of Acrylamide Hydrogels with Crosslinkers. J. Macromol. Sci. Part A 2004, 41, 419–431.

34. Karadağ, E.; Saraydın, D.; Güven, O. Influence of some crosslinkers on the swelling of acrylamide-crotonic acid hydrogels. Turkish J. Chem. 1997, 21, 151–161.

35. Colombo, P.; Bettini, R.; Santi, P.; Peppas, N.A. Swellable matrices for controlled drug delivery: Gel-layer behaviour, mechanisms and optimal performance. Pharm. Sci. Technolo. Today 2000, 3, 198–204.

36. Karadağ, E.; Ödemiş, H.; Kundakçi, S.; Üzüm, Ö.B. Swelling characterization of acrylamide/zinc acrylate/xanthan gum/sepiolite hybrid hydrogels and Its application in sorption of janus green B from aqueous solutions. Adv. Polym. Technol. 2016, 35, 248–259.

37. Dai, H.; Zhang, Y.; Ma, L.; Zhang, H.; Huang, H. Synthesis and response of pineapple peel carboxymethyl cellulose-g-poly (acrylic acid-co-acrylamide)/graphene oxide hydrogels. Carbohydr. Polym. 2019, 215, 366–376.

38. Siouffi, A.; Guiochon, G. Chromatography: Thin-Layer (PLANAR)|Theory of Thin-Layer (Planar) Chromatography. Encycl. Sep. Sci. 2000, 915–930.

39. Varaprasad, K.; Jayaramudu, T.; Sadiku, E.R. Removal of dye by carboxymethyl cellulose, acrylamide and graphene oxide via a free radical polymerization process. Carbohydr. Polym. 2017, 164, 186–194.

40. Borsali, R.; Pecora, R. Soft-Matter Characterization; Springer Science & Business Media: Berlin/Heidelberg, Germany, 2008; ISBN 140204464X.

41. Robinson, G.; Ross-Murphy, S.B.; Morris, E.R. Viscosity-molecular weight relationships, intrinsic chain flexibility, and dynamic solution properties of guar galactomannan. Carbohydr. Res. 1982, 107, 17–32.

42. Gudeman, L.F.; Peppas, N.A. pH-sensitive membranes from poly (vinyl alcohol)/poly (acrylic acid) interpenetrating networks. J. Membr. Sci. 1995, 107, 239–248.

43. Fekete, T.; Borsa, J.; Takács, E.; Wojnarovits, L. Synthesis of carboxymethylcellulose/acrylic acid hydrogels with superabsorbent properties by radiation-initiated crosslinking. Radiat. Phys. Chem. 2015, 124.

44. Wong, R.; Ashton, M.; Dodou, K. Effect of Crosslinking Agent Concentration on the Properties of Unmedicated Hydrogels. Pharmaceutics 2015, 7, 305–319.

45. Varma, D.M.; Gold, G.T.; Taub, P.J.; Nicoll, S.B. Injectable carboxymethylcellulose hydrogels for soft tissue filler applications. Acta Biomater. 2014, 10, 4996–5004.

46. Kalia, S.; Choudhury, A.R. Synthesis and rheological studies of a novel composite hydrogel of xanthan, gellan and pullulan. Int. J. Biol. Macromol. 2019, 137, 475–482.

47. Micic, M.; Suljovrujic, E. Network parameters and biocompatibility of p(2-hydroxyethyl methacrylate/itaconic acid/oligo(ethylene glycol) acrylate) dual-responsive hydrogels. Eur. Polym. J. 2013, 49, 3223–3233.

48. Lorenzo, E.; Katime, I. Some Mechanical Properties of Poly[(acrylic acid)-co-(itaconic acid)] Hydrogels. Macromol. Mater. Eng. 2003, 288, 607–612.

49. Lejardi, A.; Hernandez, R.; Gonzalez, M.C.; Santos, J.; Etxeberria, A.; Sarasua, J.-R.; Mijangos, C. Novel hydrogels of chitosan and poly(vinyl alcohol)-g-glycolic acid copolymer with enhanced rheological properties. Carbohydr. Polym. 2014, 103, 267–273.

50. Braihi, A. Viscoelastic and Rheological Properties of Carboxymethyl Cellulose/Starch/Graphite Oxide as Superabsorbent Hydrogel Nano Composites (SHNCs). Int. J. Mater. Sci. Appl. 2015, 4, 30–36.

51. Manaila, E.; Craciun, G.; Ighigeanu, D.; Lungu, B.; Dumitru, M.; Stelescu, M. Electron Beam Irradiation: A Method for Degradation of Composites Based on Natural Rubber and Plasticized Starch. Polymers 2021, 13, 1950.

52. Suhartini, M.; Yoshii, F.; Nagasawa, N.; Mitomo, H. Radiation Yield and Radicals Produced in Irradiated Poly (Butylene Succinate). Atom Indon. 2011, 30.

53. Nagasawa, N.; Yagi, T.; Kume, T.; Yoshii, F. Radiation crosslinking of carboxymethyl starch. Carbohydr. Polym. 2004, 58, 109–113.

54. Ritger, P.L.; Peppas, N.A. A simple equation for description of solute release I. Fickian and non-fickian release from non-swellable devices in the form of slabs, spheres, cylinders or discs. J. Control. Release 1987, 5, 23–36.

55. Mahmudi, N.; Sen, M.; Rendevski, S.; Güven, O. Radiation synthesis of low swelling acrylamide based hydrogels and determination of average molecular weight between cross-links. Nucl. Instrum. Methods Phys. Res. Sect. B Beam Interact. Mater. Atoms 2007, 265, 375–378.

56. Canal, T.; Peppas, N.A. Correlation between mesh size and equilibrium degree of swelling of polymeric networks. J. Biomed. Mater. Res. 1989, 23, 1183–1193.

9 Chapter

INSTANTANEOUS DEGELLING THERMORESPONSIVE HYDROGEL

1. INTRODUCTION: AN OVERVIEW

Biodegradable polymers are widely used in the pharmaceutical business. These materials are commonly used as the main platform for delivering therapeutic agents. Their slow disintegration in the body enables a continuous release of drugs over a long period without requiring the removal of the delivery vehicle. Polymers with the ability to react to outside influences were created to enable specific medication release when the appropriate stimulus is present. These materials contain functional groups that are capable of rapidly responding to slight variations in temperature, pH, or light. This response causes a physical change in the polymer, which in turn activates the release of drugs [1].

Polymeric hydrogels are polymers that can absorb significant quantities of water and form a gel as a result of crosslinking between the polymer chains. Thermoresponsive hydrogels with intelligent properties were created to undergo a sol-gel transition in response to changes in temperature [2]. This property allows for the introduction of a liquid solution into a biological environment, which then undergoes a temperature-induced transformation, resulting in the formation of a gel at the site of injection. The gelling temperatures of these polymers and their biocompatibility are determined by their molecular architectures [3,4].

The article extensively discussed the biocompatible and biodegradable properties of copolymers made from poly (lactic-co-glycolic acid) (PLGA) and poly (ethylene glycol) (PEG). These polymers have the ability to generate hydrogels that respond to changes in temperature when placed in water [5]. Triblock copolymers, such as those with the PLA-PEG-PLGA or PEG-PLGA-PEG morphology, have the ability to dissolve in water and undergo reversible gelation when heated [6,7,8]. The gelation properties of the hydrogels generated from these copolymers are influenced by the total molecular weight of the polymer, the ratio between the polymer blocks, and the concentration of the polymer in the solution [3,9,10,11]. Therefore, it is crucial to carefully plan and design each component of the polymer structure in order to attain the desired qualities of the gel.

Driven by the vital function of cysteine–cysteine bonds and their cleavage in biological processes, disulfide (S–S) bonds have been utilized in many functional materials because of their redox reactivity. Materials that include disulfide bonds can be specifically broken apart under conditions that involve reduction, especially when exposed to the reducing environment inside cells. This process gives polymers the ability to respond to changes in redox potential. Disulfide bonds were introduced into polyurethanes to provide self-healing properties [12], in electrochemical polymers and devices [13,14], and in drug carriers to enable reduction-specific crosslinking or drug binding [15,16,17] for the controlled release of antitumor drugs due to the abundance of reducing agents in the tumor microenvironment [18]. They are also used to reduce toxicity in gene delivery platforms [19,20,21,22], and in supramolecular polymer applications [23]. By introducing disulfide links into biocompatible thermoreversible hydrogels, it is possible to create materials that are responsive to both temperature and reduction.

This study presents a description of a triblock copolymer called PEG-PLA-PEG, which contains one disulfide link per molecule. The polymer solution exhibits thermoresponsiveness, becoming completely soluble in water at low temperatures and forming a gel between 32 and 40 °C. The polymer contains a disulfide bond that can be cleaved by strong reducing agents, causing the hydrogel to collapse immediately. This proof of concept was demonstrated through the breaking of the disulfide bond in the presence of hydride. A synthetic analogue lacking a disulfide bond exhibited stability as a hydrogel, even when exposed to severe reductive conditions. This work is the first known report of a thermoresponsive hydrogel that can instantly collapse under regulated settings, namely when exposed to decreasing conditions (Figure 1).

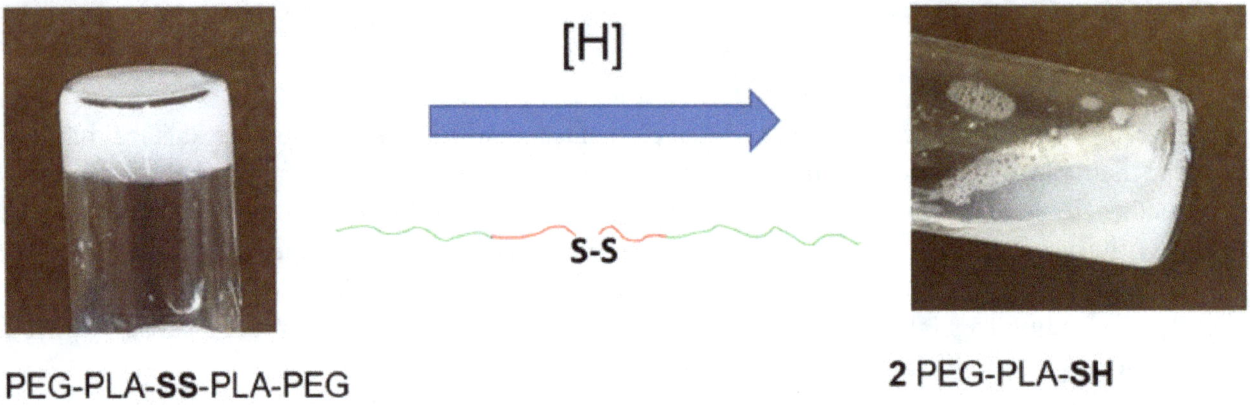

Figure 1. PEG-PLA-PEG triblock copolymers with embedded disulfide bonds immediately collapse upon addition of reducing agent NaBH4. Disulfide-bearing polymer hydrogels afford controlled hydrogel collapsibility due to S–S bond cleavage.

2. RESULTS

2.1. Rational Design

Jeong et al. first reported the synthesis of triblock copolymers containing a central poly (lactic acid) (PLA) block bearing two sidechains of poly (ethylene glycol) (PEG) [24]. The synthetic method reported there comprised of two steps. First, poly(ethylene glycol) methyl ether Mn = 550 (mPEG 550) was used as a macroinitiator of lactide polymerization (via ring-opening polymerization in the presence of stannous octoate catalyst) to form a diblock copolymer. Two hydroxyl termini of two distinct PLA blocks were subsequently linked by reaction with diisocyanate, forming urethane linkages (Figure 2A). Whereas the original research focused on polymers with sol–gel properties, which required cooling to move from sol form to gel morphology, subsequent reports focused on the development of sol–gel transitions, which occurred upon heating of the polymer solution. The enhancement of sol-to-gel properties from relying on cooling (gel at low temperatures and sol at high temperatures) to being responsive to heating (sol at room temperature and gel at elevated temperature) was crucial, as the practical application of such gels relies on the incorporation of active agents in a room-temperature solution, which upon injection and exposure to physiological temperatures forms a semi-solid gel capable of extended release of the active agent. These sol–gel thermosensitive hydrogels were based on an amorphous PLA polymer block (as opposed to crystalline poly (L-lactic acid) polymer blocks employed in the original work). Gel properties, biodegradability, and drug release were all subsequently optimized, leading to over two decades of research and application of injectable sol–gel biodegradable polymer solutions [6,11,25].

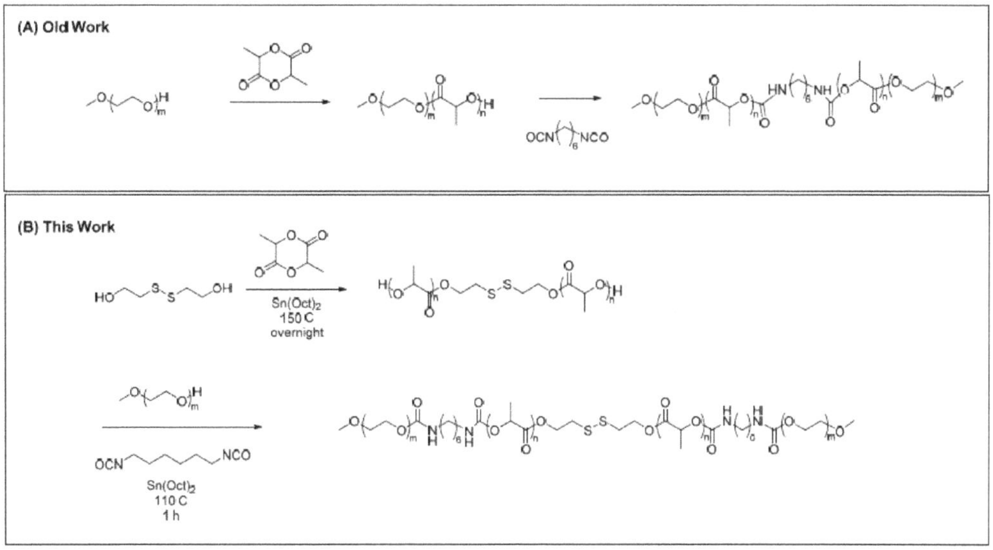

Figure 2. (A) Synthesis by Jeong et al. of a PEG-PLA-PEG triblock copolymer with urethane linkages [24]. (B) Synthesis of PEG-PLA-PEG with urethane linkages between polymer blocks and one embedded disulfide bond.

The PEG-PLA-PEG triblock copolymer described here is an analog of the original work performed by Jeong and co-workers. Its chemical structure bears the modular hydrophobic and hydrophilic units at the appropriate ratios for effecting sol–gel transitions near physiological temperatures, with the added benefit of one embedded disulfide bond. In doing so, the 'smart' hydrogel was imbued with secondary responsiveness to reduction agents in addition to its thermal responsivity. The hydrogel resultant thereof was therefore capable of injection as a liquid at ambient temperatures, gelling upon moderate heating, and subsequent collapse of the 3D gel structure upon exposure to strong reducing agents [26]. The following is a description of the synthesis and characterization of the novel triblock copolymer with a comprehensive study of the gel properties of aqueous solutions prepared thereof.

2.2. Synthesis

In a similar fashion as the original work on PEG-PLA-PEG triblock copolymers, several synthetic steps were employed in order to incorporate a disulfide bond into the triblock copolymer structure. In the first synthetic step, 2-hydroxyethyl disulfide was used as the initiator of DL-lactide polymerization. In doing so, an amorphous DL-poly(lactic acid) (DL-PLA) was obtained bearing one embedded disulfide bond (PLA(SS)). In order to complete the triblock copolymer synthesis, mPEG with hydroxyl termini were linked to either terminus of the amorphous PLA diol by a 2:2:1 molar reaction between mPEG diisocyanate and PLA(SS), affording urethane bridges between PEG and PLA polymer blocks with an overall PEG-PLA-SS-PLA-PEG copolymer structure (Figure 2B).

2.3. Characterization

The formation of PLA(SS) and PEG-PLA-SS-PLA-PEG polymers was confirmed spectroscopically by 1H NMR. Polymer molecular weights (MW) were estimated by size-exclusion chromatography (SEC). The presence of a disulfide bond was confirmed by elemental analysis (CHNS).

2.3.1. 1H NMR

The structures of both PLA(SS) and PEG-PLA-SS-PLA-PEG were confirmed by 1H NMR (Figure 3) by observing all available proton peaks. For PLA(SS), in addition to characteristic multiplets for lactic acid at 5.2 and 1.5 ppm [3], a downfield shift of 2-hydroxyethyl disulfide CH2 protons was observed upon polymerization in the PLA(SS) spectrum, indicating the formation of the ester bond at the termini of the disulfide reagent. No other peaks were observed in the spectrum, indicating a pure substance.

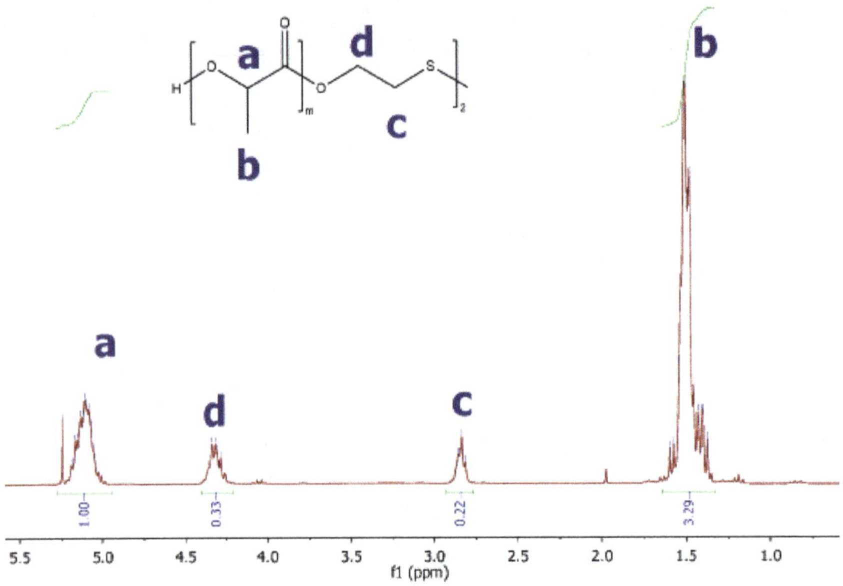

(a)

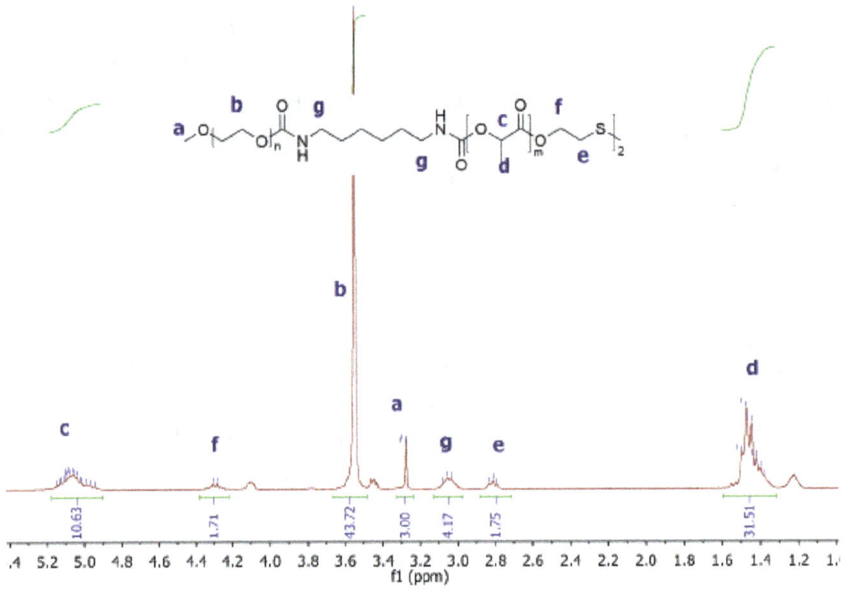

(b)

Figure 3. (a) 1H NMR spectra of PLA(SS) with peak assignments.; (b) 1H NMR spectra of PEG-PLA-SS-PLA-PEG with peak assignments.

Upon linkage to PEG, no chemical shifts were observed for peaks from PLA(SS), confirming the stability of the PLA block of the copolymer to the reaction conditions. Urethane linkages were confirmed by a peak at δ 3.05 with 2x integration relative to the CH2 peak at the disulfide, as expected. A characteristic PEG singlet was observed at δ 3.55, and all integrations were calculated relative to CH3 of mPEG. No other peaks were observed in the spectrum, indicating a pure substance (Figure 3).

2.3.2. Molecular Weight Determination

Weight average (Mw) and number average (Mn) molecular weights of PLA(SS) and PEG-PLA-SS-PLA-PEG were estimated by SEC by comparing to polystyrene standards. Polydispersity (PDI) values were calculated from these measured values. Mn was further calculated by 1H NMR peak integrations using known integration values (CH2 of 2-hydroxyethyl disulfide for PLA(SS), CH3 of mPEG for triblock PEG-PLA-PEG copolymer). Mn values calculated by NMR closely reflected the expected chain lengths based on feed ratios of starting materials, indicating full conversion from PLA(SS) to triblock copolymer with PEG. Estimated values provided by SEC indicated increased polymer MW upon copolymerization with PEG by urethane bridge, and only a slight increase in PDI values indicated polymers with high purity (Table 1).

Table 1. Molecular weight determination of disulfide PLA and disulfide-containing PEG-PLA-PEG copolymer.

Polymer	PLA(SS)	PEG-PLA-SS-PLA-PEG
Expected MW [1,2]	1.53	1.96
Mn (^{1}H NMR) [1,3]	1.46	3.11
Mn (SEC) [1,4]	2.65	4.24
Mw (SEC) [1,4]	2.85	5.97
PDI [5]	1.07	1.41

[1]Values given in kDa. [2] Based on feed ratios of starting materials. [3] Calculated based on peak integrations of known value. [4] Calculated from size-exclusion chromatography (SEC). [5] Calculated from SEC as Mw/Mn.

2.4. Rheometry Studies

The PEG-PLA-SS-PLA-PEG triblock copolymer formed a free-flowing solution in double-distilled water at room temperature (25% w/w). Upon incubation at physiological temperature, the solution formed a hydrogel which did not flow upon inversion of the test tube [3,4]. In order to evaluate the flow behavior of the copolymer solution and determine the temperature at which the sol–gel transition occurred, the viscosity of copolymer solutions was measured as a function of temperature (Figure 4). Upon heating, an increase in viscosity was observed beginning at ~32 °C with a maximum 70-fold increase in viscosity at 36 °C. The high viscosity values were stable through physiologically relevant temperatures. Upon cooling of the hydrogel, the opposite trend was observed, indicating reversibility of the thermoresponsive system. This flow behavior indicates a polymeric solution capable of injectability at ambient temperatures (<30 °C) with temperature-induced gelling occurring upon exposure to physiological temperatures.

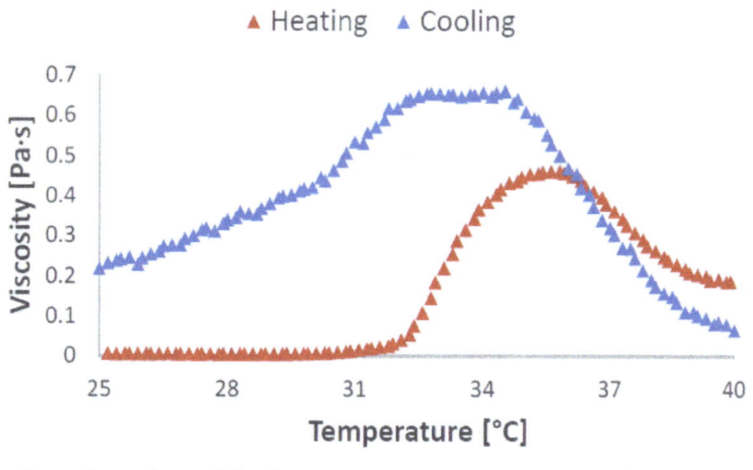

Figure 4. Heating (red) and cooling (blue) viscosity curves of 25% w/w PEG-PLA-SS-PLA-PEG aqueous solutions.

2.5. Degradation Studies

Degradation of injectable implants such as triblock copolymer hydrogels is an important parameter for the determination of the hydrogel's potential applicability to the clinic [27]. The stability of the hydrogel under physiological conditions affects the therapeutic window and controlled release profile of incorporated drugs, and biodegradation of the hydrogel matrix excludes the necessity for post-injection removal of the material, allowing it to slowly degrade and be eliminated by the body.

In the case of the PEG-PLA-PEG copolymer with an embedded disulfide bond, two modes of degradation are possible. One degradable bond is the repeating ester bond throughout the PLA polymer block, known to be susceptible to hydrolytic cleavage [28]. Secondly, the disulfide bond may degrade upon exposure to reducing agents. Hydrolytic degradation of the ester bonds was expected to occur slowly over time, allowing for slow degradation of the hydrogel matrix. The disulfide bond, however, was expected to remain intact under in vitro physiological conditions, degrading only upon exposure to reducing agent. Due to its position in the center of the polymer chain, cleavage of the disulfide bond should lead to an immediate release of water retained in the hydrogel matrix due to rupture of the triblock copolymer structure. In order to investigate these effects, degradation of PEG-PLA-SS-PLA-PEG triblock copolymer was studied in water or in the presence of reducing agents such as sodium borohydride (NaBH4) or L-glutathione.

In order to confirm the redox reactivity of the disulfide bond embedded in the PEG-PLA-PEG triblock copolymer structure, NaBH4, a strong hydride-based reducing agent, was added to the hydrogel matrix. The hydride was expected to reduce the disulfide bond, affording diblock PEG-PLA copolymers with thiol end groups (Figure 5). Indeed, a reduction in polymer molecular weight to half was observed upon reaction with hydride, indicating the cleavage of the reduction-sensitive disulfide bond in the triblock copolymer structure (Figure 5). A chemical analog of the triblock copolymer was prepared using 1,6-hexanediol as the initiator of PLA synthesis, resulting in PEG-PLA-PEG triblock copolymer without a disulfide bond (experimental details and spectral data available as Supplementary Materials, Figures S1 and S2). The molecular weight of this polymer was only slightly reduced upon exposure to the same hydride reagent, thereby confirming the role of the disulfide bond in a molecular weight reduction in the original polymer (Figure 5).

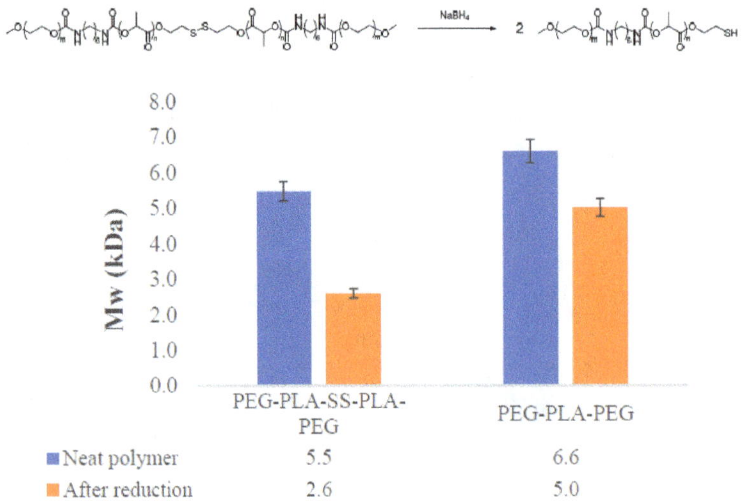

Figure 5. Sensitivity of the embedded disulfide bond in PEG-PLA-PEG triblock copolymer was confirmed by reaction with NaBH4 and subsequent reduction in molecular weight by half. A chemical analog lacking the disulfide bond retained most of its molecular weight under the same conditions, indicating the role of disulfide bond cleavage in molecular weight reduction. Error bars represent the standard deviation of experiments performed in triplicate.

In order to evaluate the in vitro degradation of PEG-PLA-SS-PLA-PEG hydrogels, polymer molecular weight was monitored for two weeks at physiological temperature (Figure 6). The hydrogel remained stable for this period, after which the molecular weight was sufficiently reduced to inhibit the water retention capabilities of the hydrogel, resulting in the release of water from the 3D hydrogel structure. A hydrogel containing 1% L-glutathione displayed no effect on the degradation rate of the polymer. We hypothesize that the rigidity of the hydrogel prevented the thiol reducing agent from reducing the disulfide bond embedded within the triblock copolymer structure. The near-constant PDI and slow Mw reduction in the polymer under either condition indicates a hydrolytically degradable polymeric hydrogel with stability in the presence of thiol reducing agents (Figure 6).

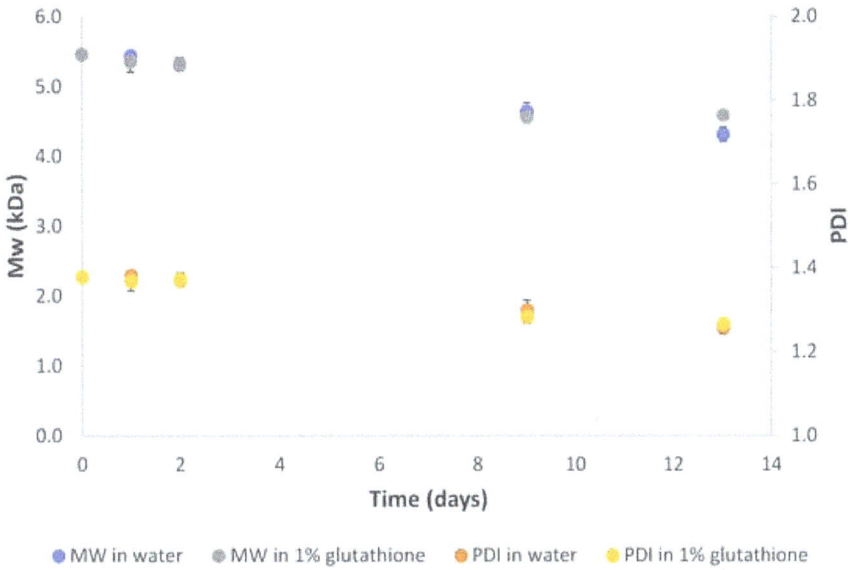

Figure 6. Degradation of PEG-PLA-SS-PLA-PEG hydrogel in distilled water or 1% glutathione solution displayed a slow hydrolytic degradation for two weeks with near-constant PDI values.

3. DISCUSSION

Polymer systems that undergo phase transitions in response to environmental stimuli such as temperature and pH have been widely investigated for drug delivery and tissue engineering applications [29]. The aqueous copolymer solution presented in this work represents a programmable biodegradable responsive polymer gel in which its gelling properties can be additionally controlled by a cleavable S–S bond triggered to cleave and change the gel properties. The polymer is a non-crosslinked triblock copolymer of polyethylene glycol (PEG) blocks and a hydroxy acid aliphatic polyester block that possesses thermo-responsive properties where the polymer is fully water soluble at temperatures between 0 °C and about 30 °C and gels at body temperature (37 °C). In addition to hydrolytically degradable ester bonds, the polymer possesses, as mentioned above, a cleavable bond that breaks upon application of an external reductive trigger that is faster than the hydrolysis of the ester bonds of the polyester block. The cleavage of this bond results in the collapse of the gel to dissolution in water with no remaining gel properties (Figure 7). This class of polymer may have medical use as a device for drug carrying or cell support and delivery. For drug or cell delivery, the rate of de-gelation can be programmed so that after its formation at body temperature, the gel may gradually or instantly lose/loosen its gel properties to allow faster release from the gel of the entrapped cells or drugs. The change in gel properties, after initial gelation at the site of action, can be tailored according to the need for a specific application, either immediately or gradually, based on the presence of reducing agents. Of specific interest is the delivery of therapeutic stem cells carrying anticancer cargo at cancerous tissue. The cells may be delivered to the site of treatment as a solution that instantly gels on the target tissue due to temperature change and starts to lose its gel properties to allow the cells to move towards cancerous tissue due to the increased neoplastic concentration of reducing agents [18].

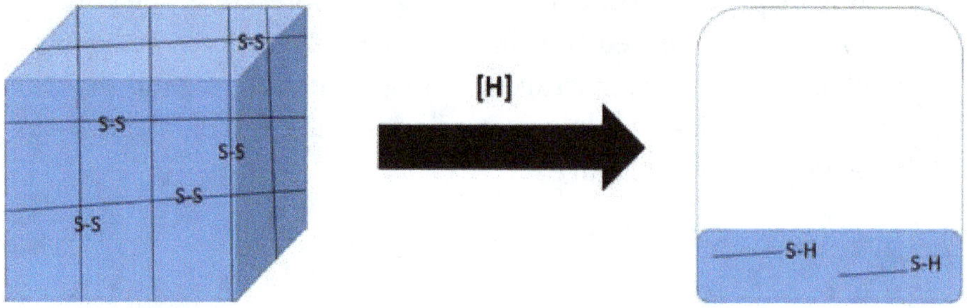

Figure 7. Three-dimensional structure of a physically crosslinked hydrogel may be instantaneously collapsed by exposure to reducing conditions.

Several important parameters are involved in the development of these systems, including robust synthetic methods, defined stimuli-responsive gelation, and degradation for application in vivo. The physical crosslinking mechanism of PEG-PLA-PEG copolymers hydrogel formation permits the additional advantage of reversibility, as no chemical bonds must be cleaved in order to re-obtain the polymer in solution [30]. However, to date, there have been no reports of non-crosslinked thermoresponsive polymer gels that possess all of the following qualities: (1) are water-soluble below physiological temperatures, (2) form a semi-solid continuous gel at physiological temperatures, and (3) possess breakable bonds that permit an immediate programmable change in gel properties following exposure to an external trigger. The PEG-PLA-PEG triblock copolymers with embedded disulfide bond described here represent such material, with gelling properties suitable for physiological application and susceptibility to water leakage upon exposure to reducing agents. The proof of concept afforded by cleavage with a hydride reagent represents an interesting option for the development of reduction-sensitive hydrogels capable of instantaneous collapse upon exposure to reducing agents.

4. CONCISE SUMMARY

Injectable platforms using thermoresponsive hydrogels that rapidly collapse when exposed to powerful reducing agents offer a fascinating method for controlled drug delivery. At normal temperatures, the polymer dissolves in water (25%) and creates a solution that flows easily. However, when heated, it transforms into a hydrogel. The presence of a disulfide link inside the polymer structure causes a controlled collapse of the hydrogel when NaBH4 is added. The hydrogel remains stable for a duration exceeding two weeks. Over time, the ester bonds in hydrogels undergo hydrolytic breakdown, causing the hydrogel to collapse and form harmless by-products. The hydrogels discussed in this context have the capacity to be utilized as carriers for the gradual release of medicinal substances, with the ability to trigger quick release upon coming into contact with reducing agents.

5. MATERIALS AND METHODS

5.1. Materials

Stannous octoate, hexamethylene diisocyanate, and poly(ethylene glycol) monomethyl ether were purchased from Sigma Aldrich (Rehovot, Israel). Lactide was purchased from Purac Biochem BV (Gorinchem, The Netherlands). Solvents were purchased from Bio-Lab Ltd. (Jerusalem, Israel).

5.2. General Methods

Chemical reactions were performed in dry glassware under N2 gas. 1H NMR spectra were obtained on a Varian 300 MHz spectrometer with CDCl3 as the solvent and tetramethylsilane as the shift reference. The molecular weights of the polymers were estimated using a gel permeation chromatography (GPC) system consisting of a Waters 1515 Isocratic HPLC pump with a Waters 2410 Refractive Index Detector and a Rheodyne (Cotati, CA, USA) injection valve with a 20 mL loop (Waters, Milford, MA, USA). Samples were eluted with CHCl3 through a linear Styragel HR4E column (7.8 × 300 mm i.d.; Waters) at a flow rate of 1 mL/min. The molecular weights were determined relative to polystyrene standards (Polyscience, Warrington, PA, USA) using a Breeze computer program.

5.3. Synthesis

PEG-PLA-PEG triblock copolymer with disulfide bond was prepared in two steps. First, 50 μL of a 10% solution of stannous octoate in dichloromethane (DCM) was added to a melt of 2-hydroxyethyl disulfide (0.13 g, 0.82 mmol) and D,L-lactide (2.4 g, 16 mmol). The solvent was allowed to evaporate, and the vial was purged with N2. The mixture was stirred at 120 °C for 2 h, followed by overnight stirring at 150 °C. The crude polymer was taken up in DCM and precipitated into ether to afford PLA disulfide PLA(SS) as a pure substance. 1H NMR (300 MHz, CDCl3, δ): 5.24–5.05 (m, LA), 4.36–4.28 (m, CH2 α-ester), 2.87–2.82 (m, CH2 α-disulfide), 1.59–1.37 (m, LA). Anal. Calcd: C, 46.74; H, 5.91; S, 3.80. Found: C, 48.00; H, 5.70; S, 4.42.

In the next step, hexamethylene diisocyanate (0.11 mL, 0.68 mmol) was added to a melt of PLA(SS) (1.0 g, 0.70 mol) and poly(ethylene glycol) methyl ether Mn = 550 (0.34 mL, 0.68 mmol). One drop of stannous octoate was added, and the mixture was stirred at 110 °C for one hour. The crude product was dissolved in cold water, filtered, and lyophilized to afford PEG-PLA(SS)-PEG as a pure substance. Proton nuclear magnetic resonance (1H NMR) (300 MHz, CDCl3, δ): 5.15–4.94 (m, LA), 4.30–4.28 (m, CH2 α-ester), 4.11–4.08 (m, PEG α-urethane), 3.55 (s, PEG), 3.28 (s, CH3), 3.07–3.03 (m, CH2 α-urethane), 2.83–2.79 (t, J = 6 Hz, CH2 α-disulfide), 1.50–1.38 (m, LA). Anal. Calcd: C, 50.53; H, 7.12; N, 2.05; S, 1.88. Found: C, 51.09; H, 7.22; N, 1.99; S, 2.27.

5.4. Rheological Studies

Sol–gel transitions were measured on 25% w/w polymer solutions in double-distilled water (DDW) by rotational tests on a Physica MCR 101 rheometer (Anton Paar, Austria) as a function of temperature from 25 to 45 °C.

5.5. Degradation Studies

25% w/w polymer solutions were prepared in DDW or in 1% solutions of either reduced L-glutathione or NaBH4. Solutions were incubated at 37 °C, and molecular weight was monitored by GPC for two weeks.

AT A GLANCE

Responsive polymeric hydrogels are extensively used in medical settings as injectable, biocompatible, and biodegradable materials that may release medicines in a regulated manner. This article presents a thermoresponsive polymer hydrogel that has covalent disulfide links. Upon reaching physiological temperatures, the cold aqueous polymer solution transforms into a hydrogel and gradually breaks down through hydrolytic cleavage of ester bonds. The presence of disulfide groups enables the rapid breaking of the redox-sensitive bond present in the polymer structure, resulting in the quick collapse of the hydrogel if desired. The PEG-PLA-SS-PLA-PEG copolymer was created by attaching PEG to PLA-SS-PLA through urethane bonds. The resulting copolymer formed a hydrogel above 32 °C and was a fluid solution at room temperature. The hydrogel's rapid collapsibility was demonstrated when it reacted with NaBH4, a potent reducing agent. However, the hydrogel exhibited stability even in the presence of glutathione solution, where it underwent gradual breakdown through hydrolytic processes. The polymeric hydrogel has the ability to degrade either over a lengthy period of time or immediately. This makes it a desirable option as a biocompatible material for controlled drug release.

REFERENCES

1. Priya James, H.; John, R.; Alex, A.; Anoop, K.R. Smart polymers for the controlled delivery of drugs—A concise overview. Acta Pharm. Sin. B 2014, 4, 120–127.
2. Jeong, B.; Kim, S.W.; Bae, Y.H. Thermosensitive sol–gel reversible hydrogels. Adv. Drug Deliv. Rev. 2012, 64, 154–162.
3. Steinman, N.Y.; Haim-Zada, M.; Goldstein, I.A.; Goldberg, A.H.; Haber, T.; Berlin, J.M.; Domb, A.J. Effect of PLGA block molecular weight on gelling temperature of PLGA-PEG-PLGA thermoresponsive copolymers. J. Polym. Sci. Part A Polym. Chem. 2019, 57, 35–39.
4. Steinman, N.Y.; Bentolila, N.Y.; Domb, A.J. Effect of Molecular Weight on Gelling and Viscoelastic Properties of Poly (caprolactone)–b-Poly (ethylene glycol)–b-Poly (caprolactone)(PCL–PEG–PCL) Hydrogels. Polymers 2020, 12, 2372.
5. Maeda, T. Structures and Applications of Thermoresponsive Hydrogels and Nanocomposite-Hydrogels Based on Copolymers with Poly (Ethylene Glycol) and Poly (Lactide-Co-Glycolide) Blocks. Bioengineering 2019, 6, 107.
6. Jeong, B.; Bae, Y.H.; Kim, S.W. In situ gelation of PEG-PLGA-PEG triblock copolymer aqueous solutions and degradation thereof. J. Biomed. Mater. Res. 2000, 50, 171–177.
7. Shim, M.S.; Lee, H.T.; Shim, W.S.; Park, I.; Lee, H.; Chang, T.; Kim, S.W.; Lee, D.S. Poly(D,L-lactic acid-co-glycolic acid)-b-poly(ethylene glycol)-b-poly (D,L-lactic acid-co-glycolic acid) triblock copolymer and thermoreversible phase transition in water. J. Biomed. Mater. Res. 2002, 61, 188–196.
8. Lee, D.S.; Shim, M.S.; Kim, S.W.; Lee, H.; Park, I.; Chang, T. Novel thermoreversible gelation of biodegradable PLGA-block-PEO-block-PLGA triblock copolymers in aqueous solution. Macromol. Rapid Commun. 2001, 22, 587–592.
9. Qiao, M.; Chen, D.; Ma, X.; Liu, Y. Injectable biodegradable temperature-responsive PLGA–PEG–PLGA copolymers: Synthesis and effect of copolymer composition on the drug release from the copolymer-based hydrogels. Int. J. Pharm. 2005, 294, 103–112.
10. Jeong, B.; Bae, Y.H.; Kim, S.W. Biodegradable thermosensitive micelles of PEG-PLGA-PEG triblock copolymers. Colloids Surf. B Biointerfaces 1999, 16, 185–193.
11. Jeong, B.; Bae, Y.H.; Kim, S.W. Thermoreversible gelation of PEG- PLGA- PEG triblock copolymer aqueous solutions. Macromolecules 1999, 32, 7064–7069.
12. Jian, X.; Hu, Y.; Zhou, W.; Xiao, L. Self-healing polyurethane based on disulfide bond and hydrogen bond. Polym. Adv. Technol. 2018, 29, 463–469.
13. Naskar, K.; Dey, A.; Maity, S.; Ray, P.P.; Ghosh, P.; Sinha, C. Biporous Cd (II) Coordination Polymer via in Situ Disulfide Bond Formation: Self-Healing and Application to Photosensitive Optoelectronic Device. Inorg. Chem. 2020, 59, 5518–5528.

14. Naoi, K.; Kawase, K.i.; Mori, M.; Komiyama, M. Electrochemistry of Poly (2, 2′-dithiodianiline): A New Class of High Energy Conducting Polymer Interconnected with S–S Bonds. J. Electrochem. Soc. 1997, 144, L173.

15. Liu, L.; Liu, P. Synthesis strategies for disulfide bond-containing polymer-based drug delivery system for reduction-responsive controlled release. Front. Mater. Sci. 2015, 9, 211–226.

16. Zhang, P.; Wu, J.; Xiao, F.; Zhao, D.; Luan, Y. Disulfide bond based polymeric drug carriers for cancer chemotherapy and relevant redox environments in mammals. Med. Res. Rev. 2018, 38, 1485–1510.

17. Wang, Y.-C.; Wang, F.; Sun, T.-M.; Wang, J. Redox-responsive nanoparticles from the single disulfide bond-bridged block copolymer as drug carriers for overcoming multidrug resistance in cancer cells. Bioconj. Chem. 2011, 22, 1939–1945.

18. Wong, D.Y.-K.; Hsiao, Y.-L.; Poon, C.-K.; Kwan, P.-C.; Chao, S.-Y.; Chou, S.-T.; Yang, C.-S. Glutathione concentration in oral cancer tissues. Cancer Lett. 1994, 81, 111–116.

19. Hoon Jeong, J.; Christensen, L.V.; Yockman, J.W.; Zhong, Z.; Engbersen, J.F.J.; Jong Kim, W.; Feijen, J.; Wan Kim, S. Reducible poly(amido ethylenimine) directed to enhance RNA interference. Biomaterials 2007, 28, 1912–1917.

20. Taranejoo, S.; Chandrasekaran, R.; Cheng, W.; Hourigan, K. Bioreducible PEI-functionalized glycol chitosan: A novel gene vector with reduced cytotoxicity and improved transfection efficiency. Carbohydr. Polym. 2016, 153, 160–168.

21. Davoodi, P.; Srinivasan, M.P.; Wang, C.-H. Synthesis of intracellular reduction-sensitive amphiphilic polyethyleneimine and poly(ε-caprolactone) graft copolymer for on-demand release of doxorubicin and p53 plasmid DNA. Acta Biomater. 2016, 39, 79–93.

22. Wang, L.-H.; Wu, D.-C.; Xu, H.-X.; You, Y.-Z. High DNA-Binding Affinity and Gene-Transfection Efficacy of Bioreducible Cationic Nanomicelles with a Fluorinated Core. Angew. Chem. Int. Ed. 2016, 55, 755–759.

23. Yang, H.; Bai, Y.; Yu, B.; Wang, Z.; Zhang, X. Supramolecular polymers bearing disulfide bonds. Polym. Chem. 2014, 5, 6439–6443.

24. Jeong, B.; Bae, Y.H.; Lee, D.S.; Kim, S.W. Biodegradable block copolymers as injectable drug-delivery systems. Nature 1997, 388, 860–862.

25. Jeong, B.; Bae, Y.H.; Kim, S.W. Drug release from biodegradable injectable thermosensitive hydrogel of PEG–PLGA–PEG triblock copolymers. J. Control Release 2000, 63, 155–163.

26. Li, X.; Wang, Y.; Chen, J.; Wang, Y.; Ma, J.; Wu, G. Controlled release of protein from biodegradable multi-sensitive injectable poly (ether-urethane) hydrogel. ACS Appl. Mater. Interfaces 2014, 6, 3640–3647.

27. Doppalapudi, S.; Jain, A.; Khan, W.; Domb, A.J. Biodegradable polymers—An overview. Polym. Adv. Technol. 2014, 25, 427–435.

28. Ramot, Y.; Haim-Zada, M.; Domb, A.J.; Nyska, A. Biocompatibility and safety of PLA and its copolymers. Adv. Drug Deliv. Rev. 2016, 107, 153–162.

29. Hogan, K.J.; Mikos, A.G. Biodegradable thermoresponsive polymers: Applications in drug delivery and tissue engineering. Polymer 2020, 211, 123063.

30. Ghahremankhani, A.A.; Dorkoosh, F.; Dinarvand, R. PLGA-PEG-PLGA Tri-Block Copolymers as In Situ Gel-Forming Peptide Delivery System: Effect of Formulation Properties on Peptide Release. Pharm. Dev. Technol. 2008, 13, 49–55.

10 Chapter

CONVERTING COMMERCIAL COPPER SULFIDE INTO INJECTABLE HYDROGELS FOR LOCALIZED PHOTOTHERMAL THERAPY

1. INTRODUCTION: AN OVERVIEW

The use of local therapy in tumor treatment has gained significant interest recently due to its benefits of targeted selectivity, controllability, and minimal systemic side effects [1,2]. Local therapy primarily consists of various methods such as local ablation (including radiofrequency ablation, irreversible electroporation, high-intensity focused ultrasound, local laser ablation, cryoablation, and chemical ablation), local phototherapy (including photothermal therapy and photodynamic therapy), local radioisotope therapy, local radiotherapy, and local chemotherapy. Several studies have demonstrated the efficacy of local therapy in various clinical scenarios, including tissue and function preservation [26], situations where surgery is not medically necessary [27], local treatment of metastatic tumors [28,29], and salvage treatment of recurrent tumors [30]. Therefore, local therapy is occasionally necessary to enhance the quality of life and increase the survival duration of patients with malignancies. Simultaneously, certain studies have documented the efficacy of combining local therapy with other therapeutic modalities, such as immunotherapy [31,32], hence highlighting the extensive potential of local therapy.

PTT, a growing local therapy, utilizes photoabsorbers to convert near-infrared (NIR) light energy into heat energy, resulting in the death of tumor cells through necrosis [16,33]. Near-infrared (NIR) photothermal therapy (PTT) is mostly performed within two specific ranges of wavelengths known as the first NIR window (NIR-I) and the second NIR window (NIR-II). The wavelength range of NIR-I spans from 750 nm to 1000 nm, while the range of NIR-II extends from 1000 nm to 1350 nm [34]. In comparison to the extensively researched NIR-I light, NIR-II light possesses greater penetration capability and better thermal safety power due to its less absorption and scattering in tissue. This characteristic grants it a significant advantage in effectively treating tumors located deeper within the body [34,35,36,37]. Furthermore, besides the significant progress of photothermal therapy (PTT) in basic research [38,39], it is highly promising that PTT, when administered locally, has also made successful advancements in a clinical trial for prostate cancer treatment [1]. This highlights the immense potential of PTT-based local therapy in clinical applications.

So far, numerous biomaterials have been created to serve as photoabsorbers. These include organic dyes [40,41,42], organic nanoparticles [43,44,45], noble metal materials [46,47,48], carbon materials [49,50], black phosphorus [51,52], and metal oxide and sulfides [53,54]. Metal sulfides, including CuS, Bi2S3, WS2, CoS, NiS, and FeS, have a high photothermal conversion efficiency due to the surface plasmon resonance effect [55,56]. Consequently, they have been extensively employed in photothermal therapy (PTT) in recent years [57,58,59,60,61,62]. Specifically, CuS, which has a high level of absorption in the near-infrared (NIR-II) biological window, has been widely utilized in NIR-II photothermal therapy (PTT) [63,64].

The existing photoabsorbers are mostly produced utilizing either bottom-up or top-down techniques, such as combining thermal oxidation etching with liquid exfoliation to create a solvent-dispersible system [65,66]. Nevertheless, these approaches encounter certain shared challenges, including intricate procedures, substantial time and energy requirements, limited exploitation of raw materials, and the absence of a universal strategy [67].

Commercial metal sulfides are a preferable alternative to the utilization of metal sulfides through intricate synthesis techniques. Commercial metal sulfides provide the benefit of being well-established industrial products with effective quality control and affordable prices. However, they have drawbacks such as the raw materials having a large particle size, being insoluble in water, and lacking suitability for biological applications. Our research has suggested a smart approach called the "turning solid into gel" strategy [68]. This involves distributing solid materials in alginate-Ca2+ hydrogel (ACH), which can convert the solid materials into an injectable hydrogel, hence increasing their bioavailability. Hence, it is intriguing to create adaptable commercial hydrogels based on metal sulfides that can serve as innovative photoabsorbers without the need for intricate synthesis.

In this study, we presented a straightforward and effective ACH platform for loading commercial CuS as a representative sample for local tumor NIR-II PTT (Figure 1). This platform offers a range of advantages for extremely efficient photothermal therapy (PTT), including an uncomplicated synthesis process, 100% loading efficiency, excellent biocompatibility, affordability, exceptional photothermal capacity, and remarkable adaptability. The CuS hydrogel (CSH) can be easily created by combining and agitating the components. CSH offers favorable syringeability and strong absorption in the near-infrared range (1064 nm) according to in vitro tests. Subsequently, CSH was utilized for in vivo photothermal therapy (PTT) investigations. The findings validate that this hydrogel exhibits excellent efficacy in eradicating tumor cells when exposed to gentle laser radiation, while also demonstrating minimal toxicity in both laboratory and living organism settings. As far as we know, this is the first instance where commercial CuS has been effectively used for extremely efficient photothermal therapy (PTT) in living organisms.

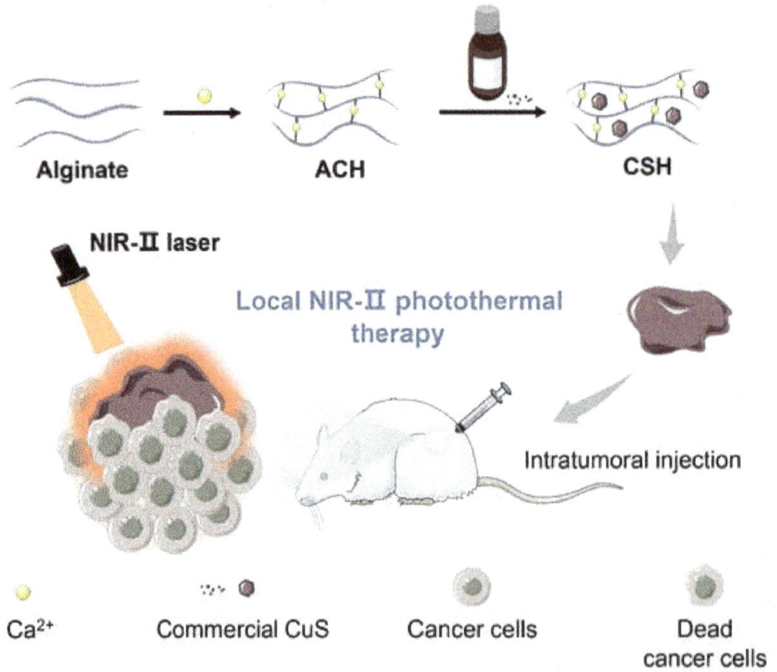

Figure 1. Schematic representation of the synthesis of CSH as a PTT agent for local NIR-II PTT in vivo.

2. RESULTS AND DISCUSSION

2.1. Synthesis and Characterization of CSH

Firstly, alginate solution and Ca2+ were mixed to produce ACH within 1 min based on their strong coordination interaction (Figure S1). Then, CSH was obtained by dispersing commercial CuS powder into ACH. To investigate the loading capacity of ACH, increasing concentrations of CSH were employed. The maximum loading capacity was 480 mg CuS/mL (Figure 2A). The long-term stability of CSH was also monitored (Table S1 and Figure S2). All concentrations of CSH were stable for more than 4 days, and concentrations of 90 mg CuS/mL and below were still stable after 14 days. Considering that excellent syringeability is essential to potential biological applications, we investigated the maximum loading capacity of CSH capable of fluently being injected with different diameters of syringe needles. The results showed that the maximum injectable concentrations for 0.45, 0.5, 0.6, and 1.2 mm syringe needles were 120, 240, 480, and 480 mg CuS/mL, respectively, and a "TMU" pattern could be written by a 0.45 mm syringe with 20 mg CuS/mL CSH (Figure 2B), which proved its excellent syringeability due to the shear-dependent and reversible gel–sol transition (Figure S3) [69].

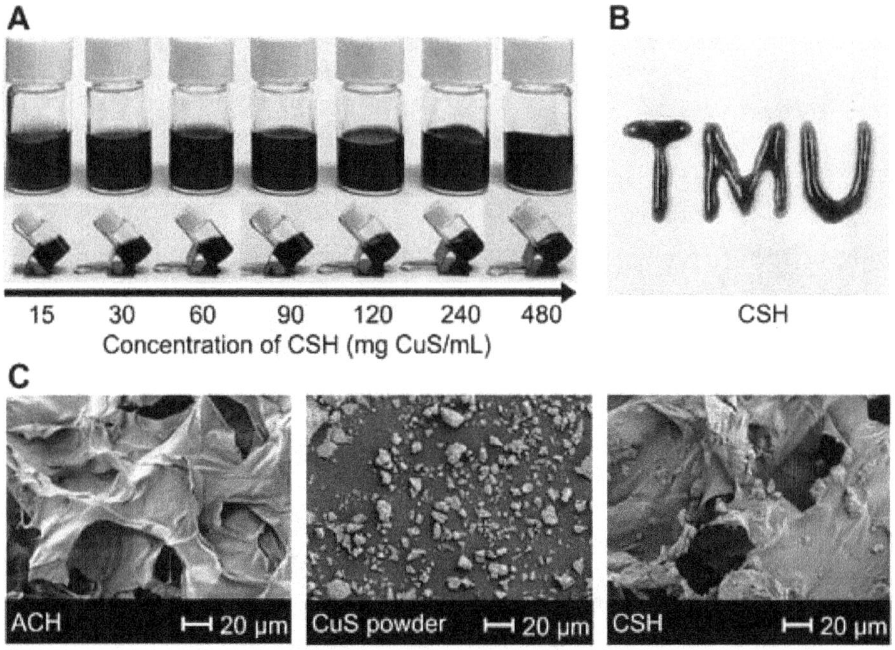

Figure 2. (A) Standing and oblique photos of different concentrations of CSH taken immediately after the preparation. (B) "TMU" formed by CSH (20 mg CuS/mL) through a 0.45 mm syringe needle. (C) SEM images of ACH, CuS particles, and CSH (20 mg CuS/mL).

Rheological experiments showed that the storage moduli (G') of ACH and CSH were higher than their loss moduli (G"), demonstrating that ACH and CSH were in a gel state with a relatively weak mechanical strength and flexible shape, which made them easily injectable (Figure S4). As the essential components of CSH, the swelling ratio and degradation behavior of ACH were further investigated. The swelling test showed that ACH could reach swelling equilibrium in 10 min in PBS (pH = 7.4). The swelling ratio of ACH was as high as 13,342.6% (Figure S5), which suggested that the internal cross-linking points of ACH were relatively few, the cross-linking density was low, and the water absorption capacity was strong. According to the ACH degradation curve (Figure S6), the degradation rate of ACH in PBS (pH = 7.4) was 51% after 7 days, which showed its excellent degradability.

The scanning electron microscope (SEM) images of ACH, CSH, and commercial CuS powder were characterized, and they indicated that the CuS particles were dispersed in the ACH with a porous structure (Figure 2C).

2.2. Photothermal Performance of CSH In Vitro

To evaluate the photothermal efficiency of CSH in vitro, different concentrations of CSH were treated with NIR-II laser irradiation (1064 nm, 1 W/cm2) for 5 min, and an infrared thermal camera was used to record the temperature elevations. Under NIR-II laser irradiation, CSH showed good photothermal capacity (Figure 3A). The temperature enhancement of CSH with different concentrations increased from 17.3 °C to 38.1 °C, while the temperature increase of ACH and PBS was just 6.4 °C. The thermal images also demonstrate the outstanding photothermal ability of CSH (Figure 3B). After undergoing the heating–cooling process three times, the heating capacity of CSH did not significantly change (Figure 3C), which indicates that CSH has good photothermal stability under NIR-II laser irradiation. Therefore, not only can the prepared CSH efficiently transform NIR laser energy to heat energy, but it can also remain stable after repeated laser illumination.

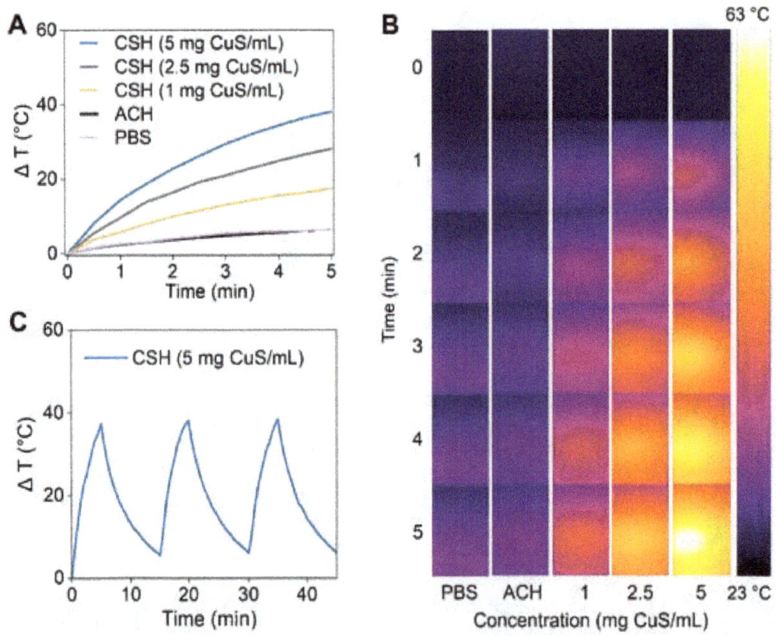

Figure 3. (A) Photothermal heating curves of PBS, ACH, and CSH irradiated by NIR-II laser (1064 nm, 1 W/cm2). (B) The thermal images of PBS, ACH, and CSH under NIR-II laser irradiation taken by an infrared thermal camera. (C) The photothermal stability of CSH.

2.3. Cytotoxicity and Cellular Uptake of CSH

CSH, which had a great photothermal efficacy under 1064 nm irradiation, super-large loading capacity, and excellent stability, was capable of being used for further studies. To evaluate its cytotoxicity, different concentrations of CSH were added to 4T1 cells in 96-well plates, and the cells were continued to be cultured for 24 h. Then, the cell viabilities were calculated through a standard MTT assay [3-(4,5-Dimethylthiazol-2-yl)-2,5-diphenyltetrazolium bromide, MTT]. The cell viability was as high as 88.8% after incubation with a high concentration of CSH (1 mg CuS/mL), which indicated the low cytotoxicity of CSH (Figure 4A). The cellular uptake experiment proved that CuS particles in CSH could not be uptaken by cells due to their big size (Figure S7), which illustrates that the mechanism of PTT based on CHS is deduced heat conduction instead of the direct interaction of cell and CSH.

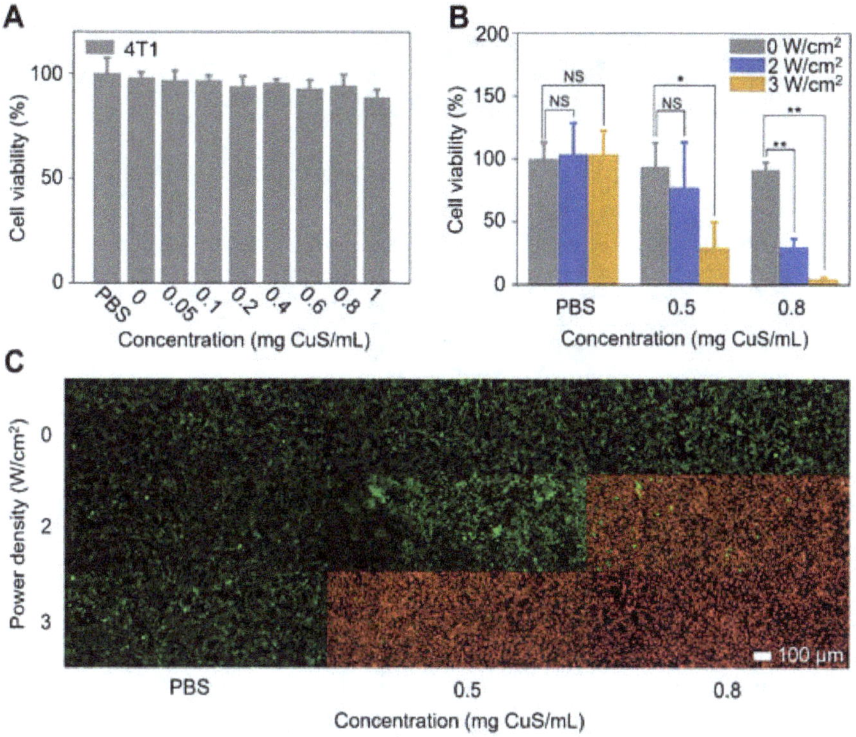

Figure 4. **(A) Cytotoxicity of 4T1 cells incubated with different concentrations of CSH or PBS. (B) Viability of 4T1 cells incubated with PBS or CSH (0.5, 0.8 mg CuS/mL) and irradiated by NIR-II laser (1064 nm: 0, 2, or 3 W/cm2) (* $p < 0.05$, ** $p < 0.01$). (C) Dead/live cell staining test of 4T1 cells treated with PBS or CSH (0.5, 0.8 mg CuS/mL) and irradiated by NIR-II laser (1064 nm: 0, 2, or 3 W/cm2).**

2.4. In Vitro PTT of CSH

Due to the low cytotoxicity of CSH, PTT of 4T1 cells using the hydrogel was investigated with an MTT assay and live and dead cell staining. As the standard procedure, 4T1 cells were cultured in a 96-well plate at 37 °C for 24 h, and different concentrations of CSH (0.5 and 0.8 mg CuS/mL) or PBS were added and incubated with the cells at 37 °C. After 1 h, the cells were irradiated by a 1064 nm laser (0, 2, or 3 W/cm2) for 5 min. The 4T1 cell viabilities showed CSH-concentration- and laser-power density-dependent deceases. After being treated with both 0.8 mg CuS/mL of CSH and 1064 nm laser irradiation (3 W/cm2), 4T1 cell viability dropped to less than 4%. However, the viability of 4T1 cells treated with only CSH or laser irradiation remained approximately 100% (Figure 4B). The fluorescent images of live and dead cells, which were stained by calcein acetoxymethyl ester (calcein AM) and propidium iodide (PI), respectively, also showed that the 4T1 cells were significantly destructed after the combined treatments (Figure 4C). These results prove that CSH has an excellent photothermal effect on tumor cells with 1064 nm laser irradiation.

2.5. Intratumoral Retention Test of CSH

In order to assess the intratumoral retention ability of CSH, computer tomography (CT) scans were carried out in vitro and in vivo under a voltage of 120 kV (clinical use). Although CSH was considered to have a weak CT value attenuation (Figure 5A,B), after intratumoral injection, it could still be found at the tumor site with CT scans. During the 2 days of CT monitoring that followed, no significant change was found in the CT value or in the morphology of CSH (Figure 5C,D), proving its good retention ability at tumor sites.

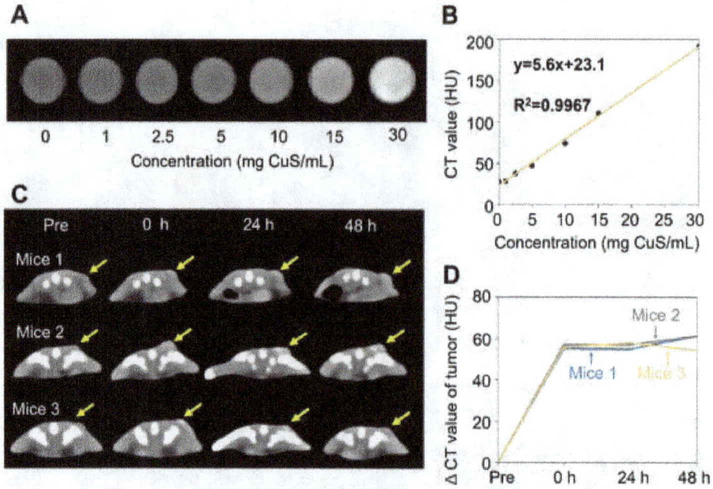

Figure 5. (A) CT images of CSH at different concentrations. (B) CT value curve of CSH. (C) CT scan of BALB/c mice before and 0, 24, and 48 h after being intratumorally injected with 20 mg CuS/mL of CSH (n = 3). (D) CT value (Hounsfield, HU) changing curves on the tumor site of BALB/c mice.

2.6. In Vivo PTT of CSH

To minimize the damage to surrounding tissues, a mild laser power (0.3 W/cm2) was employed for in vivo PTT. CSH with a concentration of 20 mg CuS/mL, which can cause a significant temperature rise in vitro, was used to guarantee effective tumor ablation (Figure S8).

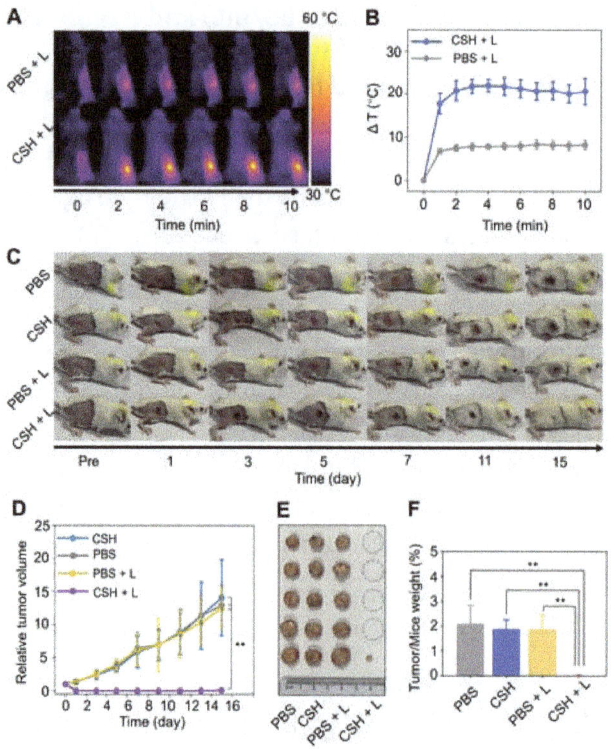

Figure 6. (A) Thermal images of tumor-bearing mice treated with only PBS or both CSH and NIR-II laser irradiation. (B) Photothermal heating curves of tumor sites taken with an infrared thermal camera. (C) Tumor-monitoring photography of mice in various groups. (D) Relative tumor volume curves of mice in different groups (n = 5 in each group, ** $p < 0.01$). (E) Excised tumors from the mice on the 15th day of the observation period. (F) Ratio of final tumor weight to final body weight of mice in different groups ($p < 0.01$).**

To evaluate the in vivo photothermal tumor therapy efficacy of CSH in the NIR-II bio-window (1064 nm), BALB/c mice were grouped according to different treatments (n = 5) as follows: (1) only PBS; (2) only CSH; (3) PBS + laser; and (4) CSH + laser. In comparison with the control, there was a noticeable temperature increase at the tumor site after being injected with CSH and irradiated with a 1064 nm laser (Figure 6A,B).

Tumor sizes were measured every 2 days to evaluate the anti-tumor capacity. The results showed that the tumor growth of mice treated with both CSH and laser irradiation was effectively inhibited, and the tumors were eliminated after PTT (Figure 6C). There was tumor recurrence in only one mouse in the combined treatment group, and the recurred tumor was significantly smaller than that in the other groups. In contrast, the tumors in the other groups multiplied, and the final tumor volumes after 15 days of growth were about 12.9, 14.1, and 12.5 times larger than the initial tumor volume in groups 1, 2, and 3, respectively (Figure 6D). The tumors were dissected and photographed on the 15th day (Figure 6E). The dissected tumors were weighed, and the ratio of tumor weight to mouse body weight in each group was calculated (Figure 6F), which further revealed that the tumors were obviously suppressed by CSH-based PTT. These results illustrate that CSH can wreck tumors entirely due to its high thermal efficiency in vivo.

2.7. In Vivo Toxicity of CSH

To assess the systemic toxicity of CSH in vivo, the weight monitoring, blood biochemistry analysis, and H&E staining of major organs of the mice were accomplished.

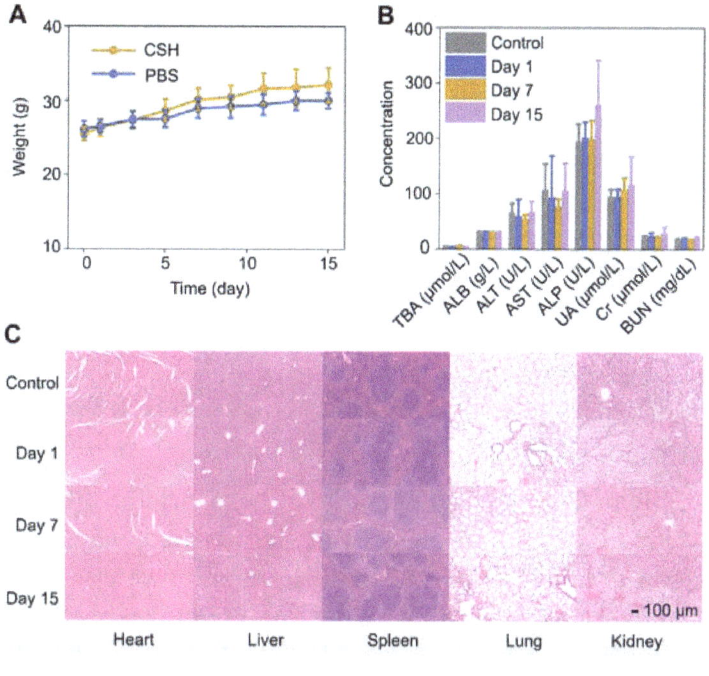

Figure 7. (A) Weight-changing curves of mice after being subcutaneously injected with PBS and 20 mg CuS/mL of CSH (n = 5 in each group). (B) Blood biochemical indexes of mice measured after being treated with PBS for 15 days and CSH (20 mg CuS/mL) for 1, 7, and 15 days (n = 5 in each group). (C) H&E-stained images of major organs of Kunming mice acquired after being treated with PBS for 15 days and CSH (20 mg CuS/mL) for 1, 7, and 15 days.

The weight monitoring results displayed no noticeable difference in body weight among the mice with various treatments (Figure 7A). The blood biochemistry analysis indicated that the liver and kidney function indexes of the mice were entirely within the normal range (Figure 7B), and no evident inflammatory lesion or organ damage was found in all major organs of the mice (Figure 7C). All of the above results confirm that CSH has good biocompatibility, which makes it a promising PTT agent with good biosafety and photothermal efficacy in vivo.

3. CONCLUSIONS

In conclusion, according to the "turning solid into gel" strategy, a robust metal sulfide hydrogel system was established to load commercial metal sulfide powders for high-efficiency tumor PTT. As a representative metal sulfide, commercial CuS powder was studied in depth. The obtained CSH was verified to have good stability, favorable syringeability, potent photothermal efficacy, and excellent retention capability at the injection site. Due to the deeper tissue penetration of NIR-II light, further studies were investigated using 1064 nm laser irradiation. The follow-up experimentations in vitro and in vivo showed the CSH to have negligible toxicity and a high photothermal killing effect on tumor cells under the irradiation of the 1064 nm laser. Therefore, as a new method of photothermal agent preparation, transforming commercial sulfides into injectable hydrogels can help to save costs, improve accuracy, and raise efficiency without worrying about toxicity, all of which give it great hope for clinical transformation.

4. MATERIALS AND METHODS

4.1. Materials

CaCl2 and sodium alginate (200 ± 20 mPa s) were obtained from Aladdin Biochemical Technology Co., Ltd. (Shanghai, China). CuS was purchased from Sigma-Aldrich trade Co., Ltd. (Shanghai, China). Fetal Bovine Serum (FBS) was provided by Lanzhou Minhai Bio-Engineering Co., Ltd. Dulbecco's Modified Eagle Medium (DMEM) was obtained from ThermoFisher Instrument Co., Ltd. (Suzhou, China). MTT was bought from Aladdin Biochemical Technology Co., Ltd. (Shanghai, China). Calcein AM and PI were provided by Dojindo Chemical L.L.C. (Shanghai, China). DMSO was purchased from Concord Technology Co., Ltd. (Tianjin, China). Ultrapure water was bought from Wahaha Group Co., Ltd. (Hangzhou, China).

4.2. Synthesis of ACH and CSH

Typically, 0.5 mL of sodium alginate (10 mg/mL) and 0.05 mL of CaCl2 (10 mg/mL) were mixed with 0.45 mL H2O to prepare ACH. Then, commercial CuS powder was added to ACH, and the system was stirred for 15 min to obtain CSH.

4.3. Stability Assessment of CSH

The homogeneous stability of CSH was evaluated for 2 weeks. Briefly, different concentrations of CSH (15, 30, 60, 90, 120, 240, 480 mg CuS/mL) were placed in vials, respectively. If the stability time was less than 10 min or the sulfide could not be dispersed in ACH, the mixture was regarded as overloaded. The stable concentrations of CSH were monitored for 14 days and photographed at different time points (10 min, 20 min, 30 min, 1 h, 2 h, 3 h, 4 h, 6 h, 8 h, 10 h, 12 h, 14 h, 16 h, 18 h, 20 h, 24 h, and then every day). During the monitoring period, if the hydrogel was found to be layered, it was regarded as precipitation.

4.4. Syringeability of CSH

The syringeability of CSH was evaluated using four sizes of syringe needles (26 G, 0.45 mm; 25 G, 0.5 mm; 23 G, 0.6 mm; 18 G, 1.2 mm). Different concentrations of CSH were extruded through the various sizes of syringe needles. For every size of syringe, the maximum injectable concentration was recorded, and a "TMU" (an abbreviation of "Tianjin Medical University") was written by its maximum injectable concentration. Then, a "TMU" was formed by CSH (20 mg CuS/mL) through a 0.45 mm syringe needle, which was used for in vivo PTT.

4.5. Characterization

Rheology experiments of ACH and CSH (20 mg CuS/mL) were conducted on a DHR-2 rheometer (TA Instruments), with a strain amplitude of 1% and an angular frequency of 10 rad/s for dynamic oscillatory time sweep measurements.

The swelling ratio and degradation behavior of ACH were investigated according to a previous study [70]. To calculate the swelling ratio, lyophilized ACH was weighed (recorded as M0), immersed in PBS (pH = 7.4), and incubated in an incubator shaker at a shaking speed of 100 rpm at 37 °C. The swelled ACH was removed, and the surface water was wiped out. Then, the collected ACH was weighed at a specific time interval (recorded as Mt). The swelling ratio (%, w/w) was calculated using the equation (Mt - M0)/M0 × 100. The experiments were carried out in triplicate to obtain an average value. The degradation behavior of ACH was assessed in PBS (pH = 7.4). The lyophilized ACH was weighed (recorded as Mi) and completely immersed in PBS, and then it was degraded in an incubator shaker at 37 °C and 100 rpm. After different time intervals (1, 3, or 7 days), ACH was washed with ultrapure water to remove PBS, freeze-dried, and weighed (recorded as Mf). The degradation rate (%, w/w) was calculated using the equation (Mi - Mf)/Mi × 100.

Field-emission scanning electron microscopy (FE-SEM) images of ACH, CuS powder, and CSH (20 mg CuS/mL) were acquired under a 2 kV accelerating voltage on a Gemini SEM 300 (ZEISS, Germany) microscope.

4.6. Photothermal Performance In Vitro

In order to evaluate the photothermal efficacy of CSH in vitro, PBS or different concentrations of CSH (0, 1, 2.5, and 5 mg CuS/mL) with a volume of 1 cm3 were placed in cuvettes with a base area of 1 cm2. Then, cuvettes were irradiated with a 1064 nm laser (1 W/cm2) for 5 min, and temperature elevations were recorded using an infrared thermal camera. In order to test its photothermal stability, CSH (5 mg CuS/mL, 1 mL) was put into a cuvette and irradiated using NIR-II (1064 nm) laser with a power density of 1 W/cm2 for 5 min, and then the system was cooled for 10 min to bring the temperature close to room temperature; the process was repeated three times.

4.7. Cell Culture and Animals

The growth and metastasis of 4T1 cells in BALB/c mice are similar to those of human breast cancer, making the cells a relatively classical and widely used cell line to test the therapeutic effects on tumors [71]. Therefore, the 4T1 cell line was used to study CSH in vitro and in vivo. 4T1 cells were cultured in a culture medium with 90% DMEM and 10% FBS. Cells were cultured in a humidified incubator (5% CO2 and 37 °C), and the culture medium was refreshed at 1–2 day intervals. Kunming mice and BALB/c mice were purchased from Beijing HFK Bioscience Co., Ltd. (Beijing, China). All animal experiments were performed according to the protocols established by the Animal Care and Use Committee of Tianjin Medical University, and all experimental operations were approved by the Animal Care and Use Committee.

4.8. Cytotoxicity and Cellular Uptake Assay

To determine the potential cytotoxic effects of CSH, 4T1 cells (1 × 104 per well) were cultured in 96-well plates with 200 μL of cell culture medium per well for 24 h. Then, after the exchange of the cell medium, PBS or different concentrations of CSH (0, 0.05, 0.1, 0.2, 0.4, 0.6, 0.8, 1 mg CuS/mL) were added to the wells. After 24 h incubation, cell viabilities were evaluated via a standard MTT test. After the cells were washed with PBS, new cell medium and MTT (10 μL, 5 mg/mL) were added and incubated with the cells for 4 h; then, the supernatant was discarded, and 120 μL of DMSO per well was added. Finally, the wells' absorptions at 490 nm were measured using a microplate reader.

The cellular uptake mechanism of CSH was also investigated. In brief, 4T1 cells (1 × 104 per well) were cultured in 96-well plates. After 24 h, PBS or different concentrations of CSH (0, 0.05, 0.1, 0.2, 0.4, 0.6, 0.8, 1 mg CuS/mL) were added and co-incubated with the cells for another 24 h. Then, the cells were washed with PBS, and 120 μL of PBS per well was added. Finally, the cells were observed under a microscope.

4.9. In Vitro Photothermal Cytotoxicity Study

The photothermal cell killing ability of CSH under 1064 nm laser irradiation was evaluated using the MTT assay. 4T1 cells (1.4 × 104 per well) were incubated in a 96-well plate for 24 h. After being washed with PBS, the 4T1 cells were treated with PBS or different concentrations of CSH (0.5 or 0.8 mg CuS/mL) for 1 h, and they were irradiated with varying densities of power of 1064 nm laser (0, 2, or 3 W/cm2) for 5 min. Then, cell viabilities were measured using the MTT assay, and the absorption of each well at 490 nm was recorded using a microplate reader.

4.10. Live/Dead Cells Staining Test

To further investigate the PTT efficacy in vitro, 4T1 cells (1.4 × 104 per well) were incubated in 96-well plates for 24 h, and CSH (0.5, 0.8 mg CuS/mL) or PBS was added and co-incubated with 4T1 cells for 1 h. Then, 4T1 cells were exposed to NIR-II laser (1064 nm, 0, 2, or 3 W/cm2) for 5 min, thoroughly washed with PBS twice, and stained with calcein AM and PI. Fluorescent images were recorded with an inverted luminescence microscope.

4.11. Intratumoral Retention Test of CSH

In vitro and in vivo CT scans were carried out via a clinical X-ray CT (SOMATOM Force, Siemens healthineers, Erlangen, Germany) under a clinical voltage (120 kV) [72]. CSHs with different concentrations (0, 1, 2.5, 5, 10, 15, 30 mg CuS/mL) were prepared, and then CT images of CSH were collected. For in vivo CT imaging, 50 μL of CSH (20 mg CuS/mL) was intratumorally injected into BALB/c mice (n = 3). Then, the mice were scanned pre-injection and after injection at different time points (0 h, 24 h, and 48 h). CT values were measured using Radiant DICOM Viewer software.

4.12. Anti-Tumor Assessment In Vivo

To ensure biosafety, a mild laser power (0.3 W/cm2) and CSH with a concentration of 20 mg CuS/mL were used for in vivo PTT. To verify the in vitro heating effect, CSH (20 mg CuS/mL) or PBS was made into 50 μL droplets, and they were irradiated using a 1064 nm laser (0.3 W/cm2) for 5 min. Thermal images of them were taken, and photothermal heating curves were obtained. Then, to explore the anti-tumor ability of CSH with 1064 nm laser irradiation, tumor-bearing BALB/c mice were divided into 4 groups (n = 5) as follows: (1) only PBS, (2) only CSH, (3) PBS + laser, and (4) CSH + laser. Mice in Group 1 were intratumorally injected with PBS (50 μL). Mice in Group 2 were intratumorally injected with CSH (20 mg CuS/mL, 50 μL). Mice in Group 3 were intratumorally injected with PBS and exposed to 1064 nm laser irradiation (0.3 W/cm2) for 10 min. Mice in Group 4 were intratumorally injected with CSH (20 mg CuS/mL, 50 μL) and exposed to 1064 nm laser irradiation (0.3 W/cm2) for 10 min. The hyperthermia effect on tumor site was carefully recorded using an infrared thermal camera. Then, tumor sizes were measured and recorded every 2 days. Tumor volume was calculated using the following formula: V = a × b2/2, where a and b mean the longest and shortest diameters, respectively. The relative volume of the tumors was the ratio of the day's volume to the initial volume. Photos of tumors in all groups were taken every 2 days, and the tumors were removed and weighed on day 15 after the treatment.

4.13. Statistics

The differences between groups were studied using one-way ANOVA, and "p" value < 0.05 was considered as statistically significant. All analyses were conducted using GraphPad Prism 8.0.2 software.

4.14. In Vivo Biosafety Analysis

In order to assess the biosafety of CSH in living organisms, Kunming mice were subjected to weight monitoring, blood biochemistry analysis, and H&E staining. In order to track changes in body weight, Kunming mice ($n = 5$) were subcutaneously injected with either CSH (50 μL, 20 mg CuS/mL) or PBS. The mice's body weights were then recorded every two days for a period of 15 days. In order to conduct blood biochemistry study and H&E staining, a group of Kunming mice were injected subcutaneously with PBS ($n = 5$) or CSH (50 μL, 20 mg CuS/mL) ($n = 15$). Five mice from the hydrogel-injected group were dissected on the 1st, 7th, and 15th days, while mice from the PBS group were dissected only on the 15th day. Following the dissection of the mice, their principal organs (namely, the heart, lung, spleen, liver, and kidney) were extracted and treated with hematoxylin and eosin staining. Additionally, blood samples were obtained. The blood samples were subjected to centrifugation at a speed of 3000 revolutions per minute for a duration of 10 minutes in order to separate and collect the serum that settled on top of the samples. Subsequently, the blood biochemistry biomarkers were examined, encompassing albumin (ALB), total bile acid (TBA), aspartate aminotransferase (AST), alanine aminotransferase (ALT), and alkaline phosphatase (ALP) to assess liver function, as well as uric acid (UA), urea nitrogen (BUN), and serum creatinine (Cr) to evaluate kidney function.

AT A GLANCE

Photothermal therapy (PTT) is an effective localized treatment that is becoming increasingly significant in the treatment of tumors. In order to optimize the effectiveness of photothermal therapy (PTT), several near-infrared photoabsorbers have been created. Metal sulfides have garnered significant attention due to their notable stability and exceptional photothermal conversion efficiency. Nevertheless, the current techniques for synthesizing metal-sulfide-based photoabsorbers are plagued by the disadvantages of intricate procedures, limited usage of raw materials, and lack of universality. In this study, we have presented a versatile and adaptable approach that can convert commercially available metal sulfides into injectable hydrogels for localized photothermal therapy (PTT). We selected copper sulfide (CuS) as a representative example because to its strong absorption in the second-window near-infrared range (1064 nm). We conducted a comprehensive study to examine its features both in vitro and in vivo. A CuS hydrogel with excellent syringeability was produced by dispersing commercially available CuS particles, which act as photoabsorbers, within an alginate-Ca^{2+} hydrogel. This synthesis technique demonstrates the exceptional advantages of an extremely straightforward synthesis procedure, achieving a loading efficiency of 100%, displaying excellent biocompatibility, being cost-effective, possessing outstanding photothermal capacity, and exhibiting broad applicability. The in vitro investigations demonstrated that the hydrogel possesses advantageous photothermal heating capability, and it clearly eradicated tumor cells when exposed to 1064 nm laser irradiation. Following direct injection into the tumor, the hydrogel containing CuS particles of significant size exhibited a high degree of accumulation inside the tumor tissues. This led to the successful implementation of localized photothermal therapy (PTT) with the use of a low-intensity laser (0.3 W/cm^2). The formulated approach for the production of CuS hydrogel offers an innovative method to employ commercially available metal sulfides for various biological uses.

REFERENCES

1. Rastinehad, A.R.; Anastos, H.; Wajswol, E.; Winoker, J.S.; Sfakianos, J.P.; Doppalapudi, S.K.; Carrick, M.R.; Knauer, C.J.; Taouli, B.; Lewis, S.C.; et al. Gold nanoshell-localized photothermal ablation of prostate tumors in a clinical pilot device study. Proc. Natl. Acad. Sci. USA 2019, 116, 18590–18596.
2. Cai, X.; Liu, B. Aggregation-Induced Emission: Recent Advances in Materials and Biomedical Applications. Angew. Chem. Int. Ed. 2020, 59, 9868–9886.
3. Kim, T.H.; Koh, Y.H.; Kim, B.H.; Kim, M.J.; Lee, J.H.; Park, B.; Park, J.-W. Proton beam radiotherapy vs. radiofrequency ablation for recurrent hepatocellular carcinoma: A randomized phase III trial. J. Hepatol. 2021, 74, 603–612.
4. Lee, S.; Kang, T.W.; Song, K.D.; Lee, M.W.; Rhim, H.; Lim, H.K.; Kim, S.Y.; Sinn, D.H.; Kim, J.M.; Kim, K.; et al. Effect of Microvascular Invasion Risk on Early Recurrence of Hepatocellular Carcinoma After Surgery and Radiofrequency Ablation. Ann. Surg. 2021, 273, 564–571.
5. Zhou, W.; Gao, Y.; Tong, Y.; Wu, Q.; Zhou, Y.; Li, Y. Anlotinib enhances the antitumor activity of radiofrequency ablation on lung squamous cell carcinoma. Pharmacol. Res. 2021, 164, 105392.
6. Ting, F.; Tran, M.; Böhm, M.; Siriwardana, A.; Van Leeuwen, P.J.; Haynes, A.M.; Delprado, W.; Shnier, R.; Stricker, P.D. Focal irreversible electroporation for prostate cancer: Functional outcomes and short-term oncological control. Prostate Cancer Prostatic Dis. 2016, 19, 46–52.
7. Martin, R.C.G., II; Kwon, D.; Chalikonda, S.; Sellers, M.; Kotz, E.; Scoggins, C.; McMasters, K.M.; Watkins, K. Treatment of 200 Locally Advanced (Stage III) Pancreatic Adenocarcinoma Patients With Irreversible Electroporation Safety and Efficacy. Ann. Surg. 2015, 262, 486–494.
8. Zhao, J.; Wen, X.; Tian, L.; Li, T.; Xu, C.; Wen, X.; Melancon, M.P.; Gupta, S.; Shen, B.; Peng, W.; et al. Irreversible electroporation reverses resistance to immune checkpoint blockade in pancreatic cancer. Nat. Commun. 2019, 10, 899.
9. Feijoo, E.R.; Sivaraman, A.; Barret, E.; Sanchez-Salas, R.; Galiano, M.; Rozet, F.; Prapotnich, D.; Cathala, N.; Mombet, A.; Cathelineau, X. Focal High-intensity Focused Ultrasound Targeted Hemiablation for Unilateral Prostate Cancer: A Prospective Evaluation of Oncologic and Functional Outcomes. Eur. Urol. 2016, 69, 214–220.
10. Natarajan, S.; Raman, S.; Priester, A.M.; Garritano, J.; Margolis, D.J.; Lieu, P.; Macairan, M.L.; Huang, J.; Grundfest, W.; Marks, L.S. Focal Laser Ablation of Prostate Cancer: Phase I Clinical Trial. J. Urol. 2016, 196, 68–75.
11. Llovet, J.M.; De Baere, T.; Kulik, L.; Haber, P.K.; Greten, T.F.; Meyer, T.; Lencioni, R. Locoregional therapies in the era of molecular and immune treatments for hepatocellular carcinoma. Nat. Rev. Gastroenterol. Hepatol. 2021, 18, 293–313.
12. Nault, J.-C.; Sutter, O.; Nahon, P.; Ganne-Carrie, N.; Seror, O. Percutaneous treatment of hepatocellular carcinoma: State of the art and innovations. J. Hepatol. 2018, 68, 783–797.
13. Thompson, R.H.; Atwell, T.; Schmit, G.; Lohse, C.M.; Kurup, A.N.; Weisbrod, A.; Psutka, S.P.; Stewart, S.B.; Callstrom, M.R.; Cheville, J.C.; et al. Comparison of Partial Nephrectomy and Percutaneous Ablation for cT1 Renal Masses. Eur. Urol. 2015, 67, 252–259.
14. Ahmed, M.; Brace, C.L.; Lee, F.T., Jr.; Goldberg, S.N. Principles of and Advances in Percutaneous Ablation. Radiology 2011, 258, 351–369.

15. Ahmed, M.; Solbiati, L.; Brace, C.L.; Breen, D.J.; Callstrom, M.R.; Charboneau, J.W.; Chen, M.-H.; Choi, B.I.; de Baere, T.; Dodd, G.D., III; et al. Image-guided Tumor Ablation: Standardization of Terminology and Reporting Criteria-A 10-Year Update. Radiology 2014, 273, 241–260.

16. Li, X.; Lovell, J.F.; Yoon, J.; Chen, X. Clinical development and potential of photothermal and photodynamic therapies for cancer. Nat. Rev. Clin. Oncol. 2020, 17, 657–674.

17. Liu, Y.; Bhattarai, P.; Dai, Z.; Chen, X. Photothermal therapy and photoacoustic imaging via nanotheranostics in fighting cancer. Chem. Soc. Rev. 2019, 48, 2053–2108.

18. Liu, C.; Cao, Y.; Cheng, Y.; Wang, D.; Xu, T.; Su, L.; Zhang, X.; Dong, H. An open source and reduce expenditure ROS generation strategy for chemodynamic/photodynamic synergistic therapy. Nat. Commun. 2020, 11, 1735.

19. Xu, C.; Pu, K. Second near-infrared photothermal materials for combinational nanotheranostics. Chem. Soc. Rev. 2021, 50, 1111–1137.

20. Aaronson, D.S.; Yamasaki, I.; Gottschalk, A.; Speight, J.; Hsu, I.C.; Pickett, B.; Roach, M.; Shinohara, K. Salvage permanent perineal radioactive-seed implantation for treating recurrence of localized prostate adenocarcinoma after external beam radiotherapy. BJU Int. 2009, 104, 600–604.

21. Chao, Y.; Xu, L.; Liang, C.; Feng, L.; Xu, J.; Dong, Z.; Tian, L.; Yi, X.; Yang, K.; Liu, Z. Combined local immunostimulatory radioisotope therapy and systemic immune checkpoint blockade imparts potent antitumour responses. Nat. Biomed. Eng. 2018, 2, 611–621.

22. Formenti, S.C.; Rudqvist, N.-P.; Golden, E.; Cooper, B.; Wennerberg, E.; Lhuillier, C.; Vanpouille-Box, C.; Friedman, K.; de Andrade, L.F.; Wucherpfennig, K.W.; et al. Radiotherapy induces responses of lung cancer to CTLA-4 blockade. Nat. Med. 2018, 24, 1845–1851.

23. Kerkmeijer, L.G.W.; Groen, V.H.; Pos, F.J.; Haustermans, K.; Monninkhof, E.M.; Smeenk, R.J.; Kunze-Busch, M.; de Boer, J.C.J.; van Zijp, J.V.D.V.; van Vulpen, M.; et al. Focal Boost to the Intraprostatic Tumor in External Beam Radiotherapy for Patients With Localized Prostate Cancer: Results From the FLAME Randomized Phase III Trial. J. Clin. Oncol. 2021, 39, 787–796.

24. Rodriguez-Ruiz, M.E.; Vanpouille-Box, C.; Melero, I.; Formenti, S.C.; Demaria, S. Immunological Mechanisms Responsible for Radiation-Induced Abscopal Effect. Trends Immunol. 2018, 39, 644–655.

25. Norouzi, M.; Nazari, B.; Miller, D.W. Injectable hydrogel-based drug delivery systems for local cancer therapy. Drug Discov. Today 2016, 21, 1835–1849.

26. Salagierski, M.; Wojciechowska, A.; Zajac, K.; Klatte, T.; Thompson, R.H.; Cadeddu, J.A.; Kaouk, J.; Autorino, R.; Ahrar, K.; Capitanio, U.; et al. The Role of Ablation and Minimally Invasive Techniques in the Management of Small Renal Masses. Eur. Urol. Oncol. 2018, 1, 395–402.

27. Nieuwenhuizen, S.; Dijkstra, M.; Puijk, R.S.; Geboers, B.; Ruarus, A.H.; Schouten, E.A.; Nielsen, K.; de Vries, J.J.J.; Bruynzeel, A.M.E.; Scheffer, H.J.; et al. Microwave Ablation, Radiofrequency Ablation, Irreversible Electroporation, and Stereotactic Ablative Body Radiotherapy for Intermediate Size (3–5 cm) Unresectable Colorectal Liver Metastases: A Systematic Review and Meta-analysis. Curr. Oncol. Rep. 2022, 24, 793–808.

28. Han, K.; Kim, J.H.; Yang, S.G.; Park, S.H.; Choi, H.-K.; Chun, S.-Y.; Kim, P.N.; Park, J.; Lee, M. A Single-Center Retrospective Analysis of Periprocedural Variables Affecting Local Tumor Progression after Radiofrequency Ablation of Colorectal Cancer Liver Metastases. Radiology 2021, 298, 212–218.

29. Nam, J.; Son, S.; Ochyl, L.J.; Kuai, R.; Schwendeman, A.; Moon, J.J. Chemo-photothermal therapy combination elicits anti-tumor immunity against advanced metastatic cancer. Nat. Commun. 2018, 9, 1074.
30. Valerio, M.; Ahmed, H.U.; Emberton, M.; Lawrentschuk, N.; Lazzeri, M.; Montironi, R.; Nguyen, P.L.; Trachtenberg, J.; Polascik, T.J. The Role of Focal Therapy in the Management of Localised Prostate Cancer: A Systematic Review. Eur. Urol. 2014, 66, 732–751.
31. Yakkala, C.; Denys, A.; Kandalaft, L.; Duran, R. Cryoablation and immunotherapy of cancer. Curr. Opin. Biotechnol. 2020, 65, 60–64.
32. Chen, Q.; Chen, M.C.; Liu, Z. Local biomaterials-assisted cancer immunotherapy to trigger systemic antitumor responses. Chem. Soc. Rev. 2019, 48, 5506–5526.
33. Wei, W.F.; Zhang, X.Y.; Zhang, S.; Wei, G.; Su, Z.Q. Biomedical and bioactive engineered nanomaterials for targeted tumor photothermal therapy: A review. Mat. Sci. Eng. C Mater. 2019, 104, 109891.
34. Lin, H.; Gao, S.S.; Dai, C.; Chen, Y.; Shi, J.L. A Two-Dimensional Biodegradable Niobium Carbide (MXene) for Photothermal Tumor Eradication in NIR-I and NIR-II Biowindows. J. Am. Chem. Soc. 2017, 139, 16235–16247.
35. Cao, Z.Y.; Feng, L.Z.; Zhang, G.B.; Wang, J.X.; Shen, S.; Li, D.D.; Yang, X.Z. Semiconducting polymer-based nanoparticles with strong absorbance in NIR-II window for in vivo photothermal therapy and photoacoustic imaging. Biomaterials 2018, 155, 103–111.
36. Guo, B.; Sheng, Z.; Hu, D.; Liu, C.; Zheng, H.; Liu, B. Through Scalp and Skull NIR-II Photothermal Therapy of Deep Orthotopic Brain Tumors with Precise Photoacoustic Imaging Guidance. Adv. Mater. 2018, 30, 1802591.
37. Tang, Z.M.; Zhao, P.R.; Ni, D.L.; Liu, Y.Y.; Zhang, M.; Wang, H.; Zhang, H.; Gao, H.B.; Yao, Z.W.; Bu, W.B. Pyroelectric nanoplatform for NIR-II-triggered photothermal therapy with simultaneous pyroelectric dynamic therapy. Mater. Horizons 2018, 5, 946–952.
38. Zhu, Y.; Wang, Y.; Williams, G.R.; Fu, L.; Wu, J.; Wang, H.; Liang, R.; Weng, X.; Wei, M. Multicomponent Transition Metal Dichalcogenide Nanosheets for Imaging-Guided Photothermal and Chemodynamic Therapy. Adv. Sci. 2020, 7, 2000272.
39. Hu, K.; Xie, L.; Zhang, Y.; Hanyu, M.; Yang, Z.; Nagatsu, K.; Suzuki, H.; Ouyang, J.; Ji, X.; Wei, J.; et al. Marriage of black phosphorus and Cu2+ as effective photothermal agents for PET-guided combination cancer therapy. Nat. Commun. 2020, 11, 2778.
40. Wu, F.; Lu, Y.; Mu, X.; Chen, Z.; Liu, S.; Zhou, X.; Liu, S.; Li, Z. Intriguing H-Aggregates of Heptamethine Cyanine for Imaging-Guided Photothermal Cancer Therapy. ACS Appl. Mater. Interfaces 2020, 12, 32388–32396.
41. Wang, Q.; Dai, Y.; Xu, J.; Cai, J.; Niu, X.; Zhang, L.; Chen, R.; Shen, Q.; Huang, W.; Fan, Q. All-in-One Phototheranostics: Single Laser Triggers NIR-II Fluorescence/Photoacoustic Imaging Guided Photothermal/Photodynamic/Chemo Combination Therapy. Adv. Funct. Mater. 2019, 29, 1901480.
42. Zhou, J.; Li, M.; Hou, Y.; Luo, Z.; Chen, Q.; Cao, H.; Huo, R.; Xue, C.; Sutrisno, L.; Hao, L.; et al. Engineering of a Nanosized Biocatalyst for Combined Tumor Starvation and Low-Temperature Photothermal Therapy. ACS Nano 2018, 12, 2858–2872.

43. Jiang, Y.; Li, J.; Zhen, X.; Xie, C.; Pu, K. Dual-Peak Absorbing Semiconducting Copolymer Nanoparticles for First and Second Near-Infrared Window Photothermal Therapy: A Comparative Study. Adv. Mater. 2018, 30, 1705980.

44. Jung, H.S.; Verwilst, P.; Sharma, A.; Shin, J.; Sessler, J.L.; Kim, J.S. Organic molecule-based photothermal agents: An expanding photothermal therapy universe. Chem. Soc. Rev. 2018, 47, 2280–2297.

45. Li, J.; Rao, J.; Pu, K. Recent progress on semiconducting polymer nanoparticles for molecular imaging and cancer phototherapy. Biomaterials 2018, 155, 217–235.

46. Cheng, Y.; Chang, Y.; Feng, Y.; Jian, H.; Tang, Z.; Zhang, H. Deep-Level Defect Enhanced Photothermal Performance of Bismuth Sulfide-Gold Heterojunction Nanorods for Photothermal Therapy of Cancer Guided by Computed Tomography Imaging. Angew. Chem. Int. Ed. 2018, 57, 246–251.

47. Tang, W.; Dong, Z.; Zhang, R.; Yi, X.; Yang, K.; Jin, M.; Yuan, C.; Xiao, Z.; Liu, Z.; Cheng, L. Multifunctional Two-Dimensional Core-Shell MXene@Gold Nanocomposites for Enhanced Photo-Radio Combined Therapy in the Second Biological Window. ACS Nano 2019, 13, 284–294.

48. Wang, Q.; Wang, H.; Yang, Y.; Jin, L.; Liu, Y.; Wang, Y.; Yan, X.; Xu, J.; Gao, R.; Lei, P.; et al. Plasmonic Pt Superstructures with Boosted Near-Infrared Absorption and Photothermal Conversion Efficiency in the Second Biowindow for Cancer Therapy. Adv. Mater. 2019, 31, 1904836.

49. Gu, Z.; Zhu, S.; Yan, L.; Zhao, F.; Zhao, Y. Graphene-Based Smart Platforms for Combined Cancer Therapy. Adv. Mater. 2019, 31, 1800662.

50. Liu, H.; Li, C.; Qian, Y.; Hu, L.; Fang, J.; Tong, W.; Nie, R.; Chen, Q.; Wang, H. Magnetic-induced graphene quantum dots for imaging-guided photothermal therapy in the second near-infrared window. Biomaterials 2020, 232, 119700.

51. Liang, X.; Ye, X.; Wang, C.; Xing, C.; Miao, Q.; Xie, Z.; Chen, X.; Zhang, X.; Zhang, H.; Mei, L. Photothermal cancer immunotherapy by erythrocyte membrane-coated black phosphorus formulation. J. Control. Release 2019, 296, 150–161.

52. Choi, J.R.; Yong, K.W.; Choi, J.Y.; Nilghaz, A.; Lin, Y.; Xu, J.; Lu, X. Black Phosphorus and its Biomedical Applications. Theranostics 2018, 8, 1005–1026.

53. Zhou, Z.; Wang, X.; Zhang, H.; Huang, H.; Sun, L.; Ma, L.; Du, Y.; Pei, C.; Zhang, Q.; Li, H.; et al. Activating Layered Metal Oxide Nanomaterials via Structural Engineering as Biodegradable Nanoagents for Photothermal Cancer Therapy. Small 2021, 17, 2007486.

54. Huang, X.; Zhang, W.; Guan, G.; Song, G.; Zou, R.; Hu, J. Design and Functionalization of the NIR-Responsive Photothermal Semiconductor Nanomaterials for Cancer Theranostics. Acc. Chem. Res. 2017, 50, 2529–2538.

55. Liu, Y.; Ji, X.; Liu, J.; Tong, W.W.L.; Askhatova, D.; Shi, J. Tantalum Sulfide Nanosheets as a Theranostic Nanoplatform for Computed Tomography Imaging-Guided Combinatorial Chemo-Photothermal Therapy. Adv. Funct. Mater. 2017, 27, 1703261.

56. Cheng, H.; Wen, M.; Ma, X.; Kuwahara, Y.; Mori, K.; Dai, Y.; Huang, B.; Yamashita, H. Hydrogen Doped Metal Oxide Semiconductors with Exceptional and Tunable Localized Surface Plasmon Resonances. J. Am. Chem. Soc. 2016, 138, 9316–9324.

57. Cheng, L.; Liu, J.; Gu, X.; Gong, H.; Shi, X.; Liu, T.; Wang, C.; Wang, X.; Liu, G.; Xing, H.; et al. PEGylated WS2 Nanosheets as a Multifunctional Theranostic Agent for in vivo Dual-Modal CT/Photoacoustic Imaging Guided Photothermal Therapy. Adv. Mater. 2014, 26, 1886–1893.

58. Guan, G.; Wang, X.; Huang, X.; Zhang, W.; Cui, Z.; Zhang, Y.; Lu, X.; Zou, R.; Hu, J. Porous cobalt sulfide hollow nanospheres with tunable optical property for magnetic resonance imaging-guided photothermal therapy. Nanoscale 2018, 10, 14190–14200.
59. Liu, J.; Zheng, X.; Yan, L.; Zhou, L.; Tian, G.; Yin, W.; Wang, L.; Liu, Y.; Hu, Z.; Gu, Z.; et al. Bismuth Sulfide Nanorods as a Precision Nanomedicine for in Vivo Multimodal Imaging-Guided Photothermal Therapy of Tumor. ACS Nano 2015, 9, 696–707.
60. Yang, K.; Yang, G.; Chen, L.; Cheng, L.; Wang, L.; Ge, C.; Liu, Z. FeS nanoplates as a multifunctional nano-theranostic for magnetic resonance imaging guided photothermal therapy. Biomaterials 2015, 38, 1–9.
61. Yang, W.; Guo, W.; Le, W.; Lv, G.; Zhang, F.; Shi, L.; Wang, X.; Wang, J.; Wang, S.; Chang, J.; et al. Albumin-Bioinspired Gd:CuS Nanotheranostic Agent for In Vivo Photoacoustic/Magnetic Resonance Imaging-Guided Tumor-Targeted Photothermal Therapy. ACS Nano 2016, 10, 10245–10257.
62. Lei, Z.; Zhang, W.; Li, B.; Guan, G.; Huang, X.; Peng, X.; Zou, R.; Hu, J. A full-spectrum-absorption from nickel sulphide nanoparticles for efficient NIR-II window photothermal therapy. Nanoscale 2019, 11, 20161–20170.
63. Zhang, C.; Sun, W.; Wang, Y.; Xu, F.; Qu, J.; Xia, J.; Shen, M.; Shi, X. Gd-/CuS-Loaded Functional Nanogels for MR/PA Imaging-Guided Tumor-Targeted Photothermal Therapy. ACS Appl. Mater. Interfaces 2020, 12, 9107–9117.
64. Sun, H.; Zhang, Y.; Chen, S.; Wang, R.; Chen, Q.; Li, J.; Luo, Y.; Wang, X.; Chen, H. Photothermal Fenton Nanocatalysts for Synergetic Cancer Therapy in the Second Near-Infrared Window. ACS Appl. Mater. Interfaces 2020, 12, 30145–30154.
65. Liu, S.; Pan, X.T.; Liu, H.Y. Two-Dimensional Nanomaterials for Photothermal Therapy. Angew. Chem. Int. Ed. 2020, 59, 5890–5900.
66. Hu, J.-J.; Cheng, Y.-J.; Zhang, X.-Z. Recent advances in nanomaterials for enhanced photothermal therapy of tumors. Nanoscale 2018, 10, 22657–22672.
67. Wu, Y.; Liang, Y.; Liu, Y.; Hao, Y.; Tao, N.; Li, J.; Sun, X.; Zhou, M.; Liu, Y.N. A Bi2S3-embedded gellan gum hydrogel for localized tumor photothermal/antiangiogenic therapy. J. Mater. Chem. B 2021, 9, 3224–3234.
68. Sun, S.-K.; Wu, J.-C.; Wang, H.; Zhou, L.; Zhang, C.; Cheng, R.; Kan, D.; Zhang, X.; Yu, C. Turning solid into gel for high-efficient persistent luminescence-sensitized photodynamic therapy. Biomaterials 2019, 218, 119328.
69. Ding, X.; Wang, Y. Weak Bond-Based Injectable and Stimuli Responsive Hydrogels for Biomedical Applications. J. Mater. Chem. B 2017, 5, 887–906.
70. Chandel, A.K.S.; Nutan, B.; Raval, I.H.; Jewrajka, S.K. Self-Assembly of Partially Alkylated Dextran-graft-poly[(2-dimethylamino)ethyl methacrylate] Copolymer Facilitating Hydrophobic/Hydrophilic Drug Delivery and Improving Conetwork Hydrogel Properties. Biomacromolecules 2018, 19, 1142–1153.
71. Jiang, K.; Song, X.; Yang, L.; Li, L.; Wan, Z.; Sun, X.; Gong, T.; Lin, Q.; Zhang, Z. Enhanced antitumor and anti-metastasis efficacy against aggressive breast cancer with a fibronectin-targeting liposomal doxorubicin. J. Control. Release 2018, 271, 21–30.
72. Henschke, C.I.; Yankelevitz, D.F.; McCauley, D.I.; Rifkin, M.; Fiore, E.S.; Austin, J.H.M.; Pearson, G.D.N.; Shiau, M.C.; Kopel, S.; Klippenstein, D.; et al. CT screening for lung cancer: Diagnoses resulting from the New York Early Lung Cancer Action Project. Radiology 2007, 243, 239–249.

11 Chapter

HYDROGELS LOADED WITH QUERCETIN/ HYDROXYPROPYL-B-CYCLODEXTRIN INCLUSION COMPLEX FOR FASTER WOUND HEALING

1. INTRODUCTION: AN OVERVIEW

Eventually, every individual may encounter an injury that results in harm to the skin or other tissues, and it is crucial to provide appropriate care for it. When selecting a dressing for a wound, it is important to examine various aspects such as the wound's color and depth, the level of infection, the amount of fluid discharge, and the state of the skin surrounding the wound [1]. An optimal dressing should possess the following characteristics: it should preserve a moist environment, shield against bacterial infiltration, expedite the development of epithelialization, offer thermal insulation, and be easily removable without leaving any residue after the healing process [2].

Modern wound dressings have been designed to enhance biological processes, including antibacterial, antifungal, and antiviral capabilities, in contrast to conventional wound dressings that only provide coverage for wounds. Medicinal plants are commonly used to create herbal medications that are added to wound dressing materials in order to enhance their healing effects [3,4]. Quercetin, a fascinating constituent of the flavonoid group, can be found in onions, apples, berries, tea, and tomatoes [5,6]. Furthermore, it possesses physicochemical characteristics and the capacity to be absorbed, serving as a dietary source of nutrients, and primarily influencing inflammation and immunological function. Studies have indicated that consuming a daily dose of 50 to 800 mg of quercetin (which makes up 75% of the total) can provide possible health benefits due to its efficient absorption in the organs involved in metabolism and elimination [7]. Quercetin's antioxidant, anti-inflammatory, antibacterial, antiviral, anticancer, antitoxic, and immunomodulatory properties demonstrate its potential therapeutic utility in promoting wound healing.

Nevertheless, the primary obstacles that restrict its effectiveness in promoting wound healing are its inadequate solubility, limited bioavailability, and hydrophobic nature, resulting in a tendency to dissolve more easily in organic solutions [8,9]. To mitigate the drawbacks of quercetin, a drug-delivery method can be employed, such as the formation of an inclusion complex with host molecules such cyclodextrins (CDs), which is a widely utilized strategy [10].

Cyclodextrins are cyclic oligosaccharides formed from the enzymatic conversion of starch. They consist of six (α-CD), seven (β-CD), or eight (γ-CD) α-D-glucopyranose units, connected by α-1,4 glycosidic linkages. Cyclodextrins are used in the food processing industry to manufacture reduced-cholesterol products and boost the bioavailability of targeted molecules by enhancing hydrophobic-hydrophilic interactions between proteins and other molecules [11]. Typically, CDs possess a conical form, featuring a hydrophilic outside and a hydrophobic interior hollow. CDs have the ability to capture hydrophobic molecules by utilizing their hydrophobic cavity, resulting in the formation of host-guest complexes. These complexes are utilized in the pharmaceutical sector [12]. CDs are employed to enclose and preserve aromas and tastes, preventing food from losing its color and reducing unpleasant scents. They can be utilized in the packaging of food products [13]. Furthermore, CDs are employed in several medicinal contexts, including wound dressings. Specifically, inclusion complexes consisting of curcumin and hydroxypropyl-cyclodextrin have been included into bacterial cellulose hydrogels to augment the process of wound healing [14]. Hydroxypropyl-β-cyclodextrin (HP-β-CD) is primarily used as a complexing agent to boost the water solubility and bioavailability of medications that have low solubility. It also helps improve the stability of these medications [15].

Hydrogels are 3D structures made of natural or synthetic polymers that consist of over 90% water. The polymer chains undergo crosslinking, which results in the formation of a network that ensures the integrity of the hydrogel structure. This crosslinking can occur by either physical or chemical means. Hydrogels possess a porous structure that is highly attracted to water and can absorb a specific quantity of water, which is determined by characteristics such as pore size and the type of polymer used [16]. Hydrogels are commonly used for wound dressing due to their capacity to create a moist environment, absorb exudates, have good biocompatibility, low adherence to wound tissue, reduce pressure and shear stresses, protect against bacterial invasion, and provide thermal insulation [16,17]. Hydrogels possess the capacity to have a cooling and soothing effect on the skin, making them beneficial for alleviating pain caused by burns or painful wounds [18]. Polyvinyl alcohol (PVA) is a hydrophilic polymer that can dissolve in water and is capable of being broken down by natural processes. It is created by chemically converting polyvinyl acetate (PVAc) through a process called hydrolysis. This polymer exhibits favorable mechanical characteristics, exceptional thermal durability, minimal toxicity, and remarkable gas impermeability, rendering it extensively employed in various fields such as biomedical, food packaging, paper coating, textile sizing, as well as a thickening agent and emulsion stabilizer. Additionally, it can be combined with other polymers to enhance the overall properties. Polyvinyl alcohol (PVA) hydrogel wound dressings can be created by physical crosslinking using the cyclic freeze-thaw method [19]. As a result of the PVA structure, they are capable of forming hydrogen bonds between -OH groups when freezing occurs, which causes the polymer chains to become crosslinked in a crystalline zone. Nevertheless, the sub-zero circumstances, encompassing factors such as temperature, duration, and frequency of repetitions, may impact the extent of physical interconnections or the level of crystalline structure in the resulting hydrogels. At lower temperatures, the interactions between PVA chains increase, leading to a higher degree of crystal formation. At low freezing temperatures, small water crystals are created and then transformed into gas without going through the liquid phase. This process results in the construction of a structure with many small holes. On the other hand, when there are multiple cycles of freezing and thawing, more water crystals are formed and the structure becomes more tightly interconnected.

The objective of this study was to create an inclusion complex by enclosing quercetin within HP-β-CD in order to overcome its limitations. Additionally, the aim was to fabricate new hydrogels that incorporate the inclusion complex, with the intention of enhancing the rate of wound healing. The physical properties of drug-loaded hydrogels were assessed in terms of gelation, swelling, and compression. The study also examined the release of quercetin from the hydrogels and its cytotoxic effects on NCTC 929 clone cells, in order to assess the biological capabilities of the drug-loaded hydrogels.

2. RESULTS AND DISCUSSION

2.1. Characterization of Quercetin/HP-β-CD Inclusion Complex

FTIR spectroscopy was used to investigate the successful encapsulation of quercetin into HP-β-CD (Figure 1). The FTIR spectrum of quercetin (Figure 1a) showed –OH groups stretching at 3395 and 3268 cm−1; C=O aryl ketone stretching at 1663 cm−1; C=C aromatic ring stretching at 1605, 1560, and 1518 cm−1; and C–O aryl ether ring stretching, C–O phenol stretching, and C–CO–C ketone stretching and bending at 1256, 1194, and 1164 cm−1, respectively [22]. For HP-β-CD (Figure 1b), the characteristic broad peak of OH group stretching was shown at 3344 cm−1; C–H stretching was detected at 2926 cm−1; and C-O-C glucose unit stretching occurred at 1020 cm−1 [23]. The FTIR spectrum of the physical mixture of the two (Figure 1c) showed characteristic peaks of quercetin, including C=O aryl ketone stretching (1663 cm−1), C=C aromatic ring stretching (1609, 1560, and 1520 cm−1), and C–CO–C ketone stretching and bending (1165 cm−1), as well as C–O–C glucose unit stretching of HP-β-CD. When quercetin was encapsulated, its characteristic peaks seemed to be masked (Figure 1d). The C=O aryl ketone stretching (1663 cm−1), C=C aromatic ring stretching (1605, 1560, and 1518 cm−1), and C–CO–C ketone stretching and bending (1164 cm−1) peaks of quercetin were absent in the inclusion complex, which may be because the quercetin was entrapped in the HP-β-CD cavities [24]. Signals of the characteristic bands of quercetin shifted from 3282 cm−1 to 3142 cm−1 O–H bending signals showing the formation of quercetin/HP-β-CD inclusion complex [25].

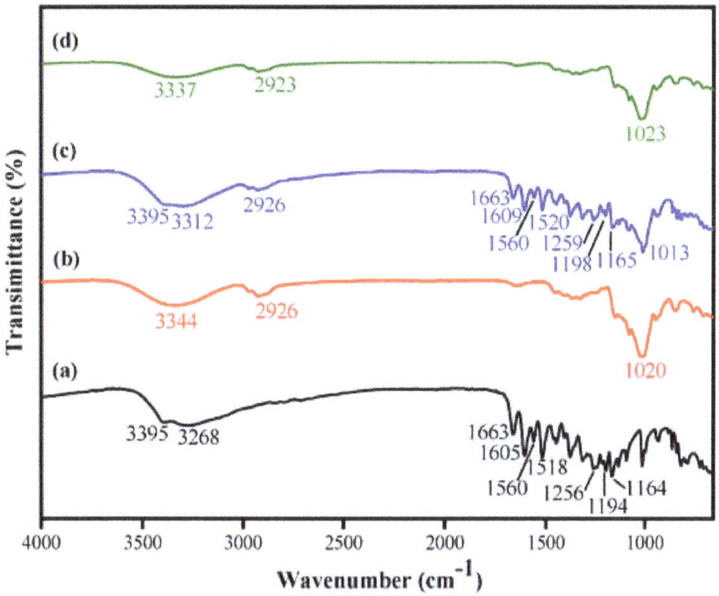

Figure 1. FTIR spectra of (a) quercetin, (b) HP-β-CD, (c) quercetin/HP-β-CD physical mixture, and (d) quercetin/HP-β-CD inclusion complex.

The XRD patterns provide further evidence of the successful encapsulation of quercetin in HP-β-CD (Figure 2). Characteristic diffraction peaks (Figure 2a) indicated that quercetin existed in a crystalline form, with peaks at three 2θ positions (10.66°, 12.34°, and 27.26°) [26], while HP-β-CD (Figure 2b) showed broad peaks of amorphous materials caused by random X-ray scattering [27]. After the complexation of quercetin and HP-β-CD, there were no noticeable crystalline peaks of quercetin in the inclusion complex (Figure 2d) because the CD cavities hindered its crystalline structure, unlike the physical mixture, which showed crystalline and amorphous behavior (Figure 2c) of quercetin and HP-β-CD, respectively. These results suggest that quercetin was successfully encapsulated into HP-β-CD cavities, leading to the loss of characteristic peaks. Similarly, Zhu et al. encapsulated l-menthol, which is crystalline (2θ = 8.0°, 14.0°, 20.5°, and 21.6°), into amorphous HP-β-CD and observed no sharp peaks of l-menthol [28]. This result agrees with Kim et al., who reported similar XRD patterns of some selective flavonoids in hydroxypropyl-β-cyclodextrin inclusion complexes, which was indicative of the amorphous nature of the complex [29].

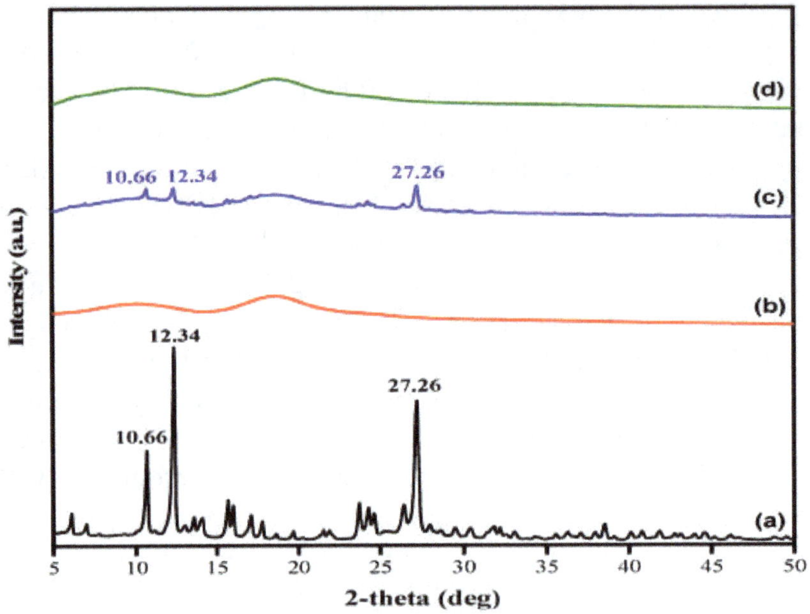

Figure 2. XRD patterns of (a) quercetin, (b) HP-β-CD, (c) quercetin/HP-β-CD physical mixture, and (d) quercetin/HP-β-CD inclusion complex.

The morphologies of quercetin, HP-β-CD, and quercetin/HP-β-CD physical mixture and inclusion complex were observed at different magnifications on SEM (Figure 3). Quercetin (Figure 3a) presented a strip-like structure with a smooth surface [30], while HP-β-CD (Figure 3b) appeared as rough spheres [31]. In the physical mixture (Figure 3c), both quercetin and HP-β-CD kept their own structure, with particles of HP-β-CD embedded with quercetin particles [32]. On the other hand, the morphology and shape of quercetin/HP-β-CD inclusion complex obtained from the solvent evaporation method were completely different from those of the original product (Figure 3d). Our results were consistent with Pradhan et al. [33] in that the morphology of Berberis anthocyanin-loaded CD inclusion complex revealed the rod shape. The encapsulation of quercetin with CD demonstrated the electrostatic and hydrogel bond interactions between quercetin and CD.

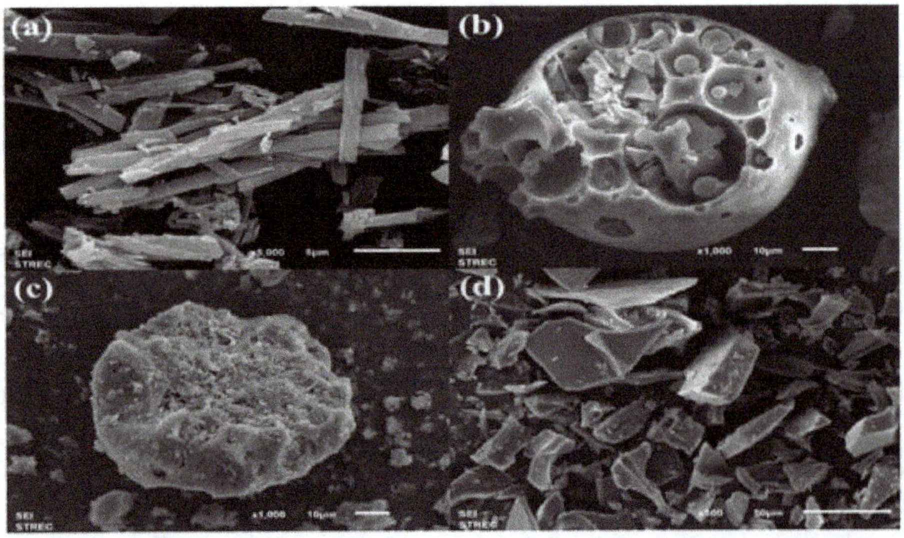

Figure 3. SEM images of (a) quercetin, (b) HP-β-CD, (c) quercetin/HP-β-CD physical mixture, and (d) quercetin/HP-β-CD inclusion complex.

2.2. Encapsulation Efficiency and Loading Capacity of Quercetin in the Inclusion Complex

Cyclodextrins are nontoxic cyclic oligosaccharides consisting of α1,4-glycosidic bonds. CDs are distinguished by hydrophilic surfaces and hydrophobic cavities. The cavities are suitable for inclusion complexes with hydrophobic drugs (quercetins) to improve the aqueous solubility and yield Van der Waals and hydrophobic forces [34,35]. Quercetin/HP-β-CD inclusion complex was formed with optimum loading and encapsulation efficacy, achieved at a mass ratio of HP-β-CD to quercetin of 4.35:1. Loading capacity (LC) is the ratio of mass of encapsulated compound to mass of the polymer (Equation (1)), and encapsulation efficacy (EE%) refers to the ratio of the mass of the encapsulated compounds to the total mass of the compounds (Equation (2)). Encapsulation is considered to be an excellent carrier of compounds when LC id ≥ 5%, and %EE in the range of 70–100% indicates excellent encapsulation capacity.

The absorbance of ~1 mg/mL inclusion complex in DMSO was 0.68917 ± 0.01715, which can be calculated back to the concentration of quercetin using the predetermined calibration curve (R2 = 0.99758). The estimated total quercetin content in the inclusion complex was 0.4554 ± 0.0089 g.

The %EE and %LC of quercetin were quantified with UV-Vis spectrophotometry at a wavelength of 379 nm and calculated using Equations (1) and (2), respectively. The %EE of the obtained quercetin/HP-β-CD inclusion complex was 90.50 ± 1.84%, and the %LC was 4.67 ± 0.13%. The %LC of the inclusion complex seemed to be quite low but was not. The molecular weight of HP-β-CD was greater than that of quercetin by around 4.6 times. We also used the excess molar ratio of HP-β-CD to prepare the inclusion complex. That explains why when we calculated %LC, which compares the drug in the complexation (or inclusion complex) in terms of weight, the molar ratio seemed to be low. This was similar to the results of Hadian et al., who prepared four formulations of geraniol/β-CD inclusion complexes by varying the mole ratio of GR:β-CD (0.44:0.13, 0.44:0.2, 0.44:0.4, and 0.44:1), and showed that %LC gradually decreased with increasing mole ratio (by 7.8 ± 0.70, 6.9 ± 0.1 6.6 ± 0.28 and 6.5 ± 0.6, respectively) [36].

2.3. Gelation of PVA Hydrogels Loaded with Quercetin/HP-β-CD Inclusion Complex

The gelation of neat and all prepared drug-loaded hydrogels in PBS (pH = 7.4) at 37 °C for 24 h was calculated using Equation (3). The experiments were repeat five times, and values were averaged. The gelation of neat PVA hydrogel was 97.40 ± 0.62%, while the gelation of drug-loaded hydrogels ranged from 74.52 ± 0.86% to 94.36 ± 0.80%, as shown in Table 1.

Table 1. Gelation value and mass content of PVA hydrogels containing various amounts of quercetin/HP-β-CD inclusion complex. Data shown as mean ± standard deviation (n = 5). IC as quercetin/HP-β-CD inclusion complex.

Sample	Gelation (%)	Mass Content (%)	
		IC	PVA
PVAIC0	97.40 ± 0.62	0.00	100.00
PVAIC0.5	94.36 ± 0.80	4.67	95.24
PVAIC1.0	92.29 ± 0.71	9.09	90.91
PVAIC1.5	89.88 ± 0.83	13.04	86.96
PVAIC2.5 85.59 ± 0.95 20.00			80.00
PVAIC5.0	74.52 ± 0.86	33.33	66.67

The percent gelation of drug-loaded hydrogel clearly gradually decreased with increasing inclusion complex content in the hydrogels. The reduction in hydrogel weight may have occurred for the following reasons. The first is that hydrogel swells when immersed in PBS, leading to a loose structure, so the quercetin can dissolve and be released out of the hydrogels [20]. The second reason is that when loading the quercetin/HP-β-CD inclusion complex into the hydrogels, the inclusion complex may disrupt the crosslink formation of the hydrogel, so some PVA that did not participate in crystallite formation could have dissolved into the solution [21].

2.4. Swelling Ratio of PVA Hydrogels Loaded with Quercetin/HP-β-CD Inclusion Complex

The swelling ratio of neat and drug-loaded hydrogels in PBS (pH = 7.4) at 37 °C at 10, 30, 60, 180, and 360 min was calculated using Equation (4). The experiment was repeated three times at each release time point, and values were averaged. The results show that the swelling ratio of hydrogels with various drug amounts was slightly higher than that of neat PVA hydrogel, but there was not much difference (Figure 4). After the first 60 min of the experiment, the swelling ratio of all hydrogels dramatically increased to around 4.3%, then gradually increased to around 5.8% at 180 min and became relatively stable at around 6.4% after immersion in PBS for 360 min. The incorporation of quercetin/HP-β-CD inclusion complex into the PVA hydrogels had no effect on the swelling ratio, which may be caused by the similar conditions of the freeze–thaw process that we used to prepare the hydrogels. Therefore, the number and temperature of freezing–thawing cycles could be the most important factors controlling the swelling ratio of hydrogels. We can also add other polymers or materials, such as gelatin or chitosan, to increase the swellability of PVA hydrogels [20].

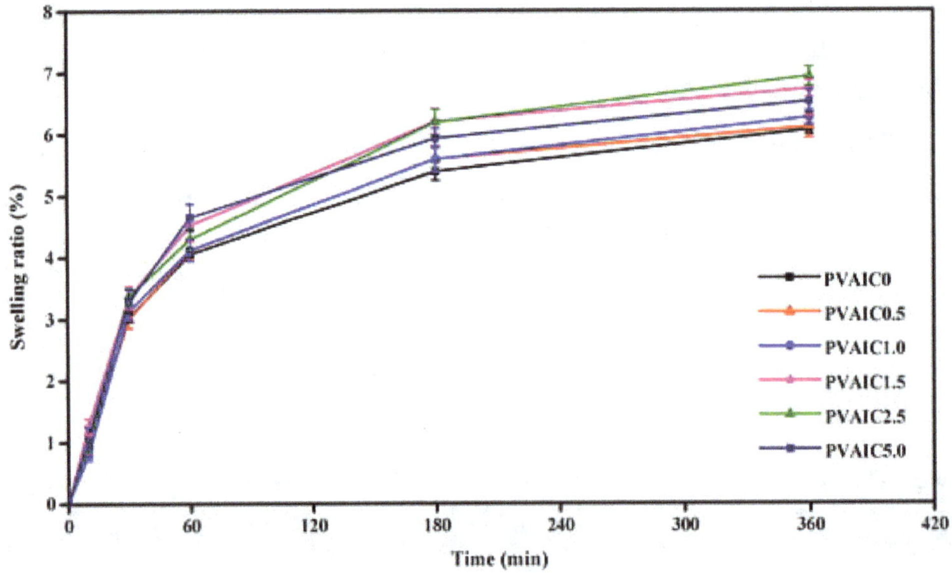

Figure 4. Swelling ratios of PVA hydrogels containing 0% (PVAIC0), 0.5% (PVAIC0.5), 1.0% (PVAIC1.0), 1.5% (PVAIC1.5), 2.5% (PVAIC2.5), and 5.0% (PVAIC5.0) quercetin/HP-β-CD inclusion complex. Data shown as mean ± standard deviation (n = 3).

2.5. Mechanical Test of PVA Hydrogels Loaded with Quercetin/HP-β-CD Inclusion Complex

The compressive strength of neat and all prepared drug-loaded hydrogels was tested, and the experiment was repeated seven times to obtain average values. The stiffness and compressive modulus of neat PVA hydrogel were 27,318 ± 831 N/m and 1.19 ± 0.07 MPa, respectively, while the values for drug-loaded hydrogels were 19,520 ± 1066 to 26,199 ± 1275 N/m and 0.88 ± 0.04 to 1.12 ± 0.08 MPa, respectively (Table 2). As the amount of inclusion complex in the hydrogels increased, stiffness and compressive modulus seemed to gradually decrease. This may have resulted from the molecules of the inclusion complex being distributed in the PVA structure, leading to a lower crystalline zone [20] and decreased crosslink density of the PVA hydrogel [37], so it was not able to withstand the compressive force compared with lower drug content hydrogels. Water content is another factor that can affect the mechanical properties of hydrogels because when the water in hydrogel evaporates, the hydrogel becomes stronger, so it will have stronger resistance against compressive force than softer or lower-water-content hydrogels [38]. Qing et al. prepared PVA hydrogel dressings that incorporated N-succinyl chitosan (NSCS) and lincomycin, and the results revealed that the introduction of 10% NSCS reduced the compression strength of pure PVA hydrogel from 0.33 to 0.25 MPa because NSCS increases the distance between PVA chains and weakens the molecular interaction force. However, when they further increased NSCS content, hydrogel bonding occurred, and the hydrogel became stronger, leading to enhanced compression strength (0.75 MPa). As NSCS content was further increased to 40 and 50%, the compression strength in turn decreased due to the low crosslinking density resulting from the decreased gel fraction [39].

Table 2. Mechanical test of PVA hydrogels containing various amounts of quercetin/HP-β-CD inclusion complex. Data shown as mean ± standard deviation (n = 7).

Sample	Stiffness (N/m)	Compressive Modulus (MPa)
PVAIC0	27,318 ± 831	1.19 ± 0.07
PVAIC0.5	26,199 ± 1275	1.12 ± 0.08
PVAIC1.0	24,878 ± 758	1.09 ± 0.10
PVAIC1.5	24,011 ± 941	0.99 ± 0.06
PVAIC2.5	21,817 ± 1001	0.92 ± 0.04
PVAIC5.0	19,520 ± 1066	0.88 ± 0.04

2.6. In Vitro PVA Hydrogels Loaded with Quercetin/HP-β-CD Inclusion Complex Dissolution and Release Kinetics Study

The characteristics of quercetin release from the prepared hydrogels were investigated using the total immersion method over a period of 48 h in PBS (pH = 7.4) and incubated at 37 °C under continuous stirring. The cumulative release of quercetin from all hydrogels was calculated using Equation (5) and reported in parts per million. As shown in Figure 5, the cumulative amount of quercetin released from the hydrogels increased rapidly with increasing immersion time for the first 5 or 8 h of the experiment, then gradually increased and reached a plateau. After a period of 48 h, the cumulative quercetin quantities released from the quercetin/HP-β-CD inclusion complex PVA hydrogels were 81.66 ± 17.42, 142.07 ± 30.43, 167.64 ± 23.30, 211.74 ± 27.78, and 286.10 ± 27.71 ppm for PVAIC0.5, PVAIC1.0, PVAIC1.5, PVAIC2.5, and PVAIC5.0, respectively. The greatest cumulative amounts of CIP released were observed when the quercetin concentration of PVAIC5.0 was used, and therefore, it can be concluded that when a greater concentration of quercetin is introduced to the nanoparticles, the drug release will increase. In the USA, the mean quercetin intake was approximately 14.90 to 16.39 mg per day [40]. Thus, the release characteristics of quercetin for the PVAIC5.0 hydrogels (close to 28.6 mg/100 g of sample) was found to be about 1.9-times higher than the oral solution of the quercetin intake per day.

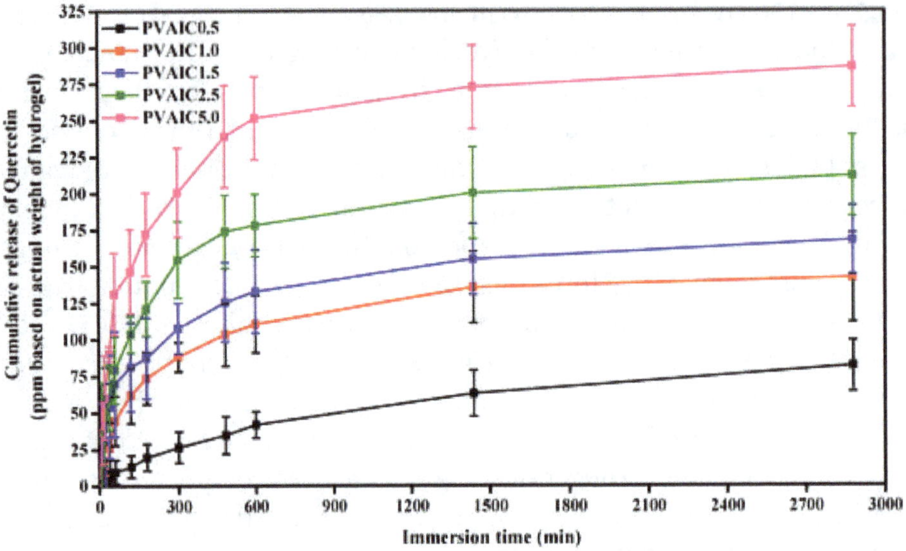

Figure 5. Release profiles of quercetin from PVA hydrogels containing 0.5% (PVAIC0.5), 1.0% (PVAIC1.0), 1.5% (PVAIC1.5), 2.5% (PVAIC2.5), and 5.0% (PVAIC5.0) of quercetin/HP-β-CD inclusion complex. Data shown as mean ± standard deviation (n = 3).

The drug-release kinetics of quercetin from PVA hydrogels loaded with quercetin/HP-β-CD inclusion complex were fitted with Zero-order and Korsmeyer–Peppas models to reveal the kinetics of drug release. Release constant (k) and release exponent (n) were calculated to identify the release mechanism, and the fitted model parameters and correlation coefficients (r2) for the release profiles are given in Table 3. An r2 coefficient of determination over 0.95 was considered to reflect the goodness of fit of the release model [41]. The release data from PVA hydrogels loaded with quercetin/HP-β-CD inclusion complex fit with the Korsmeyer–Peppas model. The release exponent (n) values can be used to indicate the release rate mechanism. The PVAIC0.5, PVAIC1.0, PVAIC1.5, PVAIC2.5, and PVAIC5.0 hydrogels had n values of 1.31, 0.77, 0.68, 0.58, and 0.77, respectively. The n values between 0.89 and 1 indicate that the release of quercetin from PVAIC1.0 to PVAIC5.0 were controlled by the mechanism of anomalous (non-Fickian) diffusion. The diffusion exponent for PVAIC0.5 is in the range of 0.89 and 1, which denotes the case II transport mechanism (zero-order kinetics).

Table 3. Modeling results of quercetin released from PVA hydrogels loaded with quercetin/HP-β-CD inclusion complex fitting with Zero-order and Korsmeyer–Peppas models.

Sample	Release Kinetic Models			
	Zero-Order		Korsmeyer–Peppas	
	k	r^2	*n*	r^2
PVAIC0.5	0.78	0.7953	1.31	0.995
PVAIC1.0	0.67	0.9127	0.77	0.9671
PVAIC1.5	1.10	0.844	0.68	0.9526
PVAIC2.5	1.49	0.7033	0.58	0.9748
PVAIC5.0	2.31	0.6468	0.77	0.9535

2.7. In Vitro Antioxidant Activity: DPPH-Radical Scavenging Ability Assay

The encapsulation of quercetin within cyclodextrin increases the antioxidant activity compared with free molecules [42]. The antioxidant activity of quercetin/HP-β-CD inclusion complex is widely used in 1,1-diphenyl-2-picrylhydrazyl (DPPH) radical scavenging assay in comparison with free counterparts and the results were shown in Figure 6. A concentration of 120 µg/mL quercetin/HP-β-CD inclusion complex was determined to have 74.80% DPPH inhibition compared with free quercetin, with 61.23% of inhibition, and ascorbic acid with 74.80%. HP-β-CD inclusion complex showed significant antioxidant activity compared with quercetin alone (Figure 6) in a dose-dependent manner. The results indicate that HP-β-CD inclusion complex is associated with the sustained release of quercetin and increases the stability of incorporated bioactive compounds such as quercetin, significantly improving its antioxidant activity [43].

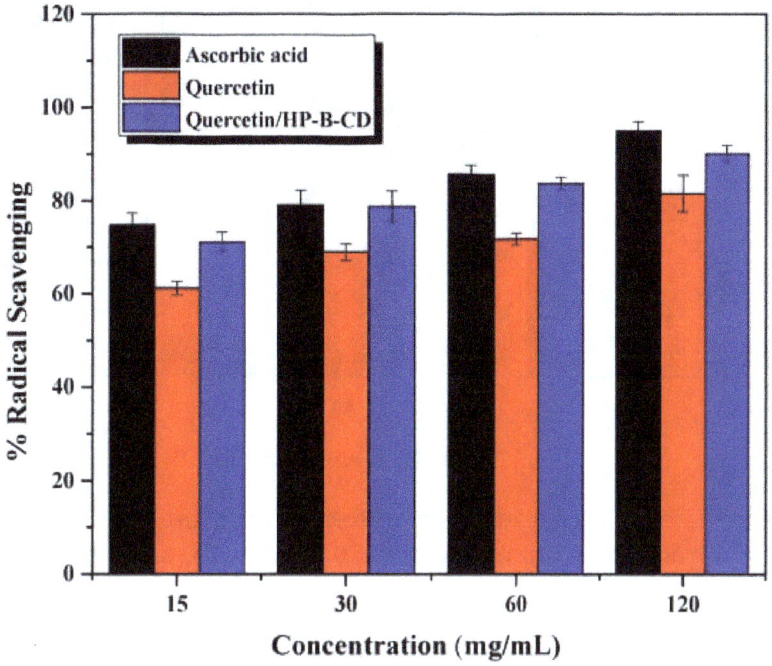

Figure 6. Scavenging activity of ascorbic acid, quercetin and quercetin/ HP-β-CD as determined by DPPH assay. Values expressed as mean ± standard deviation (n = 5).

2.8. Cytotoxicity Evaluation

The indirect cytotoxicity evaluation of all samples toward NCTC 929 clone cells investigated the potential use of drug loaded PVA hydrogels as wound dressing material. Figure 7 shows the viability of cells obtained from MTT assay after being cultured with various concentrations of media extracted from neat PVA hydrogel as blank PVA, and PVA hydrogels loaded with quercetin/HP-β-CD inclusion complex (PVAIC0.5, PVAIC1.0, PVAIC1.5, PVAIC2.5, and PVAIC5.0) compared with that obtained after cells were cultured with fresh SFM. According to the in vitro cytotoxicity standard, a reduction of cell viability by more than 30% is considered a cytotoxic effect [44]. The viability of NCTC 929 clone cells with all extraction medium concentrations, i.e., neat PVA, PVAIC0.5, PVAIC1.0, PVAIC1.5, PVAIC2.5, and PVAIC5.0, were 92.7–96.3, 76.8–88.2, 80.0–97.5, 74.7–92.9, 87.6–96.8, and 78.3–85.3%, respectively. Cell viability was higher than 70% in all prepared hydrogels loaded with quercetin/HP-β-CD inclusion complex. Therefore, all hydrogels loaded with quercetin/HP-β-CD inclusion complex with extraction medium concentrations of 0.5, 5, 10, 25, and 50 mg/mL had cell viability of more than 70%, indicating that these conditions were non-cytotoxic, and the hydrogels have potential in wound dressing applications.

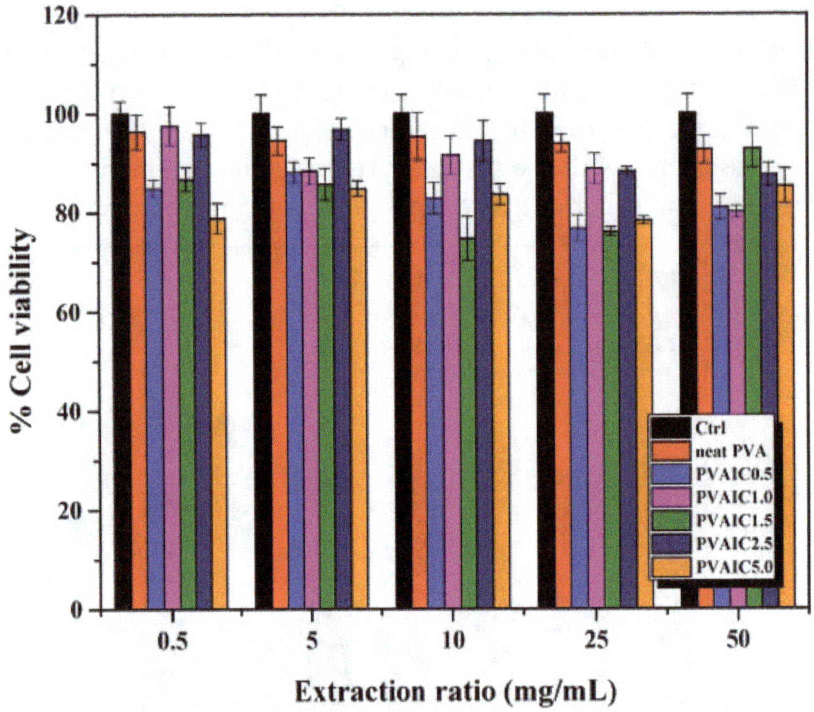

Figure 7. Indirect cytotoxicity of PVA hydrogels loaded with various amounts of quercetin/HP-β-CD inclusion complex with extraction medium: neat PVA, 0.5, 5, 10, 25, and 50 mg/mL cultured with NCTC 929 clone cells.

3. CONCLUSIONS

We successfully prepared PVA hydrogels containing quercetin/HP-β-CD inclusion complex that achieved efficient drug loading and effective encapsulation. The quercetin/HP-β-CD inclusion complex was prepared with an excess molar ratio of HP-β-CD via the solvent evaporation method. The successful encapsulation of quercetin in HP-β-CD was confirmed by FT-IR and XRD, in which characteristic peaks of quercetin could not be observed after encapsulation due to the hindering of CD cavities. As observed on SEM, the quercetin/HP-β-CD inclusion complex demonstrated rod-shaped particles. The prepared quercetin/HP-β-CD inclusion complex had 90.50 ± 1.84% encapsulation efficiency and 4.67 ± 0.13% loading capacity. It was incorporated into PVA hydrogels in various amounts (0, 0.5, 1.0, 1.5, 2.5, and 5.0% MIC/Vsolvent ratio). The hydrogels were evaluated for their potential use as wound dressing in terms of gelation, swelling, compression, release characteristics, antioxidant properties, and in vitro non-cytotoxicity. The incorporation of quercetin/HP-β-CD inclusion complex into PVA hydrogels had no effect on the swelling ratio of the hydrogels. The gelation and compressive strength of hydrogels decreased with increasing inclusion complex content, whereas the cumulative release of quercetin from hydrogels increased with higher amounts of inclusion complex.

Hydrogels loaded with quercetin/HP-β-CD inclusion complex investigated for their cytotoxicity in vitro by MTT assay were shown to be nontoxic toward mouse fibroblast NCTC 929 clone cells at the test concentrations. The results showed significantly increased antioxidant properties of quercetin/HP-β-CD inclusion complex compared with free quercetin. These findings suggest that the use of quercetin/HP-β-CD inclusion complex ensures non-cytotoxicity and potential antioxidant activity. Hence, hydrogel with quercetin/HP-β-CD inclusion complex is an attractive candidate for the encapsulation of bioactive compounds for biomedical applications.

4. MATERIALS AND METHODS

4.1. Materials

Quercetin hydrate (C15H10O7·xH2O, Mw 302.24, >96% purity) and hydroxypropyl-β-cyclodextrin (HP-β-CD, Mw ~1380–1500, >99.0% purity) were purchased from Tokyo Chemical Industry (Portland, OR, USA). Polyvinyl alcohol (PVA; Mw 89,000–98,000, 99+% hydrolyzed) was purchased from Sigma Aldrich (Saint Louis, MO, USA). Methanol (MeOH; HPLC grade) and dimethyl sulfoxide (DMSO; AR grade) were purchased from RCI Labscan (Bangkok, Thailand).

4.2. Formation of Quercetin/HP-β-CD Inclusion Complex

Lyophilized formulations of quercetin/HP-β-CD inclusion complex were prepared by freeze-drying using the neutralization method [45]. Freeze-drying is an excellent method for preserving a wide variety of heat-sensitive materials such as proteins, microbes, pharmaceutical agents, tissues, and plasma [46]. The molar ratio of quercetin to HP-β-CD, following Lan et al., was 1:4.35 [24]. First, 10 g of HP-β-CD was dissolved in 50 mL distilled water, and 0.5 g of quercetin was dissolved in 70 mL MeOH at 60 °C to obtain a homogeneous solution. Then, both solutions were mixed and refluxed at 65 °C overnight. The solution was stirred continuously at room temperature for 5 h for complete encapsulation. Next, MeOH was removed by heating the solution at 75 °C for 2 h. The solution was freeze-dried after being filtered (0.45 μm) and frozen at -20 °C. Finally, quercetin/HP-β-CD inclusion complex was obtained after freeze-drying at -50 °C for 48 h. The resulting lyophilized quercetin/HP-β-CD inclusion complex powder was stored in airtight containers and protected from light until use. Freeze-dried drug–cyclodextrin complexes were used for hydrogel preparation.

4.3. Characterization of Quercetin/HP-β-CD Inclusion Complex

Fourier-transform infrared spectrometry (FT-IR; Nicolet iS5 with iD7, Thermo Scientific, Waltham, MA, USA) was performed to reveal the chemical structure and group interaction of inclusion complex and analyzed in ATR mode with a framework region of 650 to 4000 cm−1 with 64 scans at resolution of 4 cm−1. Samples were prepared as KBr pellets.

The crystalline structure of quercetin/HP-β-CD inclusion complex was examined by X-ray diffractometry (XRD; SmartLab, Rigaku, Japan) with diffraction angle of 2θ ranging from 5 to 50° at a scanning rate of 0.02°/s.

A scanning electron microscope (SEM; JEOL, JSM-6610LV, Japan) was used to observe the morphology of quercetin/HP-β-CD inclusion complex with 15 kV accelerating voltage and magnification of 5000, 1000, and 500 times after the samples were vacuum-coated with a fine layer of gold.

4.4. Determination of Encapsulation Efficiency and Loading Capacity of Quercetin/HP-β-CD Inclusion Complex

The drug-loading process was carried out through a diffusional mechanism, and the encapsulation efficiency (%EE) was used to illustrate the quantity of added drug encapsulated in the formulation. The amount of encapsulated quercetin was determined using a direct method [18] and recorded on a calibration curve at 379 nm.

Briefly, 20 mg of quercetin/HP-β-CD inclusion complex was dissolved in 20 mL DMSO. The amount of quercetin was quantified using a UV-Vis spectrophotometer at a wavelength that provided the maximum absorbance (379 nm), and the concentration was back-calculated from the predetermined quercetin calibration curve. Encapsulation efficiency (%EE) and loading capacity (%LC) were calculated from the following equations [47]:

$$\mathrm{EE}\,(\%) = \frac{W_{\text{Entrapped Quercetin}}}{W_{\text{Quercetin loaded}}} \times 100 \quad (1)$$

$$\mathrm{LC}\,(\%) = \frac{W_{\text{Entrapped Quercetin}}}{W_{\text{Sample}}} \times 100 \quad (2)$$

4.5. Preparation of Quercetin/HP-β-CD Inclusion Complex-Loaded Hydrogels

Quercetin/HP-β-CD inclusion complex was loaded into 10% w/v PVA hydrogels. Briefly, 2 g of PVA was slowly dissolved in 20 mL distilled water at 80–100 °C for 40 min. Then, an accurate weight of quercetin/HP-β-CD inclusion complex was added to PVA solutions at concentrations of 0, 0.5, 1.0, 1.5, 2.5, and 5.0% w/v, which were labeled as conditions PVAIC0, PVAIC0.5, PVAIC1.0, PVAIC1.5, PVAIC2.5, and PVAIC5.0, respectively. Then, the solutions were stirred continuously for 20 min to obtain homogeneous solutions with uniform distribution. The homogeneous solutions were then sonicated in an 80 °C ultrasonic bath to remove air bubbles. Then, the viscous solutions were poured into 90 mm Petri dishes and covered with aluminum foil. The drug-loaded PVA hydrogels were physically crosslinked by repeated freeze–thaw cycles, with freezing at -20 °C for 20 h then thawing at room temperature for 4 h [48]. This freeze–thaw process was repeated for 4 cycles to obtain quercetin/HP-β-CD inclusion complex-loaded hydrogels. The Petri dishes containing prepared hydrogels were then sealed with parafilm and kept in the refrigerator.

4.6. Gelation of PVA Hydrogels Loaded with Quercetin/HP-β-CD Inclusion Complex

To investigate the gel fractions of the hydrogels used in the study, it was first necessary to ensure that the samples had achieved a constant weight. The prepared hydrogels were cut into 0.5 cm radius round and then dried in an oven at 50 °C for 48 h to obtain the dry weight (Wdry). After that, the dried hydrogels were submerged in phosphate-buffered saline (PBS, pH = 7.4) at 37 °C for 24 h with agitation (100 rpm) and then were dried again at 50 °C for 48 h to obtain the hydrogel weight after extraction (Wafter extraction):

$$\text{Gelation}\,(\%) = \frac{W_{\text{after extraction}}}{W_{\text{dry}}} \times 100 \quad (3)$$

4.7. Swelling Ratio of PVA Hydrogels Loaded with Quercetin/HP-β-CD Inclusion Complex

The swelling ratio was measured by taking the dried hydrogel (mdry) and the wet mass (mwet) of the hydrogel after immersion in a solution for a specified time [49]. First, each hydrogel was cut into circles of 0.5 cm radius that were then immersed in 5 mL of phosphate-buffered saline (PBS, pH = 7.4) at 37 °C with agitation (100 rpm) for 10, 30, 60, 180, and 360 min. Finally, the swelling ratio of the hydrogels was calculated using the following equation:

$$\text{Swelling ratio}\,(\%) = \frac{m_{\text{wet}} - m_{\text{dry}}}{m_{\text{dry}}} \times 100 \quad (4)$$

4.8. Mechanical Test of PVA Hydrogels Loaded with Quercetin/HP-β-CD Inclusion Complex

To observe the effect of loading the drug into the hydrogel on its mechanical properties, compression testing was performed. The hydrogels were cut into circles of 0.5 cm radius in all experiments. The compressive strength of neat PVA hydrogel and drug-loaded hydrogels was measured with a universal testing machine (LRX-Plus, AMETEK Lloyd Instruments Ltd., Hampshire, UK) in terms of stiffness and compressive modulus. The loading cell was set at a 500 N, with 0.05 N preload and 1 mm/min speed, and the measurement was stopped when the load reached 10 N [50].

4.9. In Vitro PVA Hydrogels Loaded with Quercetin/HP-β-CD Inclusion Complex Dissolution and Release Kinetics Study

The cumulative amount of released quercetin/HP-β-CD from hydrogels was measured using an immersion method according to Chuysinuan et al. [51] with minor modifications. The disc-shaped specimens (2.8 cm in diameter) were immersed in phosphate-buffered saline (PBS, pH = 7.4) and incubated at 37 °C under continuous stirring. The sample solution (1 mL) was withdrawn at specific time intervals, and an equal amount of fresh release medium was added. The absorbance of released quercetin was measured with a UV-Vis spectrophotometer (LAMBDA 850+, PerkinElmer, Waltham, MA, USA) at a wavelength of 370 nm. The obtained data were back calculated from the predetermined quercetin calibration curve (R2 = 0.9944) to determine the amounts of quercetin released from the hydrogel samples. The cumulative release of quercetin from each hydrogel at different time points (with measurement carried out in triplicate at each time point) was reported as parts per million of the amounts of quercetin released (WQuercetin released) divided by the weight of hydrogels (Whydrogel):

$$\text{Cumulative release } (\%) = \frac{W_{\text{Quercetin released}}}{W_{\text{hydrogel}}} \times 10^6 \quad (5)$$

A release kinetics study was used to ascertain the release mechanism, and the quercetin release data were fitted to Korsmeyer–Peppas and zero-order models. The first 60% of the drug release data was used to fit the models.

The Korsmeyer–Peppas model is a simple model that describes drug release from polymer nanoparticle systems [52,53]:

$$M_t/M_\infty = kt^n$$

where Mt/M∞ is the quercetin release fraction at time t, k is a constant incorporating geometric structural features, and n is the release exponent indicating the release rate mechanism. The value of n indicates the mechanism of the release; a value around 0.45 indicates case I (Fickian) diffusion, between 0.45 and 0.89 indicates anomalous (non-Fickian) diffusion, and between 0.89 and 1 indicates case II transport (zero-order kinetics).

A zero-order kinetic model is used to describe drugs that are released slowly with a constant concentration and can be characterized as ideal kinetic model in that it maintains constant drug levels during the delivery process [54,55]:

$$M_t/M_\infty = kt$$

where Mt/M∞ is the quercetin release fraction at time t, with zero-order rate constant, and t is investigation time.

4.10. In Vitro Antioxidant Characterization

The antioxidant activity of free quercetin and quercetin/HP-β-CD inclusion complex was evaluated using 2, 2-diphenyl-1-picrylhydrazyl (DPPH) free radical scavenging assay following [3]. First, quercetin and quercetin/HP-β-CD in various concentrations (15, 30, 60, and 120 μg/mL) were dissolved in methanol. Then, the samples (1 mL) were mixed with 3 mL DPPH solution (0.1 mM in methanol). The mixture was incubated for 30 min in the dark. Finally, the absorbance of the reaction mixture was measured using a microplate reader at 517 nm. Radical scavenging activity (%) was calculated using the following equation. Ascorbic acid was used as a standard and all analyses were performed in triplicate:

$$\text{DPPH radical scavenging activity } (\%) = \frac{(\text{Absorbance of control} - \text{Absorbance of tested sample})}{\text{Absorbance of control}} \times 100 \quad (6)$$

4.11. Cytotoxicity Evaluation

CDs are selected for pharmaceutical formulations to enhance the solubility, bioavailability, and stability of many drugs. However, it is important to consider their structure and cytotoxic potential for more efficient drug delivery [56]. The biocompatibility of PVA hydrogels loaded with different concentrations of quercetin/HP-β-CD inclusion complex was assessed to determine their suitability for biomedical applications. The evaluation focused on the indirect cytotoxicity towards NCTC 929 clone cells (ATCC-CCL-1, Rockville, MD, USA, 17th passage) and followed the ISO10993-5 standard test method [57]. The cells were initially grown in Dulbecco's Modified Eagle Medium (DMEM; GIBCO, Waltham, MA, USA) supplemented with 10% fetal bovine serum (FBS; GIBCO, USA) and 1% antibiotic-antimycotic agent (GIBCO, Waltham, MA, USA). Next, the cells were placed into 96-well tissue-culture polystyrene (TCPS) plates (SPL Lifescience, Pochon, Korea) with a density of 8000 cells per well. The plates were then kept in an incubator at a temperature of 37 °C, with a humidified environment consisting of 95% air and 5% CO2. Afterwards, the hydrogel samples underwent sterilization through a 30-minute exposure to UV radiation on each side. Subsequently, they were placed in a 96-well TCPS plate and immersed in serum-free medium (SFM), which consisted of DMEM and a 1% antibiotic-antimycotic agent. The samples were incubated for 24 hours to generate sample extraction at five different ratios of extraction medium (0.5, 5, 10, 25, and 50 mg/mL). NCTC 929 clone cells were cultivated individually in culture medium in wells of TCPS plates at a density of 8000 cells per well for 24 hours to enable cell attachment to the well surface. Subsequently, the medium was substituted with an extraction medium, and mouse fibroblast L929 cells were then cultured for an additional 24 hours.

The cell viability in each extraction media was assessed using the 3-(4,5-dimethylthiazol-2-yl)-2,5-diphenyltetrazolium bromide (MTT) test. Following the treatment, the medium was extracted, and the samples were rinsed with PBS. Subsequently, MTT solution with a concentration of 0.5 mg/mL was introduced, and the samples were left to incubate for a duration of 3 hours. Subsequently, the MTT solution was extracted from the well and substituted with DMSO (Labscan, Bangkok, Thailand) to facilitate the dissolution of the formazan crystals. Cell viability was assessed by measuring the absorbance of the solutions at 570 nm using a microplate reader (BioTek Instruments, Winooski, VT, USA). The cell viability of cultures treated with fresh serum-free medium (SFM) was employed as a control.

AT A GLANCE

The focus of this work was to create a hydrogel made of polyvinyl alcohol (PVA) that contains a complex of quercetin and cyclodextrin. The goal was to improve the stability and solubility of the complex. The solvent evaporation method was used to encapsulate quercetin in hydroxypropyl-β-cyclodextrin (HP-β-CD). The quercetin/HP-β-CD inclusion complex, which was produced, demonstrated an encapsulation efficiency (%EE) of 90.50 ± 1.84% and a loading capacity (%LC) of 4.67 ± 0.13%. The successful encapsulation of the complex was validated using FT-IR and XRD analysis. The quercetin/HP-β-CD inclusion complex was evenly distributed in thick solutions of PVA at different concentrations (0.5, 1.0, 1.5, 2.5, and 5.0% w/v ratio). The solution containing the drug-loaded polymer was physically crosslinked by subjecting it to numerous freeze-thaw cycles, resulting in the formation of a hydrogel.

The total amount of quercetin released from the hydrogels that were created rose as the concentrations of the inclusion complex increased. The incorporation of the inclusion complex into the PVA hydrogels did not affect their swelling ratio, but it resulted in a decrease in gelation and compressive strength as the concentration of the inclusion complex increased. The cytotoxicity of quercetin/HP-β-CD inclusion complex hydrogels was assessed using the MTT test and reported as the percentage of cell viability. The results demonstrate the biocompatibility of the tested material with NCTC 929 clone cells. The inhibitory effectiveness was assessed using the 2, 2-diphenyl-1-picrylhydrazyl (DPPH) free radical scavenging assay. The results indicate that quercetin/HP-β-CD inclusion complex hydrogels exhibit a greater degree of antioxidant activity compared to free quercetin. Our study's findings suggest that the quercetin/HP-β-CD inclusion complex hydrogels we produced have the necessary characteristics and can be considered as a delivery mechanism for quercetin in wound-healing applications.

REFERENCES

1. Cuzzell, J. Choosing a wound dressing. Geriatr. Nurs. 1997, 18, 260–265.
2. Dhivya, S.; Padma, V.V.; Santhini, E. Wound dressings—A review. BioMedicine 2015, 5, 22.
3. Maver, T.; Kurecic, M.; Smrke, D.; Stana-Kleinschek, K.; Maver, U. Plant-derived medicines with potential use in wound treatment. In Herbal Medicine, 1st ed.; Builders, P., Ed.; IntechOpen: London, UK, 2019; pp. 121–150.
4. Shedoeva, A.; Leavesley, D.; Upton, Z.; Fan, C. Wound Healing and the Use of Medicinal Plants. J. Evid. Based Complement. Altern. Med. 2019, 2019, 2684108.
5. Bhatt, K.; Flora, S.J.S. Oral co-administration of α-lipoic acid, quercetin and captopril prevents gallium arsenide toxicity in rats. Environ. Toxicol. Pharmacol. 2009, 28, 140–146.
6. Heinonen, I.M.; Meyer, A.S. Antioxidants in fruits, berries and vegetables. In Fruit and Vegetable Processing, 1st ed.; Jongen, W., Ed.; Woodhead Publishing Limited: Cambridge, OH, USA, 2002; pp. 23–51.
7. Li, Y.; Yao, J.; Han, C.; Yang, J.; Chaudhry, M.T.; Wang, S.; Liu, H.; Yin, Y. Quercetin, Inflammation and Immunity. Nutrients 2016, 8, 167.
8. Nathiya, S.; Durga, M.; Devasena, T. Quercetin, encapsulated quercetin and its application—A review. Int. J. Pharm. Pharm. Sci. 2014, 6, 20–26.
9. Razmara, R.S.; Daneshfar, A.; Sahraei, R. Solubility of Quercetin in Water + Methanol and Water + Ethanol from (292.8 to 333.8) K. J. Chem. Eng. Data 2010, 55, 3934–3936.
10. Liu, M.; Dong, L.; Chen, A.; Zheng, Y.; Sun, D.; Wang, X.; Wang, B. Inclusion complexes of quercetin with three β-cyclodextrins derivatives at physiological pH: Spectroscopic study and antioxidant activity. Spectrochim. Acta A Mol. Biomol. Spectrosc. 2013, 115, 854–860.
11. Singh, P.; Kumar, S. Chapter 2—Microbial enzyme in food biotechnology. In Enzymes in Food Biotechnology, 1st ed.; Kuddus, M., Ed.; Academic Press: London, UK, 2019; pp. 19–28.
12. Del Valle, E.M.M. Cyclodextrins and their uses: A review. Process Biochem. 2004, 39, 1033–1046.
13. Kfoury, M.; Hădărugă, N.G.; Hădărugă, D.I.; Fourmentin, S. 4—Cyclodextrins as encapsulation material for flavors and aroma. In Encapsulations; Grumezescu, A.M., Ed.; Academic Press: Cambridge, MA, USA, 2016; Volume 2, pp. 127–192.

14. Gupta, A.; Keddie, D.J.; Kannappan, V.; Gibson, H.; Khalil, I.R.; Kowalczuk, M.; Martin, C.; Shuai, X.; Radecka, I. Production and characterisation of bacterial cellulose hydrogels loaded with curcumin encapsulated in cyclodextrins as wound dressings. Eur. Polym. J. 2019, 118, 437–450.
15. Esmaeilpour, D.; Hussein, A.A.; Almalki, F.A.; Shityakov, S.; Bordbar, A.K. Probing inclusion complexes of 2-hydroxypropyl-β-cyclodextrin with mono-amino mono-carboxylic acids: Physicochemical specification, characterization and molecular modeling. Heliyon 2020, 6, e03360.
16. Zhang, M.; Zhao, X. Alginate hydrogel dressings for advanced wound management. Int. J. Biol. Macromol. 2020, 162, 1414–1428.
17. Boonkaew, B.; Barber, P.M.; Rengpipat, S.; Supaphol, P.; Kempf, M.; He, J.; John, V.T.; Cuttle, L. Development and Characterization of a Novel, Antimicrobial, Sterile Hydrogel Dressing for Burn Wounds: Single–Step Production with Gamma Irradiation Creates Silver Nanoparticles and Radical Polymerization. J. Pharm. Sci. 2014, 103, 3244–3253.
18. Weller, C.; Weller, C.; Team, V. 4—Interactive dressings and their role in moist wound management. In Advanced Textiles for Wound Care, 2nd ed.; Rajendran, S., Ed.; Woodhead Publishing: Cambridge, UK, 2019; pp. 105–134.
19. Tsou, Y.-H.; Khoneisser, J.; Huang, P.-C.; Xu, X. Hydrogel as a bioactive material to regulate stem cell fate. Bioact. Mater. 2016, 1, 39–55.
20. Figueroa-Pizano, M.D.; Vélaz, I.; Martínez-Barbosa, M.E. A Freeze-Thawing Method to Prepare Chitosan-Poly(vinyl alcohol) Hydrogels without Crosslinking Agents and Diflunisal Release Studies. J. Vis. Exp. 2020, 155, e59636.
21. Hassan, C.; Peppas, N. Structure and Morphology of Freeze/Thawed PVA Hydrogels. Macromolecules 2000, 33, 2472–2479.
22. Catauro, M.; Papale, F.; Bollino, F.; Piccolella, S.; Marciano, S.; Nocera, P.; Pacifico, S. Silica/quercetin sol-gel hybrids as antioxidant dental implant materials. Sci. Technol. Adv. Mater. 2015, 16, 035001.
23. Da Silva Júnior, W.F.; Bezerra de Menezes, D.L.; de Oliveira, L.C.; Koester, L.S.; Oliveira de Almeida, P.D.; Lima, E.S.; de Azevedo, E.P.; da Veiga Júnior, V.F.; Neves de Lima, Á.A. Inclusion Complexes of β and HPβ-Cyclodextrin with α, β Amyrin and In Vitro Anti-Inflammatory Activity. Biomolecules 2019, 9, 241.
24. Lan, Q.; Di, D.; Wang, S.; Zhao, Q.; Gao, Y.; Chang, D.; Jiang, T. Chitosan-N-acetylcysteine modified HP-β-CD inclusion complex as a potential ocular delivery system for anti-cataract drug: Quercetin. J. Drug Deliv. Sci. Technol. 2020, 55, 101407.
25. Pereira, B.A.; Silva, M.A.; Barroca, J.M.; Marques, M.P.M.; Braga, S.S. Physicochemical Properties, Antioxidant Action and Practical Application in Fresh Cheese of the Solid Inclusion Compound γ-Cyclodextrin•quercetin, in Comparison with β-Cyclodextrin•quercetin. Arab. J. Chem. 2020, 13, 205–215.
26. Aytac, Z.; Ipek, S.; Durgun, E.; Uyar, T. Antioxidant electrospun zein nanofibrous web encapsulating quercetin/cyclodextrin inclusion complex. J. Mater. Sci. 2018, 53, 1527–1539.
27. Han, D.; Han, Z.; Liu, L.; Wang, Y.; Xin, S.; Zhang, H.; Yu, Z. Solubility Enhancement of Myricetin by Inclusion Complexation with Heptakis-O-(2-Hydroxypropyl)-β-Cyclodextrin: A Joint Experimental and Theoretical Study. Int. J. Mol. Sci. 2020, 21, 766.

28. Zhu, G.; Xiao, Z.; Zhu, G.; Zhou, R.; Niu, Y. Encapsulation of l-menthol in hydroxypropyl-β-cyclodextrin and release characteristics of the inclusion complex. Pol. J. Chem. Technol. 2016, 18, 110–116.

29. Kim, S.J. Study of flavonoid/hydroxypropyl-β-cyclodextrin inclusion complexes by UV-Vis, FT-IR, DSC, and X-Ray diffraction analysis. Prev. Nutr. Food Sci. 2020, 25, 449–456.

30. Lv, R.; Qi, L.; Zou, Y.; Zou, J.; Luo, Z.; Shao, P.; Tamer, T.M. Preparation and structural properties of amylose complexes with quercetin and their preliminary evaluation in delivery application. Int. J. Food Prop. 2019, 22, 1445–1462.

31. Ding, Y.; Pang, Y.; Vara Prasad, C.; Wang, B. Formation of inclusion complex of enrofloxacin with 2-hydroxypropyl-β-cyclodextrin. Drug Deliv. 2020, 27, 334–343.

32. Pralhad, T.; Rajendrakumar, K. Study of freeze-dried quercetin–cyclodextrin binary systems by DSC, FT-IR, X-ray diffraction and SEM analysis. J. Pharm. Biomed. 2004, 34, 333–339.

33. Pradhan, C.P.; Mandal, A.; Dutta, A.; Sarkar, R.; Kundu, A.; Saha, S. Delineating the behavior of Berberis anthocyanin/β-cyclodextrin inclusion complex in vitro: A molecular dynamics approach. LWT 2022, 157, 113090.

34. Choipang, C.; Buntum, T.; Chuysinuan, P.; Techasakul, S.; Supaphol, P.; Suwantong, O. Gelatin scaffolds loaded with asiaticoside/2-hydroxypropyl-β-cyclodextrin complex for use as wound dressings. Polym. Adv. Technol. 2021, 32, 1187–1193.

35. Al-Qubaisi, M.S.; Rasedee, A.; Flaifel, M.H.; Eid, E.E.M.; Hussein-Al-Ali, S.; Alhassan, F.H.; Salih, A.M.; Hussein, M.Z.; Zainal, Z.; Sani, D.; et al. Characterization of thymoquinone/hydroxypropyl-β-cyclodextrin inclusion complex: Application to anti-allergy properties. Eur. J. Pharm. Sci. 2019, 133, 167–182.

36. Hadian, Z.; Maleki, M.; Abdi, K.; Atyabi, F.; Mohammadi, A.; Khaksar, R. Preparation and Characterization of Nanoparticle β-Cyclodextrin:Geraniol Inclusion Complexes. Iran J. Pharm. Res. 2018, 17, 39–51. [PubMed]

37. Afshar, M.; Dini, G.; Vaezifar, S.; Mehdikhani, M.; Movahedi, B. Preparation and characterization of sodium alginate/polyvinyl alcohol hydrogel containing drug-loaded chitosan nanoparticles as a drug delivery system. J. Drug Deliv. Sci. Technol. 2020, 56, 101530.

38. Chou, S.-F.; Woodrow, K.A. Relationships between mechanical properties and drug release from electrospun fibers of PCL and PLGA blends. J. Mech. Behav. Biomed. Mater. 2017, 65, 724–733.

39. Qing, X.; He, G.; Liu, Z.; Yin, Y.; Cai, W.; Fan, L.; Fardim, P. Preparation and properties of polyvinyl alcohol/N–succinyl chitosan/lincomycin composite antibacterial hydrogels for wound dressing. Carbohydr. Polym. 2021, 261, 117875.

40. Sampson, L.; Rimm, E.; Hollman, P.C.; de Vries, J.H.; Katan, M.B. Flavonol and flavone intakes in US health professionals. J. Am. Diet. Assoc. 2002, 102, 1414–1420.

41. Cai, B.; Engqvist, H.; Bredenberg, S. Development and evaluation of a tampering resistant transdermal fentanyl patch. Int. J. Pharm. 2015, 488, 102–107.

42. Thanyacharoen, T.; Chuysinuan, P.; Techasakul, S.; Nooeaid, P.; Ummartyotin, S. Development of a gallic acid-loaded chitosan and polyvinyl alcohol hydrogel composite: Release characteristics and antioxidant activity. Int. J. Biol. Macromol. 2018, 107, 363–370.

43. Nalini, T.; Basha, S.K.; Sadiq, A.M.; Kumari, V.S. In vitro cytocompatibility assessment and antibacterial effects of quercetin encapsulated alginate/chitosan nanoparticle. Int. J. Biol. Macromol. 2022, 219, 304–311.

44. Srivastava, G.K.; Alonso-Alonso, M.L.; Fernandez-Bueno, I.; Garcia-Gutierrez, M.T.; Rull, F.; Medina, J.; Coco, R.M.; Pastor, J.C. Comparison between direct contact and extract exposure methods for PFO cytotoxicity evaluation. Sci. Rep. 2018, 8, 1425.

45. Figueiras, A.; Carvalho, R.A.; Ribeiro, L.; Torres-Labandeira, J.J.; Veiga, F.J.B. Solid-state characterization and dissolution profiles of the inclusion complexes of omeprazole with native and chemically modified β-cyclodextrin. Eur. J. Pharm. Biopharm. 2007, 67, 531–539.

46. Kulkarni, A.; Dias, R.; Ghorpade, V. Freeze dried multicomponent inclusion complexes of quercetin: Physicochemical evaluation and pharmacodynamic study. Marmara Pharm. J. 2019, 23, 403–414.

47. Niu, Y.; Deng, J.; Xiao, Z.; Kou, X.; Zhu, G.; Liu, M.; Liu, S. Preparation and slow release kinetics of apple fragrance/β-cyclodextrin inclusion complex. J. Therm. Anal. Calorim. 2020, 143, 3775–3781.

48. Ahsan, A.; Farooq, M.A. Therapeutic potential of green synthesized silver nanoparticles loaded PVA hydrogel patches for wound healing. J. Drug Deliv. Sci. Technol. 2019, 54, 101308.

49. Raghuwanshi, V.S.; Garnier, G. Characterisation of hydrogels: Linking the nano to the microscale. Adv. Colloid Interface Sci. 2019, 274, 102044.

50. Fan, W.; Zhang, Z.; Liu, Y.; Wang, J.; Li, Z.; Wang, M. Shape memory polyacrylamide/gelatin hydrogel with controllable mechanical and drug release properties potential for wound dressing application. Polymer 2021, 226, 123786.

51. Chuysinuan, P.; Thanyacharoen, T.; Thongchai, K.; Techasakul, S.; Ummartyotin, S. Preparation of chitosan/hydrolyzed collagen/hyaluronic acid based hydrogel composite with caffeic acid addition. Int. J. Biol. Macromol. 2020, 162, 1937–1943.

52. Korsmeyer, R.W.; Gurny, R.; Doelker, E.; Buri, P.; Peppas, N.A. Mechanisms of solute release from porous hydrophilic polymers. Int. J. Pharm. 1983, 15, 25–35.

53. Brunner, E. Reaktionsgeschwindigkeit in heterogenen Systemen. Z. Phys. Chem. 1904, 47, 56–102.

54. Rasetti-Escargueil, C.; Grangé, V. Pharmacokinetic profiles of two tablet formulations of piroxicam. Int. J. Pharm. 2005, 295, 129–134.

55. Freitas, M.N.; Marchetti, J.M. Nimesulide PLA microspheres as a potential sustained release system for the treatment of inflammatory diseases. Int. J. Pharm. 2005, 295, 201–211.

56. Staedler, D.; Idrizi, E.; Kenzaoui, B.H.; Juillerat-Jeanneret, L. Drug combinations with quercetin: Doxorubicin plus quercetin in human breast cancer cells. Cancer Chemother. Pharmacol. 2011, 68, 1161–1172.

57. Kudłacik-Kramarczyk, S.; Drabczyk, A.; Głąb, M.; Alves-Lima, D.; Lin, H.; Douglas, T.E.L.; Kuciel, S.; Zagórska, A.; Tyliszczak, B. Investigations on the impact of the introduction of the Aloe vera into the hydrogel matrix on cytotoxic and hydrophilic properties of these systems considered as potential wound dressings. Mater. Sci. Eng. C. 2021, 123, 111977.

12
Chapter

THE SYNTHESIS OF A NEW TYPE OF THERMOSENSITIVE HYDROGEL, BASED ON P(OEGMA-CO-MEO2MA) MICROGELS, AND THEIR UTILIZATION AS A THREE-DIMENSIONAL CELL SCAFFOLD

1. INTRODUCTION: AN OVERVIEW

Biomedical researchers and clinical workers have shown significant interest in using tissue engineering technology to create three-dimensional (3D) cell models for tumor research, drug screening, and tissue healing in recent years [1,2,3]. This involves utilizing in vitro 3D cell culture techniques. Scaffold materials are crucial in this context. Scaffold materials serve as carriers for seed cells and bioactive growth factors. They create a suitable microenvironment for cell growth, reproduction, metabolism, and other physiological functions. Additionally, they stimulate cells to construct a 3D model [4,5,6,7]. An ideal scaffold material should possess a three-dimensional porous network structure that promotes the implantation of cells and facilitates the transportation of nutrients and metabolic waste. Simultaneously, it should possess optimal biocompatibility, degradability, and mechanical strength that align with the characteristics of the extracellular matrix.

Hydrogel is a soft substance that has a three-dimensional network structure. It is created using many ways of physical or chemical crosslinking, as documented in references [8,9,10,11]. Hydrogels possess physical features that closely resemble those of extracellular matrices, which have been extensively employed as scaffold materials for 3D cell culture [12,13,14]. Specifically, hydrogel scaffolds made from thermosensitive polymers have garnered significant interest [15,16,17]. Thermosensitive polymers typically exhibit a liquid state at normal room temperature and can be uniformly blended with cells, medicines, growth factors, and other substances. At the low critical dissolution temperature (LCST) or volume phase transition temperature (VPTT), and in the presence of suitable ions, the thermosensitive polymer undergoes in situ gelation to create injectable 3D hydrogels. Therefore, it has the ability to embed cells in their original location without causing damage and stimulate more cell proliferation. Cooling can be used to liquefy cells or complex cell clumps, making them easy to release.

Poly(N-isopropylacrylamide) (PNIPAM) is a well researched thermosensitive material that exhibits a lower critical solution temperature (LCST) or a volume phase transition temperature (VPTT) of 32 °C. It is commonly employed as a scaffold material [18,19,20]. Ekerdt and his team created a thermosensitive hyaluronic acid (HA)-PNIPAM brush polymer using thiol-ene click chemistry. This polymer could be mixed with human embryonic stem cells (hESCs) or human induced pluripotent stem cells (hiPSCs) at a temperature of 4 °C. When the temperature was raised to 37 °C, the mixture quickly formed a hydrogel in its original location. After five days of cultivation, the cells grew rapidly and formed regular multicellular spheroids. Importantly, the cells maintained their pluripotency even after multiple passages. In addition, this particular hydrogel demonstrated the capability to release the cells following the processes of freezing, liquefaction, and centrifugation [22]. Liang et al. produced a dendritic thermosensitive polymer by combining PNIPAM, dimeric glycerol, and polyethylene glycol (PEG). This polymer could potentially serve as a scaffold for the cultivation of hiPSCs, which could be released via the process of chilling, liquefaction, and centrifugation [23].

In order to enhance and refine the characteristics of the created thermosensitive hydrogel, spherical thermosensitive polymers, known as microgels, were additionally employed to construct hydrogel scaffolds. Microgels are nanoscale hydrogel particles that are spherical in shape and range in size from 10 nm to a few microns. They possess several advantages such as straightforward preparation, rapid reaction, and convenient modification, making them highly appealing in the field of biomedicine [24,25,26,27]. Gan et al. have effectively fabricated this particular thermosensitive hydrogel scaffold utilizing PNIPAM-based thermosensitive microgels, marking the first instance of its construction. The resulting hydrogel possesses an interconnected porosity architecture that greatly facilitates the transportation of oxygen, nutrients, and biological metabolites. HEK 293T cells, derived from human embryonic kidney, were successfully enclosed within the hydrogel and exhibited robust growth [28]. In addition, we utilized PNIPAM-based microgels as scaffold materials to encapsulate aggregates of HepG2 cells, a type of human hepatocellular carcinoma. This approach allowed us to rapidly generate multicellular spheroids of HepG2 cells [29]. Furthermore, Shen et al. employed NIPAM and acrylamide (AAm) copolymer microgels, which have a higher affinity for water, as scaffold materials to cultivate mouse melanoma cells [30]. It was discovered that the cells exhibited progressive growth in a uniformly scattered state. Greater particle size of the microgels leads to enhanced cellular activity. The Dai group employed PNIPAM-based microgels with a slight negative charge and a slight positive charge as cell scaffolds to investigate the cultivation and characteristics of mouse embryonic mesenchymal stem cells (MSCs), correspondingly [31,32].

Although PNIPAM thermosensitive polymers offer various benefits, the resulting thermosensitive hydrogels experience significant shrinkage over time, greatly restricting cell growth and activity. Despite attempts to mitigate the shrinking of hydrogels by including negatively charged acrylic acid (AA) [33], mixing PEG [29,34], and modifying the particle size of the microgels [35], the outcome remains unsatisfactory. Furthermore, PNIPAM polymers possess a certain degree of toxicity and provide challenges in terms of degradation. Thus, it is imperative to discover a thermosensitive gel material that exhibits little shrinkage, as well as exceptional biocompatibility and biodegradability. Researchers have shown considerable interest in innovative thermosensitive polymers based on oligo(ethylene glycol) methacrylate (OEGMA) and 2-(2-methoxyethoxy) ethyl acrylate (MEO2MA) in recent years. This is due to their exceptional biocompatibility and biodegradability [36,37].

OEGMA and MEO2MA are two forms of PEG-based macromolecular monomers that have comparable structures. The LCST (lower critical solution temperature) of POEGMA homopolymer is approximately 90 °C, and it typically functions as a hydrophilic chain segment. On the other hand, PMEO2MA homopolymer has an LCST of around 19 °C and typically serves as a hydrophobic chain segment. The copolymers derived from these substances exhibit exceptional thermosensitive properties. By manipulating the relative amounts of OEGMA and MEO2MA during the polymerization process, the LCST or VPTT can be adjusted within the range of 19 to 90 °C. Multiple studies have utilized these innovative thermosensitive polymers for the cultivation of diverse cell types. Anderson et al. created a thermosensitive polymer, namely a brush-like P(OEGMA-co-MEO2MA) polymer, by manipulating the proportion of OEGMA and MEO2MA. It served as a matrix material for investigating the impact on adhesion and morphology of L-929 murine fibroblasts [38,39]. The study revealed that the incorporation of thermosensitive polymers with a lower fraction of hydrophilic OEGMA was more conducive to cell development and adherence. Furthermore, Bakaic et al. [40] developed a hydrogel scaffold that can be injected into the body and has the ability to respond to changes in temperature and pH. This was achieved by chemically crosslinking neutral or charged aldehyde functionalized POEGMA copolymers with hydrazide functionalized POEGMA copolymer precursors. The researchers discovered that the charged hydrogel exhibited greater support for the two-dimensional (2D) adhesion of 3T3 mouse fibroblasts and the three-dimensional (3D) stability and proliferation of ARPE-19 human retinal epithelial cells.

This publication examines the viability of utilizing P(OEGMA-co-MEO2MA) spherical microgels for the construction of thermosensitive hydrogel 3D cell scaffolds, marking the first instance of such an application. Initially, we employed OEGMA and MEO2MA to synthesize a range of P(OEGMA-co-MEO2MA) microgels with varying VPTT and sizes via free radical polymerization. Subsequently, thermosensitive hydrogels were synthesized using the thermal induction approach, incorporating unique P(OEGMA-co-MEO2MA) microgels, while subjecting the mixture to heating and the presence of salts. The MCF-7 human breast cancer cells were utilized as the cellular model for embedding and co-culturing within a new thermosensitive hydrogel. As far as we know, there are no documented instances of P(OEGMA-co-MEO2MA)-based microgels being used as scaffolds for 3D cell culture. Thus, our study has the potential to offer novel insights for the advancement of thermosensitive hydrogel scaffolds.

2. RESULTS AND DISCUSSION

2.1. The Synthesis of P(OEGMA-co-MEO2MA) Thermosensitive Microgels

In this work, to investigate the influence of the properties of microgels on subsequent hydrogel formation, five types of P(OEGMA-co-MEO2MA) microgel samples, with different thermosensitivities or sizes, were synthesized by a free radical polymerization reaction [26].

As shown in Table 1, we first synthesized M1-M3 samples through changing the mass ratio of thermosensitive monomers OEGMA and MEO2MA. We measured the particle sizes of microgel aqueous solutions at different temperatures using a nanometer particle size meter. It can be seen from Figure 1A that the particle sizes of all the three P(OEGMA-co-MEO2MA) microgels decreased gradually with increase in temperature, which indicated that all of them had great thermosensitivity, similar to linear and branched P(OEGMA-co-MEO2MA) polymers. At lower temperatures, due to the strong hydrogen bonding of oxygen atoms in the microgel, the microgel particle has strong hydrophilicity and high water content, so the particle size is larger, which is manifested as volume swelling. When the temperature gradually increases, the hydrogen bond is gradually destroyed, and the microgel gradually shows strong hydrophobic effect derived from alkyl groups, so the internal water content decreases, the volume shrinks, and the particle size decreases. The temperature at which the slope of the curve changes the most is defined as the VPTT. As shown in the figure, the VPTT of M1, M2, and M3 were about 21 °C, 2 °C, and 2 °C, respectively, which means that with the increase of the relative proportion of OEGMA in the monomers, the final microgel has a higher VPTT. This is because, compared with MEO2MA, OEGMA has more ether oxygen bonds in the molecular structure, which can form more hydrogen bonds with water molecules; so, the more OEGMA content, the stronger the hydrophilicity of the formed P(OEGMA-co-MEO2MA) microgel. In addition, we found that the sizes of M3 were larger than others at most temperatures, which may also be due to the higher content of OEGMA increasing the hydrophilicity and swelling of this microgel. This is in agreement with what was reported in the literature [41].

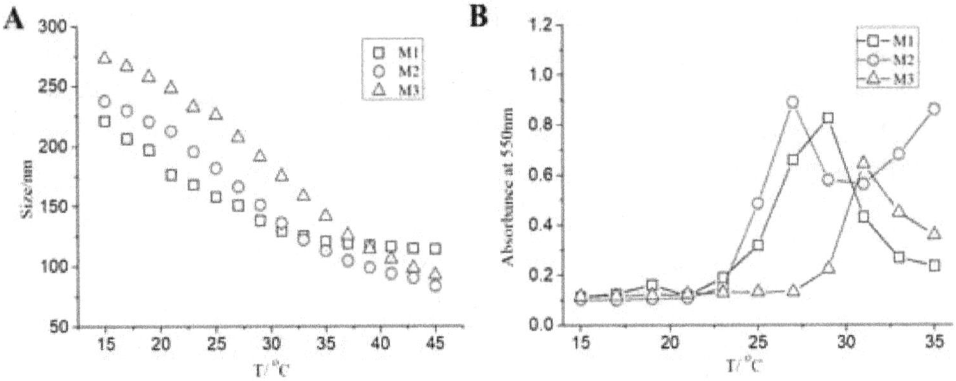

Figure 1. Hydrodynamic radius (A) and absorbance (B) of P(OEGMA-co-MEO2MA) microgel samples M1-M3 as a function of temperature.

Table 1. The feeding amount of each sample during the preparation process.

Sample	n (MEO_2MA): n (OEGMA)	m/ g				
		MEO_2MA	OEGMA	SDS	BIS	KPS
M1	95% : 3%	2.5002	0.1259	0.0284	0.0431	0.0812
M2	90% : 8%	2.3664	0.3368	0.0284	0.0431	0.0812
M3	85% : 13%	2.2371	0.5462	0.0284	0.0431	0.0812
M4	85% : 13%	2.2371	0.5462	0.0284	0.0431	0.0812
M5	85% : 13%	2.2371	0.5462	0.0284	0.0431	0.0812

In order to further study the thermosensitivity of the above prepared P(OEGMA-co-MEO2MA) microgels, we used phosphate buffer solution (PBS) to configure a certain concentrated dilute solution of microgel samples, and measured the change of absorbance at different temperatures by using an UV-Vis spectrophotometer. As shown in Figure 1B, when the temperature was lower, the absorbance of all the three microgel solutions was basically the same, but when the temperature exceeded a certain higher critical value, it mutated suddenly. The reason for this phenomenon is that with increase of temperature, hydrophilic microgels gradually become hydrophobic. When a critical temperature is reached, microgels gather together to form small aggregates, resulting in increased light refraction, thus decreasing light transmittance and increasing absorbance. Actually, the critical temperature is the VPTT. Similarly, we found that the VPTT of three samples was 21 °C, 23 °C and 27 °C, respectively. With increase of the relative proportion of OEGMA, the VPTTs of the samples M1, M2, M3 increased gradually, which was consistent with previous results.

In addition, the effect of the dosage of surfactant sodium dodecyl sulfate (SDS) on P(OEGMA-co-MEO2MA) microgel properties was also investigated. We synthesized microgel samples M4 and M5, and compared them with sample M3. The SDS dosages in their synthesis processes were reduced (Table 1). We first also measured the particle sizes of microgel solutions at different temperatures. As shown in Figure 2A, the particle sizes of M4 and M5 also decreased gradually with increase of temperature, which indicated the presence of excellent thermosensitivity. From M3, M4 to M5, with reduction of the SDS dosages, the sizes of the obtained microgels increased at measured temperatures, especially at lower temperatures. For instance, at 15 °C, the sizes of M3, M4 and M5 were 273.3 nm, 363.5 nm, and 406.7 nm, respectively. Moreover, we found that all these three types of microgel samples had the same VPTT, around 27 °C. The results of absorbance measurements at different temperatures, shown in Figure 2B, also indicates that the VPTT remains constant regardless of the SDS dosage. These phenomena indicated that with the decrease of SDS dosage, the relative particle size of the microgel generally increased, but the VPTT remained basically unchanged. This is because SDS in the reaction process does not directly participate in the polymerization reaction, but plays the role of stabilizing the parent particles of microgel by forming micelles. So, the SDS dosage will not affect the thermosensitivities and VPTTs of obtained microgels. The lower the SDS content, the larger the size of the micelle formed from the SDS, so the size of the final P(OEGMA-co-MEO2MA) microgel particle is larger.

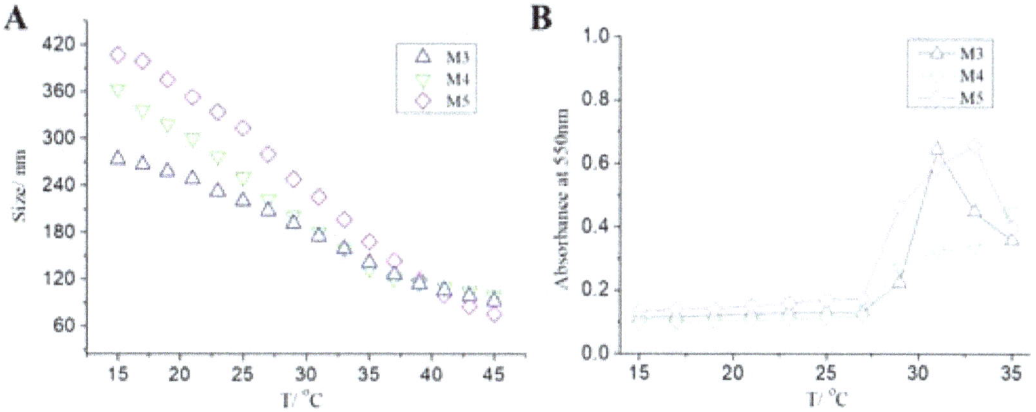

Figure 2. Hydrodynamic radius (A) and absorbance (B) of P(OEGMA-co-MEO2MA) microgel samples M3-M5 as a function of temperature.

In summary, three types of thermosensitive P(OEGMA-co-MEO2MA) microgels, with different VPTTs, were synthesized by free radical emulsion polymerization through change in the mass ratio of OEGMA and MEO2MA. With increase of the content of OEGMA, the VPTT of the obtained microgels also increased. In addition, two types of thermosensitive P(OEGMA-co-MEO2MA) microgels, with different sizes and the same VPTT, were also synthesized by polymerization through change of the SDS dosage. With the decrease of the dosages of SDS, the sizes of the obtained microgels increased.

2.2. The Preparation of P(OEGMA-co-MEO2MA) Thermosensitive Microgels-Based Hydrogel

Next, we studied the feasibility of using P(OEGMA-co-MEO2MA) microgels to construct thermosensitive hydrogels through the thermal induction method. Referring to our previously reported method [29], we configured the PBS solutions of microgel samples with a concentration of 3.0 wt%, and placed them at 37 °C, which was higher than their VPTT. Then, we observed the formation of hydrogels by the inverted method and photographed the changes of these hydrogels.

As shown in Figure 3, after maintaining for 10 min at 37 °C, all three types of P(OEGMA-co-MEO2MA) microgels could gel and form white hydrogels. This result was consistent with that of the traditional PNIPAM thermosensitive polymer. When the temperature was higher than their VPTT, the microgels, in their hydrophobic contraction state, would aggregate under the synergistic influence of hydrophobicity and the shielding effect of ions from PBS, thus forming hydrogel macroscopically. The gelation times were about 5 min to 10 min. Meanwhile, we found that the hydrogels formed from M1 and M2 showed obvious shrinkage after 2 h, and the shrinkage degree increased with time. After 48 h, the volume of the hydrogels formed from M1 and M2 was less than 1/2 of the initial value. This phenomenon of volume shrinkage has been reported several times regarding traditional PNIPAMs and other thermosensitive hydrogels, because of the dehydration of thermosensitive materials at higher temperature [19,20]. As an ideal scaffold material, the volume contraction should be avoided as far as possible in order to maintain sufficient cell growth space and ensure the circulation of nutrients and metabolic waste [4,6].

Figure 3. Photographs of the P(OEGMA-co-MEO2MA) hydrogels formed from M1–M3 microgels at different times.

Encouragingly, we found that the volume change of the hydrogel formed by the prepared sample M3 with a higher VPTT was less than 5% and no significant shrinkage was observed. So, we also observed and recorded the formation of hydrogels from M4 and M5 microgels with the same VPTT and larger sizes. It can be seen from Figure 4 that the hydrogels formed from M4 and M5 also showed obvious shrinkage after 2 h, and the shrinkage degree increased with time, which was hardly surprising.

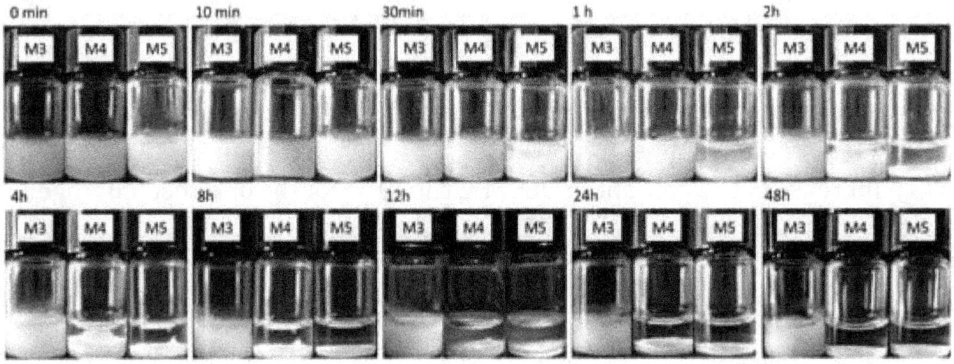

Figure 4. Photographs of the P(OEGMA-co-MEO2MA) hydrogels formed from M3–M5 microgels at different times.

We further investigated the quantitative relation between the volume shrinkage degree Vt/V0 of the hydrogel (the ratio of the volume of the formed hydrogel at a certain time point to the initial volume) and time (Figure 5). We found that the hydrogels formed from M1 or M2 contracted rapidly in the first 8 h, when the volume was less than half of the initial volume. After 48 h, the volumes of the samples were about 20% and 30% of the initial volume, respectively. The shrinkage degree was very large. However, the volume of hydrogel formed from M3 still had a volume of more than 98% of the initial volume after 48 h without significant shrinkage. We speculated that the reason why the hydrogel formed by M3 did not shrink significantly may be related to its containing longer OEGMA chain fragments with excellent hydrophilic properties, which could hinder the shrinkage of the hydrogel. In our previous report [29,34], we found that adding a small amount of PEG to the prepared PNIPAM microgel significantly reduced the shrinkage of the resulting hydrogel. Due to the presence of a large number of PEG fragments in OEGMA, we thought that OEGMA may have a similar effect to PEG in reducing hydrogel shrinkage.

We also compared the change of Vt/V0 of the hydrogels formed from M3, M4, and M5 microgels. It was shown that, as the size of P(OEGMA-co-MEO2MA) microgel increased, the shrinkage of hydrogel also increased. For instance, after 48 h, the volume of the hydrogel samples formed from M3, M4, and M5 were about 98%, 19% and 17% of the initial volume, respectively. We think the reason was that the larger the particle size of the microgel particles with the same VPTT, the higher the water content. Therefore, when at a certain temperature higher than their VPTT, the shrinkage degree of microgels is greater, resulting in more solvents being extruded from the microgels, which would also cause more volume shrinkage of the macroscopic hydrogels.

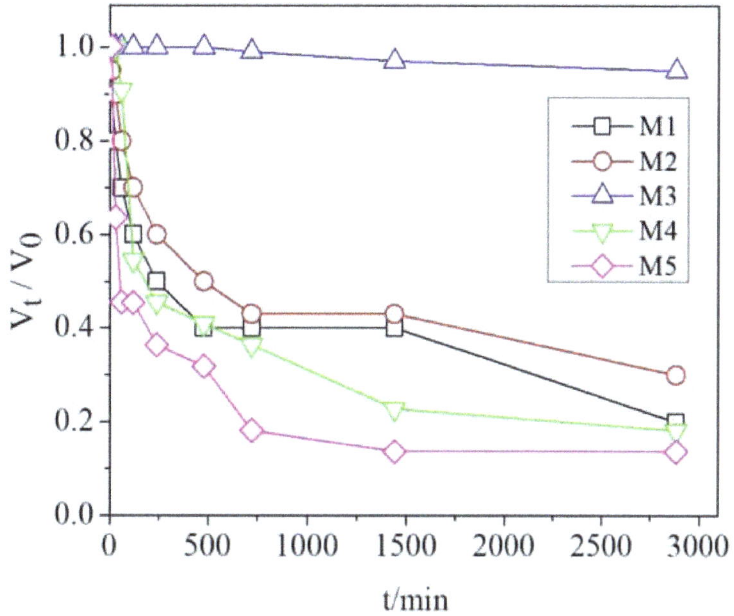

Figure 5. The change of Vt/V0 (shrinkage degree) of P(OEGMA-co-MEO2MA) hydrogels formed from microgels as a function of time.

These results indicated that, by adjusting the thermosensitivity (VPTT) and particle size of these synthesized P(OEGMA-co-MEO2MA) microgels, the properties of the obtained thermosensitive hydrogels can be adjusted, especially volume shrinkage. We also found that the obtained P(OEGMA-co-MEO2MA) thermosensitive hydrogels showed a small shrinkage degree only by controlling the synthesis conditions of P(OEGMA-co-MEO2MA) microgels; while for the commonly used PNIPAM hydrogels, complex operations, such as physical blending or copolymerization, were required to achieve the same effect. Therefore, the P(OEGMA-co-MEO2MA) thermosensitive hydrogel showed better application prospects in the biomedical field. In following work, we will continue to conduct in-depth studies on P(OEGMA-co-MEO2MA) microgel-based hydrogels, especially the gelation mechanism, gelation conditions and rheological behavior.

2.3. P(OEGMA-co-MEO2MA) Thermosensitive Hydrogels for 3D Cell Culture

A hydrogel without shrinkage is more suitable for cell embedding and 3D growth. Therefore, we selected microgel sample M3 as the scaffold material for subsequent 3D cell culture. A type of tumor cell, MCF-7 human breast cancer cell, was chosen as the model cell to explore the feasibility of using this hydrogel to construct a 3D cell model [29,30,33,34,35]. After a certain concentration of microgel dispersion, cells were evenly mixed at room temperature, and then transferred to 48-well culture plates and placed in a cell incubator at 37 °C for culturing. About 1 h later, a stable hydrogel was formed and the cells were encapsulated in situ in it. Part of the DMEM cell culture medium was added to continue this culture process.

Firstly, we investigated the viability of cells in the P(OEGMA-co-MEO2MA) hydrogel scaffold using MTT assay. As shown in Figure 6, the absorbance values measured by the MTT method increased significantly in the first two days of cell culture, which may be related to a rapid increase in the number of living cells in this hydrogel scaffold. This result also indirectly reflected the favorable biocompatibility of this hydrogel scaffold material. The increase of cell viability slowed down after the third day, which may be related to the accumulation of metabolic waste, and the reduction of nutrients and relative living space caused by the increase in the number of cells.

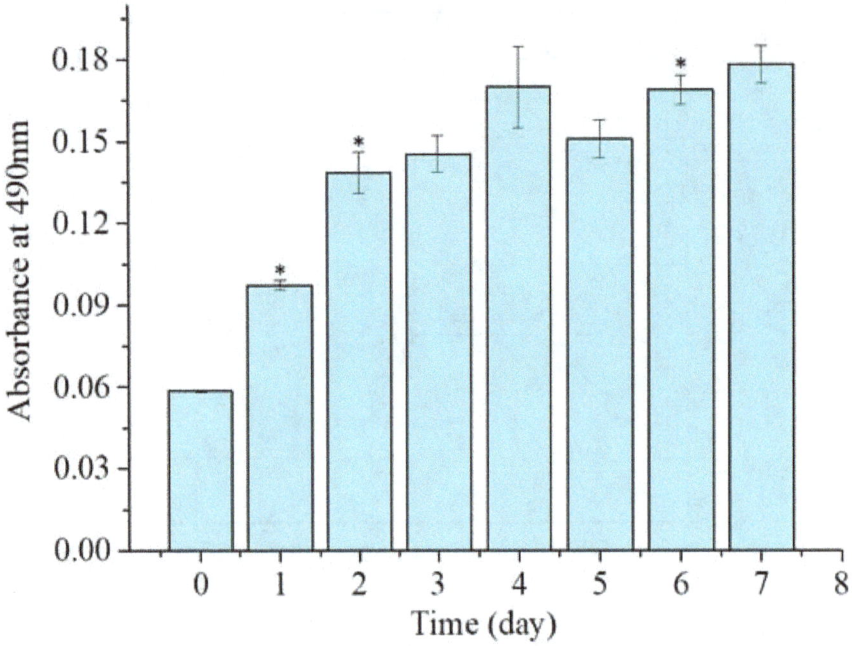

Figure 6. Viability of MCF-7 cells in formed thermosensitive P(OEGMA-co-MEO2MA) hydrogel as assessed by MTT assay. The data shown are the mean of three independent experiments. Error bars indicate the standard deviations. An asterisk (*) indicates a significant difference between this group of data and the previous group ($p < 0.05$).

We further observed the growth and morphology of cells in P(OEGMA-co-MEO2MA) thermosensitive hydrogel scaffolds by AO/EB staining, as shown in Figure 7. On the day of cell embedding (day 0), we found that dispersed MCF-7 cells did not adhere to the hydrogel scaffold, but grew uniformly in a single spherical shape in the scaffold. It is well known that cells can only stick to surfaces that are hydrophobic enough, so this cell behavior in the scaffold may be caused by the hydrophilic environment inside the scaffold. After 1 day of culture, some smaller cell clusters could be observed, which may have formed by cell division or cell aggregation. As the culture time was further extended, cells grew faster and the number of cells increased, and most of them showed a green color in the observed field of view, indicating that the cells were still alive and the hydrogel scaffold could maintain the cell growth. When cultured for 7 days, we observed that most of the cells were able to form small multicellular spheres, which may be due to interaction between cells in adjacent cell clusters as the cells continued to spread and grow within the hydrogel scaffold, thus aggregating to form multicellular spheres. Therefore, we can speculate that this type of thermosensitive P(OEGMA-co-MEO2MA) hydrogel scaffold can be used to construct tumor multicellular sphere models and may have wide application prospects in drug screening and tumor research. Besides, based on our previous research experience [29], this type of hydrogel scaffold should also be able to be used for 3D culture of many other types of cells, which may extend its range of applications.

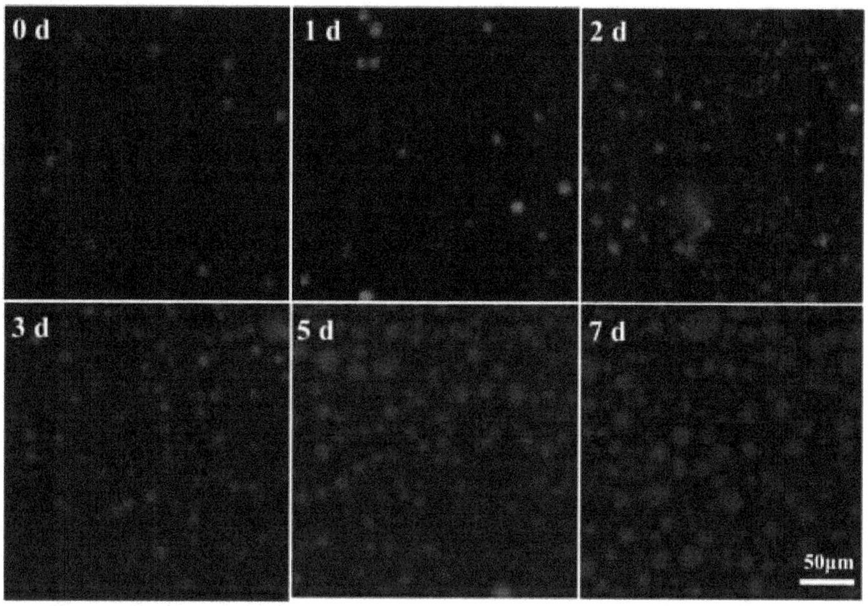

Figure 7. Fluorescence images of MCF-7 cells cultured in the formed thermosensitive P(OEGMA-co-MEO2MA) hydrogel. The cells were stained with acridine orange (AO) and ethidium bromide (EB) beforehand.

3. CONCISE SUMMARY

We synthesized five types of thermosensitive microgel samples utilizing MEO2MA and OEGMA as monomers. These monomers have high biocompatibility and biodegradability. The synthesis was done using free radical polymerization, resulting in microgels with varying VPTTs (volume phase transition temperatures) or particle sizes. The VPTTs and particle sizes were examined using a nanoscale particle size meter and UV spectrophotometer. The VPTT of the microgels rose as the content of OEGMA increased. The diameters of the produced microgels rose as the dosage of SDS decreased. Thermal induction can cause the aggregation of many types of prepared microgels, resulting in the formation of three-dimensional hydrogels. All of these hydrogels exhibited the typical contraction observed in thermosensitive hydrogels. Nevertheless, we observed that as the VPTT increased or the particle size decreased, the contraction became less apparent. Out of all the hydrogels, the hydrogel generated from M3 maintained a volume of over 98% of its initial volume after 48 hours without any noticeable shrinkage. These hydrogels have the potential to serve as a framework for the 3D cultivation of MCF-7 model cells and have demonstrated outstanding biocompatibility. Cells have the ability to proliferate inside this framework and typically organize themselves into multicellular spherical structures. This method presents a straightforward approach to construct thermosensitive hydrogel 3D cell scaffolds, which have significant potential for various applications in the biomedical area, including tumor research, drug screening, and tissue engineering.

4. MATERIALS AND METHODS

4.1. Materials

Oligo(ethylene glycol) methacrylate (OEGMA, Mn = 300), 2-(2-methoxyethoxy) ethyl acrylate (MEO2MA), N,N′-methylene diacrylamide (BIS), sodium dodecyl sulfate (SDS), were obtained from Tianjin Heowns Biochemical Technology Co, LTD; potassium persulfate (KPS) was obtained from Tianjin Damao Reagent Factory; Phosphate buffer (PBS), DMEM medium, fetal bovine serum, streptomycin mixture, trypsin, thiazole blue (MTT), acridine orange (AO), ethidium bromide (EB) were obtained from Wuhan Rutgers Biotechnology Co; MCF-7 breast cancer cells were purchased from the Stem Cell Bank of the Chinese Academy of Sciences.

4.2. Synthesis and Characterization of P(OEGMA-co-MEO2MA) Thermosensitive Microgel

In this manuscript, we first used MEO2MA and OEGMA as polymerization monomers, BIS as crosslinker, SDS as surfactant, and KPS as initiator to generate P(OEGMA-co-MEO2MA) thermosensitive microgels by free radical polymerization at 70 °C [26] (Scheme 1). The content of BIS was 2%, the mass of KPS were 0.0812 g, respectively. Three types of microgel samples M1, M2 and M3 were synthesized by changing the molar ratios of MEO2MA and OEGMA as 95 : 3, 90 : 8, and 85 : 13, respectively. In contrast to the synthesis of M3, another two types of microgel samples, M4 and M5, were synthesized by reducing the SDS dosage, while keeping the other substances constant. The detail feeding amount of each sample is in Table 1.

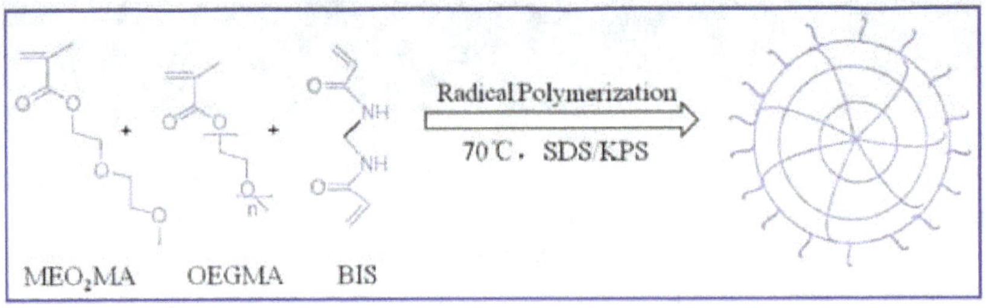

Scheme 1. The synthesis of P(OEGMA-co-MEO2MA) thermosensitive microgel.

The typical preparation process for sample M1 was as follows: 2.5002 g MEO2MA, 0.1259 g OEGMA, 0.0431 g BIS and 0.0284 g SDS were weighed accurately, dissolved in 100 mL deionized water, and transferred to a three-necked flask. After being passed into N2 for 1 h, KPS (dissolved in 5 mL of deionized water) was added to initiate polymerization under magnetic stirring. After 5 h, the reaction was terminated and transferred to a dialysis belt and dialysis for 7 days. Then, the sample was freeze-dried and preserved.

In order to test the thermosensitivities of these samples, the changes of particle size and absorbance (turbidity) with temperature were measured by a nanometer particle size meter and an ultraviolet spectrophotometer, respectively.

4.3. Preparation and Characterization of P(OEGMA-co-MEO2MA) Microgels-Based Thermosensitive Hydrogel

The thermosensitive P(OEGMA-co-MEO2MA) hydrogels were prepared through the thermal induction method [29].

A certain amount of P(OEGMA-co-MEO2MA) microgel samples were weighed and dissolved in phosphate buffered saline (PBS) to obtain a final solution with mass fraction of 3.0 wt%. Then, they were placed in a physiological temperature environment, which was higher than their VPTTs. The formation process of thermosensitive hydrogel from microgels and its stability and shrinkage were observed by the inverted method and recorded by camera. The gelation time was defined as the time point at which the inverted hydrogel no longer flowed visually.

To quantitatively analyze the shrinkage degree of the obtained hydrogels, we calculated the ratio of the volume of the obtained hydrogel at certain time point (Vt) to the initial volume (V0) by measuring with a ruler. The thermosensitive hydrogel sample with minimum shrinkage was screened out for the next cell experiments.

4.4. P(OEGMA-co-MEO2MA) Thermosensitive Hydrogels for 3D Cell Culture

4.4.1. Cell Culture

The model cells used in this experiment were human breast cancer cells (MCF-7), which were cultured in a cell culture incubator using DMEM medium. The main components of the culture medium were 90% DMEM high glucose medium, 10% fetal bovine serum, and 100 unit/mL penicillin/streptomycin double antibodies. The culture temperature was 37 °C and the carbon dioxide concentration was 5%.

4.4.2. Cells Embedded and Cultured in Hydrogel Scaffold

A certain concentration of P(OEGMA-co-MEO2MA) microgel solution was prepared and mixed well with an equal volume of the uniform cell suspension to form a mixed solution, in which the cell concentration was about 1 × 105 cells/mL and the microgel concentration was 3.0 wt%. To a 48-well culture plate, 0.5 mL of the above cell/gel mixture was added and placed in a cell culture incubator. With increasing temperature, the solution gelled gradually. After the morphology of the gel in the culture plate was fixed, 0.5 mL of the culture solution was added to each well and recorded as day 0. The culture solution was changed regularly to ensure a nutritious environment for cell growth. The whole process is shown in Scheme 2.

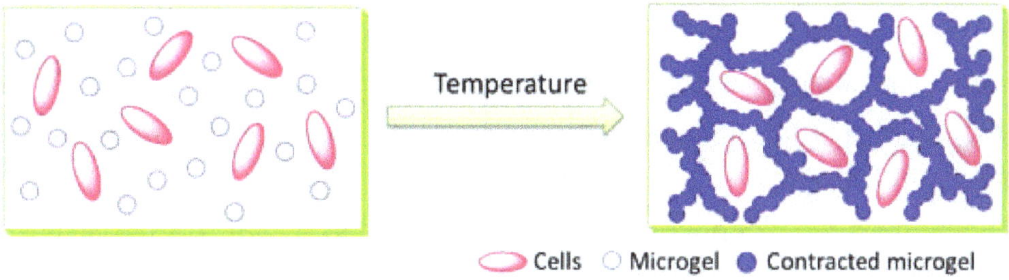

Scheme 2. The preparation of the P(OEGMA-co-MEO2MA) thermosensitive hydrogel scaffold.

4.4.3. Cell Viability (MTT Assay)

MTT assay was used to detect the viability of cells in the scaffolds. From day 0 of cell culture, three wells were selected each day and 100.0 μL of MTT solution was added to the wells. After continuing to incubate for 4 h, the gel pieces were removed, and 10.0 mL of DMSO was added. Then, the mixture was shaken for 10 min in the dark to dissolve the formed formazan. The absorbance of the solution at 490 nm was measured using a UV-Vis spectrophotometer, and the MTT curve of cells was plotted.

4.4.4. Cell Activity and Morphology (AO/EB Staining)

Starting from day 0 of cell culture, three wells were chosen after a specific number of days of culturing. Then, a single droplet of AO/EB double staining solution was introduced into each of the selected wells. Following an additional incubation period of 1 hour, the cells were extracted and the morphology and growth condition of the cells within the scaffold were examined using an inverted fluorescent microscope.

AT A GLANCE

Thermosensitive hydrogel scaffolds have garnered significant interest in the field of three-dimensional (3D) cell growth. Developing a thermosensitive hydrogel material with low shrinkage, great biocompatibility, and biodegradability is highly essential. Five distinct thermosensitive microgels were synthesized through free radical polymerization, utilizing 2-methyl-2-propenoic acid-2-(2-methoxyethoxy) ethyl ester (MEO2MA) and oligoethylene glycol methyl ether methacrylate (OEGMA) as thermosensitive monomers. These microgels vary in their volume phase transition temperature (VPTT) and particle sizes. The VPTT and particle sizes were examined using a nanoscale particle size meter and an ultraviolet spectrophotometer. This study examines the practicality of utilizing P(OEGMA-co-MEO2MA) microgels to create a thermosensitive hydrogel using the thermal induction approach, which has not been previously described. The thermosensitive hydrogel, which exhibited the best performance, was selected for the in situ embedding and three-dimensional (3D) culturing of MCF-7 breast cancer cells. The experimental results obtained from the AO/EB and MTT techniques demonstrate that the initial scaffold material exhibits remarkable biocompatibility, and cells proliferate fast within the 3D scaffold while maintaining a high capacity for proliferation. Simultaneously, there is an inclination to cluster together and create multicellular spheres. Hence, this initial P(OEGMA-co-MEO2MA) thermosensitive hydrogel can act as a remarkably biocompatible and readily functionalized 3D cell culture platform with significant prospects in the biomedical field.

REFERENCES

1. Khademhosseini, A.; Langer, R. A decade of progress in tissue engineering. Nat. Protoc. 2016, 11, 1775–1781.
2. Esdaille, C.J.; Washington, K.S.; Laurencin, C.T. Regenerative engineering: A review of recent advances and future directions. Regen. Med. 2021, 16, 495–512.
3. Atala, A.; Kasper, F.K.; Mikos, A.G. Engineering complex tissues. Sci. Transl. Med. 2012, 4, 160rv12.
4. Gaharwar, A.K.; Singh, I.; Khademhosseini, A. Engineered biomaterials for in situ tissue regeneration. Nat. Rev. Mater. 2020, 5, 686–705.
5. Reddy, M.S.B.; Ponnamma, D.; Choudhary, R.; Sadasivuni, K.K. A comparative review of natural and synthetic biopolymer composite scaffolds. Polymers 2021, 13, 1105.
6. Koons, G.L.; Diba, M.; Mikos, A.G. Materials design for bone-tissue engineering. Nat. Rev. Mater. 2020, 5, 584–603.
7. Ebhodaghe, S.O. Natural polymeric scaffolds for tissue engineering applications. J. Biomater. Sci. Polym. Ed. 2021, 32, 2144–2194.
8. Bashir, S.; Hina, M.; Iqbal, J.; Rajpar, A.H.; Mujtaba, M.A.; Alghamdi, N.A.; Wageh, S.; Ramesh, K.; Ramesh, S. Fundamental concepts of hydrogels: Synthesis, properties, and their applications. Polymers 2020, 12, 2702.
9. Jian, X.; Feng, X.; Luo, Y.; Li, F.; Tan, J.; Yin, Y.; Liu, Y. Development, preparation, and biomedical applications of DNA-based hydrogels. Front. Bioeng. Biotechnol. 2021, 9, 661409.
10. Ahmed, E.M. Hydrogel: Preparation, characterization, and applications: A review. J. Adv. Res. 2015, 6, 105–121.
11. Feng, X.; Luo, Y.; Li, F.; Jian, X.; Liu, Y. Development of natural-drugs-based low-molecular-weight supramolecular gels. Gels 2021, 7, 105.
12. Spicer, C.D. Hydrogel scaffolds for tissue engineering: The importance of polymer choice. Polym. Chem. 2020, 11, 184–219.
13. Portnov, T.; Shulimzon, T.R.; Zilberman, M. Injectable hydrogel-based scaffolds for tissue engineering applications. Rev. Chem. Eng. 2017, 33, 91–107.
14. Radulescu, D.-M.; Neacsu, I.A.; Grumezescu, A.-M.; Andronescu, E. New insights of scaffolds based on hydrogels in tissue engineering. Polymers 2022, 14, 799.

15. Zhang, Y.; Yu, J.-K.; Ren, K.; Zuo, J.; Ding, J.; Chen, X. Thermosensitive hydrogels as scaffolds for cartilage tissue engineering. Biomacromolecules 2019, 20, 1478–1492.

16. Suntornnond, R.; An, J.; Chua, C.K. Bioprinting of thermoresponsive hydrogels for next generation tissue engineering: A review. Macromol. Mater. Eng. 2017, 302, 1600266.

17. Doberenz, F.; Zeng, K.; Willems, C.; Zhang, K.; Groth, T. Thermoresponsive polymers and their biomedical application in tissue engineering–A review. J. Mater. Chem. B 2020, 8, 607–628.

18. Dou, Q.Q.; Liow, S.S.; Ye, E.Y.; Lakshminarayanan, R.; Loh, X.J. Biodegradable thermogelling polymers: Working towards clinical applications. Adv. Healthc. Mater. 2014, 3, 977.

19. Nagase, K.; Yamato, M.; Kanazawa, H.; Okano, T. Poly(N-isopropylacrylamide)-based thermoresponsive surfaces provide new types of biomedical applications. Biomaterials 2018, 153, 27–48.

20. Tang, L.; Wang, L.; Yang, X.; Feng, Y.; Li, Y.; Feng, W. Poly(N-isopropylacrylamide)-based smart hydrogels: Design, properties and applications. Prog. Mater. Sci. 2020, 115, 100702.

21. Ekerdt, B.; Fuentes, C.M.; Lei, Y.; Adil, M.M.; Ramasubramanian, A.; Segalman, R.A.; Schaffer, D.V. Thermoreversible hyaluronic acid-PNIPAAm hydrogel systems for 3D stem cell culture. Adv. Health Mater. 2018, 7, e1800225.

22. Kim, G.; Jung, Y.; Cho, K.; Lee, H.J.; Koh, W.-G. Thermoresponsive poly(N-isopropylacrylamide) hydrogel substrates micropatterned with poly(ethylene glycol) hydrogel for adipose mesenchymal stem cell spheroid formation and retrieval. Mater. Sci. Eng. C 2020, 115, 111128.

23. Liang, W.; Bhatia, S.; Reisbeck, F.; Zhong, Y.; Singh, A.K.; Li, W.; Haag, R. Thermoresponsive hydrogels as microniches for growth and controlled release of induced pluripotent stem cells. Adv. Funct. Mater. 2021, 31, 2010630.

24. Li, F.; Luo, Y.; Feng, X.; Guo, Y.; Zhou, Y.; He, D.; Xie, Z.; Zhang, H.; Liu, Y. Two-dimensional colloidal crystal of soft microgel spheres: Development, preparation and applications. Colloids Surfaces B Biointerfaces 2022, 212, 112358.

25. Liu, Y.; Guan, Y.; Zhang, Y. Facile assembly of 3D binary colloidal crystals from soft microgel spheres. Macromol. Rapid Commun. 2014, 35, 630–634.

26. Guan, Y.; Zhang, Y.J. PNIPAM microgels for biomedical applications: From dispersed particles to 3D assemblies. Soft Matter 2011, 7, 6375–6384.

27. Gan, T.; Zhang, Y.; Guan, Y. In situ gelation of P(NIPAM-HEMA) microgel dispersion and its applications as injectable 3D cell scaffold. Biomacromolecules 2009, 10, 1410–1415.

28. Liu, Y.; Guan, Y.; Zhang, Y. Chitosan as inter-cellular linker to accelerate multicellular spheroid generation in hydrogel scaffold. Polymer 2015, 77, 366–376.

29. Shen, J.; Ye, T.; Chang, A.; Wu, W.; Zhou, S. A colloidal supra-structure of responsive microgels as a potential cell scaffold. Soft Matter 2012, 8, 12034–12042.

30. Shen, Z.; Bi, J.; Shi, B.; Nguyen, D.; Xian, C.J.; Zhang, H.; Dai, S. Exploring thermal reversible hydrogels for stem cell expansion in three-dimensions. Soft Matter 2012, 8, 7250–7257.

31. Shen, Z.; Mellati, A.; Bi, J.; Zhang, H.; Dai, S. A thermally responsive cationic nanogel-based platform for three-dimensional cell culture and recovery. RSC Adv. 2014, 4, 29146–29156.

32. Gan, T.; Guan, Y.; Zhang, Y. Thermogelable PNIPAM microgel dispersion as 3D cell scaffold: Effect of syneresis. J. Mater. Chem. 2010, 20, 5937–5944.

33. Cheng, D.; Wu, Y.; Guan, Y.; Zhang, Y. Tuning properties of injectable hydrogel scaffold by PEG blending. Polymer 2012, 53, 5124–5131.
34. Gu, J.; Zhao, Y.; Guan, Y.; Zhang, Y. Effect of particle size in a colloidal hydrogel scaffold for 3D cell culture. Colloids Surfaces B Biointerfaces 2015, 136, 1139–1147.
35. Vancoillie, G.; Frank, D.; Hoogenboom, R. Thermoresponsive poly(oligo ethylene glycol acrylates). Prog. Polym. Sci. 2014, 39, 1074–1095.
36. Badi, N. Non-linear PEG-based thermoresponsive polymer systems. Prog. Polym. Sci. 2016, 66, 54–79.
37. Anderson, C.R.; Abecunas, C.; Warrener, M.; Laschewsky, A.; Wischerhoff, E. Effects of methacrylate-based thermoresponsive polymer brush composition on fibroblast adhesion and morphology. Cell. Mol. Bioeng. 2016, 10, 75–88.
38. Anderson, C.R.; Gambinossi, F.; DiLillo, K.M.; Laschewsky, A.; Wischerhoff, E.; Ferri, J.K.; Sefcik, L.S. Tuning reversible cell adhesion to methacrylate-based thermoresponsive polymers: Effects of composition on substrate hydrophobicity and cellular responses. J. Biomed. Mater. Res. Part A 2017, 105, 2416–2428.
39. Bakaic, E.; Smeets, N.M.; Badv, M.; Dodd, M.; Barrigar, O.; Siebers, E.; Lawlor, M.; Sheardown, H.; Hoare, T. Injectable and degradable poly(oligoethylene glycol methacrylate) hydrogels with tunable charge densities as adhesive peptide-free cell scaffolds. ACS Biomater. Sci. Eng. 2017, 4, 3713–3725.
40. Tatry, M.-C.; Galanopoulo, P.; Waldmann, L.; Lapeyre, V.; Garrigue, P.; Schmitt, V.; Ravaine, V. Pickering emulsions stabilized by thermoresponsive oligo(ethylene glycol)-based microgels: Effect of temperature-sensitivity on emulsion stability. J. Colloid Interface Sci. 2020, 589, 96–109.
41. Keerl, M.; Pedersen, J.S.; Richtering, W. Temperature sensitive copolymer microgels with nanophase separated structure. J. Am. Chem. Soc. 2009, 131, 3093–3097.

13 Chapter

FLUORESCENT SENSORS THAT UTILIZE MOLECULARLY IMPRINTED HYDROGELS FOR THE DETECTION OF ENVIRONMENTALLY AND MEDICINALLY SIGNIFICANT ANALYTES

1. INTRODUCTION: AN OVERVIEW

The close relationship between human health and environmental pollution, which leads to a range of disorders, has made environmental monitoring of pollutants and biomedical identification of biomarkers a global priority [1,2]. Consequently, there is a growing need to create sensors that can specifically identify different substances, such as harmful pesticides and excessive mycotoxins in food samples [3,4,5,6], as well as biomarkers and tumor cells that are indicative of certain diseases [7,8,9,10]. Furthermore, the sensors that are acquired should be manufactured in a way that is cost-efficient and have the ability to be used quickly, with high sensitivity, and in portable settings. Typically, conventional receptors like enzymes, antibodies, DNA, and cells are employed to interact with the desired analyte. There has been a notable progress in the creation of sensors, particularly in relation to common operations that rely on these biomolecules [11]. Although these biosensors have a high level of specificity, they are hindered by issues such as reagent stability and availability, high cost, and complex analysis processes [12].

In this particular situation, scientists are actively searching for artificially created receptors that can specifically identify and attach to certain molecules with a strong attraction, while also being resistant to chemical and thermal changes, simple to create, and able to be used multiple times [13]. Molecular imprinting technology (MIT) is a highly promising approach for creating synthetic receptors. It involves the co-polymerization of functional monomers and a crosslinking agent in the presence of the template, which can be the target molecule or a derivative of it. Following the removal of the template, cavities are formed with dimensions, configuration, and properties that match those of the template. These cavities are then used to create a customized receptor known as a molecularly imprinted polymer (MIP) [14,15].

Due to advancements in manufacturing, the Massachusetts Institute of Technology (MIT) has been utilized for several types of target molecules, including small molecules, macromolecules, and bigger entities [16,17]. Due to its antibody-like affinity, considerable selectivity, and great biocompatibility, the synthesized MIP has a wide range of applications including analyte purification, biosensing, bioimaging, drug delivery, and cancer therapy [17,18,19,20,21,22]. Nevertheless, the adoption of the MIP for industrial applications is progressing at a sluggish pace. Despite the potential of MIP-based sensors, their utilization in the industry remains limited primarily due to the challenges associated with manufacturing and practical implementation [23]. Thus, the commercialization of sensors based on molecularly imprinted polymers (MIPs) depends on the simplified manufacturing process and portable nature to enhance their use in diverse scenarios [24].

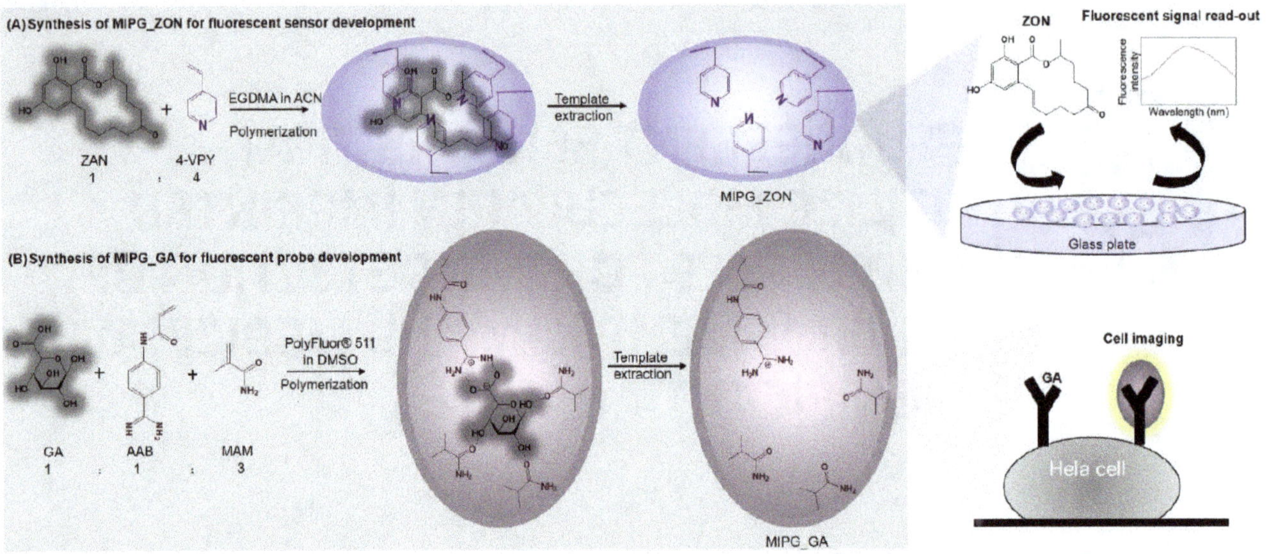

Scheme 1. Schematic illustration of molecular imprinting technology for two typical fluorescent sensors development. (A) The molecular imprinting process for the synthesis of MIPG_ZON, which can be pasted on the glass substrate in order to work as a sensor for ZON detection by reading out the fluorescence signal of ZON. (B) The molecular imprinting process for the synthesis of fluoresce-incorporated MIPG_GA as a fluorescent probe which can be directly used for cancer cell imaging.

In order to produce MIP-based sensors that are convenient for future commercial applications, two fluorescent sensors were created. These sensors utilize either the analyte's inherent fluorescence property or the receptor's fluorescence effect as the sensing mechanism (Scheme 1). A molecularly imprinted polymeric hydrogel (MIPG) was created to target zearalenone (ZON), a mycotoxin produced by various species of Fusarium molds that contaminates cereals such as corn, barley, oats, wheat, and sorghum [25]. ZON is known to have harmful effects on humans and animals, including nephrotoxic, neurotoxic, carcinogenic, estrogenic, and immunosuppressive effects [26]. MIPG_ZON was synthesized by employing an optimized quantity of 4-vinylpyridine (4-VPY) as the functional monomer and ethylene glycol dimethacrylate (EGDMA) as the cross-linking monomer to target zearalanone (ZAN). ZAN is an analogue of ZON but lacks a double bond, which could potentially disrupt the polymerization reaction. Subsequently, the MIPG_ZON compound was applied to the glass plate in order to identify the presence of ZON in corn juice. Simultaneously, a MIPG (Molecularly Imprinted Polymer Gel) containing fluorescein was created to specifically detect glucuronic acid (GA), a tiny monosaccharide that serves as the epitope of hyaluronic acid. Hyaluronic acid is a biomarker found on the surface of malignant or infectious cells [27,28,29]. The MIPG_GA was synthesized by cross-linking fluorescein dimethacrylate (PolyFluor® 511) utilizing two functional monomers, namely (4-acrylamidophenyl)(amino)methaniminium acetate (AAB) and methacrylamide (MAM). AAB interacts strongly with the carboxylic group of GA, while MAM forms hydrogen bonds with GA. The conclusive use of MIP_GA in cancer cell imaging demonstrated the significant potential of our Molecularly Imprinted Polymer (MIP) as a fluorescent probe for biomedical investigation.

2. RESULTS AND DISCUSSION

2.1. Characterization of Synthesized MIPGs_ZON

Three MIPGs_ZON using the functional monomer 4-VPY at different proportions were synthesized (Table 1), with the expectation that the "π–π" interaction between ZON and the monomer would lead to strong affinity. According to our preliminary study (Figure S3), the double bond in the chemical structure of ZON interferes with the imprinting process, which brought no difference between MIPG and NIPG. To address this problem, an analogue ZAN was used as a template for MIPGs_ZON synthesis. The hydrodynamic diameter of the ZAN imprinted polymers was analyzed by the dynamic light scattering (DLS) on a Zeta-sizer NanoZS (results are summarized in Table 1), which shows a very homogenous size for each MIPG_ZON.

Table 1. Composition of ZAN imprinted polymers synthesized in 400 μL anhydrous ACN and their corresponding sizes.

Polymer	ZAN (mmol)	4-VPY (mmol)	EGDMA (mmol)	ABDV (mg)	Size (nm)	PDI
MIPG1	0.03	0.06	0.6	3.1	345 ± 5	0.163
NIPG1	-	0.06	0.6	3.1	455 ± 16	0.160
MIPG2	0.03	0.12	0.6	3.2	443 ± 11	0.137
NIPG2	-	0.12	0.6	3.2	483 ± 6	0.199
MIPG3	0.03	0.18	0.6	3.4	468 ± 12	0.156
NIPG3	-	0.18	0.6	3.4	503 ± 9	0.085

2.2. Equilibrium Binding Studies on MIPGs_ZON

In Figure 1A, our hydrogel is a dense three-dimensional fiber network as observed under the scanning electron microscope (SEM). The relevant component analysis was carried out by a Fourier transform infrared spectrometer (FTIR, Figure S4). By magnifying the SEM image, the thickness of the fibers is found to be at the nanometer level. Figure 1B represents the binding behavior of the three MIPGs_ZON ranging from 0.1 to 5 mg/mL towards 2.5 μM ZON in ACN, indicating that the ratio ZAN: 4-VPY: EGDMA of 1:4:20 corresponding to MIPG2 gave the best binding specificity, especially for low polymer concentrations where 0.1 mg of MIP can bind to 0.5 nmol of ZON, leading to an imprinting factor of 3.0. Afterwards, 1 mg of MIPG2 and NIPG2 were incubated overnight with ZON at concentrations varying from 0.5 to 10 μM in 1 mL ACN. As shown in Figure 1C, the Langmuir-type isotherm equation was used for plotting ZON bound versus free in case of MIPG2 for the specific binding analysis, in which the binding capacity Bmax was found to be 5.8 nmol/mg, and the dissociation constant Kd was 7.9 μM with R2 = 0.99. At the same time, the Freundlich model was used in NIPG2 cases for the non-specific adsorption analysis, it was found that Bmax = 4.3 nmol/mg and Kd = 7.1 μM with R2 = 0.98. Therefore, MIPG2 presents higher binding capacity while a similar affinity towards ZON, with respect to NIPG2. As shown in Figure 1D, the corresponding Scatchard plot, ZON bound/free versus ZON bound is represented by a straight line, indicating a homogeneous MIPG2. The variables N and Ka in the Scatchard equation (B/F = KaN–KaB) correspond to the total number of binding sites and the association constant, which were found to be ~ 5.1 × 105 M-1 and 4.0 nmol/mg, respectively.

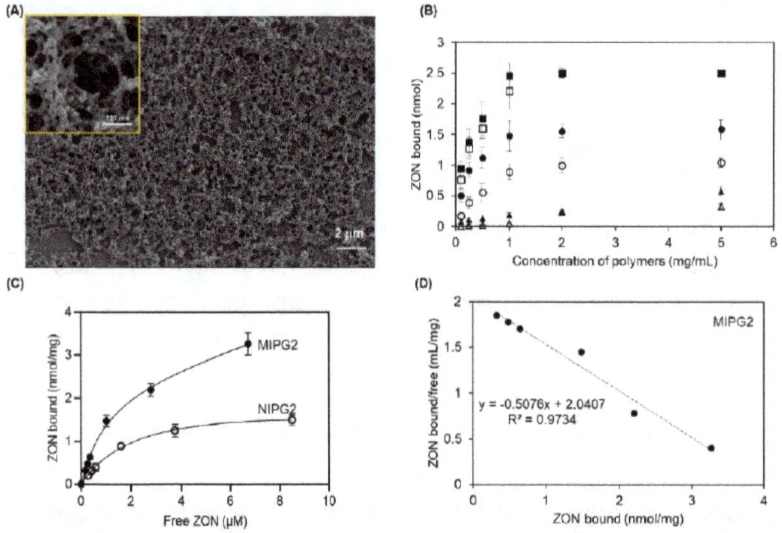

Figure 1. (A) SEM image of the MIPGs_ZON. (B) Equilibrium binding isotherms for 2.5 μM ZON on MIPGs_ZON at concentrations ranging from 0.1 to 5 mg/mL in 1 mL ACN. Data are means from three independent experiments. The error bars represent standard deviations. (C) Determination of the binding capacity of 1 mg MIPG2 and NIPG2 towards ZON varying from 0.5 to 10 μM in 1 mL ACN, respectively. Data are means from three independent experiments. The error bars represent standard deviations. (D) Corresponding Scatchard plot of MIPG2.

2.3. Application of MIPGs_ZON-Based Fluorescent Sensor for Real Sample Tests

In order to study the anti-interference ability of the MIPG2, the structurally similar reagents tetracycline (TC), amoxicillin (AMO), and levofloxacin (LEV) were used for the selectivity test. As shown in Figure 2A, MIPG2 exhibited lower binding towards other structural analogs with respect to ZON, demonstrating a great potential in anti-interference detecting.

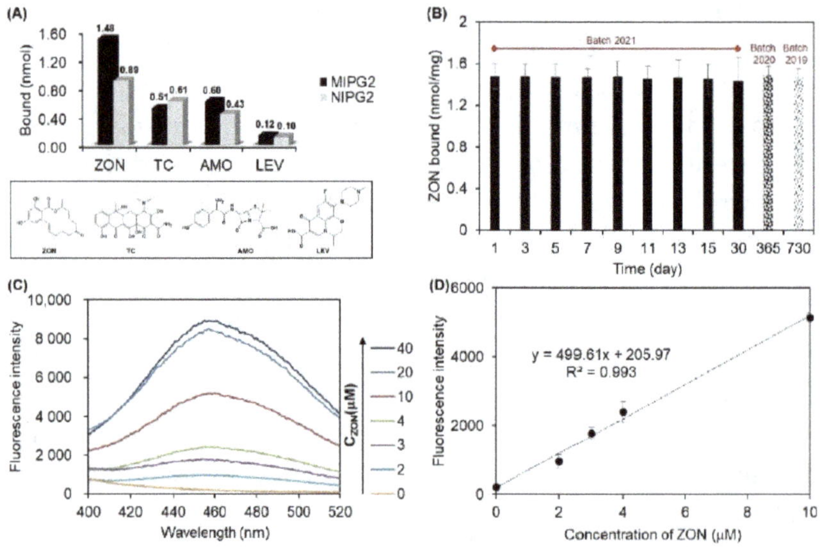

Figure 2. (A) Selectivity test of 1 mg MIPG2 and NIPG2 towards ZON and three structurally similar reagents tetracycline (TC), amoxicillin (AMO), and levofloxacin (LEV) at 2.5 μM, respectively. The chemical structure of three structurally similar reagents are given below. (B) Detection record of 1 mg MIPG2 (of the same batch) towards 2.5 μM ZON during 30 days after each template extraction, as well

as 1 mg MIPG2 (of a different batch) towards 2.5 μM ZON in 1 and 2 years, respectively. (C) Analyte-dependent fluorescence emission spectra of MIPG2 with respect to ZON ranging from 0 to 40 μM in corn juice. (D) Lear fitting plot of fluorescence intensity of MIPG2 towards ZON in corn juice. Corn juice tests were performed on glass substrates coated with MIPG2.

On the other hand, the NIPG2 shows a high cross-reactivity towards TC and AMO. Therefore, our MIPG2 is promising to detect ZON in real samples. Furthermore, the colloidal stability and batch-to-batch reproducibility were investigated prior to the real sample tests. As shown in Figure 2B, when 1 mg MIPG2 (of the same batch) was used to detect 2.5 μM ZON repeatedly after each template extraction, the binding amount was almost constant during one month, indicating its reversibility property. Meanwhile, 1 mg MIPG2 (of a different batch) was employed to recognize 2.5 μM ZON in 1 and 2 years, the amount of ZON bound remained the same, showing its excellent stability and reproducibility for sensor development.

It is well-known that ZON is a mycotoxin produced by several species of Fusarium that may contaminate cereals, in particular corn, barley, oats, wheat, and sorghum. Thus, ZON contamination can cause toxic effects in humans and animals, including nephrotoxic, neurotoxic, carcinogenic, estrogenic, and immunosuppressive effects [25,26]. Since MIPG2 exhibits good specificity and selectivity for ZON, they were then adhered on a glass substrate in order to work as a fluorescent sensor for ZON detection in real samples. Herein, we bought some corn juice as ZON-containing practical samples, and diluted it 4 times with ACN for testing. Using the standard recovery method, ZON at concentrations ranging from 0 to 40 μM were spiked in the corn juice. According to the results of ZON detection in real samples (Figure 2C,D), our MIPG2-based optical sensor may detect ZON at the linear concentration range of 0–10 μM with a limit of detection (LOD) of 1.6 μM (calculated according to the equation: $D = 3\sigma/k$, where σ is the relative standard deviation of the blank sample, k is the slope of the calibration line), showing great potential for mycotoxin detection in food testing. Although, biological molecular recognition elements exhibit lower LOD, such as the LOD of antibody-based sensors for ZON is at the nM level [30,31], and aptamer-based sensors can detect ZON at the pM level [32,33], such biosensors are expensive and have a short service life. Compared with the chromatography method which exhibited a LOD in the nM range [5], our MIPG_ZON-based fluorescent sensor is more convenient for commercial applications. Inspired by recent quantum dots-incorporated MIPG [4], our MIPG_ZON promises to be smaller and effective, and the optimization could be established on this protocol.

2.4. Synthesis and Binding Characterization of MIPG_GA

The aim of this part is to synthesize a fluorescent MIPG_GA, which is capable of specifically binding towards GA (an epitope of hyaluronic acid which is usually overexpressed on cancer cells) for cell imaging. To achieve this goal, the particle size of MIPG_GA must be controlled at the nanometer scale, and the MIPG must exhibit good compatibility with water. Hence, our MIPG_GA was generated using AAB and MAM as functional monomers to create specific binding sites (Scheme 1). According to our intensive research [28,29,34], AAB is able to strongly interact with the carboxylic group and MAM is a co-monomer that establishes a hydrogen interaction with the target molecule. As shown in Figure 3A,B, the hydrodynamic size of MIPG_GA and NIPG_GA was found to be 520 and 585 nm, respectively. The SEM image (Figure 3C) shows the three-dimensional nanofiber structure of the hydrogel (FTIR analysis in Figure S5). Using these polymers to evaluate the binding performance towards GA, different concentrations of the polymers ranging from 0.1 to 2 mg/mL were incubated with 1 mM GA. After the subtraction of unbound GA from the total amount of the analyte added by the well-established Dubois method, the amount of bound GA was determined for each concentration of polymers (Figure 3D). Although the NIPG_GA bound significant amount of analyte was due to the presence of AB, which might be largely distributed on the polymer surface and very accessible to GA. It was found that MIPG_GA bound specifically to GA as the binding to NIPG_GA was lower.

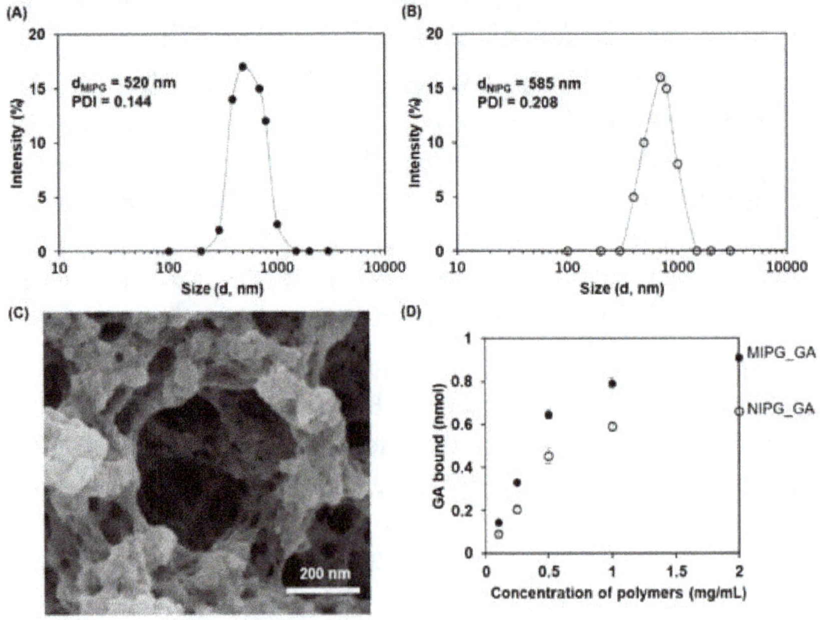

Figure 3. Dynamic light scattering measurements of 0.5 mg/mL (A) MIPG_GA and (B) NIPG_GA in H2O (mean value of two experiments). (C) SEM image of MIPG_GA. (D) Equilibrium binding isotherm for 1 mM GA on MIPG_GA and NIPG_GA at the concentrations ranging from 0.1 to 2 mg/mL in water. Data are means of three independent experiments. The error bars represent standard deviations.

2.5. Application of MIPG_GA-Based Fluorescent Probe for Cell Imaging

The objective of this part is to observe the interaction between fluorescent polymers and Hela cells before or after the enzyme treatment, in order to demonstrate the potential of our MIPG_GA to detect cancer cells in clinical uses. The images shown in Figure 4A,B correspond to the fluorescence intensity analysis for Hela cells incubated with MIPG_GA or NIPG_GA, respectively, indicating that the MIPG_GA specifically bound towards the epitope of hyaluronic acid on Hela cells. Our fluorescent probe is more convenient for commercializing due to its excellent stability, though less sensitive with respect to the hyaluronic acid-binding protein (HABP, Figure S6) [35,36,37]. Moreover, compared with nano-sized MIP containing quantum dots [28,29], our MIPG_GA is easier to fabricate using a broad-spectrum cross-linker FAM-based monomer for signaling, showing the universal utility of this strategy.

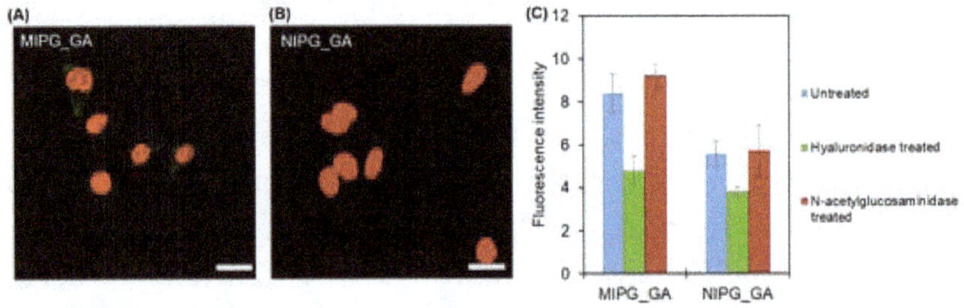

Figure 4. Confocal microscope images of fixed Hela cells incubated MIPG_GA (A) and NIPG_GA (B), that exhibit green fluorescence under proper filter sets of confocal microscopy. Nuclei stained with PI (red). Scale bar: 50 μm. (C) Summarized histogram results showing the binding performance of polymers towards untreated, hyaluronidase treated, and N-acetylglucosaminidase treated Hela cells. Data are means of three independent experiments. The error bars represent standard deviations.

Furthermore, according to Figure 4C, the MIPG_GA obtained from fluorescence microscope images exhibited more binding towards Hela cells with respect to the NIPG_GA. Particularly, hyaluronidase effectively removes GA on the surface of Hela cells by breaking down hyaluronic acids. This is why the fluorescence intensity for hyaluronidase treated Hela cells became lower than the untreated cells. On the other hand, N-acetylglucosaminidase plays a role to hydrolyze the non-reducing terminal of N-acetyl-D-glucosamine, whereby more GA is obtained. Thus, the fluorescence intensity of N-acetylglucosaminidase treated Hela cells increased. In addition, the MIPG_GA exhibited almost the same cell imaging fluorescence intensity in 2 years, indicating its excellent stability.

3. CONCISE SUMMARY

To summarize, a fluorescence sensor based on MIP (Molecularly Imprinted Polymer) was effectively created and utilized to detect and track ZON (Zearalenone) in actual samples. Simultaneously, a fluorescent probe based on MIP (Molecularly Imprinted Polymer) was created and employed for direct cell imaging. These two examples demonstrate the aesthetic appeal of molecular imprinting technology, showcasing its potential and adaptability in creating fluorescence sensors for practical use. The primary achievement of this study is the identification of a practical and feasible approach for utilizing MIPGs in real-life scenarios. This demonstrates the significant potential of these fluorescence sensors in detecting specific substances of interest in food and biological samples.

4. MATERIALS AND METHODS

4.1. Reagents

The (4-acrylamidophenyl)(amino)methaniminium acetate (AAB) was synthesized as previously described [34]. The 4-vinylpyridine (4-VPY), methacrylamide (MAM), ethylene glycol dimethacrylate (EGDMA), azo-bis-dimethylvaleronitrile (ABDV), azobisisobutyronitrile (AIBN), zearalanone (ZAN), zearalenone (ZON), hyaluronidase, N-acetylglucosaminidase, hyaluronic acid-binding protein (HABP), biotin, streptavidin, and fluorescein-5-isothiocyanate (FITC) was purchased from Sigma-Aldrich (Shanghai, China). Fluorescein dimethacrylate (PolyFluor® 511) was obtained from Polysciences (Hirschberg an der Bergstrasse, Germany). Glucuronic acid (GA), tetracycline (TC), amoxicillin (AMO), and levofloxacin (LEV) came from Sinopharm Chemical Reagent Co., Ltd. (Shanghai, China). All the solvents used in the experiments as well as acids and bases are of analytical grade and anhydrous for acetonitrile (ACN) and dimethylsulfoxide (DMSO). The materials for cells culturing are: Phosphate buffer saline (PBS), penicillin/streptomycin, high glucose Dulbecco's Modified Eagle's Medium (DMEM) supplied with L-glutamine, fetal bovine serum (FBS), trypsin, ethylenediaminetetraacetic acid, and were obtained from Thermo Fisher Scientific (Shanghai, China).

4.2. Synthesis of MIPGs_ZON and MIPG_GA

MIPGs_ZON were synthesized in the presence of the ZAN using different functional monomer ratios. The 4-VPY was distilled under reduced pressure before use. First of all, 10 mg (0.03 mmol) ZAN, 6.4 or 12.7 or 19.2 μL (0.06, 0.12, 0.18 mmol) 4-VPY, and 113.1 μL (0.60 mmol) EGDMA was mixed in 400 μL of ACN. After sonication for 5 min, the initiator ABDV (1 mol% with respect to the number of double bonds) was added. Then, the prepolymerization mixture was purged with nitrogen for 5 min and placed in a water bath maintained at 40 °C overnight. The non-imprinted polymeric hydrogels (NIPGs) are synthesized in the same way but without the template. For template extraction, the polymers were washed 3 times with methanol/acetic acid (7/3), once with methanol, 3 times methanol containing 2.8% ammonia, and 3 times with methanol for 1 h at room temperature. The final polymers were obtained after drying in a vacuum overnight.

MIPG_GA was synthesized by mixing 4.85 mg (0.025 mmol) GA, 4.73 mg (0.025 mmol) AAB (1 h pre-incubation with GA), 6.38 mg (0.075 mmol) MAM, and 234 mg (0.5 mmol) PolyFluor® 511 in 1903 μL DMSO. Afterwards, 1.01 mg (0.0062 mmol) AIBN was added and before nitrogen purging for 5 min. The mixture was then heated overnight at 50 °C. The next day, a precipitated polymer was obtained. The NIPGs were synthesized in the same way but without the template. For template extraction, the polymers are washed 3 times with 1 M hydrochloric acid, once with methanol, 3 times methanol containing 2.8% ammonia, and 3 times with methanol for 1 h at room temperature. The final polymers were obtained after drying in a vacuum overnight.

The sizes of synthesized polymers were measured by dynamic light scattering (DLS) analysis on a Zetasizer NanoZS at 25 °C, in ACN for MIPs_ZON and in H2O for MIP_GA. Scanning electron microscopy (SEM) imaging was carried out on a Quanta FEG 250 scanning electron microscope (FEI Europe), by spraying gold. Prior to the binding studies, the dried MIPGs were added into a dialysis bag for checking whether there was any leakage of the template one day later. If there is any leakage, a second round of washing will be carried out. Afterwards, the MIPG2_ZON and MIPG_GA were analyzed by Fourier transform infrared spectrometer (FTIR), respectively. In addition, 5 mg/mL MIPG2_ZON in ACN and 5 mg/mL MIPG_GA in DMSO images were taken, in order to prove the gel format.

4.3. Equilibrium Binding Studies

For MIPGs_ZON, the equilibrium binding experiments with ZON were performed in anhydrous acetonitrile. The particles of 10 mg MIPGs_ZON and NIPGs_ZON were suspended in 1 mL ACN with intensive sonication (ultrasonic power: 80 W, frequency: 40 KHz, 20 min), in order to work as stock solutions, respectively. Then, polymer concentrations at 0.1, 0.25, 0.5, 1, 2, 5 mg/mL were incubated with 2.5 μM ZON in 1 mL ACN overnight. The samples were then centrifuged at 17,500 rpm for 20 min and a 700 μL aliquot of the supernatant was taken for unbound ZON determination. Then, the amount of bound ZON was calculated by subtracting the amount of unbound from the total. A calibration curve of ZON (Ex = 280 nm, Em = 455 nm, slit: 5) was obtained on F-7000 spectrofluorometer (Hitachi High-Technologies, Tokyo, Japan), as shown in Figure S1. To investigate the binding capacity, 1 mg of polymers were incubated with ZON varying from 0.5 to 10 μM in 1 mL ACN, with results plotted with the Langmuir-type isotherm equation. The selectivity study was performed by testing the binding of 1 mg polymers towards 2.5 μM ZON, and three structurally similar reagents: Tetracycline (TC), amoxicillin (AMO), and levofloxacin (LEV).

For MIPG_GA, the recognition properties were evaluated using the method of Dubois [28]. Dubois's method is a colorimetric method widely used to determine monosaccharide based on the preconversion of sugars into furfural derivatives upon heating with strong acids, followed by the formation of a colored complex with phenol. Prior to the binding assay, a calibration curve was generated with standard solutions of GA at concentrations ranging from 0.1 to 5 mM in water (Figure S2). Afterwards, the polymers at concentrations varying from 0.1 to 2 mg/mL were incubated with 1 mM GA in 1 mL water overnight. The samples were then centrifuged at 17,500 rpm for 1 h, in order to provide the supernatant for GA determination by Dubois's method.

4.4. Application of MIPG_ZON-Based Fluorescent Sensor for Real Sample Tests

Initially, 100 mg MIPG2_GA and NIPG2_GA were mixed with 480 mg poly (vinyl alcohol) in water and smeared on the glass plates, respectively. Herein, poly (vinyl alcohol) were used as an adhesion agent due to its excellent transparent feature, exhibiting no interference to ZON excitation light. After 30 min of shaking in an oven at 90 °C, the glass plates were left for 1 h at room temperature to cool down. The glass plates coated with MIPG2_ZON and NIPG2_ZON were stored in the fridge at 4 °C until use. For the real sample tests, the spiked samples were prepared using ZON at concentrations from 0–40 μM dissolved in DMSO. The prepared ZON samples were then mixed with corn juice at a ratio of 1:9 for the determination of recovery. For fluorescence intensity measurements after 1 h incubation, the Cary Eclipse fluorescence spectrophotometer (Varian, Palo Alto, CA, USA) was used.

4.5. Application of MIPG_GA-Based Fluorescent Probe for Cell Imaging

Cells were grown in a 5% CO2 incubator at 37 °C using DMEM supplemented with 10% premium FBS and antibiotics (100 units/mL penicillin and 100 μg/mL streptomycin) in cell culture dishes. For imaging studies, Hela cells were grown on cover slips and then incubated with 0.5 mg/mL polymers or 50 ng/mL FAM-HABP (synthesized in two steps, first with a biotinylated HABP, followed by incubation with streptavidin-FITC [37]) for 1 h. The cell monolayers were then washed 3 times with PBS (pH 7.4) and immediately measured by a Leica TSC SP5 confocal scanning laser microscopy (CSLM) system (Ex = 480, Em = 515).

AT A GLANCE

Two fluorescent sensors were created in this study using molecularly imprinted polymeric hydrogels (MIPGs) to detect zearalenone (ZON) and glucuronic acid (GA). The sensors rely on the self-fluorescence property of the analyte for ZON detection and the fluorescence effect of the receptor for GA detection. While there have been notable improvements in MIPG-fluorescent sensors that have better stability than natural receptor-sensors, there is a growing need to create sensing devices that are affordable, user-friendly, and portable for commercial use. The commercial potential of MIPG-fluorescent sensors is focused on in this study. The MIPG_ZON sensor is created utilizing zearalanone, which is an analogue of ZON, as a template. This sensor has excellent detection capabilities, even in corn samples, with a limit of detection of 1.6 μM. Simultaneously, a fluorescein-labeled molecularly imprinted polymer gel (MIPG_GA) is produced and employed for direct visualization of cancer cells, demonstrating notable specificity and selectivity. Finally, our combined application outcomes provide fresh prospects for the detection of ecologically and medicinally significant analytes using MIPG-fluorescent sensors.

REFERENCES

1. Patel, B.R.; Noroozifar, M.; Kerman, K. Review-Nanocomposite-Based Sensors for Voltammetric Detection of Hazardous Phenolic Pollutants in Water. J. Electrochem. Soc. 2020, 167, 037568.
2. Sow, W.T.; Ye, F.; Zhang, C.; Li, H. Smart materials for point-of-care testing: From sample extraction to analyte sensing and readout signal generator. Biosens. Bioelectron. 2020, 170, 112682.
3. Beloglazova, N.; Lenain, P.; Tessier, M.; Goryacheva, I.; Hens, Z.; De Saeger, S. Bioimprinting for multiplex luminescent detection of deoxynivalenol and zearalenone. Talanta 2019, 192, 169–174.
4. Shao, M.; Yao, M.; Saeger, S.D.; Yan, L.; Song, S. Carbon Quantum Dots Encapsulated Molecularly Imprinted Fluorescence Quenching Particles for Sensitive Detection of Zearalenone in Corn Sample. Toxins 2018, 10, 438.
5. Lhotska, I.; Gajdosova, B.; Solich, P.; Satinsky, D. Molecularly imprinted vs. reversed-phase extraction for the determination of zearalenone: A method development and critical comparison of sample clean-up efficiency achieved in an on-line coupled SPE chromatography system. Anal. Bioanal. Chem. 2018, 410, 3265–3273.

6. Li, G.; Zhang, K.; Fizir, M.; Niu, M.; Sun, C.; Xi, S.; Hui, X.; Shi, J.; He, H. Rational design, preparation and adsorption study of a magnetic molecularly imprinted polymer using a dummy template and a bifunctional monomer. New J. Chem. 2017, 41, 7092–7101.
7. Li, Q.; Shinde, S.; Grasso, G.; Caroli, A.; Abouhany, R.; Lanzillotta, M.; Pan, G.; Wan, W.; Rurack, K.; Sellergren, B. Selective detection of phospholipids using molecularly imprinted fluorescent sensory core-shell particles. Sci. Rep. 2020, 10, 9924.
8. Li, R.; Feng, Y.; Pan, G.; Liu, L. Advances in Molecularly Imprinting Technology for Bioanalytical Applications. Sensors 2019, 19, 177.
9. Ma, Y.; Yin, Y.; Ni, L.; Miao, H.; Wang, Y.; Pan, C.; Tian, X.; Pan, J.; You, T.; Li, B.; et al. Thermo-responsive imprinted hydrogel with switchable sialic acid recognition for selective cancer cell isolation from blood. Bioact. Mater. 2021, 6, 1308–1317.
10. Wang, J.; Dai, J.; Xu, Y.; Dai, X.; Zhang, Y.; Shi, W.; Sellergren, B.; Pan, G. Molecularly Imprinted Fluorescent Test Strip for Direct, Rapid, and Visual Dopamine Detection in Tiny Amount of Biofluid. Small 2019, 15.
11. Bhalla, N.; Jolly, P.; Formisano, N.; Estrela, P. Introduction to biosensors. Biosens. Technol. Detect. Biomol. 2016, 60, 1–8.
12. Morales, M.A.; Halpern, J.M. Guide to Selecting a Biorecognition Element for Biosensors. Bioconjugate Chem. 2018, 29, 3231–3239.
13. Fuchs, Y.; Soppera, O.; Haupt, K. Photopolymerization and photostructuring of molecularly imprinted polymers for sensor applications-A review. Anal. Chim. Acta 2012, 717, 7–20.
14. Haupt, K.; Rangel, P.X.M.; Bui, B.T.S. Molecularly Imprinted Polymers: Antibody Mimics for Bioimaging and Therapy. Chem. Rev. 2020, 120, 9554–9582.
15. Xu, J.; Miao, H.; Wang, J.; Pan, G. Molecularly Imprinted Synthetic Antibodies: From Chemical Design to Biomedical Applications. Small 2020, 16.
16. Poma, A.; Turner, A.P.F.; Piletsky, S.A. Advances in the manufacture of MIP nanoparticles. Trends Biotechnol. 2010, 28, 629–637.
17. Xu, J.; Ambrosini, S.; Tamahkar, E.; Rossi, C.; Haupt, K.; Bui, B.T.S. Toward a Universal Method for Preparing Molecularly Imprinted Polymer Nanoparticles with Antibody-like Affinity for Proteins. Biomacromolecules 2016, 17, 345–353.
18. Xu, S.; Wang, L.; Liu, Z. Molecularly Imprinted Polymer Nanoparticles: An Emerging Versatile Platform for Cancer Therapy. Angew. Chem. Int. Ed. 2021, 60, 3858–3869.
19. Cheong, W.J.; Yang, S.H.; Ali, F. Molecular imprinted polymers for separation science: A review of reviews. J. Sep. Sci. 2013, 36, 609–628.
20. Saylan, Y.; Yilmaz, F.; Ozgur, E.; Derazshamshir, A.; Yavuz, H.; Denizli, A. Molecular Imprinting of Macromolecules for Sensor Applications. Sensors 2017, 17, 898.
21. Vaneckova, T.; Bezdekova, J.; Han, G.; Adam, V.; Vaculovicova, M. Application of molecularly imprinted polymers as artificial receptors for imaging. Acta Biomater. 2020, 101, 444–458.
22. Korde, B.A.; Mankar, J.S.; Phule, S.; Krupadam, R.J. Nanoporous imprinted polymers (nanoMIPs) for controlled release of cancer drug. Mater. Sci. Eng. C 2019, 99, 222–230.

23. Wackerlig, J.; Schirhagl, R. Applications of Molecularly Imprinted Polymer Nanoparticles and Their Advances toward Industrial Use: A Review. Anal. Chem. 2016, 88, 250–261.

24. Lowdon, J.W.; Dilien, H.; Singla, P.; Peeters, M.; Cleij, T.J.; van Grinsven, B.; Eersels, K. MIPs for commercial application in low-cost sensors and assays—An overview of the current status quo. Sens. Actuators B 2020, 325, 128973.

25. Urraca, J.L.; Marazuela, M.D.; Merino, E.R.; Orellana, G.; Moreno-Bondi, M.C. Molecularly imprinted polymers with a streamlined mimic for zearalenone analysis. J. Chromatogr. A 2006, 1116, 127–134.

26. Lucci, P.; Derrien, D.; Alix, F.; Perollier, C.; Bayoudh, S. Molecularly imprinted polymer solid-phase extraction for detection of zearalenone in cereal sample extracts. Anal. Chim. Acta 2010, 672, 15–19.

27. Xuan-Anh, T.; Acha, V.; Haupt, K.; Bernadette Tse Sum, B. Direct fluorimetric sensing of UV-excited analytes in biological and environmental samples using molecularly imprinted polymer nanoparticles and fluorescence polarization. Biosens. Bioelectron. 2012, 36, 22–28.

28. Demir, B.; Lemberger, M.M.; Panagiotopoulou, M.; Rangel, P.X.M.; Timur, S.; Hirsch, T.; Bui, B.T.S.; Wegener, J.; Haupt, K. Tracking Hyaluronan: Molecularly Imprinted Polymer Coated Carbon Dots for Cancer Cell Targeting and Imaging. ACS Appl. Mater. Interfaces 2018, 10, 3305–3313.

29. Panagiotopoulou, M.; Salinas, Y.; Beyazit, S.; Kunath, S.; Duma, L.; Prost, E.; Mayes, A.G.; Resmini, M.; Bui, B.T.S.; Haupt, K. Molecularly Imprinted Polymer Coated Quantum Dots for Multiplexed Cell Targeting and Imaging. Angew. Chem. Int. Ed. 2016, 55, 8244–8248.

30. Thongrussamee, T.; Kuzmina, N.S.; Shim, W.B.; Jiratpong, T.; Eremin, S.A.; Intrasook, J.; Chung, D.H. Monoclonal-based enzyme-linked immunosorbent assay for the detection of zearalenone in cereals. Food Addit. Contam. Part A 2008, 25, 997–1006.

31. Wang, D.; Zhang, Z.; Zhang, Q.; Wang, Z.; Zhang, W.; Yu, L.; Li, H.; Jiang, J.; Li, P. Rapid and sensitive double-label based immunochromatographic assay for zearalenone detection in cereals. Electrophoresis 2018, 39, 2125–2130.

32. Caglayan, M.O.; Ustundag, Z. Detection of zearalenone in an aptamer assay using attenuated internal reflection ellipsometry and it's cereal sample applications. Food Chem. Toxicol. 2020, 136, 111081.

33. Sun, S.; Xie, Y. An enhanced enzyme-linked aptamer assay for the detection of zearalenone based on gold nanoparticles. Anal. Methods 2021, 13, 1255–1260.

34. Nestora, S.; Merlier, F.; Beyazit, S.; Prost, E.; Duma, L.; Baril, B.; Greaves, A.; Haupt, K.; Bui, B.T.S. Plastic Antibodies for Cosmetics: Molecularly Imprinted Polymers Scavenge Precursors of Malodors. Angew. Chem. Int. Ed. 2016, 55, 6252–6256.

35. Kolapalli, S.P.; Kumaraswamy, S.B.; Mortha, K.K.; Thomas, A.; Das Banerjee, S. UNIVmAb reactive albumin associated hyaladherin as a potential biomarker for colorectal cancer. Cancer Biomark. 2021, 30, 55–62.

36. Li, H.; Guo, L.; Li, J.W.; Liu, N.; Qi, R.; Liu, J. Expression of hyaluronan receptors CD44 and RHAMM in stomach cancers: Relevance with tumor progression. Int. J. Oncol. 2000, 17, 927–932.

37. Rangel, P.X.M.; Lacief, S.; Xu, J.; Panagiotopoulou, M.; Kovensky, J.; Bui, B.T.S.; Haupt, K. Solid-phase synthesis of molecularly imprinted polymer nanolabels: Affinity tools for cellular bioimaging of glycans. Sci. Rep. 2019, 9, 3923.

14 Chapter

THE IMPACT OF CRYOCONCENTRATED BLUEBERRY JUICE AS A FUNCTIONAL INGREDIENT IN THE PRODUCTION OF COMMERCIAL CONFECTIONARY HYDROGELS

1. INTRODUCTION: AN OVERVIEW

Confectionery hydrogel products are highly popular and in demand among consumers, particularly youngsters and the elderly, because of their visually appealing appearance, texture, and mouthfeel [1]. More precisely, the confectionary hydrogel goods are produced using substances that cause gelling, such as sucrose, glucose syrup, acids, flavorings, and colorants [2]. Therefore, by varying the processing method and combination of ingredients, it is feasible to produce hydrogel products with various textures. These textures include gelatin gels with a soft consistency [3], gummy candies with a firm, soft, and chewy structure [4], and marshmallows or aerated gummy candies (which contain air bubbles) with a soft and elastic texture [5].

Conversely, the increasing consciousness and understanding (both at an individual and communal level) among customers have prompted food manufacturers to substitute dangerous components found in numerous food products that pose risks to human health. Confectionery hydrogel products have faced scrutiny because of their elevated sugar levels and limited nutritional value. This combination has been linked to obesity, diabetes, cardiovascular disorders, and hypertension [6]. Consumers are currently seeking confectionary hydrogel foods that are healthier, meaning they have lower sugar content and better levels of natural antioxidants, compared to the existing commercial items [7]. Consequently, scientists are striving to enhance the nutritional content and customer perception of these goods. The predominant approaches involve replacing sugar with low-calorie sweeteners, such as sorbitol, isomaltulose, and stevia [6,8,9]. Therefore, hydrogel products have been enriched with a range of bioactive ingredients, including tea extract [7,10], hibiscus extract [11], betalain-rich extract [12], watermelon juice [13], pomegranate juice [14], and cryoconcentrated blueberry juice [15,16].

Cryoconcentrated juices, processed using cryoconcentration technology at low temperatures, preserve their natural sugars, bioactive components, and antioxidant activity [17]. Additionally, they exhibit appealing colors and offer notable health advantages [18,19,20]. It is important to note that these bioactive compounds may lose their therapeutic properties during storage due to their instability under different environmental conditions [21,22]. Hence, it is crucial to ascertain their stability under various storage circumstances. Kinetic models have been employed to forecast the impact of storage on the stability of bioactive substances [21]. Tutunchi et al. [22] and Rodríguez-Sánchez et al. [23] found that the breakdown of betanin and betaxanthins in gummy candy follows first-order kinetics. Nevertheless, there is a scarcity of research on the deterioration rates of bioactive substances in hydrogels made from confectionery products under storage conditions. This is of great significance to the food industry, as they are keen on creating confectionery products that offer health benefits.

The objective of this study was to assess the physical and textural characteristics, stability of the bioactive compound, and antioxidant activity in commercially available hydrogel products made from confectionery foods that were fortified with cryoconcentrated blueberry juice. The products were stored at two different temperatures (4 °C and 25 °C) for a period of 35 days.

2. RESULTS AND DISCUSSION

2.1. Characterization of Hydrogels Samples

2.1.1. Physicochemical Properties

Table 1 summarizes the physicochemical properties of the samples.

Physicochemical Parameter	GG	AGG	GM	AGM
TSS (°Brix)	16.8 ± 0.0 [a]	21.0 ± 0.1 [b]	61.0 ± 1.0 [cd]	60.1 ± 0.2 [c]
pH	5.0 ± 0.0 [a]	5.2 ± 0.0 [c]	5.1 ± 0.0 [b]	5.1 ± 0.0 [b]
Moisture (%)	83.1 ± 0.1 [d]	78.9 ± 0.2 [c]	38.7 ± 0.4 [a]	38.3 ± 2.6 [ab]
Water activity	0.988 ± 0.001 [d]	0.981 ± 0.003 [c]	0.889 ± 0.009 [ab]	0.877 ± 0.008 [a]
Density (kg/m^3)	1083.2 ± 15.9 [b]	408.6 ± 5.3 [a]	1202.1 ± 33.1 [c]	406.5 ± 14.2 [a]
Gas hold-up (ε, %)	ND	62.3 ± 0.9 [a]	ND	66.2 ± 1.5 [b]

[a-d]: Different superscripts within the same row indicate significant differences at $p \le 0.05$. GG: gelatin gel; AGG: aerated gelatin gel; GM: gummy; AGM: aerated gummy; TSS: total soluble solid; ND: not determined.

Initially, GG and AGG reached a TSS content of around 17 °Brix and 21 °Brix, respectively. Rubio-Arraez et al. [3] reported a similar TSS value (20 °Brix) in gels with citrus juice and non-cariogenic sweeteners. Thus, this result means that CBJ provides the TSS necessary to obtain gelatin gel without sweeteners. While GM and AGM exhibited lower TSS values (≈60 °Brix) than that indicated by studies in gummies and marshmallows [10,11,24,25], which reported values between 70 °Brix and 80 °Brix in the hydrogel samples. This difference may be due to the partial substitution of the syrup for CBJ, resulting in GM and AGM with a lower amount of added sugar. For pH, the values ranged from 5.0 to 5.2. These values are lower than those previously reported for similar products, ranging from 6.0 and 6.9 [5,24,25].

This decrease in pH value can be related to the CBJ added in the samples, which provides organic acids, such as malic, citric, and shikimic acids [26]. In the same way, the moisture content and aw in the GG and AGG samples were in agreement with the values reported by Rubio-Arraez et al. [3]. For GM and AGM hydrogels, the results were significantly higher than those stated by Mardani et al. [25] and Periche et al. [5,24], who observed a moisture content value close to 16% to 24% and aw of approximately 0.75 to 0.85. In contrast, Mandura et al. [10] and Šeremet et al. [7] reported high moisture content (34%) for gummy with a white tea infusion as an ingredient. Therefore, these differences could be attributed to TSS and phenolic compounds since the phenolic groups can hold more water due to interactions between the hydroxyl groups of polyphenols, proteins, sugars, and water [27].

For density, the aerated samples (AGG and AGM) showed a reduction of 35% compared to the non-aerated samples (GG and GM) due to the incorporation of air bubbles [15]. This incorporated air (ε) corresponds to 62% and 66% for AGG and AGM, respectively. These results are in line with the values reported by Casas-Forero et al. [15] and Mardani et al. [25] for aerated gelatin gel and marshmallows, respectively. Furthermore, air is the "main ingredient" of aerated confectionery, and it provokes a significant decrease in density [28].

2.1.2. Rheological and Mechanical Properties

Figure 1 shows the rheological behavior of the hydrogel samples.

First, all samples exhibited similar behavior. Thus, at low shear rates (<10 s−1), the viscosity decreased with an increasing shear rate, indicating that the samples had a pseudoplastic fluid and shear thinning behavior. Then, at high shear rates (>10 s−1), the viscosity no longer changed with the increase in shear rate (Newtonian fluid) [6]. Similarly, Figure 1a (relationship between apparent viscosity and shear rate) shows that GG had the lowest viscosity (0.006 Pa·s). In contrast, AGG significantly increased its viscosity (0.083 Pa·s), showing values close to those found for GM and AGM (≈0.05 Pa·s). This increase in viscosity is mainly related to an increase in the amount of gelatin added in the AGG, GM, and AGM samples, which was significantly higher than GG hydrogel, with an increase of 2.6, 2.0, and 2.0 times, respectively. This behavior agrees with that reported by Casas-Forero et al. [15]. Furthermore, as described above, aeration in confectionery changes its physical properties, in particular, an increase in viscosity and a decrease in fluidity [28].

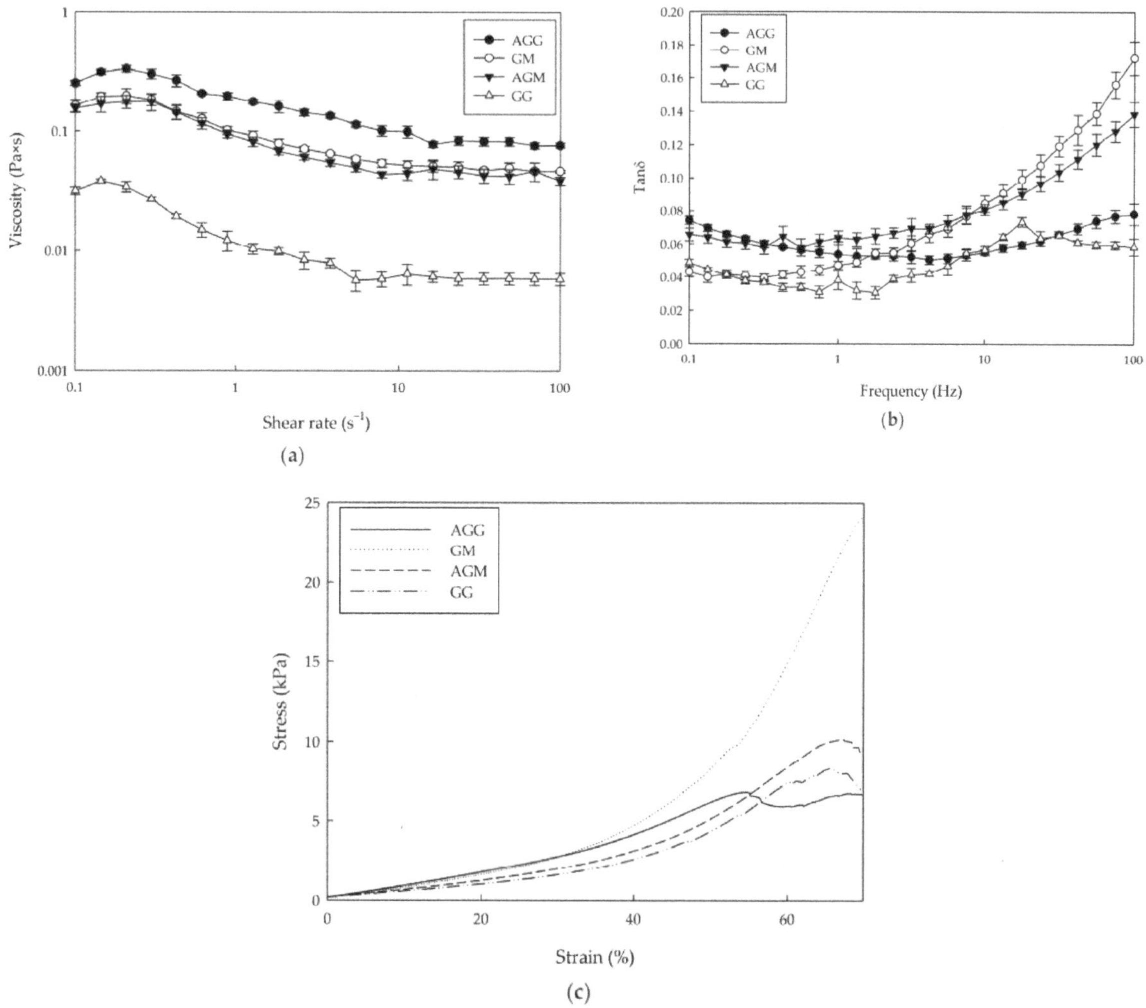

Figure 1. Rheological and mechanical properties of hydrogel samples: (a) The changes in viscosity during shear rate; (b) The changes in tan δ (G″/G′) during frequency scanning; (c) Stress and strain curves in uniaxial compression.

In the same way, the value of tan δ (G″/G′) is a parameter of the viscoelastic behavior of the materials [29]. As shown in Figure 1b, the tan δ of all samples was less than 0.5 in the entire frequency range, indicating that an elastic behavior predominated over the viscous, which is characteristic of these types of hydrogel products [6].

Additionally, stress-deformation curves of the samples during uniaxial compression are depicted in Figure 1c. All samples exhibited a sigmoid stress–strain behavior, characteristic of foam materials [30]. As can be seen, GG had fracture stress and strain of 8 kPa and 65%, respectively, exhibiting the characteristics of a soft material. The results are consistent with the literature, which reports 4 kPa to 8 kPa and 60% values for fracture stress and strain, respectively [31,32]. AGG was characterized by a weaker and less ductile texture with 6 kPa for fracture stress and 55% for fracture strain. This difference between GG and AGG is attributed to the air bubbles, which reduce the matrix content per unit of sectional [32]. This is in line, with Hartel et al. [28] who indicated that the texture properties of aerated confectionery are largely dependent on the air phase. For GM and AGM, fracture stress and strain values in GM (23 kPa and 69%, respectively) were higher than achieved by AGM (10 kPa and 67%, respectively). This means that GM had a firm and elastic texture, while AGM was a product with a soft elastic texture. Likewise, AGM showed fracture stress close to those obtained in a commercial marshmallow (9 kPa) [33], indicating that it is possible to develop hydrogel foods with CBJ as a source of natural sugar and bioactive compounds without affecting its texture, which would be a potential commercial advantage.

2.1.3. Microstructural Features

Micrographs of the confectionery are shown in Figure 2.

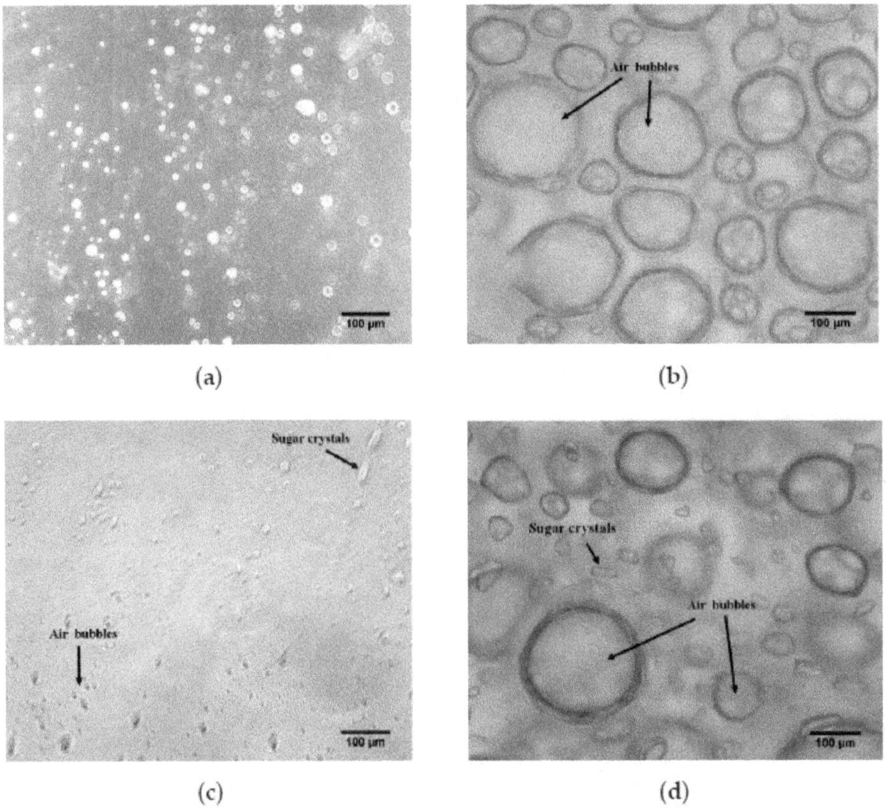

Figure 2. Optical micrograph images of gelatin-based confectionery at 10X: (a) GG; (b) AGG; (c) GM; (d) AGM.

GG exhibited a homogeneous structure with a small strand formation (Figure 2a), which is characteristic of the microstructure of bovine gelatin [12]. GM preserved the uniform structure of GG with the formation of some air bubbles and any tiny sugar crystals (Figure 2c). A similar microstructure was reported by de Moura et al. [11] and Kumar et al. [34] in jelly candies with hibiscus extract-encapsulated and vegan gummy candy, respectively. In the case of the aerated samples, AGG and AGM showed small and large air bubbles. Large bubbles with a marked spherical shape were characterized in AGG hydrogels (Figure 2b), while AG showed small bubbles with non-spherical shapes and the formation of tiny sugar crystals (Figure 2d). The difference in bubble size between AGG and AGM can be attributed to the gelatin content, as evidenced by Casas-Forero et al. [15] in the study of aerated gelatin gels that increased small bubbles as the amount of gelatin decreased.

2.2. Stability of Hydrogel Samples during Storage

2.2.1. Stability of TBC Content and Antioxidant Activity

Total bioactive compounds (TBC) and antioxidant activity (AA) values in the initial sample and hydrogel samples are shown in Table 2.

Table 2. Total bioactive compounds (TBC) and antioxidant activity (AA) of the hydrogel samples.

Sample	TBC			AA	
	TPC	TAC	TFC	DPPH	FRAP
CBJ	773.3 ± 8.7 ^{e}	22.3 ± 0.7 ^{e}	566.8 ± 7.6 ^{e}	4585.4 ± 8.5 ^{e}	4442.0 ± 61.9 ^{d}
GG	233.4 ± 1.6 ^{a}	3.9 ± 0.1 ^{cd}	136.1 ± 2.9 ^{cd}	782.9 ± 3.0 ^{b}	1067.8 ± 13.5 ^{a}
AGG	251.0 ± 3.6 ^{cd}	3.5 ± 0.2 ^{a}	133.7 ± 3.7 ^{c}	759.1 ± 1.1 ^{a}	1057.3 ± 45.9 ^{a}
GM	247.7 ± 0.9 ^{c}	3.7 ± 0.2 ^{abc}	119.6 ± 3.9 ^{a}	911.3 ± 2.2 ^{d}	1075.1 ± 12.6 ^{ab}
AGM	238.5 ± 3.9 ^{ab}	3.7 ± 0.1 ^{ab}	123.9 ± 4.7 ^{ab}	870.5 ± 4.8 ^{c}	1083.3 ± 31.8 ^{ab}

$^{a-e}$: Different superscripts within the same column indicate significant differences at $p \le 0.05$. CBJ: cryoconcentrated blueberry juice; GG: gelatin gel; AGG: aerated gelatin gel; GM: gummy; AGM: aerated gummy.

The fortification of food with juice rich in bioactive compounds can improve the nutritional and functional values of these products [35]. First, CBJ had a TBC content close to 770 mg GAE/100 g for TPC, 22 mg C3G/100 g for TAC, and 560 mg CEQ/100 g for TFC, and the AA values were approximately 4585 µmol TE/100 g for DPPH and 4440 µmol TE/100 g for FRAP.

On day 0, hydrogel samples displayed a TBC content ranging from 230 to 250 mg GAE/100 g for TPC, 3.5 to 3.9 mg C3G/100 g for TAC, and 120 to 136 mg CEQ/100 g for TFC. These slight variations in TBC content between the samples could be due to processing conditions since all samples were fortified with CBJ at 30% (w/w). Compared to other studies, the TPC obtained was higher than those reported in white tea-based candies (170–180 mg GAE/100 g) [7], pomegranate juice-based candies (72–159 mg GAE/100 g) [14], and jelly candies with rosemary extract (227 mg GAE/100 g) [8]. In contrast, Rivero et al. [9] reported polyphenol content of over 50% in candies with raspberry juice powder (490–550 mg GAE/100 g). These differences between studies can be related to the formulation, and the process conditions used.

In terms of AA, all the samples exhibited similar FRAP values (1050–1085 µmol TE/100 g), while DPPH values varied between samples. Thus, in DPPH, GG and AGG had values of 780 and 760 µmol TE/100 g, respectively, while GM and AGM increased the DPPH results, reaching values of 910 and 870 µmol TE/100 g, respectively. The values are comparable to those reported by Rivero et al. [9], Mandura et al. [10], and Hani et al. [36], who studied the AA of gummy with red pitaya fruit puree, white tea-based candies, and jellies containing honey and propolis, respectively.

Additionally, we observed a proportional trend between CBJ and hydrogel samples. A decrease varying from ≈70% to ≈85% for TBC and DPPH values were observed, while FRAP decreased in all the samples

(44.7% for GG, 24.5% for AGG, 39.5% for GM, and 19.1% for AGM). These variations are in agreement with the results reported by Cedeño-Pinos et al. [8], who suggested that the reduction in the AA level in gelatin-based candies may be due to interactions between ingredients, the presence of amino acids, or the interaction of antioxidants with the gelling matrix.

Figure 3 shows the degradation of total bioactive compounds of the hydrogel samples and CBJ during storage.

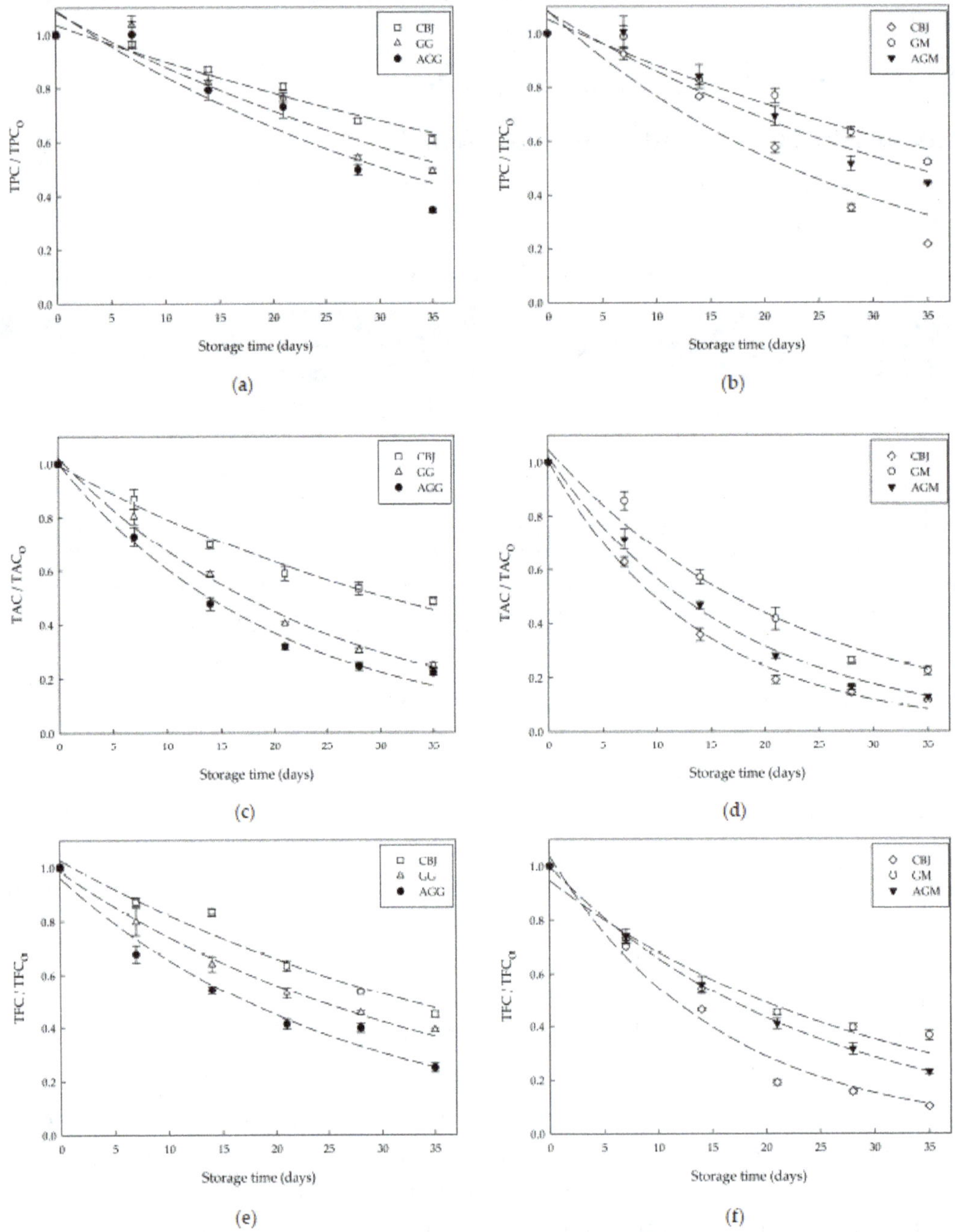

Figure 3. Degradation of total bioactive compounds of the gelatin-based confectionery and CBJ during storage: (a) TPC at 4 °C; (b) TPC at 25 °C; (c) TAC at 4 °C; (d) TPC at 25 °C; (e) TFC at 4 °C, (f) TPC at 25 °C.

The CBJ showed a significant reduction in TBC by increasing the temperature from 4 to 25 °C during storage. This decrease in TBC was between 40% and 50% for CBJ stored at 4 °C and about 80% for CBJ stored at 25 °C. This result was higher than the previously reported value by Orellana-Palma et al. [19] for cryoconcentrated Calafate juice, who reported a loss between 4% and 25% in the TBC after 35 days of storage at 4 °C. Regarding anthocyanin stability studies in blueberry juice, Barba et al. [37] indicated a decrease of about 9% after 7 days of storage at 4 °C in juice treated with PEF. Furthermore, Cortellino and Rizzolo [38] and Zhang et al. [39] observed in thermally treated juice a loss of 60% after 150 days at 20 °C and 80% after ten days at 4 °C, respectively. These differences in the TBC degradation rate during storage can be related to pretreatments before juice storage that could inactivate enzymes, such as polyphenol oxidase and peroxidase, which are considered responsible for the decay of phenols in berry-derived foods [38].

Figure 3a,b show that TPC gradually decreased as the storage time increased. After 35 days, the polyphenols loss was 50% and 65% for GG and AGG, respectively, whereas GM and AGM showed a lower loss with 48% and 55% values. This difference between samples could be related to the moisture content and water activity, since a more significant volume of water favors reactant mobility, contributing to phenolic compounds degradation [40]. Maier et al. [41] also observed a decrease in the phenolic content from 243.6 mg/kg to 82.6 mg/kg in gelatin gels enriched with grape pomace extract during storage. Furthermore, Tutunchi et al. [22] indicated a loss of TPC of 25% to 35% for gummy candy with red beet extract powder at 28 days. Meanwhile, Šeremet et al. [7] reported fluctuations in the TPC during the storage of white tea-based candies.

On the other hand, TAC decreased markedly over 35 days, with a total decrease of 75%, 78%, 77%, and 87% for GG, AGG, GM, and AGM, respectively (Figure 3c,d). These results were consistent with the findings of de Moura et al. [11] and Maier et al. [41]. They reported a reduction of close to 70% in anthocyanin content in jelly candy with hibiscus extract and gelatin gels enriched with grape pomace extract, respectively. Tavares et al. [42] mentioned that anthocyanins are unstable and highly susceptible to degradation under conditions such as pH, temperature, light, oxygen, enzymes, ascorbic acid, and copigments. Likewise, Chen et al. [21] and Teribia et al. [43] suggested that anthocyanin degradation could result from condensation and polymerization reactions with other phenolic compounds present in the sample. Hence, this anthocyanin decrease can lead to undesirable color changes and a reduction in antioxidant activity.

Like the results of the TPC and TAC, the storage also had an unfavorable influence on the TFC (Figure 3e,f). At the end of 35 days of storage, the flavonoids content in GG and AGG stored at 4 °C decreased by approximately 60% and 75%, respectively, and for the GM and AGM samples stored at 25 °C, this reduction was up to 62% and 77%, respectively. In general terms, the stability of TBC in gelatin-based foods was slightly higher than CBJ mainly when stored at 25 °C, possibly due to the interaction of polyphenols with gelatin that led to lower availability of polyphenols for hydrolysis and oxidation reactions [22,23]. Similarly, Kia [44] indicated that the gelatin triple helix could retain antioxidant compounds due to crosslinking and the formation of three-dimensional networks. On the other hand, the TBC stability in samples containing air bubbles inside the food matrix (AGG and AGM) was significantly ($p \leq 0.05$) lower than those samples non-aerated (GG and GM). This can be related to oxygen reacting with antioxidant compounds through hydrogen atom donation of the hydroxyl group to a free radical, leading to a decrease in the bioactive compounds [45].

Figure 4 shows a significant reduction ($p \leq 0.05$) in the values of DPPH and FRAP during the storage period at 4 and 25 °C.

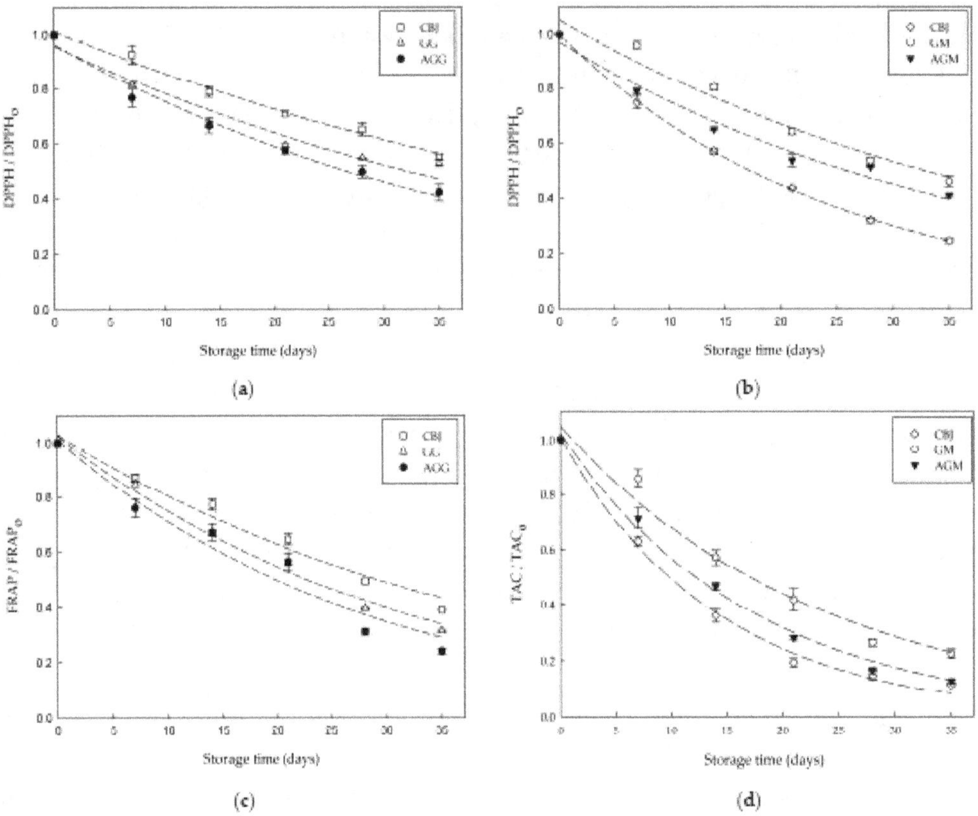

Figure 4. Antioxidant activity stability of hydrogel samples and CBJ during storage: (a) DPPH at 4 °C; (b) DPPH at 25 °C; (c) FRAP at 4 °C; (d) FRAP at 25 °C.

DPPH values of hydrogel samples decreased by approximately 54% by day 35. Meanwhile, a higher loss was observed in the FRAP assay, which decreased between 65% and 75% by day 35. This reduction might be due to the degradation of bioactive compounds, mainly anthocyanins [21]. The decrease in AA during storage has been reported in other studies, including a decrease in AA of 64% in jelly prepared with citrus juice [3], 51% in white tea-based candies [10], 21% to 37% in gummy candy with red beet extract powder [22], 15% in vegan gummy candies with betalains nanoliposomes [34], and 55% to 70% in gummy candy with red beet extract [46].

On the other hand, k and t1/2 are summarized in Table 3.

Table 3. Kinetic parameters of total bioactive compounds and antioxidant activity in the CBJ and hydrogel samples stored under different conditions.

Assay	Kinetic Parameters	4 °C			25 °C		
		CBJ	GG	AGG	CBJ	GM	AGM
TPC	k (10^{-3} day^{-1})	13.0 ± 0.6 [a]	18.6 ± 0.1 [c]	24.9 ± 1.4 [e]	36.8 ± 0.7 [f]	16.4 ±0.6 [b]	21.3 ± 1.3 [d]
	$t_{1/2}$ (day)	53.3 ± 2.7 [f]	37.2 ± 0.3 [d]	27.9 ± 1.6 [b]	18.8 ± 0.4 [a]	42.4 ± 1.5 [e]	32.6 ± 1.9 [c]
	R^2	0.94	0.87	0.86	0.89	0.92	0.90
TAC	k (10^{-3} day^{-1})	22.3 ± 1.2 [a]	40.9 ± 1.1 [b]	47.9 ± 1.7 [c]	67.7 ± 2.2 [ed]	43.6 ± 2.4 [b]	60.1 ± 7.2 [d]
	$t_{1/2}$ (day)	31.1 ± 1.8 [e]	16.9 ± 0.4 [d]	14.5 ± 0.5 [c]	10.3 ± 0.3 [a]	15.9 ± 0.9 [cd]	11.5 ± 0.1 [b]
	R^2	0.97	0.99	0.97	0.97	0.98	0.99
TFC	k ($10^{-3}$$day^{-1}$)	21.8 ± 0.5 [a]	28.4 ± 0.5 [b]	38.8 ± 1.6 [d]	66.7 ± 0.9 [f]	33.1 ± 0.1 [c]	41.9 ± 1.7 [e]

Assay	Kinetic Parameters	4 °C			25 °C		
		CBJ	GG	AGG	CBJ	GM	AGM
	$t_{1/2}$ (day)	31.8 ± 0.7 [e]	24.4 ± 0.4 [d]	17.9 ± 0.7 [b]	10.4 ± 0.2 [a]	20.9 ± 0.8 [c]	16.5 ± 0.6 [b]
	R^2	0.96	0.98	0.95	0.97	0.90	0.99
DPPH	k (10^{-3} day^{-1})	16.3 ± 0.6 [a]	21.2 ± 0.4 [b]	25.6 ± 0.6 [c]	40.2 ± 0.7 [d]	21.3 ± 0.6 [b]	26.2 ± 0.6 [c]
	$t_{1/2}$ (day)	42.5 ± 1.7 [d]	32.8 ± 0.6 [c]	27.1 ± 0.6 [b]	17.2 ± 0.3 [a]	32.6 ± 0.9 [c]	26.5 ± 0.6 [b]
	R^2	0.98	0.90	0.96	0.98	0.96	0.97
FRAP	k (10^{-3} day^{-1})	24.7 ± 0.6 [a]	31.7 ± 0.2 [b]	37.9 ± 1.7 [d]	54.2 ± 1.5 [f]	34.1 ± 0.3 [c]	41.7 ± 0.6 [e]
	$t_{1/2}$ (day)	28.0 ± 0.7 [e]	21.9 ± 0.2 [d]	18.2 ± 0.8 [c]	12.8 ± 0.3 [a]	20.3 ± 0.2 [c]	16.6 ± 0.2 [b]
	R^2	0.96	0.98	0.94	0.97	0.97	0.99

[a–f]: Different superscripts within the same column indicate significant differences at $p \leq 0.05$. CBJ: cryoconcentrated blueberry juice; GG: gelatin gel; AGG: aerated gelatin gel; GM: gummy; AGM: aerated gummy. TPC: Total polyphenol content, TAC: Total anthocyanin content, TFC: Total flavonoid content, k: kinetic degradation rate, $t_{1/2}$: half-life time, R^2: correlation index.

The TBC and AA degradation was appropriately fitted to a first-order kinetic model (R2 > 0.9), which is similar to the previous studies on the degradation of bioactive compounds in blueberries [41,47]. As expected, in CBJ samples, the storage temperature significantly affected the k values of TBC and AA. The results indicate that CBJ stored at 25 °C degrades three times faster than CBJ stored at 4 °C. Anthocyanins showed the most remarkable changes, decreasing t1/2 from 31 days to 10 days. As reported by Tavares et al. [42], storage temperature influences anthocyanin degradation due to their high thermosensitivity.

By comparing samples stored at 4 °C, we observed that the matrix did not have a protective effect against the degradation of TBC and AA, as the value of k was approximately 1.3 and 1.6 times higher for GG and AGG compared to CBJ, respectively. These differences in the value of k could be related to the available water content since CBJ had a high TSS (45 °Brix), exhibited a lower aw (0.911) than the GG and AGG samples (0.988 and 0.981, respectively). Thus, these conditions reduce the degradation of phenolic compounds in CBJ by decreasing the mobility of the reactants within the food [22,40]. In samples stored at 25 °C, a protective effect of gelatin was noticeable on the degradation of TBC and AA. A 52% and 38% lower k were observed in GM and AGM than in CBJ, respectively. Similar results were reported by Tutunchi et al. [22], who indicated that betanin was more stable in gummy candy than a beverage. In this regard, Rodriguez-Sánchez et al. [23] suggested that interactions with proteins may protect some pigments such as betaxanthins from hydrolysis and oxidation. Likewise, Muhamad et al. [48] indicated that sugar has a protective effect on anthocyanins, which could be assigned to its steric interference with reaction products of anthocyanin–phenolic polymerization [49]. The anthocyanins stability results showed a t1/2 of 16 and 12 days for GM and AGM, respectively. Tutunchi et al. [22] also observed a similar t1/2 in gummy candies enriched with red beet extract powder.

2.2.2. Stability of Color

The CIELAB parameters (L*, a*, b*, and ΔE) of gelatin-based foods and CBJ stored for 35 days are shown in Table 4.

Table 4. Color parameters of CBJ and gelatin-based confectionery during storage.

Sample	Time (Days)	L*	a*	b*	ΔE
4 °C					
CBJ	0	8.0 ± 0.2 [d]	24.6 ± 0.9 [f]	4.9 ± 0.2 [c]	-
	7	8.0 ± 0.0 [d]	22.5 ± 0.0 [e]	4.6 ± 0.1 [c]	2.1 ± 1.0 [a]
	14	7.1 ± 0.2 [c]	21.6 ± 0.0 [d]	4.3 ± 0.0 [b]	3.1 ± 1.0 [b]
	21	6.9 ± 0.2 [bc]	21.0 ± 0.0 [c]	4.3 ± 0.0 [b]	3.8 ± 0.9 [c]
	28	6.7 ± 0.2 [b]	19.3 ± 0.1 [b]	3.2 ± 0.1 [a]	5.7 ± 0.9 [d]
	35	6.3 ± 0.2 [a]	17.4 ± 0.6 [a]	3.2 ± 0.0 [a]	7.6 ± 1.5 [e]
GG	0	18.1 ± 0.6 [d]	7.6 ± 0.4 [f]	1.1 ± 0.0 [d]	-
	7	17.2 ± 0.3 [cd]	7.1 ± 0.2 [e]	1.1 ± 0.1 [d]	1.2 ± 0.4 [a]
	14	16.6 ± 0.3 [bc]	5.8 ± 0.2 [d]	0.8 ± 0.1 [c]	2.4 ± 0.2 [b]
	21	16.2 ± 0.5 [b]	4.5 ± 0.1 [c]	0.7 ± 0.0 [c]	3.8 ± 0.3 [c]
	28	15.9 ± 0.6 [ab]	4.1 ± 0.1 [b]	0.6 ± 0.0 [b]	4.3 ± 0.1 [d]
	35	15.3 ± 0.2 [a]	3.8 ± 0.1 [a]	0.5 ± 0.0 [a]	4.8 ± 0.4 [e]
AGG	0	60.2 ± 0.5 [e]	10.2 ± 0.2 [f]	10.0 ± 0.1 [d]	-
	7	58.7 ± 0.7 [d]	9.6 ± 0.1 [e]	9.9 ± 0.3 [d]	1.7 ± 0.3 [a]
	14	56.8 ± 0.3 [c]	8.4 ± 0.2 [d]	9.2 ± 0.0 [c]	4.0 ± 0.0 [b]
	21	54.8 ± 0.1 [b]	7.9 ± 0.0 [c]	8.6 ± 0.1 [b]	6.0 ± 0.5 [c]
	28	53.4 ± 0.1 [a]	5.2 ± 0.0 [b]	8.5 ± 0.0 [b]	8.6 ± 0.2 [d]
	35	51.9 ± 1.4 [a]	4.6 ± 0.0 [a]	7.7 ± 0.0 [a]	10.3 ± 0.9 [e]
25 °C					
CBJ	0	8.0 ± 0.2 [f]	24.6 ± 0.9 [f]	4.9 ± 0.2 [e]	-
	7	6.8 ± 0.1 [e]	18.4 ± 0.1 [e]	3.9 ± 0.1 [d]	6.4 ± 1.0 [a]
	14	5.6 ± 0.0 [d]	13.5 ± 0.5 [d]	1.3 ± 0.0 [c]	11.9 ± 1.3 [b]
	21	4.6 ± 0.0 [c]	10.7 ± 0.1 [c]	1.0 ± 0.0 [b]	14.8 ± 1.0 [c]
	28	3.6 ± 0.0 [b]	9.0 ± 0.0 [b]	0.9 ± 0.0 [a]	16.7 ± 1.0 [d]
	35	3.0 ± 0.0 [a]	7.6 ± 0.0 [a]	0.9 ± 0.0 [a]	18.1 ± 1.0 [e]
GM	0	22.1 ± 0.6 [e]	4.0 ± 0.1 [f]	9.4 ± 0.4 [d]	-
	7	21.0 ± 0.4 [d]	2.8 ± 0.1 [e]	9.7 ± 0.1 [c]	1.7 ± 0.3 [a]
	14	19.3 ± 0.1 [c]	2.5 ± 0.1 [d]	8.7 ± 0.3 [b]	3.3 ± 0.5 [b]
	21	18.6 ± 0.4 [b]	2.4 ± 0.1 [c]	8.8 ± 0.1 [b]	3.9 ± 0.3 [c]
	28	17.9 ± 0.2 [a]	2.3 ± 0.0 [b]	8.7 ± 0.3 [b]	4.6 ± 0.4 [d]
	35	17.3 ± 0.5 [a]	1.7 ± 0.0 [a]	8.1 ± 0.2 [a]	5.5 ± 0.3 [e]
AGM	0	56.4 ± 1.6 [f]	11.3 ± 0.5 [f]	9.4 ± 0.4 [d]	-
	7	53.4 ± 0.8 [e]	11.2 ± 0.1 [e]	8.9 ± 0.1 [c]	3.0 ± 1.0 [a]
	14	47.8 ± 0.6 [d]	10.9 ± 0.2 [d]	8.8 ± 0.1 [c]	8.6 ± 2.2 [b]
	21	46.0 ± 1.0 [c]	9.9 ± 0.3 [c]	8.7 ± 0.3 [bc]	10.5 ± 0.9 [c]
	28	44.2 ± 0.5 [b]	9.1 ± 0.0 [b]	8.4 ± 0.1 [b]	12.4 ± 1.6 [d]
	35	40.7 ± 0.3 [a]	8.8 ± 0.1 [a]	7.9 ± 0.1 [a]	15.9 ± 1.7 [e]

[a-f]: Different superscripts within the same column indicate significant differences at $p \leq 0.05$. CBJ: cryoconcentrated blueberry juice; GG: gelatin gel; AGG: aerated gelatin gel; GM: gummy; AGM: aerated gummy. TPC: Total polyphenol content, TAC: Total anthocyanin content, TFC: Total flavonoid content, k: kinetic degradation rate, $t_{1/2}$: half-life time, R^2: correlation index.

On day 0, CBJ had an intense reddish color mainly due to anthocyanins (L* ≈ 7.9, a* ≈ 24.56, and b* ≈ 4.91), and its addition explains the marked color of the samples, as can be seen in Figure 5.

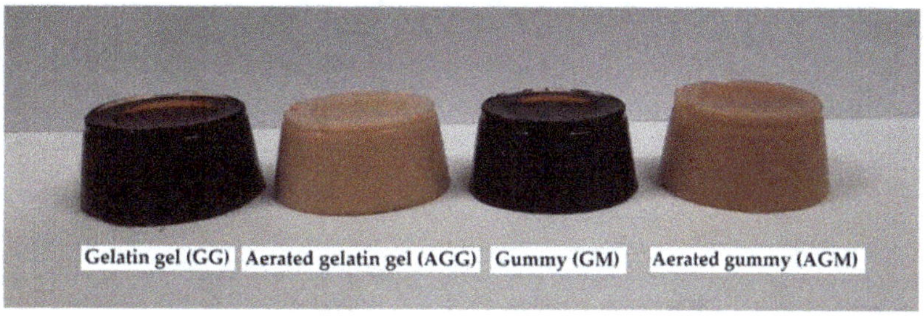

Figure 5. Photographs of the gelatin-based confectionery at day 0.

GG had low L* and higher a* compared to GM, indicating that GG presents a darker reddish color and may be related to a lower gelatin content in the formulation [24], highlighting the intense color of the CBJ. Meanwhile, the aerated samples (AGG and AGM) exhibited an opaque appearance with higher values of L* ($p \leq 0.05$) compared to the non-aerated samples (GG and GM), which is mainly attributed to the air bubbles as reported by Zúñiga et al. [32] and Casas-Forero et al. [15]. In addition, the aeration produces an evident decrease in the darkness of the samples.

In terms of storage, a general decrease in L* values was observed, indicating a progressive darkening. On day 35, CBJ, GG, and AGG showed a decrease of 21%, 16%, and 14%, respectively, stored at 4 °C, while CBJ, GM, and AGM, which were kept at 25 °C, exhibited a reduction of 62%, 22%, and 28%, respectively. Thus, these results denote a protective effect of gelatin on color stability, as indicated by Otálora et al. [12] with gummy candies. Likewise, several authors have reported higher color stability in confectionery, such as gels and candies, made with gelatin as a gelling agent, compared to pectin [7,36,41]. Similarly, the coordinates of a* and b* decreased significantly with time storage ($p \leq 0.05$), indicating that the samples lost their characteristic red coloration, which could be attributed to the degradation of anthocyanins in the chalcones [50].

The ΔE increased with time, reaching values greater than 5 units at 35 days. According to de Moura et al. [11], the human eye can distinguish the color change between samples with ΔE values greater than 3.5. The color difference was more noticeable in the CBJ stored at 25 °C (ΔE ≈18 units), whereas GG and GM exhibited lower values (ΔE ≈5 units). Comparable results were reported by Rivero et al. [9], de Moura et al. [11], and Yan et al. [51].

From a practical point of view, the present study can be useful in the food industry and molecular gastronomy, since the findings obtained are the use of the starting point for the cryoconcentrates used in gelling materials, which in turn allows multiple possibilities for food innovations.

3. CONCISE SUMMARY

The results of this study indicate that it is feasible to enhance the nutritional value of gelatin-based confectionary by including cryoconcentrated juice, without negatively affecting their physical and chemical qualities or texture. The nutritional value of CBJ has been enhanced by the inclusion of bioactive substances that possess a high antioxidant potential. On day 0, gelatin-based diets exhibited elevated Total Bacterial Count (TBC) and Amino Acid (AA) levels. Concerning the stability investigation, there was a notable reduction in the total bacterial count (TBC) and ascorbic acid (AA) content in the CBJ and gelatin-based foods during storage at both 4 °C and 25 °C. However, gelatin had a protective effect against the loss of bioactive chemicals and color alterations. In addition, in order to maintain TBC, it is anticipated that GG and G can be held for a minimum of 21 days at 4 °C and 25 °C, respectively. However, in aerated products (AGG and AGM), the storage duration would be reduced to 15 days at both 4 °C and 25 °C. These data provide evidence that CBJ may be regarded as a pioneering and effective component. Moreover, additional research is necessary to assess the bioactivity of gelatin-based confectionary with cryoconcentrates in order to validate their possible advantageous health impacts.

4. MATERIALS AND METHODS

4.1. Materials

Blueberries (Vaccinium corymbosum L.) were purchased from a local market in Chillán (36°36′24″ S 72°06′12″ W, XVI Región del Ñuble, Chile), and the fruits were stored at 4 °C until processing. Gelatin from bovine powder skin type B (G9382), sucrose, Folin–Ciocalteu reagent, gallic acid, cyanidin-3-glucoside, catequin, 1,1-diphenyl-2-picryl-hydrazyl (DPPH), 2,4,6-tris(2-pyridyl)-S-triazine (TPTZ), and 6-hydroxyl-2,5,7,8-tetramethyl-2-carboxylic acid (Trolox) were purchased from Sigma-Aldrich (St. Loius, MO, USA). Glucose syrup 42 DE was acquired locally from Furet Ltd. (Chillán, Chile). Standards and enzymes used to simulate in vitro digestion, such as human salivary α-amylase (1031), porcine pepsin (P6887), porcine pancreatin (H2625), and bovine bile (B3883), were purchased from Sigma-Aldrich (St. Loius, MO, USA). Calcium chloride (CaCl2), sodium carbonate (Na2CO3), ferric chloride hexahydrate (FeCl3·6H2O), sodium acetate (CH3COONa), sodium hydroxide (NaOH), hydrochloric acid (HCl), sodium nitrite (NaNO2), aluminum chloride (AlCl3), and potassium chloride (KCl) were obtained from Merck (Darmstadt, Germany). Ultrapure and distilled water were used for the preparation of all aqueous solutions.

4.2. General Experimental Procedure

Figure 6 shows the experimental procedure of the hydrogels fortified with CBJ, the characterization of hydrogels samples, storage stability, determination of total bioactive compounds (TBC) and antioxidant activity (AA), color measurement, and half-life time analysis.

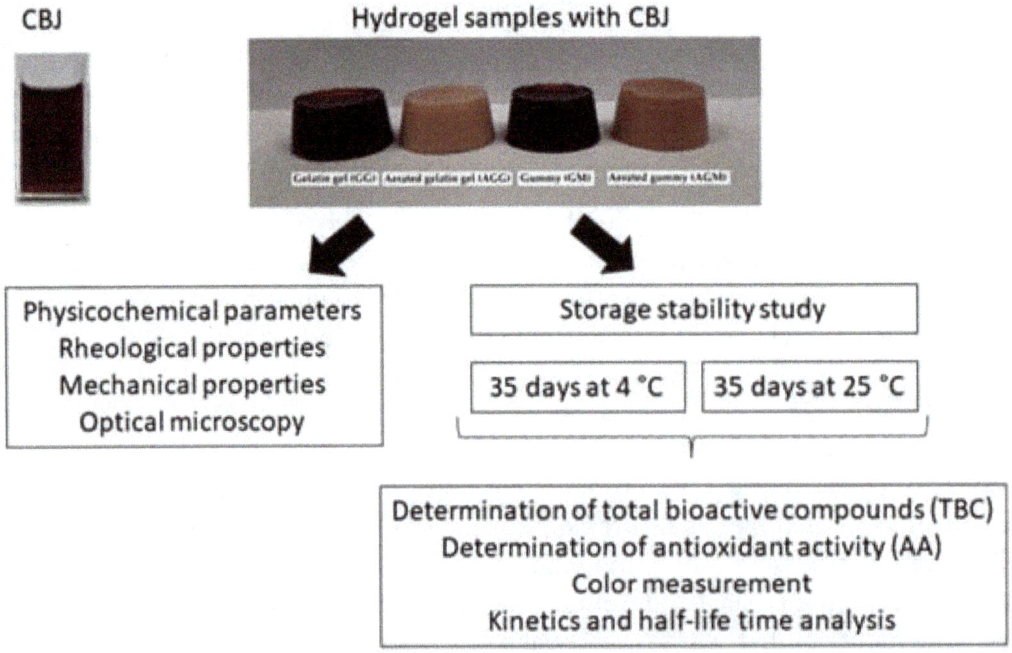

Figure 6. General experimental procedure.

4.3. CBJ Preparation

Blueberries were washed with tap water to remove dirt, and later, the juice was extracted in a tabletop juicer (JE2001, Nex, Barcelona, Spain). Immediately, the juice was filtered through a nylon cloth (0.8 mm mesh) to remove skin and seeds. Then, the juice was concentrated through three cryoconcentration cycles using the centrifugal method described previously by Casas-Forero et al. [16]. Hence, CBJ corresponds to the cryoconcentrated sample acquired in the final cycle. After the third cycle, the CBJ reached values of 45 °Brix and a 4.1 pH. Thus, the cryoconcentrated juice was stored until use as an ingredient in the elaboration of hydrogel samples.

4.4. Preparation of Hydrogel Products

For the commercial hydrogels products based on confectionery foods, four types of samples were prepared, gelatin gel (GG), aerated gelatin gel (AGG), gummy (GM), and aerated gummy (AGM). For GG and AGG samples, the hydrogels with CBJ were performed according to Casas-Forero et al. [15], with slight modifications. Separately, the gelatin powder (3 g for GG and 8 g for AGG) was hydrated in 70 mL distilled water at room temperature for 10 min. Then, the solutions were mechanically stirred (300 rpm) for 10 min at 60 °C. After, the solutions were cooled and maintained for 5 min at 40 °C, and later, 30 g of CBJ was added to each solution, and the solution (with CBJ) was stirred (300 rpm) for 5 min at 40 °C. Later, the gel solution with CBJ was poured into plastic vessels (approximately 34 mm inner diameter and 15 mm height). For AGG samples, air bubbles were incorporated using an Oster® mixer (Oster 2532, 250-watt, 6-speed, Rianxo, Spain) for 9 min, and then, the aerated gel solution with CBJ was deposited into plastic vessels. Finally, the GG and AGG samples were maintained overnight at 4 °C until the formation of a solid gel.

For GM and AGM samples, separately, 6 g of gelatin powder was hydrated in 25 mL distilled water for 10 min at room temperature. Then, the solution was heated at 60 °C and mechanically stirred at 300 rpm until complete dissolution (≈10 min). The gelatin solution was cooled and maintained until at 40 °C. In parallel, a syrup solution was prepared with 350 mL of distilled water, 350 mL of sucrose, and 300 mL of glucose syrup. Thus, once the solution was homogenized with constant agitation, it was heated for 5 min at 110 °C, and then, it was cooled to 50 °C. After, 30 g of CBJ and 45 g of syrup were mixed with the gelatin solution, and the mixture was stirred (300 rpm) for 5 min at 40 °C. Thus, for GM, the gel solution with CBJ and syrup was placed into plastic vessels. For AGM, the aeration process was performed for 4 min using an Oster® mixer (Oster 2532, 250-watt, 6-speed, Rianxo, Spain). Finally, the aerated gel solution with CBJ and syrup was placed into plastic vessels. Thus, the GM and AGM were solidified overnight at room temperature.

4.5. Characterization of Hydrogels Samples

4.5.1. Physicochemical Parameters

Total soluble solids (TSS) expressed as °Brix were analyzed using a digital PAL-1 refractometer (PAL-3, ≈1 mL, range: 0–93 °Brix, precision: ±0.1 °Brix, Atago Inc., Tokyo, Japan). The pH was measured using a digital pH meter (HI 2210, Hanna Instruments, Woonsocket, RI, USA). The moisture content was obtained gravimetrically by vacuum drying at 60 °C in a vacuum oven (3618–1CE, Lab Line Instruments Inc., Melrose Park, IL, USA) for 24 h (AOAC, 20.013, 2000). The water activity (aw) was determined using a dew-point hygrometer (AquaLab Model 4TE, Decagon Devices Inc., Pullman, WA, USA). The density was measured by the flotation method according to Zúñiga and Aguilera [31], and the results were expressed as kg/m3. Gas hold-up (ε) of AGG and AGM was obtained by comparing the density of the aerated sample (ρAGG or ρAGM) with the respective gas-free sample (ρGG or ρGM) density. The ε was calculated according to Equation (1).

$$\varepsilon(\%) = \left(1 - \frac{\rho_{AGG}\ or\ \rho_{AGM}}{\rho_{GG}\ or\ \rho_{GM}}\right) * 100 \quad (1)$$

where ρAGG, ρAGM, ρGG, and ρGM are the density of aerated gelatin gel (AGG), aerated gummy (AGM), gelatin gel (GG), and gummy (GM), respectively.

4.5.2. Rheological Properties

The rheological properties were evaluated by a rotational-type rheometer (Physica MCR300, Anton Paar GmbH, Stuttgart, Germany) using a parallel plate geometry (50 mm diameter) with a 1 mm gap. The apparent viscosity of the samples was determined at 25 °C with a shear rate of 0.1 to 100 s-1 [27]. The frequency sweep measurement was performed from 0.1 to 100 Hz at a constant stress of 3.0 Pa within the linear viscoelastic region at 25 °C [52].

4.5.3. Mechanical Properties

The uniaxial compression test was carried out using a texture analyzer (TAXT plus100, Stable Micro Systems Ltd., Surrey, UK) with a load cell of 5.0 kg. The samples were compressed with a 50 mm diameter cylindrical aluminum probe (P50) at a constant speed of 1 mm/s up to a compression strain of 70% [53]. The true stress (σH) and true strain (εT) were calculated from force-time curves by Equations (2) and (3), respectively.

$$\sigma_H = F\left[\frac{h_o - h}{h_o\ A}\right] \quad (2)$$

$$\varepsilon_T = -\ln\left[\frac{h_o}{h_o - h}\right] \quad (3)$$

where F is the compression force (N), A is the cross-sectional area of the sample (m2), and ho and h are the initial and final height after compression (m), respectively.

4.5.4. Optical Microscopy

The samples were cut into thick slices of 3 mm with a razor blade and the slices were placed on a slide. Photomicrographs were acquired using an Olympus Trinocular Microscope (Olympus Co., Tokyo, Japan) coupled to a digital camera, Olympus LC 20 (Olympus Co., Munster, Germany) with an objective lens Nikon 10× [12].

4.6. Storage Stability Study (SSS)

Specifically, GG and AGG were maintained at 4 °C in the dark in a refrigerated incubator (FOC 215E, Velp Scientific Inc., Milano, Italy), and GM and AGM were stored at 25 °C in the dark in a thermostatic chamber (Memmert UF110, Memmert, Schwabach, Germany). For each sample and the control (CBJ), total bioactive compounds (TBC), antioxidant activity (AA), and color were analyzed at 0, 7, 14, 21, 28, and 35 days of storage.

4.7. Determination of Total Bioactive Compounds (TBC)

Total polyphenol content (TPC) was measured according to the Folin–Ciocalteu method described by Waterhouse [54], with minor modifications. Gallic acid (GA) was used as standard. An amount of 100 μL of the sample was mixed with 500 μL of 10-fold diluted Folin–Ciocalteu reagent. Then, the solution was vigorously mixed with 1500 μL of Na2CO3 (20% w/v). After 90 min in the dark at room temperature (incubation), the absorbance was recorded at 760 nm. TPC was calculated as milligrams of GA equivalents (GAE) per 100 g of sample (mg GAE/100 g).

Total monomeric anthocyanin content (TAC) was quantified by the differential pH method according to Lee et al. [55], with some modifications. Cyanidin-3-O-glucoside (C3G) was used as standard. An amount of 200 μL of the sample was mixed with 800 μL of KCl (pH 1.0, 0.025 M) and 800 μL of CH3COONa (pH 4.5, 0.4 M) buffers. After 30 min in the dark at room temperature (incubation), the absorbance was measured at 510 and 700 nm. TAC was calculated as milligrams of C3G equivalents per 100 g of sample (mg C3G/100 g).

Total flavonoid content (TFC) was determined using the aluminum chloride colorimetric method described by Dewanto et al. [56], with modifications. Catequin (C) was used as standard. An amount of 250 μL of the sample was mixed with 1000 μL of distilled water and 75 μL of NaNO2 (5% w/v). After 6 min in the dark at room temperature (incubation), 75 μL of AlCl3 (10% w/v), 500 μL of NaOH (1 M), and 600 μL of distilled water were added to the solution, and later, the absorbance was measured at 510 nm. TFC was calculated as milligrams of C equivalents (CEQ) per 100 g of sample (mg CEQ/100 g).

4.8. Determination of Antioxidant Activity (AA)

The DPPH assay was assessed using the method reported by Brand-Williams et al. [57], with minor modifications. An amount of 150 μL of the sample was mixed with 2850 μL of DPPH methanolic solution. The mixture was kept in the dark at room temperature for 30 min (incubation), and the absorbance was measured at 515 nm.

The ferric reducing antioxidant power (FRAP) assay was performed according to Benzie and Strain [58], with some modifications. Briefly, FRAP reagent was prepared with 50 mL of CH3COONa buffer (pH 3.6, 300 mM), 5.0 mL of TPTZ (10 mM in HCl (40 mM)), and 5.0 mL of FeCl3·6H2O (20 mM) (10:1:1 ratio), and then the solution was incubated at 37 °C. An amount of 150 μL of the sample was mixed with 2850 μL of FRAP reagent. The solution was kept in the dark at 37 °C for 30 min (incubation), and the absorbance was measured at 593 nm.

For DPPH and FRAP assays, Trolox (T) was used as the standard curve, and the results were expressed as μmol Trolox equivalents (TE) per 100 g of sample (μM TE/100 g).

TBC (TPC and TAC) and AA (DPPH and FRAP) values were measured by spectrophotometric analysis (T70 UV/Vis spectrophotometer, Oasis Scientific Inc., Greenville, SC, USA).

4.9. Color Measurement

Color analysis was performed through CIELab coordinates (L*: darkness–lightness, a*: green-red axis, b*: blue-yellow axis) using a colorimeter (CM-5, Konica Minolta, Osaka, Japan), with illuminant D65 and an observer angle of 10°, where the samples were filled in a glass cuvette, and thus, the CIELab values were measured [59].

4.10. Kinetics and Half-Life Time Analysis

First-order kinetics was used to describe the degradation TBC and AA content [22]. The reaction rate constant (k) and half-life time (t1/2) were calculated according to Equations (4) and (5), respectively.

$$\frac{C_t}{C_o} = e^{-k\,t} \tag{4}$$

$$t_{1/2} = \frac{\ln 2}{k} \tag{5}$$

where Co is the initial TBC and AA content, and Ct is the TBC and AA content at time t (days).

4.11. Statistical Analysis

The experiments were reproduced three times and measurements were conducted in duplicate. The data were reported as the mean value plus or minus the standard deviation (SD).

A one-way analysis of variance (ANOVA) was performed for each variable to identify differences between samples, and least significant difference (LSD) tests were performed for the comparison of means at a significance level of 5% ($p \leq 0.05$) using Statgraphics Centurion XVI software (v. 16.2.04, StatPoint Technologies Inc., Warrenton, VA, USA).

AT A GLANCE

Hydrogels has the ability to absorb and retain substances within the spaces between their polymer network, thanks to their three-dimensional cross-linked structure. As a result, these matrices find utility in various technical applications. This study examines the physical and chemical characteristics, texture, bioactive components, and stability of antioxidant activity in commercial hydrogels that have been enhanced with cryoconcentrated blueberry juice (CBJ) and stored for a period of 35 days. CBJ was included into many commercial hydrogels, including gelatin gel (GG), aerated gelatin gel (AGG), gummy (GM), and aerated gummy (AGM). The samples exhibited a range of 230 to 250 mg GAE/100 g, 3.5 to 3.9 mg C3G/100 g, and 120 to 136 mg CEQ/100 g for total polyphenol, anthocyanin, and flavonoid content, respectively. Among the samples, GG and GM had the slowest degradation rate for bioactive components, whereas AGM had the greatest degradation rate. The GG and GM samples can be maintained for a maximum of 21 days without any noticeable alterations, however the AGG and AGM samples showed indications of changes after around 15 days. CBJ has significant potential as a functional ingredient because of its high nutritional value. It can be used to enhance various hydrogel samples and protect their components during in vitro digestion, thereby improving their bioaccessibility after digestion.

REFERENCES

1. Yang, Z.; Chen, L.; McClements, D.J.; Qiu, C.; Li, C.; Zhang, Z.; Ming, M.; Tian, Y.; Zhu, K.; Jin, Z. Stimulus-responsive hydrogels in food science: A review. Food Hydrocoll. 2022, 124, 107218.
2. Klein, M.; Poverenov, E. Natural biopolymer-based hydrogels for use in food and agriculture. J. Sci. Food Agric. 2020, 100, 2337–2347.
3. Rubio-Arraez, S.; Capella, J.V.; Castelló, M.L.; Ortolá, M.D. Physicochemical characteristics of citrus jelly with non-cariogenic and functional sweeteners. J. Food Sci. Technol. 2016, 53, 3642–3650.
4. Ge, H.; Wu, Y.; Woshnak, L.; Mitmesser, S. Effects of hydrocolloids, acids and nutrients on gelatin network in gummies. Food Hydrocoll. 2021, 113, 106549.
5. Periche, Á.; Castelló, M.L.; Heredia, A.; Escriche, I. Stevia rebaudiana, oligofructose and isomaltulose as sugar replacers in marshmallows: Stability and antioxidant properties. J. Food Process. Preserv. 2016, 40, 724–732.

6. Dai, H.; Li, X.; Du, J.; Ma, L.; Yu, Y.; Zhou, H.; Guo, T.; Zhang, Y. Effect of interaction between sorbitol and gelatin on gelatin properties and its mechanism under different citric acid concentrations. Food Hydrocoll. 2020, 101, 105557.
7. Šeremet, D.; Mandura, A.; Cebin, A.; Martinić, A.; Galić, K.; Komes, D. Challenges in confectionery industry: Development and storage stability of innovative white tea-based candies. J. Food Sci. 2020, 85, 2060–2068.
8. Cedeño-Pinos, C.; Martínez-Tomé, M.; Murcia, M.A.; Jordán, M.J.; Bañón, S. Assessment of rosemary (Rosmarinus officinalis L.) extract as antioxidant in jelly candies made with fructan fibres and stevia. Antioxidants 2020, 9, 1289.
9. Rivero, R.; Archaina, D.; Sosa, N.; Schebor, C. Development and characterization of two gelatin candies with alternative sweeteners and fruit bioactive compounds. LWT-Food Sci. Technol. 2021, 141, 110894.
10. Mandura, A.; Šeremet, D.; Ščetar, M.; Vojvodić Cebin, A.; Belščak-Cvitanović, A.; Komes, D. Physico-chemical, bioactive, and sensory assessment of white tea-based candies during 4-months storage. J. Food Process. Preserv. 2020, 44, e14628.
11. de Moura, S.C.; Berling, C.L.; Garcia, A.O.; Queiroz, M.B.; Alvim, I.D.; Hubinger, M.D. Release of anthocyanins from the hibiscus extract encapsulated by ionic gelation and application of microparticles in jelly candy. Food Res. Int. 2019, 121, 542–552.
12. Otálora, M.C.; de Jesús Barbosa, H.; Perilla, J.E.; Osorio, C.; Nazareno, M.A. Encapsulated betalains (Opuntia ficus-indica) as natural colorants. Case study: Gummy candies. LWT-Food Sci. Technol. 2019, 103, 222–227.
13. Marinelli, V.; Lucera, A.; Incoronato, A.L.; Morcavallo, L.; Del Nobile, M.A.; Conte, A. Strategies for fortified sustainable food: The case of watermelon-based candy. J. Food Sci. Technol. 2021, 58, 894–901.
14. Cano-Lamadrid, M.; Calín-Sánchez, Á.; Clemente-Villalba, J.; Hernández, F.; Carbonell-Barrachina, Á.A.; Sendra, E.; Wojdyło, A. Quality parameters and consumer acceptance of jelly candies based on pomegranate juice "Mollar de Elche". Foods 2020, 9, 516.
15. Casas-Forero, N.; Orellana-Palma, P.; Petzold, G. Comparative study of the structural properties, color, bioactive compounds content and antioxidant capacity of aerated gelatin gels enriched with cryoconcentrated blueberry juice during storage. Polymers 2020, 12, 2769.
16. Casas-Forero, N.; Moreno-Osorio, L.; Orellana-Palma, P.; Petzold, G. Effects of cryoconcentrate blueberry juice incorporation on gelatin gel: A rheological, textural and bioactive properties study. LWT-Food Sci. Technol. 2021, 138, 110674.
17. Orellana-Palma, P.; Guerra-Valle, M.; Gianelli, M.P.; Petzold, G. Evaluation of freeze crystallization on pomegranate juice quality in comparison with conventional thermal processing. Food Biosci. 2021, 41, 101106.
18. da Silva Haas, I.C.; de Espindola, J.S.; de Liz, G.R.; Luna, A.S.; Bordignon-Luiz, M.T.; Prudêncio, E.S.; de Gois, J.S.; Fedrigo, I.M.T. Gravitational assisted three-stage block freeze concentration process for producing enriched concentrated orange juice (Citrus sinensis L.): Multi-elemental profiling and polyphenolic bioactives. J. Food Eng. 2022, 315, 110802.
19. Orellana-Palma, P.; Tobar-Bolaños, G.; Casas-Forero, N.; Zúñiga, R.N.; Petzold, G. Quality attributes of cryoconcentrated calafate (Berberis microphylla) juice during refrigerated storage. Foods 2020, 9, 1314.

20. Qin, F.G.; Ding, Z.; Peng, K.; Yuan, J.; Huang, S.; Jiang, R.; Shao, Y. Freeze concentration of apple juice followed by centrifugation of ice packed bed. J. Food Eng. 2021, 291, 110270.
21. Chen, J.Y.; Du, J.; Li, M.L.; Li, C.M. Degradation kinetics and pathways of red raspberry anthocyanins in model and juice systems and their correlation with color and antioxidant changes during storage. LWT-Food Sci. Technol. 2020, 128, 109448.
22. Tutunchi, P.; Roufegarinejad, L.; Hamishehkar, H.; Alizadeh, A. Extraction of red beet extract with β-cyclodextrin-enhanced ultrasound assisted extraction: A strategy for enhancing the extraction efficacy of bioactive compounds and their stability in food models. Food Chem. 2019, 297, 124994.
23. Rodríguez-Sánchez, J.A.; Cruz, M.T.C.; Barragán-Huerta, B.E. Betaxanthins and antioxidant capacity in Stenocereus pruinosus: Stability and use in food. Food Res. Int. 2017, 91, 63–71.
24. Periche, Á.; Heredia, A.; Escriche, I.; Andrés, A.; Castelló, M.L. Optical, mechanical and sensory properties of based-isomaltulose gummy confections. Food Biosci. 2014, 7, 37–44.
25. Mardani, M.; Yeganehzad, S.; Ptichkina, N.; Kodatsky, Y.; Kliukina, O.; Nepovinnykh, N.; Naji-Tabasi, S. Study on foaming, rheological and thermal properties of gelatin-free marshmallow. Food Hydrocoll. 2019, 93, 335–341.
26. Mikulic-Petkovsek, M.; Schmitzer, V.; Slatnar, A.; Stampar, F.; Veberic, R. Composition of sugars, organic acids, and total phenolics in 25 wild or cultivated berry species. J. Food Sci. 2012, 77, C1064–C1070.
27. Sadahira, M.S.; Rodrigues, M.I.; Akhtar, M.; Murray, B.S.; Netto, F.M. Influence of pH on foaming and rheological properties of aerated high sugar system with egg white protein and hydroxypropylmethylcellulose. LWT-Food Sci. Technol. 2018, 89, 350–357.
28. Hartel, R.W.; Joachim, H.; Hofberger, R. (Eds.) Aerated Confections. In Confectionery Science and Technology, 1st ed.; Springer Nature: Cham, Switzerland, 2018; pp. 301–327.
29. Yang, F.; Zhang, M.; Bhandari, B.; Liu, Y. Investigation on lemon juice gel as food material for 3D printing and optimization of printing parameters. LWT-Food Sci. Technol. 2018, 87, 67–76.
30. Laurindo, J.B.; Peleg, M. Mechanical measurements in puffed rice cakes. J. Texture Stud. 2007, 38, 619–634.
31. Zúñiga, R.N.; Aguilera, J.M. Structure–fracture relationships in gas-filled gelatin gels. Food Hydrocoll. 2009, 23, 1351–1357.
32. Zúñiga, R.N.; Kulozik, U.; Aguilera, J.M. Ultrasonic generation of aerated gelatin gels stabilized by whey protein β-lactoglobulin. Food Hydrocoll. 2011, 25, 958–967.
33. Kaletunc, G.; Normand, M.D.; Johnson, E.A.; Peleg, M. "Degree of elasticity" determination in solid foods. J. Food Sci. 1991, 56, 950–953.
34. Kumar, S.S.; Chauhan, A.S.; Giridhar, P. Nanoliposomal encapsulation mediated enhancement of betalain stability: Characterisation, storage stability and antioxidant activity of Basella rubra L. fruits for its applications in vegan gummy candies. Food Chem. 2020, 333, 127442.
35. Oliveira, A.; Amaro, A.L.; Pintado, M. Impact of food matrix components on nutritional and functional properties of fruit-based products. Curr. Opin. Food Sci. 2018, 22, 153–159.
36. Hani, N.; Romli, S.; Ahmad, M. Influences of red pitaya fruit puree and gelling agents on the physico-mechanical properties and quality changes of gummy confections. Int. J. Food Sci. 2015, 50, 331–339.

37. Barba, F.; Jäger, H.; Meneses, N.; Esteve, M.; Frígola, A.; Knorr, D. Evaluation of the quality changes of blueberry juice during refrigerated storage after processing of high pressure and pulsed electric fields. Innov. Food Sci. Emerg. Technol. 2012, 14, 18–24.

38. Cortellino, G.; Rizzolo, A. Storage stability of novel functional drinks based on ricotta cheese whey and fruit juices. Beverages 2018, 4, 67.

39. Zhang, L.; Wu, G.; Wang, W.; Yue, J.; Yue, P.; Gao, X. Anthocyanin profile, color and antioxidant activity of blueberry (Vaccinium ashei) juice as affected by thermal pretreatment. Int. J. Food Prop. 2019, 22, 1035–1046.

40. Mar, J.M.; Silva, L.S.; Rabelo, M.S.; Muniz, M.P.; Nunomura, S.M.; Correa, R.F.; Kinupp, V.F.; Campelo, P.H.; Bezerra, J.A.; Sanches, E. Encapsulation of amazonian blueberry juices: Evaluation of bioactive compounds and stability. LWT-Food Sci. Technol. 2020, 124, 109152.

41. Maier, T.; Fromm, M.; Schieber, A.; Kammerer, D.R.; Carle, R. Process and storage stability of anthocyanins and non-anthocyanin phenolics in pectin and gelatin gels enriched with grape pomace extracts. Eur. Food Res. Technol. 2009, 229, 949–960.

42. Tavares, I.M.; Sumere, B.R.; Gómez-Alonso, S.; Gomes, E.; Hermosín-Gutiérrez, I.; da Silva, R.; Lago-Vanzela, E.S. Storage stability of the phenolic compounds, color and antioxidant activity of jambolan juice powder obtained by foam mat drying. Food Res. Int. 2020, 128, 108750.

43. Teribia, N.; Buve, C.; Bonerz, D.; Aschoff, J.; Hendrickx, M.; van Loey, A. Impact of processing and storage conditions on color stability of strawberry puree: The role of PPO reactions revisited. J. Food Eng. 2021, 294, 110402.

44. Kia, E.M.; Langroodi, A.M.; Ghasempour, Z.; Ehsani, A. Red beet extract usage in gelatin/gellan based gummy candy formulation introducing Salix aegyptiaca distillate as a flavouring agent. J. Food Sci. Tech. 2020, 57, 3355–3362.

45. Tarone, A.; Cazarin, C.; Junior, M. Anthocyanins: New techniques and challenges in microencapsulation. Food Res. Int. 2020, 133, 109092.

46. Amjadi, S.; Ghorbani, M.; Hamishehkar, H.; Roufegarinejad, L. Improvement in the stability of betanin by liposomal nanocarriers: Its application in gummy candy as a food model. Food Chem. 2018, 256, 156–162.

47. da Rosa, J.R.; Nunes, G.L.; Motta, M.H.; Fortes, J.P.; Weis, G.C.C.; Hecktheuer, L.H.R.; Muller, E.I.; de Menezes, C.R.; da Rosa, C.S. Microencapsulation of anthocyanin compounds extracted from blueberry (Vaccinium spp.) by spray drying: Characterization, stability and simulated gastrointestinal conditions. Food Hydrocoll. 2019, 89, 742–748.

48. Muhamad, I.I.; Jusoh, Y.M.; Nawi, N.M.; Aziz, A.A.; Padzil, A.M.; Lian, H.L. Advanced natural food colorant encapsulation methods: Anthocyanin plant pigment. In Natural and Artificial Flavoring Agents and Food Dyes, 1st ed.; Grumezescu, A., Holban, A.M., Eds.; Elsevier: Amsterdam, The Netherlands, 2018; pp. 495–526.

49. Vukoja, J.; Pichler, A.; Kopjar, M. Stability of anthocyanins, phenolics and color of tart cherry jams. Foods 2019, 8, 255.

50. Sinela, A.; Rawat, N.; Mertz, C.; Achir, N.; Fulcrand, H.; Dornier, M. Anthocyanins degradation during storage of Hibiscus sabdariffa extract and evolution of its degradation products. Food Chem. 2017, 214, 234–241.

51. Yan, B.; Davachi, S.M.; Ravanfar, R.; Dadmohammadi, Y.; Deisenroth, T.W.; van Pho, T.; Odorisio, P.A.; Darji, R.H.; Abbaspourrad, A. Improvement of vitamin C stability in vitamin gummies by encapsulation in casein gel. Food Hydrocoll. 2021, 113, 106414.

52. Huang, T.; Zhao, H.; Fang, Y.; Lu, J.; Yang, W.; Qiao, Z.; Lou, Q.; Xu, D.; Zhang, J. Comparison of gelling properties and flow behaviors of microbial transglutaminase (MTGase) and pectin modified fish gelatin. J. Texture Stud. 2019, 50, 400–409.

53. Li, X.; Liu, X.; Lai, K.; Fan, Y.; Liu, Y.; Huang, Y. Effects of sucrose, glucose and fructose on the large deformation behaviors of fish skin gelatin gels. Food Hydrocoll. 2020, 101, 105537.

54. Waterhouse, A.L. Determination of total phenolics. Curr. Protoc. Food Anal. Chem. 2002, 6, I1.1.1–I1.1.8.

55. Lee, J.; Durst, R.; Wrolstad, R. Determination of total monomeric anthocyanin pigment content of fruit juices, beverages, natural colorants, and wines by the pH differential method: Collaborative study. J. AOAC Int. 2005, 88, 1269–1278.

56. Dewanto, V.; Wu, X.; Adom, K.K.; Liu, R.H. Thermal processing enhances the nutritional value of tomatoes by increasing total antioxidant activity. J. Agric. Food Chem. 2002, 50, 3010–3014.

57. Brand-Williams, W.; Cuvelier, M.E.; Berset, C.L.W.T. Use of a free radical method to evaluate antioxidant activity. LWT-Food Sci. Technol. 1995, 28, 25–30.

58. Benzie, I.F.; Strain, J.J. The ferric reducing ability of plasma (FRAP) as a measure of "antioxidant power": The FRAP assay. Anal. Biochem. 1996, 239, 70–76.

59. Orellana-Palma, P.; Guerra-Valle, M.; Zúñiga, R.N. Centrifugal filter-assisted block freeze crystallization applied to blueberry juice. Processes 2021, 9, 421.

60.

15

Chapter

STABILITY AND SENSORY ANALYSIS OF NANO MATRIX SOFT CONFECTIONARY FOR ORAL SUPPLEMENTATION OF VITAMIN D

1. INTRODUCTION: AN OVERVIEW

Vitamin D, also known as calciferol, is a type of lipid-soluble vitamin that consists of a mixture of several steroidal derivatives. The derivatives mentioned are cholecalciferol (vitamin D3), which is derived from cholesterol, calcidiol (partially active hydroxylated form of cholecalciferol), calcitriol (dihydroxylated active form), ergocalciferol (D2), and its mono and dihydroxylated derivatives [1]. Vitamin D can be produced within the body (endogenous) or obtained from external sources (exogenous). When the human skin epidermis is exposed to ultraviolet-B (UVB) radiation found in sunlight, it naturally generates vitamin D3 from 7-dehydrocholesterol. Exogenous vitamin D is obtained from food sources, such as dairy products, oily fish, liver, egg yolk, etc., as well as from supplements. Due to the restricted availability of natural dietary sources that contain exogenous vitamin D, it is becoming more common to recommend supplementation to address its insufficiency. Vitamin D, in essence, denotes vitamin D3. The main form of vitamin D that circulates in the body is called 25-hydroxyvitamin D. It is transformed into an active form called 1,25-dihydroxyvitamin D within the human body [2].

Individuals with serum vitamin D levels below 20 ng/mL or 50 nmol/L (as determined by the Institute of Medicine in 2011) are classified as vitamin D deficient. Based on a thorough examination of the available research, the Institute of Medicine (IOM) determined in 2011 that a blood concentration of 25(OH)D at or above 20 ng/mL is sufficient for maintaining good bone health [3,4,5]. Vitamin D insufficiency is linked to various clinical symptoms across different age groups. Vitamin D insufficiency is widely recognized as a significant factor in the development of musculoskeletal problems, mostly affecting bone health. In the absence of vitamin D, only 10-15% of calcium from the diet and around 60% of phosphorus are effectively absorbed. The user's text is "[2]." Musculoskeletal fatigue is a consequence of a lack of vitamin D, and there is a correlation between low levels of vitamin D and conditions such as rheumatoid arthritis, multiple sclerosis, and arthritis. Additionally, vitamin D deficiency is closely linked to various cardiovascular risk factors, including myocardial heart infarction, congestive heart failure, diabetic cardiovascular disease, and peripheral arterial disease. Several studies have indicated that vitamin D deficiency has a role in the heightened risk of cerebrovascular accident (CVA) and is associated with an increased risk of depression and schizophrenia [7,8,9]. Vitamin D governs the function of B and T-lymphocytes and controls the process of cell differentiation and proliferation [10,11]. Several studies have recognized that women are a vulnerable demographic when it comes to vitamin D insufficiency, particularly pregnant women who face even higher risks. This deficiency can contribute to several pregnancy problems, such as gestational diabetes and preeclampsia [12].

The problem of vitamin D insufficiency is widespread globally, affecting around 1 billion individuals [13]. Multiple reports indicate that a significant proportion, ranging from 40% to 100%, of older individuals in the United States and European countries have a deficiency of vitamin D. In addition, other investigations have demonstrated a deficiency in vitamin D levels regardless of age, gender, or geographical location [2, 13, 14, 15]. Based on a 2015 meta-analysis report, the prevalence of vitamin D deficiency was 35% higher in obese individuals, regardless of their age or demographic characteristics [16]. There is a correlation between vitamin D insufficiency and populations with higher levels of melanin in their skin, as well as communities in Middle East nations that use considerable skin care [13,17].

The Institute of Medicine (IOM) has released guidelines recommending a daily intake of 600 IU of vitamin D for those between the ages of one and 70. The Recommended Dietary Allowance (RDA) for vitamin D is 800 IU per day for persons over 70 years of age and 400 IU per day for newborns up to one year of age [18]. The Endocrine Society in the United States has suggested achieving serum 25(OH) D concentrations of more than 30 ng/mL for the treatment and prevention of vitamin D insufficiency. Ideally, the recommended range is 40-60 ng/mL [4]. In order to achieve a serum 25(OH)D level of 30 ng/mL, it is recommended that deficient people get either 50,000 IU of vitamin D3 weekly or 6000 IU daily for a duration of 8 weeks. The Endocrine Society recommends a daily maintenance dose of 1000 IU for patients up to 18 years old and 1500-2000 IU for patients between the ages of 18 and 50 [4,19].

Constraints associated with supplementary solid oral administration: The oral route is often considered the most dependable method for administering active pharmaceutical ingredients (APIs). The majority of medications, over 60%, are in the form of solid oral dosage forms. This is because they are easy to produce, there are many different substances that may be used in their formulation, the development process is less expensive, they allow for more precise dosing compared to liquid medications, and they are easy to handle and dispense. Due to its non-invasive nature, patients experience greater comfort compared to alternative methods. Oral formulations have significant drawbacks, such as poor solubility in the stomach and intestines, leading to reduced effectiveness of many drugs. Additionally, these formulations can be quickly broken down in the stomach, potentially causing the active pharmaceutical ingredient (API) to degrade due to enzymes in the gastrointestinal (GI) tract [21]. Another significant issue is the lack of adherence to treatment among children and the elderly, primarily because of dysphagia. A survey conducted on 6158 patients found that 26% of them experienced challenges with swallowing medications [22]. A further study discovered that 37.4% of the total number of participants (n = 1051) experienced difficulties in swallowing pills and capsules [17]. supplementing, whether for prevention or treatment, is often considered less important than other urgent medical demands or essential medications. As a result, the level of adherence to supplementing decreases. Adults who have dysphagia or those who are hesitant to take drugs often disregard the additional pills. Parents sometimes overlook the administration of supplements in solid oral form since youngsters are typically unwilling to take tablets.

Constraints regarding the stability and ability to be absorbed of vitamin D: Cholecalciferol is extremely vulnerable to degradation from environmental factors such as sunshine, heat, and oxygen. When exposed to open air in its natural state, it easily undergoes isomerization or oxidation, leading to a loss of its functionality and physiological effects. Multiple studies indicate that vitamin D has low water solubility, and incorporating it into carrier oils can safeguard it against the detrimental effects of heat or oxidation [23,24,25]. Vitamin D has a restricted ability to be absorbed by the body, which reduces its efficiency when taken as a pill, and larger amounts are needed to maintain optimal levels. Research has shown that around 50% of orally taken Vitamin D3 is absorbed [26].

Lipid-based formulations have achieved commercial recognition due to their capacity to enhance the oral bioavailability and solubility of medicines that have low water solubility. They improve the absorption of medications by increasing the solubility of pharmaceuticals that are weakly soluble or insoluble in water, and by facilitating the emulsification process in the gastrointestinal tract (GIT). The use of nano-emulsions (NEs) in the food industry has gained significant attention due to the alterations in the physical and chemical properties and improved effectiveness resulting from adjustments in particle size. Reducing the size of particles in emulsion-based formulations can have several positive effects for different food and supplement applications. These include: (i) increased stability against the clumping and separation of droplets due to gravity; (ii) improved visual clarity; and (iii) enhanced absorption in the body when taken orally [27,28]. Salvia-Trujillo et al. did a study where they found that smaller lipid droplets were digested at a faster rate in simulated gastrointestinal fluid. This suggests that mixed micelles, which help solubilize lipophilic vitamins, may form more rapidly in the small intestine. NEs are highly suitable for encapsulating lipophilic nutraceuticals due to their ability to produce effective and transparent delivery systems with excellent oral bioavailability [29]. Nanoemulsions (NEs) are thermodynamically stable systems consisting of at least two immiscible liquids, with droplet sizes typically below 200 nm. These materials have excellent stability when stored and can be manufactured using low amounts of surfactants [30,31]. NEs can be synthesized utilizing high-energy emulsification technologies, such as ultrasonic generators or high-pressure homogenizers. Nanoemulsion-based chewable matrices, which are easily masticable and swallowable, have demonstrated potential as delivery systems. These matrices have the ability to entrap medication, resulting in sustained and regulated release, while also preventing rapid disintegration in the stomach.

Chewable gummy vitamins have gained popularity among both children and adults in recent times. Gummy supplements are said to address the issue of customer weariness with pills and to disguise the unpleasant taste or odor frequently associated with vitamins [32]. Currently, youngsters with all their teeth readily like gelled gummies and candies due to their pleasant taste and chewiness. These treats are often flavored with a variety of fruity and appealing scents and extracts, and they have a sweet taste. Currently, gummy supplements make approximately 20% of the top 10 vitamin D pills. This is because gummy soft chew vitamins have gained significant acceptance among individuals of all age groups. Multiple studies have demonstrated that the nutritional status of children has been enhanced by the use of chewable vitamins [32,33]. Considering the factors mentioned above, a soft confectionery system was created using nano-emulsion technology. The system consisted of a chewable matrix and was then analyzed for its textural profile, dissolving, and consumer acceptance.

2. MATERIALS

The procurement of Vitamin D (97–99% purity) was done from HiMedia Laboratories Pvt. Ltd. in Mumbai, India. The user bought cold-pressed, unadulterated maize oil from Deve Herbes, located in Delhi, India. The following substances were acquired from Sigma-Aldrich, Inc., St. Louis, MO, USA: Gelatin with a bloom strength of 160 g, Poloxamer 188, lecithin, and sucralose. Transcutol and Poloxamer 407 were acquired from Gattefosse, located in Saint-Priest, France. Tween 20, Tween 80, Tween 60, and Span 20 were purchased from Loba Chemie Pvt. Ltd., located in Mumbai, India. Methanol and water of high-performance liquid chromatography (HPLC) grade were purchased from Merck, located in Mumbai, India. The Food Safety and Standards Authority of India (FSSAI) procured sorbitol/xylitol, sucralose, citric acid, and flavors from local suppliers.

3. EXPERIMENTAL DESIGN

3.1. Preparation and Optimization of Vitamin D Nanoemulsion

The oil phase of nanoemulsion was selected based on the reported solubility of vitamin D in various oils [34,35,36,37,38,39]. Surfactant and co-surfactant were selected based on their ability to spontaneously emulsify corn oil [40]. Briefly, 100 mg of corn oil was dissolved in 3 mL of methylene chloride and then surfactant/co-surfactant (10 mL) was added and stirred at 200 rpm (at 40 °C) using magnetic stirrer to remove methylene chloride. Thereafter, 1 mL of sample was withdrawn and in it 10 mL of distilled water was added gradually, and percentage transmittance was determined at 339 nm wavelength using a UV spectrophotometer [41]. All the measurements were performed in triplicate.

Table 1. (A). Grading system of oil and Smix based on appearance and emulsification time. (B). Visual screening data of nano emulsions using different ratio of Smix and oil.

A. Grade	Parameter			Self-Emulsion Time (min)	
1.	Rapid, clear nanoemulsion			<1	
2.	Rapid, slight hazy nanoemulsion			<2	
3.	Slow, turbid emulsion			>3	
4.	No emulsification			>4	
B. Grading based on appearance					
Ratio of Smix: Oil	*Co-surfactant to surfactant ratio*				
	1:1	2:1	1:2	1:3	3:1
10:3	1	1	1	1	1
09:3	1	1	1	1	1
08:3	2	1	1	1	1
07:3	2	1	1	2	1
06:3	2	1	2	2	1
05:3	3	1	2	2	2
04:3	3	1	2	3	2
03:3	4	1	2	4	2
02:3	4	2	3	4	3
01:3	4	4	4	4	4

Nanoemulsion optimization was performed by aqueous titration method followed by pseudo-ternary phase diagram as described previously [42,43]. Corn oil containing a weighed quantity of vitamin D was admixed with the combination of surfactant and co-surfactant (Smix) in different ratios (1:1, 2:1, 1:2, 1:3 and 3:1). The mixtures were then titrated with water, vortexed, and analyzed visually for their appearance, clarity, or turbidity (Table 1) and plotted as ternary phase diagram. Clear dispersions (nanoemulsions) with a minimum concentration of Smix were selected for further evaluation [44].

3.2. Characterization of Vitamin D Nanoemulsion

The average particle size of the fabricated nanoemulsion was evaluated using Zetasizer Nano ZS90 (Malvern Instruments, Malvern, UK) via dynamic light scattering (DLS). An ambient temperature of 25 °C and scatter angle of 173° were set to carry out the measurements. To eliminate the overlapping effects of scattering, a dilution of 500 times was performed with buffer solution of equivalent pH before measurements. A disposable capillary cell of 1 mL was employed for zeta potential measurement.

To analyze morphology, a Morgagni 268D transmission electron microscope (TEM, Fei Electron Optics, Eindhoven, The Netherlands) functioning at 80 kV was employed. On a 400-mesh copper panel coated with carbon foil, one drop of each sample previously diluted 100 times was applied and allowed to dry for two minutes. A negative stain of 2% phosphotungstic acid was then applied to the sample slide and allowed to dry before examination. Observations were conducted at room temperature, using a smaller objective aperture, a lower accelerating voltage of 80 kV, and at increasing magnifications [44].

3.3. Stability Study of Vitamin D Nanoemulsion

The high-performance liquid chromatography (HPLC) analysis was employed for quantification of vitamin D. Waters e2695 separations Module HPLC system (Waters Co., Milford, MA, USA) equipped with autosampler, Waters 2489 dual wavelength UV detector and a column heater was used. The symmetry C8 column by Waters, having 100 Å pore size with 4.6 mm internal diameter, 5 µm particle size, and 250 mm length, was used for the stationary phase and Empower™ software (Version 3, feature release 4) from Waters was used to process and analyze the results.

A primary stock solution of 1 µg/mL was prepared by dissolving vitamin D in HPLC grade methanol. Hence, 50 mg vitamin D was serially diluted to obtain standard solution of desired concentration. The following process parameters were utilized for separation: temperature was maintained at 25 ± 2 °C, methanol and HPLC grade water in a ratio of 90:10 (v/v) were employed as mobile phase, 1.0 mL/min flow rate was set and maintained, sample cooler was set to 15 °C, and detection was carried out at 265 nm. The standard solution was analyzed 5 times (n = 5) for regression.

Vitamin D nanoemulsion was freshly prepared and stored in light resistant container. Since vitamin D degrades rapidly in the aqueous solution, a comparison was made with the oil solution of vitamin D prepared by mixing a weighed quantity of vitamin D in corn oil. One mL of both samples was diluted to 100 mL using HPLC grade methanol and were then analyzed on day 0, day 15 and day 30.

3.4. Soft Confectionary (Nano Gummy) Preparation

The nanoemulsion based gelled matrix system was developed using gelatin (hydrocolloids), as shown in Figure 1. The matrices were made from a blend of type A and type B gelatin with a bloom strength of 160 g. Different combinations of sweeteners (sorbitol, xylitol, sucralose), citric acid, and flavors (coffee, lime, vanilla, orange, candy) were tested.

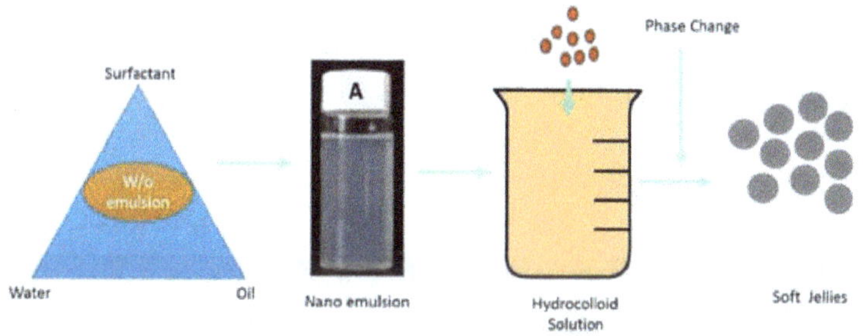

Figure 1. Formulation development process from nano emulsion optimization to gummy formation.

Briefly, gelatin types A and B in equal quantity were dissolved in a small quantity of water heated to 60 °C by stirring at 100 rpm. To this, the previously developed o/w nanoemulsion was added, and the mixture was stirred for 2 min. Sweeteners were then added followed by citric acid, flavoring agent, and corn oil (plasticizer). The influence of corn oil on the matrix properties and product acceptability was studied. The formula of the developed Nano gummy using 10% corn oil is shown in Table 2.

Table 2. Nano gummy formula with 10 wt. % corn oil.

Ingredients	Amount (wt. %)
O/W nanoemulsion	34.29
Gelatin	10.29
Sorbitol	13.20
Xylitol	30.79
Sucralose	0.38
Citric acid	0.38
Lime flavor	0.66
Corn Oil	10.00

3.5. Texture Profile Analysis of Nano Gummy

The textural properties of gummy confections were evaluated using two-bite compression tests [45]. Gummies were formulated with different corn oil concentrations (0%, 10%, 20% and 30%) and textural attributes, including hardness and stickiness, were measured. A TA.XT2i Texture Analyzer (Stable Micro Systems, Godalming, UK) was used for the evaluation. All measurements were made using a 5 kg load cell. A cylinder aluminum plate probe with a diameter of 7.5 cm, lubricated using the corn oil was lowered at a speed of 1 mm/s to compress 50% of confectionary jelly cube measuring 10 mm3 placed on the fixed bottom surface. The equipment was adjusted to zero, trigger force was kept at 0.05 N, and the plate was automatically lowered until the bottom surface of the plate just contacted the table. All the measurements were made in triplicate. Data were obtained and analyzed using Stable Micro System's Exponent software, which evaluates the force–time curve formed during the compression of the sample [46,47].

3.6. Sensory Evaluation of Nano Gummy

Sensory assessment is accomplished by documenting the outcome of the assessor's replies based on the parameters of taste, texture/appearance, aroma, and overall liking after a comprehensive study. A questionnaire was prepared, and sensory assessments were performed using a qualitative 9-point hedonic scale, with 1 being least likely and 9 being most. Participant were given three different gummies, viz. Placebo gummy (nano gummy without vitamin D), Nano gummy, and Marketed gummy (Azveston Healthcare Pvt. Ltd., Bengaluru, India) on Day 1 and again on Day 45 after storage. Volunteers were screened based on inclusion and exclusion criteria and were informed about the study objectives and recruited after providing informed consent. Gender-neutral healthy volunteers who spoke and understood English, aged between 18 and 45 years, and had no medical history or known sensitivities to the materials used in gummy fabrication were eligible to participate. Participants younger than 18 years old, smokers, those having flu or similar manifestation, those who did not speak or comprehend the language of choice, and those who had known sensitivities to the materials used in gummy production were all excluded. Oral hygiene product, food and aerated drinks were restricted 2-h before and throughout the study period. Responses were collected based on the questionnaire, including: How do you like the taste? How do you like the aroma? How do you like the appearance and texture and overall product? One-way ANOVA (regression analysis) and t-test with Welch correction, with $p < 0.05$ as significantly different, were used for the evaluation of differences between the gummy and changes over 45 days of storage.

4. RESULTS

4.1. Preparation and Optimization of Vitamin D Nanoemulsion

Based on the previously reported solubility of vitamin D, corn oil was selected as the oily phase. Results of the emulsification study (Figure 2) showed that tween 20 has maximum potential to emulsify corn oil followed by lecithin. The decreasing order of emulsification potential among was as follows: tween 20 (96%) > lecithin (87%) > tween 80 (84%) > tween 60 (84%) > transcutol (72%) > poloxamer 407 (61%) > poloxamer 188 (53%) > span 20 (49%). From the results obtained, tween 20 and lecithin were selected as surfactant and co-surfactant, respectively, for the preparation of the vitamin D nanoemulsion.

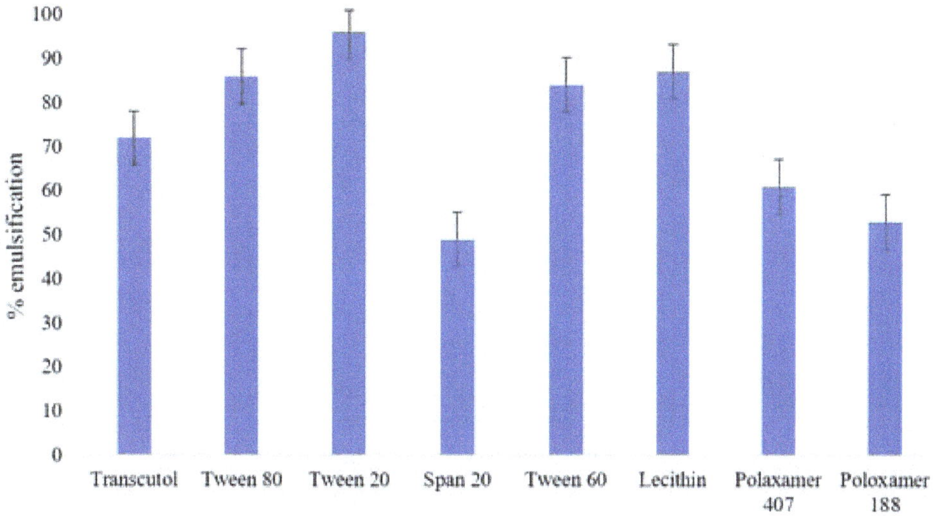

Figure 2. Comparative corn oil emulsification potential of various screened surfactants.

Data obtained from a grading system based on the visual screening, appearance, and emulsification capacity of formulations prepared using different ratios of Smix and Oil were obtained and are tabulated in Table 1. Data revealed that Smix ratio 2:1 resulted in a clear and stable nanoemulsion. Thus, Smix 2:1 was selected for pseudo ternary analysis to obtain the optimized formulation. The nanoemulsion region is shown by a colored region in the pseudo-ternary phase diagram (Figure 3). From the results, it was evident that 3% corn oil, 3% tween 20 and lecithin in a ratio of 2:1 and 94% water resulted in the optimized vitamin D nanoemulsion utilized for further evaluation.

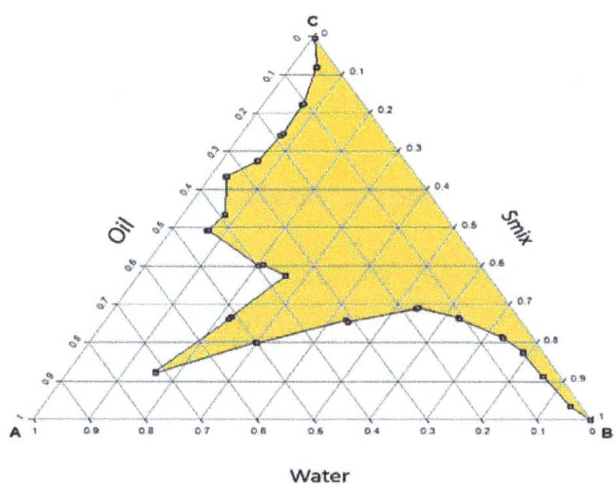

Figure 3. Pseudo-ternary Phase diagram of optimized nano-emulsion with corn oil, Smix (tween 20 + lecithin; 2:1) and water.

4.2. Characterization of Vitamin D Nanoemulsion

As illustrated in Figure 4B, the average particle size and particle size distribution of the optimized vitamin D nanoemulsion were found to be 118.6 ± 4.3 nm and 0.11 ± 0.30, respectively. The zeta-potential was found to be -27 ± 0.82 mV, indicating that the formed nanoemulsion was stable.

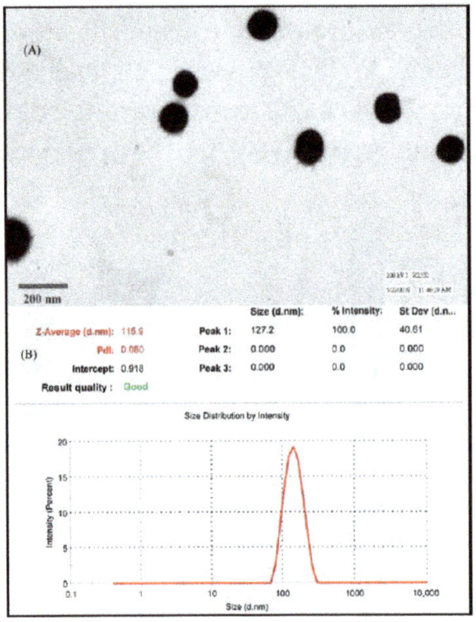

Figure 4. (A) HI-RES TEM micrograph of NE. (B) Particle size distribution of optimized NE.

TEM revealed that the vitamin D nanoemulsion was spherical (Figure 4A). The droplet size measured by TEM (127 ± 7.2 nm) was slightly larger, but not significantly different from that measured by dynamic light scattering (118.6 ± 4.3 nm). The slightly larger size observed in TEM could be due to the measurement of the hydration layer surrounding the globule, which was not measured using DLS due to difference in principle.

4.3. Stability Study of Vitamin D Nanoemulsion

A representative chromatogram of vitamin D is shown in Figure 5A,B. The standard solution of a known concentration of vitamin D (1 µg/mL) was analyzed five times for robustness. The data obtained for the regression plot of vitamin D by HPLC are shown in Table 3.

Table 3. Average retention area, retention time and peak height of stock solution.

Concentration (µg/mL)	Average Area	Retention Time	Height	Area %
1	4,132,567	7.643	284,993	100
1	4,134,052	7.620	285,379	100
1	4,142,904	7.660	286,712	100
1	4,130,736	7.663	284,132	100
1	4,140,783	7.677	285,363	100
Mean	4,136,208.4	7.652	285,315.8	100
Std. Dev.	5329.52	0.022	930.21	NA

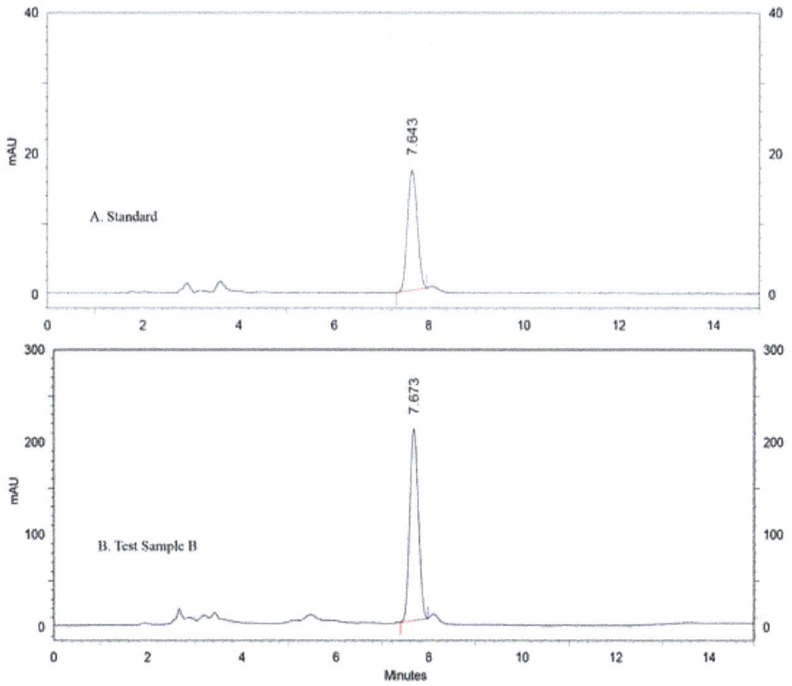

Figure 5. HPLC Chromatograph of (A) Standard solution (B) Test sample B.

The linear equation for the calibration plot was used to quantify the concentration of vitamin D on days 0, 15, and 30. Vitamin D in corn oil solution degraded by 5.49% in 15 days and 8.97% in 30 days, whereas the nanoemulsion formulation degraded by 2.94% and 7.63% on day 15 and day 30, respectively (Table 4).

Table 4. Concentration reduction and percent reduction of vitamin D on day 15 and 30.

Test Day	Sample A (IU/mL)	% Reduction	Sample B (IU/mL)	% Reduction
Day 0	289,302.01	0	38,282.13	0
Day 15	273,420.01	5.49	37,155.68	2.94
Day 30	263,360.20	8.97	35,360.46	7.63

4.4. Texture Analysis

The gummies prepared using different concentrations of corn oil as plasticizer were subjected to texture analysis. The maximum force during the initial compression was used to determine hardness (g) and the negative force with which the probe disconnects from the sample over its way up after compressing was termed as stickiness (g). The mean hardness values and mean stickiness values of each gummy type on day 1, 15, 30, and 45 are shown in Table 5. Gummies without corn oil were softest on day 1 (1540 ± 125 g) but they were extremely soft and difficult to handle and not in the range of an acceptable marketed gummy. Although gummies with 40% corn oil had the lowest stickiness value of the four (76 ± 8 gm on day 45) due the excess of oil present, they were the hardest. As shown in Figure 6, gummies with 10% corn oil had the optimum hardness value (3100 ± 200 g on day 1 and 3600 ± 210 g on day 45), and stickiness was also minimal. It was noted that the gummies prepared with 10% corn oil had excellent hardness and texture up to 30 days and this was retained up to 45 days on further storage.

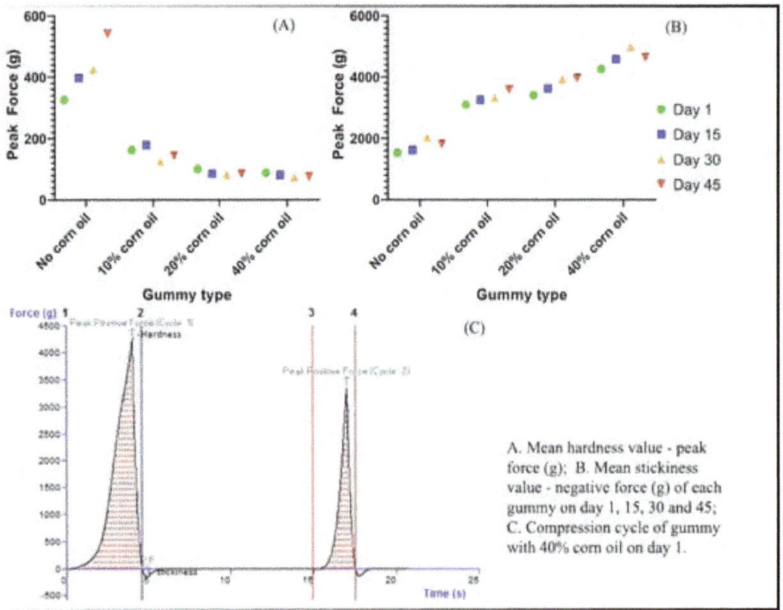

Figure 6. Graphical representation of (A) peak hardness cum (B) peak stickiness of each gummy on day 1, 15, 30 and 45 and (C) force-time curve of 40% corn oil gummy on day 1.

Table 5. (A) Mean hardness value (Peak force (g) mean ± STDV) and (B) mean stickiness value (negative force (g) mean ± STDV) of each gummy on day 1, 15, 30 and 45.

A. Mean Hardness Value (Peak Force (g) Mean ± STDV) of Each Gummy				
Oil Content in Gummy	Day 1	Day 15	Day 30	Day 45
No corn oil	1540 ± 125	1620 ± 100	2030 ± 150	1820 ± 155
10% corn oil	3100 ± 200	3260 ± 155	3330 ± 165	3600 ± 210
20% corn oil	3410 ± 265	3630 ± 230	3930 ± 195	3970 ± 265
40% corn oil	4260 ± 165	4600 ± 185	5000 ± 205	4650 ± 195
B. Mean Stickiness Value (Negative Force (g) Mean ± STDV) of Each Gummy				
No corn oil	325 ± 45	400 ± 35	425 ± 60	550 ± 60
10% corn oil	160 ± 20	180 ± 30	125 ± 25	145 ± 33
20% corn oil	100 ± 15	85 ± 15	80 ± 7	85 ± 12
40% corn oil	90 ± 14	80 ± 14	75 ± 9	76 ± 8

4.5. Sensory Analysis

The Placebo, Nano, and Marketed gummies were compared and the data obtained via sensory evaluation conducted at Day 1 and at Day 45 following storage are presented in Table 6. The highest score for taste was observed for the Marketed gummy. However, the difference between the score compared to placebo and nano gummy was insignificant ($p > 0.05$). Further, storage for a period of 45 days had no detrimental effect on the taste and the scores for all the gummies remained same. The aroma of Placebo and Nano gummies was significantly better than the Marketed gummy ($p = 0.01$; $p < 0.05$). Following 45 days storage, the loss of aroma was observed for all the gummies, but the loss was maximum for the Marketed gummy. Although the appearance and texture of the Marketed gummy was better than Nano gummy ($p = 0.029$; $p < 0.05$), the nano gummy retained both texture and appearance without any significant change ($p = 0.45$) over 45 days of storage. Overall liking of the Nano gummy was comparable to Marketed gummy and was similar on day 1 and day 45 without any significant difference ($p = 0.4$).

Table 6. Average scores from the sensory evaluation on Day 1 and day 45.

Day 1	Placebo Gummy	Nano Gummy	Marketed Gummy
Taste	7.1 ± 1.7	6.7 ± 1.6	7.3 ± 1.4
Aroma	7.4 ± 1.3	7.0 ± 0.8	5.4 ± 0.7
Appearance/texture	7.0 ± 0.6	6.8 ± 0.9	7.6 ± 1.2
Overall liking	6.9 ± 0.9	6.6 ± 0.8	7.2 ± 1.0
Day 45			
Taste	7.2 ± 1.4	6.6 ± 1.0	7.4 ± 0.7
Aroma	6.7 ± 0.7	6.4 ± 1.1	4.5 ± 0.6
Appearance/texture	7.1 ± 0.8	6.9 ± 1.2	7.4 ± 0.9
Overall liking	6.7 ± 0.6	6.5 ± 0.6	7.3 ± 1.1

5. DISCUSSION

This study investigates the capability of a new nano matrix-based candy (Nano gummy) to administer vitamin D. The new matrix not only inhibits the degradation of vitamin D, but also improves compliance. Our findings indicated that vitamin D was effectively retained in the O/W nanoemulsion and maintained its stability throughout storage. In addition, sensory analysis experiments indicated that the taste and overall appeal to consumers were successfully attained.

The Nano gummy discussed in this study was created using a two-step process. Initially, vitamin D enriched nanoemulsions were created using the process of spontaneous emulsification and aqueous titration. Corn oil was chosen as the oil phase based on solubility experiments and published literature.

A previous study supported that the in vitro digestibility and bioaccessibility of vitamin D were maximum in long chain triglycerides such as corn oil compared to medium chain triglycerides [34,35]. Studies conducted by Yang et al. [36], Qian et al. [37], and Rao et al. [38] showed similar results where long chain triacylglycerols, viz. corn oil, displayed better bioavailability characteristics of fat-soluble vitamins in contrast to indigestible oils due to the improved solubilization ability of mixed micelles produced by long chain fatty acids [39]. From the results of the emulsification study, tween 20 and lecithin with maximum potential to emulsify corn oil (96% and 87%, respectively) were selected as surfactant and co-surfactant, respectively.

Pseudo-ternary phase diagrams were drawn to locate the nanoemulsion zones, where oil, surfactant, co-surfactant, and water exist as clear and homogeneous dispersions. Ternary phase diagrams provide information on the best combinations to use for the preparation of nanoemulsion. The droplet size of the nanoemulsion is critical as it influences the solubility as well as bioavailability and it is well established that smaller size droplets result in enhanced solubility and bioavailability owing to the higher surface area [34,35].

The phase analysis demonstrated that when the surfactant: co-surfactant ratio was 2:1, the largest proportion of oil was integrated in the nanoemulsion system and smaller size nanoemulsions. The 3% corn oil, 2% tween 20, 1% lecithin, and 94% water resulted in an optimized vitamin D nanoemulsion with globule size 118.6 ± 4.3 nm and particle size distribution of 0.11 ± 0.30. The zeta potential of nanoemulsion is important because its value can be linked to colloidal dispersion stability. In a dispersion, the zeta potential demonstrates the extent of repulsion between contiguous, identically charged particles. A higher zeta potential indicates stability for particles and molecules of sufficiently small size. In cases of lower zeta potential, repulsive forces are weaker compared to attractive forces, leading to the breaking and flocculation of dispersions. Colloids having a large zeta potential (either negative or positive) are deemed electrically stable [48,49]. The optimized formulation displayed zeta-potential of −27 ± 0.82 mV, indicating that the developed nanoemulsions are stable. Using TEM, the structural and dimensional characterization of the optimized vitamin D nanoemulsion was accomplished. The nanoemulsion droplet structure was confirmed to be spherical, and each droplet had a discrete region surrounding it, which suggested vitamin D encapsulation. The size of the nanoemulsion droplets obtained by TEM was in agreement with that obtained by DLS. Blackness in the core of particles as observed in TEM image could be attributed to the solubilization of vitamin D in the oil.

During the stability study, the results showed that vitamin D nanoemulsion retained the vitamin D concentration quite well, although it was exposed to water in the emulsion. Jiang et al. [23] showcased similar behavior when encapsulating vitamin D in carrier oils and storing away from light, which prevents its otherwise easy degradation. This study confirmed the stability of vitamin D in the nanoemulsion upon storage for extended time periods. The vitamin D degradation was only 7.63% in nanoemulsion compared to 8.97% in oil following 30-day storage. The stability results do not appear to be distinctly different but are significant because in the nanoemulsion, the vitamin D, prone to hydrolysis, was in contact with water.

In the second step, the optimized nanoemulsion was mixed with hydrocolloid matrix to form the Nano gummy. It was hypothesized that the Nano gummy might improve the bioavailability of vitamin D, although this was not tested in the present research. The improved bioavailability of vitamin D from Nano gummy could be explained by two mechanisms. Firstly, gummies are meant for chewing in the mouth where they are mixed with saliva and begin to dissolve. Most likely, the released nanoemulsion from the Nano gummy in the buccal cavity will allow enhanced permeation of vitamin D and thus improved bioavailability. Such enhancement of buccal permeation by nanoemulsions has been previously reported [39,50,51]. Secondly, the released nanoemulsion from the Nano gummy will disperse in the gastric fluid. The nanoemulsion results in the enhanced solubilization of vitamin D and its subsequent absorption from GIT, resulting in improved bioavailability. Thus, the Nano gummy could significantly enhance the bioavailability of vitamin D encapsulated in nanosized emulsion droplets in hydrocolloid matrix [50].

Oil integration significantly influenced the textural qualities of gummies during the design process. The gummies containing 10% corn oil consistently demonstrate the ideal level of hardness throughout the whole investigation. The introduction of oil resulted in a significant rise in the hardness value. However, for gummies, the ideal hardness value often falls within the range of approximately 3100 g force. Following the application of a force of 3100 g, the gummies undergo a transformation, becoming more rigid and compact. Texture profile analysis can be employed to monitor alterations in texture throughout the drying process of gummy confections. The stickiness results indicated that the gummies with the least amount of oil, specifically 10% oil, had greater adhesiveness throughout the investigation, as depicted in Table 5. The viscosity increases as the oil content decreases. The stickiness of gummies with 40% oil content was further reduced, but, they exhibited increased hardness and a loss of their chewy texture. The texture of the gummies is influenced by their age, as the water content undergoes changes over time, and water is a crucial factor in preserving the texture [52]. The concentrations of additional additives were consistent throughout the experiment. Sucralose is a synthetic sweetener, whereas sorbitol and xylitol are examples of sugar alcohols. These substances function as sweeteners and create a desire among children to willingly consume the product. Different flavors enhance the fragrance and overall flavor of the product, as these variables significantly influence the overall preference for the product. The homogenization rates were restricted to 1000 rpm due to the potential risk of instability or disruption of the emulsion being integrated.

While both instrumental and sensory methodologies can be used to evaluate texture, research has shown that combining sensory analysis with instrumental analysis produces more accurate qualities for a consumer-oriented product [53]. The sensory assessment technique is extensively employed in food product development, quality control, shelf life monitoring, customer acceptance, and is considered by some experts to be one of the most sensitive and reliable approaches in specific circumstances [54]. During the 45-day storage period at room temperature, the manufactured gummy maintained its texture and was generally well-received. This indicates a product that is perceived by consumers as being stable and reliable. The consistent flavor retention throughout storage suggests that the excipients were mutually compatible and effectively encapsulated vitamin D under a nano-emulsion coating. Despite the sensory scores indicating a significant loss of aroma in the stored confectionery, this issue might be resolved by making adjustments to the flavoring ingredients.

6. CONCISE SUMMARY

The utilization of a nanoemulsion-based matrix system, comprising of edible oil, emulsifier, and hydrocolloid, has demonstrated its efficacy in creating stable and nano-enhanced vitamin D gummies that are easily accepted by consumers. The Nano gummy that was created has the ability to increase the bioavailability of vitamin D. Nevertheless, it is necessary to conduct extensive stability experiments and evaluate the results before proceeding with scaling up.

AT A GLANCE

Approximately half of the global population is affected by a lack of Vitamin D, and numerous studies have emphasized the link between Vitamin D and various clinical symptoms, such as musculoskeletal, cardiovascular, cerebrovascular, and neurological problems. The present study aimed to produce vitamin D oil-in-water (O/W) nanoemulsions and combine them into palatable gummies in order to improve bioavailability, stability, and patient adherence. The nano-emulsion was produced using the spontaneous emulsification method, with maize oil as the base and tween 20 and lecithin as the emulsifiers. The optimization process utilized pseudo-ternary phase diagrams to determine the ideal conditions. The resulting optimized nanoemulsion had an average particle size of 118.6 ± 4.3 nm and a polydispersity index (PDI) of 0.11 ± 0.30. The HPLC stability analysis revealed that the nano-emulsion effectively inhibited degradation, resulting in the retention of over 97% of active vitamin D over a 15-day period, as compared to 94.5% in the oil solution. Consistent findings were achieved after additional storage analysis. Subsequently, vitamin D gummies were created by utilizing emulsion-based gelled matrices, with gelatin serving as the hydrocolloid and different amounts of corn oil. The texture examination showed that gummies containing 10% corn oil had an ideal hardness of 3095.6 ± 201.7 g on the first day, which remained steady on day 45 with similar values of 3594.4 ± 210.6 g. The sensory evaluation conducted by 19 judges using the nine-point hedonic scale indicated that there was no significant change ($p > 0.05$) in the taste and general acceptance of the prepared gummies during a 45-day storage period. This study proposed that nanoemulsions continuously inhibit the environmental deterioration of vitamin D, which is previously recognized for its protective effects in the gastrointestinal tract by facilitating continuous release in the intestines and improving overall absorption. The soft chewable matrices were easily masticated and ingested, resulting in improved patient adherence.

REFERENCES

1. Demer, L.L.; Hsu, J.J.; Tintut, Y. Steroid hormone vitamin D: Implications for cardiovascular disease. Circ. Res. 2018, 122, 1576–1585.

2. Holick, M.F. Vitamin D deficiency. N. Engl. J. Med. 2007, 357, 266–281.
3. Del Valle, H.B.; Yaktine, A.L.; Taylor, C.L.; Ross, A.C. Dietary Reference Intakes for Calcium and Vitamin D; National Academies Press: Washington, DC, USA, 2011.
4. Holick, M.F.; Binkley, N.C.; Bischoff-Ferrari, H.A.; Gordon, C.M.; Hanley, D.A.; Heaney, R.P.; Murad, M.H.; Weaver, C.M. Evaluation, Treatment, and Prevention of Vitamin D Deficiency: An Endocrine Society Clinical Practice Guideline. J. Clin. Endocrinol. Metab. 2011, 96, 1911–1930.
5. American Geriatrics Society Workgroup on Vitamin D Supplementation for Older Adults. Recommendations abstracted from the American Geriatrics Society consensus statement on vitamin D for prevention of falls and their consequences. J. Am. Geriatr. Soc. 2014, 62, 147–152.
6. Aparna, P.; Muthathal, S.; Nongkynrih, B.; Gupta, S.K. Vitamin D deficiency in India. J. Fam. Med. Prim. Care 2018, 7, 324–330.
7. Pilz, S.; Tomaschitz, A.; Drechsler, C.; Zittermann, A.; Dekker, J.M.; Marz, W. Vitamin D Supplementation: A Promising Approach for the Prevention and Treatment of Strokes. Curr. Drug Targets 2011, 12, 88–96.
8. Milaneschi, Y.; Hoogendijk, W.J.G.; Lips, P.; Heijboer, A.; Schoevers, R.; Van Hemert, A.M.; Beekman, A.T.F.; Smit, J.H.; Penninx, B.W.J.H. The association between low vitamin D and depressive disorders. Mol. Psychiatry 2014, 19, 444–451.
9. Sarris, J.; Murphy, J.; Mischoulon, D.; Papakostas, G.I.; Fava, M.; Berk, M.; Ng, C.H. Adjunctive Nutraceuticals for Depression: A Systematic Review and Meta-Analyses. Am. J. Psychiatry 2016, 173, 575–587.
10. Borkar, V.V.; Devidayal; Verma, S.; Bhalla, A.K. Low levels of vitamin D in North Indian children with newly diagnosed type 1 diabetes. Pediatr. Diabetes 2010, 11, 345–350.
11. Zipitis, C.S.; Akobeng, A.K. Vitamin D supplementation in early childhood and risk of type 1 diabetes: A systematic review and meta-analysis. Arch. Dis. Child. 2008, 93, 512–517.
12. Holick, M.F. A call to action: Pregnant women in-deed require vitamin D supplementation for better health outcomes. J. Clin. Endocrinol. Metab. 2019, 104, 13–15.
13. Van Schoor, N.; Lips, P. Global overview of vitamin D status. Endocrinol. Metab. Clin. 2017, 46, 845–870.
14. Lips, P.; Hosking, D.; Lippuner, K.; Norquist, J.; Wehren, L.; Maalouf, G.; Ragi-Eis, S.; Chandler, J. The prevalence of vitamin D inadequacy amongst women with osteoporosis: An international epidemiological investigation. J. Intern. Med. 2006, 260, 245–254.
15. Manson, J.E.; Brannon, P.M.; Rosen, C.J.; Taylor, C.L. Vitamin D deficiency-is there really a pandemic. N. Engl. J. Med. 2016, 375, 1817–1820.
16. Pereira-Santos, M.; Costa, P.D.F.; Assis, A.D.; Santos, C.D.S.; Santos, D.D. Obesity and vitamin D deficiency: A systematic review and meta-analysis. Obes. Rev. 2015, 16, 341–349.
17. Holick, M.F. The vitamin D deficiency pandemic: Approaches for diagnosis, treatment and prevention. Rev. Endocr. Metab. Disord. 2017, 18, 153–165.
18. Pludowski, P.; Holick, M.F.; Grant, W.B.; Konstantynowicz, J.; Mascarenhas, M.R.; Haq, A.; Povoroznyuk, V.; Balatska, N.; Barbosa, A.P.; Karonova, T.; et al. Vitamin D supplementation guidelines. J. Steroid Biochem. Mol. Biol. 2018, 175, 125–135.

19. Chung, M.; Lee, J.; Terasawa, T.; Lau, J.; Trikalinos, T.A. Vitamin D With or Without Calcium Supplementation for Prevention of Cancer and Fractures: An Updated Meta-analysis for the U.S. Preventive Services Task Force. Ann. Intern. Med. 2011, 155, 827–838.
20. Schiele, J.T.; Quinzler, R.; Klimm, H.-D.; Pruszydlo, M.G.; Haefeli, W.E. Difficulties swallowing solid oral dosage forms in a general practice population: Prevalence, causes, and relationship to dosage forms. Eur. J. Clin. Pharmacol. 2013, 69, 937–948.
21. Fields, J.; Go, J.T.; Schulze, K.S. Pill Properties that Cause Dysphagia and Treatment Failure. Curr. Ther. Res. 2015, 77, 79–82.
22. Andersen, O.; Zweidorff, O.K.; Hjelde, T.; Rødland, E.A. Problems when swallowing tablets. A questionnaire study from general practice. Tidsskr. Nor. Laegeforening Tidsskr. Prakt. Med. Raekke 1995, 115, 947–949.
23. Jiang, S.; Yildiz, G.; Ding, J.; Andrade, J.; Rababahb, T.M.; Almajwalc, A.; Feng, H. Pea protein nano-emulsion and nanocomplex as carriers for protection of cholecalciferol (vitamin D3). Food Bioprocess Technol. 2019, 12, 1031–1040.
24. Zareie, M.; Abbasi, A.; Faghih, S. Thermal Stability and Kinetic Study on Thermal Degradation of Vitamin D3 in Fortified Canola Oil. J. Food Sci. 2019, 84, 2475–2481.
25. Donsì, F.; Annunziata, M.; Vincensi, M.; Ferrari, G. Design of nano-emulsion-based delivery systems of natural antimicrobials: Effect of the emulsifier. J. Biotechnol. 2012, 159, 342–350.
26. Bothiraja, C.; Pawar, A.; Deshpande, G. Ex-Vivo absorption study of a nanoparticle based novel drug delivery system of vitamin D3 (Arachitol Nano™) using everted intestinal sac technique. J. Pharm. Investig. 2016, 46, 425–432.
27. Salvia-Trujillo, L.; Fumiaki, B.; Park, Y.; McClements, D.J. The influence of lipid droplet size on the oral bioavailability of vitamin D2 encapsulated in emulsions: An in vitro and in vivo study. Food Funct. 2017, 8, 767–777.
28. McClements, D.J.; Rao, J. Food-grade nano-emulsions: Formulation, fabrication, properties, performance, biolog-ical fate, and potential toxicity. Crit. Rev. Food Sci. Nutr. 2011, 51, 285–330.
29. Golfomitsou, I.; Mitsou, E.; Xenakis, A.; Papadimitriou, V. Development of food grade O/W nano-emulsions as carriers of vitamin D for the fortification of emulsion based food matrices: A structural and activity study. J. Mol. Liq. 2018, 268, 734–742.
30. McClements, D.J. Nanoemulsions versus microemulsions: Terminology, differences, and similarities. Soft Matter 2012, 8, 1719–1729.
31. Silva, H.D.; Cerqueira, M.Â.; Vicente, A.A. Nanoemulsions for food applications: Development and characteriza-tion. Food Bioprocess Technol. 2012, 5, 854–867.
32. Kleiman-Weiner, M.; Luo, R.; Zhang, L.; Shi, Y.; Medina, A.; Rozelle, S. Eggs versus chewable vitamins: Which intervention can increase nutrition and test scores in rural China? China Econ. Rev. 2013, 24, 165–176.
33. Stewart, R.; Askew, E.; Mcdonald, C.; Metos, J.; Jackson, W.; Balon, T.; Prior, R. Antioxidant Status of Young Children: Response to an Antioxidant Supplement. J. Am. Diet. Assoc. 2002, 102, 1652–1657.
34. Ozturk, B.; Argin, S.; Ozilgen, M.; McDonald, D.J. Nanoemulsion delivery systems for oil-soluble vitamins: Influence of carrier oil type on lipid digestion and vitamin D3 bioaccessibility. Food Chem. 2015, 187, 499–506.

35. Schoener, A.L.; Zhang, R.; Lv, S.; Weiss, J.; McClements, D.J. Fabrication of plant-based vitamin D 3-fortified nano-emulsions: Influence of carrier oil type on vitamin bioaccessibility. Food Funct. 2019, 10, 1826–1835.

36. Yang, Y.; McClements, D.J. Vitamin E bioaccessibility: Influence of carrier oil type on digestion and release of emulsified α-tocopherol acetate. Food Chem. 2013, 141, 473–481.

37. Qian, C.; Decker, E.A.; Xiao, H.; McClements, D.J. Nanoemulsion delivery systems: Influence of carrier oil on β-carotene bio-accessibility. Food Chem. 2012, 135, 1440–1447.

38. Rao, J.; Decker, E.A.; Xiao, H.; McClements, D.J. Nutraceutical nano-emulsions: Influence of carrier oil compo-sition (digestible versus indigestible oil) on β-carotene bioavailability. J. Sci. Food Agric. 2013, 93, 3175–3183.

39. Hsu, C.-Y.; Wang, P.-W.; Alalaiwe, A.; Lin, Z.-C.; Fang, J.-Y. Use of Lipid Nanocarriers to Improve Oral Delivery of Vitamins. Nutrients 2019, 11, 68.

40. Imran, M.; Iqubal, M.K.; Imtiyaz, K.; Saleem, S.; Mittal, S.; Rizvi, M.M.A.; Ali, J.; Baboota, S. Topical nanostructured lipid carrier gel of quercetin and resveratrol: Formulation, optimization, in vitro and ex vivo study for the treatment of skin cancer. Int. J. Pharm. 2020, 587, 119705.

41. Hasan, N.; Imran, M.; Kesharwani, P.; Khanna, K.; Karwasra, R.; Sharma, N.; Rawat, S.; Sharma, D.; Ahmad, F.J.; Jain, G.K.; et al. Intranasal delivery of Naloxone-loaded solid lipid nanoparticles as a promising simple and non-invasive approach for the management of opioid overdose. Int. J. Pharm. 2021, 599, 120428.

42. McClements, D.J. Edible nano-emulsions: Fabrication, properties, and functional performance. Soft Matter 2011, 7, 2297–2316.

43. Vandamme, T.F.; Anton, N. Low-energy nanoemulsification to design veterinary controlled drug delivery devices. Int. J. Nanomed. 2010, 5, 867.

44. Zakkula, A.; Gabani, B.B.; Jairam, R.K.; Kiran, V.; Todmal, U.; Mullangi, R.; Babulal, B.G. Preparation and optimization of nilotinib self-micro-emulsifying drug delivery systems to enhance oral bioavailability. Drug Dev. Ind. Pharm. 2020, 46, 498–504.

45. Phillips, G.O.; Williams, P.A. Handbook of Hydrocolloids; Elsevier: Amsterdam, The Netherlands, 2009.

46. Teixeira-Lemos, E.; Almeida, A.R.; Vouga, B.; Morais, C.; Correia, I.; Pereira, P.; Guiné, R.P.F. Development and characterization of healthy gummy jellies containing natural fruits. Open Agric. 2021, 6, 466–478.

47. Marfil, P.H.; Anhê, A.C.; Telis, V.R. Texture and Microstructure of Gelatin/Corn Starch-Based Gummy Confections. Food Biophys. 2012, 7, 236–243.

48. Sari, T.P.; Mann, B.; Kumar, R.; Singh, R.R.; Sharma, R.; Bhardwaj, M.; Athira, S. Preparation and characterization of nanoemulsion encapsulating curcumin. Food Hydrocoll. 2015, 43, 540–546.

49. Gurpret, K.; Singh, S.K. Review of Nanoemulsion Formulation and Characterization Techniques. Indian J. Pharm. Sci. 2018, 80, 781–789.

50. Grossmann, R.E.; Tangpricha, V. Evaluation of vehicle substances on vitamin D bioavailability: A systematic review. Mol. Nutr. Food Res. 2010, 54, 1055–1061.

51. Kotta, S.; Khan, A.W.; Pramod, K.; Ansari, S.H.; Sharma, R.K.; Ali, J. Exploring oral nanoemulsions for bioavailability enhancement of poorly water-soluble drugs. Expert Opin. Drug Deliv. 2012, 9, 585–598.

52. Ergun, R.; Lietha, R.; Hartel, R.W. Moisture and Shelf Life in Sugar Confections. Crit. Rev. Food Sci. Nutr. 2010, 50, 162–192.

53. Rahman, M.S.; Al-Attabi, Z.H.; Al-Habsi, N.; Al-Khusaibi, M. Measurement of Instrumental Texture Profile Analysis (TPA) of Foods. In Techniques to Measure Food Safety and Quality; Springer: Cham, Switzerland, 2021; pp. 427–465.

54. Kemp, S.E. IFST PFSG committee Application of sensory evaluation in food research. Int. J. Food Sci. Technol. 2008, 43, 1507–1511.

16 Chapter

INVESTIGATING CANCER MECHANOBIOLOGY WITH BIOMIMETIC HYDROGELS: OVERVIEW, BIOMEDICAL APPLICATIONS, AND FUTURE PERSPECTIVES

1. CELLULAR MICROENVIRONMENT

The cellular microenvironment consists of a combination of extracellular matrix proteins, soluble signaling molecules, adjacent cells, and the physical features of the niche. These elements influence cell activity through both direct and indirect biomechanical and biochemical signals [1,2]. Studies have demonstrated that the characteristics of the microenvironment, such as its stiffness and composition, can influence and guide cell physiology and lineage development [3,4]. The discovery of such discoveries has motivated scientific investigations aimed at characterizing the cellular microenvironment and its connections to cellular activities through the use of in vitro tissue models capable of replicating biomechanical circumstances [5,6]. In this discussion, we will focus on the cellular niche, namely the biological and mechanical characteristics of the extracellular matrix (ECM), how cells perceive these characteristics, and the disruption of cell-ECM interactions in different disease conditions. This review will demonstrate how these variables, which engage in crosstalk with cells, contribute to cellular activity and overall health by examining numerous research publications on the issue. This review aims to comprehensively examine the current understanding and gaps in knowledge regarding the biomechanics of cell-microenvironment interactions. It will also explore how biomimetic hydrogel models can be used to bridge these knowledge gaps, as well as the practical applications of biomimetic hydrogels in the field of biomedicine.

1.1. Extracellular Matrix

Tissues can be characterized by two constituents: cellular and non-cellular. The extracellular matrix is a component that exists outside of cells and is made up of proteins, polysaccharides, growth factors, signaling molecules, proteases, and water [7] (Figure 1). The distribution of these components is heterogeneous rather than homogeneous, leading to distinct niche microenvironments for each cell, as well as tissue-specific mechanical, physical, and biochemical features [7]. These entities play a significant role in governing and moderating cellular activities. The ECM is mostly comprised of proteins, although it also contains several other components. These components consist of proteoglycans like hyaluronan, as well as fibrous proteins like as collagens, elastins, fibronectins, and laminins [8]. These proteins serve to attach cells to the extracellular matrix (ECM) through focal adhesions and facilitate the transmission of signals between cells [8].

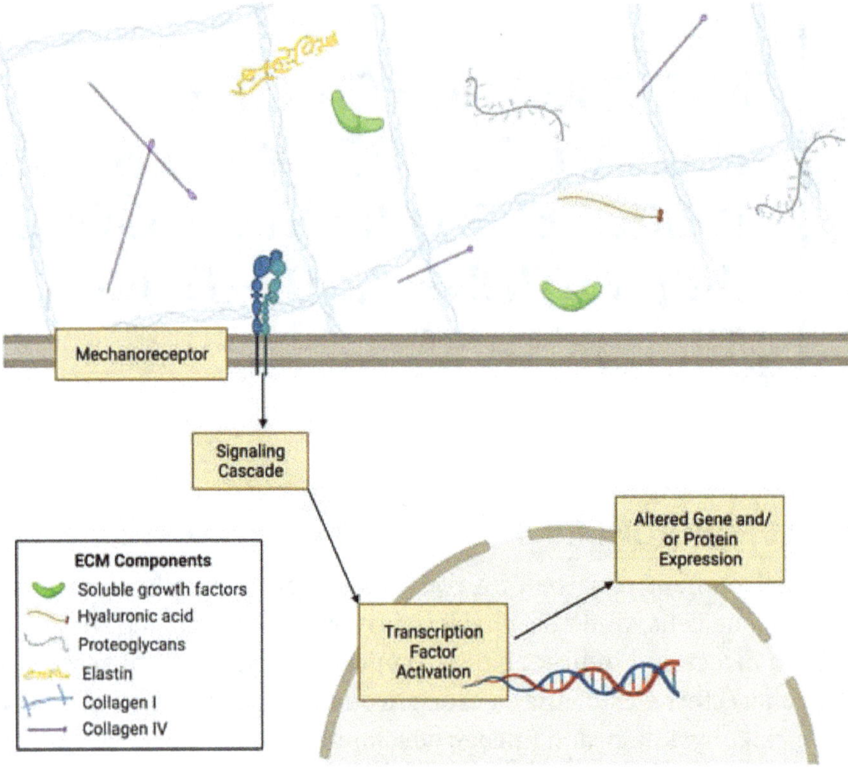

Figure 1. Extracellular matrix (ECM) and cellular mechanotransduction. ECM components such as collagens can alter mechanical properties to induce changes in cellular signaling and gene or protein expression via mechanoreceptors and mechanotransduction proteins. Legend on the bottom left shows which ECM component is represented by each symbol. (Created with BioRender.com, accessed on 29 May 2022).

The composition of the extracellular matrix (ECM) also has an impact on other physical properties, including elasticity, stiffness, porosity, static architecture, and dynamic deformations of the matrix [9]. The mechanical properties of the extracellular matrix (ECM) are greatly influenced by the concentration, assembly, and crosslinking densities of its structural collagens and elastins. These factors affect the composite strength, elasticity, and mechanical resistance of the ECM [10,11]. The features of the ECM are not fixed; they experience dynamic alterations as the ECM is constantly being modified through protein degradation, deposition, or modifications that might be intrinsic or induced by cellular activity [10]. During tissue repair, the presence of growth factors and cytokines in the extracellular matrix (ECM) triggers the activation of matrix metalloproteinases, which are responsible for altering the ECM [12]. The processes of wound healing and tissue remodeling are triggered by mechanical and pharmacological stimuli, which activate growth factors to alter the composition of the extracellular matrix (ECM) [13]. Specialized kinds of extracellular matrices (ECMs) have demonstrated their significance as regulators of tissue and cell function. The basement membrane plays a crucial role in the structure and function of blood vessels due to its involvement in the process of angiogenesis [12]. Research has demonstrated that the extracellular matrix (ECM) plays a role in controlling stem cell destiny, cell growth, cell specialization, cell movement, and tissue regrowth [11,14,15,16].

1.1.1. Cell-ECM Interactions

The cell and its ECM are involved in a dynamic reciprocity, through which cues from the ECM and cellular activities are in crosstalk to maintain a healthy state [17]. Signaling processes are one of the ways that the cell-ECM interactions are facilitated. Cell-matrix adhesion sites, or focal contacts, enable communication between cells and the ECM through physical connections of cellular integrins and cadherins to ligands in the ECM [15,16]. Focal contacts are important to cellular processes that require physical attachment to the ECM, such as migration and angiogenesis [18]. Engagement of integrins and cadherins to certain ligands can activate signaling pathways such as that of the Rho family of GTPases to stimulate structural changes in the cell or induce other processes, therefore serving as an important step in biochemical cell-ECM interactions [6,16].

Proteoglycans present in the ECM have functions such as inducing aggregation and participating in ECM structure by adhering to structural proteins [19]. Some proteoglycans reside on the surfaces of epithelial cells, where they act similar to cell-adhesion molecules and bind collagens and fibronectin to anchor cells to the ECM [20,21]. In addition to these vital functions, proteoglycans are active co-receptors that mediate cellular signaling by binding soluble ligands in the ECM and encouraging the formation of receptor complexes on cell surfaces [22,23]. Proteoglycan co-receptors are vital to various developmental processes, and the loss of co-receptor function or mutation has been implicated in diseases such as cancer and ischemic heart disease [23]. Therefore, the ECM also plays a crucial role in mediating cell–cell communication, which will be discussed more thoroughly in Section 1.2 of this review.

Mechanobiology of the Cellular Microenvironment

The ECM is involved in mechanical crosstalk with cells [24], which relies on mechanotransduction proteins that help to regulate intracellular tensile response to mechanical forces from the ECM [25]. Mechanical stimuli that cells may receive include shear stress, membrane tension, force, strain, stiffness, and drag force [25]. These stimuli are listed in Table 1 with the mechanotransduction proteins identified to be involved in recognizing the stimuli and eliciting the response in cells. While the extracellular matrix of many cancers, including colon, breast, and prostate cancer, is stiffer than that of healthy tissues [26,27], cell deformability or reduced stiffness has been correlated to increased metastatic potential and invasiveness in cancer. In a study of ovarian cancer cells of varying invasiveness, Xu et al. found that the more-invasive ovarian cancer cell line was more deformable compared to the less-invasive cell line [28]. In another study, Hayashi and Iwata confirmed that cancer cells are softer (i.e., lower Young's modulus) than normal cells using atomic force microscopy [29]. Another study reported that cisplatin treatment caused decreased stiffness and invasiveness of prostate cancer cells [30], suggesting that various cell lines may have varied mechanical properties in the cancer state. Further controversy exists within the field of glioblastoma biomechanics. Gliomas are highly variable; therefore, measurements of tumor stiffness may vary depending on location of measurement [31]. Although some GBM tissues were stiffer than healthy reference tissues, GBM tissues, on average, were less stiff than healthy tissues [31,32,33]. The controversy regarding GBM tumor stiffness could by fueled by differences in methods of measurement, as there is currently no standard practice in the field [34,35,36].

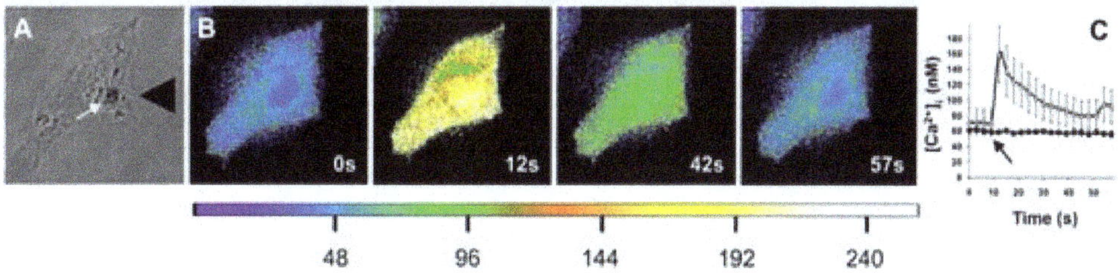

Figure 2. Application of extracellular stress leads to increased intracellular calcium concentrations. Matthews et al. applied high levels of stress to cells (A) and found that when imaged via Fura-2AM

ratio imaging, it led to a transient increase in calcium concentrations, as shown in pseudocolor images ranging from blue to yellow (B) that is quantified (C) as a function of time for control and gadolinium chloride-treated cells. (Figure reprinted/adapted with permission from Ref. [61]. Copyright 2006, National Academy of Sciences. Mechanical stimuli coming from the extracellular environment can be processed by cells through mechanotransduction pathways. The proteins of these pathways translate mechanical cues to induce biochemical and genetic responses.

Table 1. Mechanotransducers of various mechanical properties and human cellular responses.

Functional Category	Mechano-Transducers	Mechanical Signal	Examples of Cellular Responses
Cell Mechanical and Physical Properties	Integrins	Force	RhoA activation leading to increased cell stiffness [62,63]
	Focal Adhesions	Force	Actin polymerization [55]
	Yes-associated protein (YAP)	Force	Oligodendrocyte morphology and maturation [40]
	Titin	Force	Implicated in development of mechanical unloading-induced diaphragm weakness [64]
	Stress Fibers (actin filaments, myosin II, etc.)	Force	Transmit tension to other proteins, regulate assembly of filaments [65]
	Vinculin	Force	Transmit tensile force [66]
	Myosin II	Force	Increased cortical tension and cell membrane fusion promotion [67]
	Vasodilator stimulated phosphoprotein (VASP), zyxin, and Testin LIM domain protein (TES)	Force	Regulate junction dynamics [68]
	Neurogenic locus notch homolog protein 1 (NOTCH1)	Shear Stress	Altered cell morphology [69]
	Piezo1	Force	Vascular structure [70]
	Lamin A	Rigidity	Nuclear mechanics [71,72]
	Integrins	Force	Tyrosine Phosphorylation, MAPK signaling [15]
Alters Signaling Pathways	Focal Adhesions	Force	Integrin convergence [73]
	Fibronectin	Force	Altered integrin binding [74]
	T-cell receptor (TCR)	Force	T-cell calcium and IL-2 secretion [75]
	Talin	Force	Recruitment of vinculin to focal adhesion complexes [76]
	Piezo2	Force	Serotonin release [77]
	Vinculin	Force	Enhanced PI3K activation [78]
	p130Cas	Force	Activation of Cas signaling pathway [79]
	Syndecan-1	Force	Activation of pro-inflammatory and growth-stimulating pathways [80]

Functional Category	Mechano-Transducers	Mechanical Signal	Examples of Cellular Responses
	Transient Receptor Potential Cation Channel Subfamily V Member 4 (TRPV4)	Force	Reorientation and flow-mediated nitric oxide production [81]
	Ion Channels	Force	Cell signaling [82]
	von Willebrand factor—glycoprotein Ib complex (VWF-GPIb)	Shear Stress	Enhanced calcium triggering in platelets and T cells [83]
	Platelet endothelial cell adhesion molecule-1 (PECAM-1)	Shear Stress	Tyrosine kinase Src and PI3K signaling activated [84]
	G-protein coupled receptor 68 (GPR68)	Shear Stress	Component in signaling for cardiovascular pathophysiology [85]
	β-catenin	Shear Stress	Activated expression of FOXC2 transcription factor [86]
	Caveolin-1 and β1 Integrin	Stiffness	FA assembly and turnover [62]
	rho-associated, coiled-coil-containing protein kinase (ROCK) 1 and 2	Stiffness	Regulation of RhoA signaling pathways [87]
	YAP	Stiffness	Altered translocation depending on surrounding stiffness [88]
	Piezo1	Force	Ion Permeation and selection [89]
	C-X-C motif chemokine receptor (CXCR1/2)	Shear Stress	Mediates laminar shear-stress-induced endothelial cell migration [90]
	Transforming growth factor beta 1 (TGFβ1)	Shear Stress	Collagenase-dependent fibroblast migration [91]
Migration	RhoA	Force	Collective cell migration [92]
	Vinculin and metavinculin	Force	Regulation of cell adhesion and motility [66]
	NOTCH1	Shear Stress	Decreased proliferation [69]
	Caveolin 1	Rigidity	Decreased proliferation [93]
Cancer	YAP1	Shear Stress Stiffness	Cancer cell motility [54] Nuclear localization of YAP1 [94]
	TGFβ1	Shear Stress	Human melanoma cell tumor invasiveness [91]
	PI3K/Akt pathway	Stiffness	Overexpression of VEGF in hepatocarcinoma cells [63]
	TRPV4 ion channel	Stiffness	Tumor vascularization through down-regulation of Rho kinase activity [95]
	microRNAs	Stiffness	Altered expression in different stiffness conditions [96]
	Twist1	Stiffness	Induction of EMT and tumor metastasis [97]
	Myocardin related transcription factor A (MRTF-A)	Stiffness	Regulates miRNAs involved in myogenic differentiation [88]
Differentiation	Focal Adhesions	Force	Osteogenic differentiation [98] Myofibroblastic differentiation [99]
	Transient Receptor Potential Cation Channel Subfamily M Member 7 (TRPM7)	Shear Stress	Osteogenic differentiation of mesenchymal stromal cells [100]

For instance, integrins and focal adhesion proteins are mechanotransduction proteins that communicate mechanical forces to the cell cytoskeleton, and there have been some studies that show force-dependent integrin activation [37,38,39]. Yes-associated protein (YAP) and other transcription factors have been shown to translocate to the nucleus in stiffer substrates [40]. Similarly, E-cadherin is a mechanotranducer of shear stress [41,42,43]. Many mechanotransduction proteins have been studied in the context of specific cell types, including platelet endothelial cell adhesion molecule-1 (PECAM-1) in skeletal muscle cells, G protein-coupled receptors in endothelial cells, and vascular endothelial growth factor receptor 2 (VEGFR2) in chondrocytes [44,45,46,47]. Cell surface receptors also transduce mechanical signals to cells upon recognition of a ligand that sustained force from the ECM, which can then cause conformational changes in the mechanoreceptor to activate a protein signaling pathway to alter cellular processes [48]. Such membrane proteins are not the sole propagators of mechanical force; studies have shown that there are mechanotransduction systems in cells that enable the progression of force through a long distance [49]. Src and Rac1 have been shown to be activated at distances from 30-60 μm from the original area the force was applied to, via cytoskeletal mediation of the force [49,50]. Mechanical signals can also be transmitted throughout the cytoskeleton to the nucleus via proteins such as linker of nucleo- and cyto-skeleton (LINC) complex 9 to change chromatin structure and cause nuclear stiffening [49,51]. Mechanotransduction systems translate the numerous mechanical cues from the ECM into biochemical signals interpreted by the cell that lead to signaling cascades that control transcription, proliferation, migration, and many other cellular processes [52,53,54,55] (Figure 2). For instance, smooth muscle cells can migrate along gradients of substrate stiffness through durotaxis [56]. ECM biomechanics is not only vital for cellular processes, but also regulates tissue and organ-level processes such as tissue differentiation, morphogenesis, and development [57,58,59,60]. Therefore, elucidating the interplay of ECM biomechanical and biochemical signals with mechanotransduction proteins and pathways is critical to understanding diverse aspects of cellular and tissue health.

1.1.2. Cell-ECM Interactions in Cancer

Problems in mechanotransduction can result from changes in ECM mechanical properties and defects in proteins involved in mechano-sensitivity [101]. Since mechanotransduction is essential for modulating cellular homeostasis, its failure is linked to metastasis and cancer progression [96]. For instance, there are many proteins implicated in mechanotransduction in glioblastoma (GBM). Talin1 inhibition has been observed to decrease cell spreading and limit cell stiffness changes of glioma cells in response to ECM stiffness, proving its role as a mechanosensory [76]. Non-muscle myosin II depletion reduced the effect of matrix confinement on GBM cell motility [102]. Constitutive activation of RhoA GTPase caused lower sensitivity to matrix stiffness of GBM cells in toxicity assays [55]. Increased matrix stiffness was correlated to Hras, RhoA, and rho-associated, coiled-coil-containing protein kinase 1 (ROCK1) upregulation, which are mechanosensor proteins that are implicated in migration and proliferation in cancers in general [61,87]. Integrins are also particularly important in mechanotransduction by relaying signals from the ECM to the cell actin cytoskeleton and are essential to cell migration and cell-matrix adhesion in cancer [103].

Identifying proteins involved in mechanotransduction and their roles in cancer progression can be an essential part of developing therapeutic strategies to hinder cancer progression and malignancy. Certain studies have shown the potential effects of targeting mechanotransduction proteins on cancer cells. Knock-down of CD44 led to decreased structural microtubule, vimentin, and glial fibrillary acidic protein expression and decreased migration and cell stiffness [104]. Targeted inhibition of integrins in the tumor microenvironment has been shown to reduce angiogenesis and inhibit tumor growth [103].

While these studies show partial inhibition of mechanical sensitivity as decreasing invasive properties of tumor cells, other studies have shown that tumor initiating cells are mostly insensitive to mechanical cues from the ECM and that mechanically-insensitive cells have increased motility and invasiveness in vitro [28,105,106,107,108,109]. A complex approach is needed to target mechanical sensitivity in cells through mechanotransduction-based therapeutics for it to become a promising mode of cancer treatment. Therefore, extensive research in mechanotransduction and cell-ECM crosstalk is essential.

1.2. Neighboring Cells and Secreted Factors

In addition to the ECM components, cells are surrounded by heterogeneous populations of neighboring cells that are unique to the tissue and sub-location within the tissue [110]. How cells communicate and influence one another is crucial to maintaining homeostasis and coordinating processes that require several cells, such as tissue formation and regeneration.

Cells can interact through secreted signals that are recognized by membrane-bound receptors through either paracrine signaling, between cells, or autocrine signaling, which is from one cell to itself. In fact, cancer cells are often able to "override" signals from neighboring cells through autocrine pro-survival and proliferation signals. It is crucial to better understand the specific interactions between cells that promote healthy conditions or lead to disease states. For instance, Zervantonakis et al. found that fibroblasts in the tumor microenvironment of HER2 positive breast-cancer cells reduced drug sensitivity through paracrine signaling that activates mechanistic target of rapamycin (mTOR, anti-autophagic) and anti-apoptotic signals [111]. In addition to elucidating mechanisms of tumor resistance, cell–cell interaction dynamics can provide insights to developing self-assembled multicellular structures in vitro. Mueller et al. demonstrated that by utilizing pulsed light activation to control engineered photo-switchable cell-cell interactions, they were able to control the spatial organization of multicellular structures without a scaffold [112]. This highlights the importance of maintaining cell interactions in efforts to mimic physiological conditions in vitro.

Cells are also involved in communications through physical contacts with one another. This is another way that mechanical stimuli may play a role in influencing cellular processes. Physical contacts are also involved in collective migration, which has been exhibited by metastatic and invasive cancer cells [113]. The cytoskeletal tension at cell-cell contacts can serve as a significant regulator of mechanotransduction pathways, and there is a wide field of study on the mechano-sensing implications of cell-cell contacts such as focal adhesions and adherens [114]. One method in particular, physically interacting cell sequencing (PIC-seq), is a novel sequencing approach that combines cell sorting, RNA-sequencing, and computational modeling to describe complex cellular interactions in different contexts [115]. However, the heterogeneity and complexity of the interactions between cells and with the ECM need to be better recapitulated in vitro to provide a more accurate understanding of cellular communication networks.

1.3. Hydrogels as In Vitro Models of the Cellular Microenvironment

Three-dimensional hydrogel models of the cellular microenvironment have gained interest in recent years for their improved mimicry of in vivo conditions as opposed to two-dimensional cell cultures (Table 2). These hydrogel models have been extensively studied and compared to 2D culture and in vivo conditions. Cells have been shown to have differing spatio-physical properties in 3D and conventional 2D culture conditions [116,117,118,119]. Hsieh and colleagues showed that cells cultured in 2D exhibit greater drug sensitivity than in vivo, and 3D cultures exhibit chemosensitivities comparable to solid tumors provided that they have similar cell density [120]. Expanding the cellular environment from 2D to 3D has also been shown to affect proliferation and metabolism [120,121,122]. Using 3D cultures has led to significant advancements in the understanding of cell migration strategies as well, since some of these could not be observed in 2D cultures [123,124,125]. For instance, fibroblasts have several different migration strategies that are utilized in different microenvironmental conditions, demonstrating cellular plasticity [125]. Such findings prompt questions about the mechano- and bio-sensing abilities of cells in relation to their environments. In this section, we will elaborate on the different methods utilized thus far to improve cell-microenvironment mimicry in the form of hydrogel models, the cellular response to such models in contrast with conventional cell-culture methods, and factors to be considered in crafting a hydrogel model.

Table 2. 2D versus 3D cultures [117,118,119,123,126,127,128].

2-Dimensional Culture		3-Dimensional Culture	
Advantages	Disadvantages	Advantages	Disadvantages
Simple	Does not mimic in vivo structure	More like in vivo structure	Expensive
Reproducible	Fewer interactions with environment	Niches are available	Time consuming
Inexpensive	Access to unlimited amount of nutrients from media	Access to nutrients is not unlimited, varies	Less reproducible
	Less diverse phenotype and polarity	Can form organs or spheroid clusters of cells	More complex and difficult to carry out
	Altered cell morphology	Allows study of cell-cell and cell-ECM interactions	Fewer interactions with environment

1.3.1. Mimicking Cellular Microenvironment Biomechanics

There have been many diverse approaches to capturing the complexity of the ECM and the biomechanical cues directed by it via in vitro hydrogel models. Tissue specificity is a crucial factor to consider in developing an ECM-mimetic hydrogel-based cell culture system, since the cells of each tissue produce and degrade matrix constituents, leading to variable ECM composition, bioactivity, and mechanics amongst tissues [129] (Figure 3). In the bone, the ECM is calcified to provide structural support to the tissue, while in tendons it is structured to provide tensile strength [130,131].

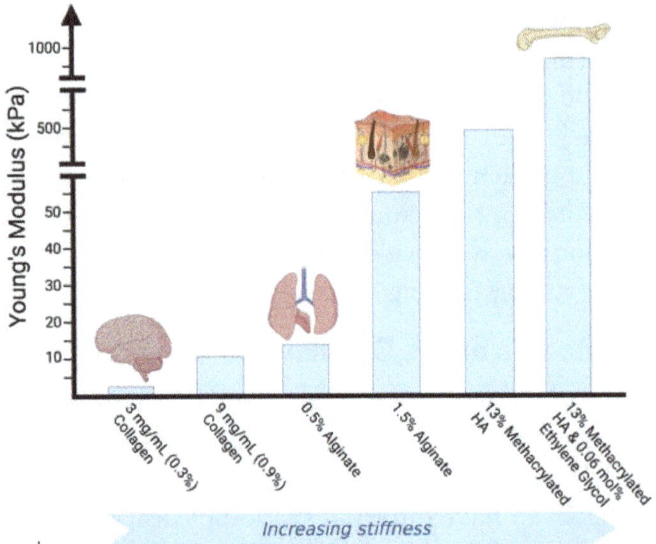

Figure 3. Young's modulus of hydrogels. The reported Young's modulus (kPa) values of various hydrogels are plotted with an image of the tissue that has similar Young's modulus. It is crucial to choose a hydrogel model that corresponds to the in vivo mechanical properties of tissue that is relevant to the work. (Created with BioRender.com, accessed on 14 July 2022).

Most natural ECM mimics in vitro have been hydrogels composed of ECM components such as collagen, fibronectin, or hyaluronic acid. Such hydrogels have been significant in studying cellular behavior in 3D environments as well as cell response to controlled mechanical properties such as stiffness, elasticity, and rigidity [63,132,133,134,135,136,137,138,139]. Photo-crosslinking, chemical crosslinking, changing fiber density, and the development of 'smart' hydrogels that are responsive to external stimuli (e.g., changes in pH, temperature, and light) have all been used to fine-tune hydrogels to mimic the cellular microenvironment mechanically [140,141,142,143,144,145,146,147,148,149,150,151]. For example, photo-crosslinking of gelatin methacryloyl (GelMa) hydrogels alters hydrogel stiffness based on the light intensity, exposure time, and concentration of photo-initiator used in photo-crosslinking [143]. Nanocomposite smart hydrogels have been produced to change volume, Young's modulus, and breaking strength based on applied chemical and physical stimuli [148,150].

1.3.2. Recapitulating Cellular Micro-environment Heterogeneity

While the above-discussed methods are useful to study the effects of semi-isolated mechanical conditions on cells, they often disregard the biological complexity of the ECM. In vivo ECMs can also serve as a medium through which cells communicate with one another via secreted growth and signaling factors in vitro. Several protocols have been developed to decellularize ECM isolated from animal or human tissue, or to harvest ECM secreted by fibroblasts in vitro [152,153,154,155]. To better capture the biological complexity of the ECM, hydrogel models have been developed to incorporate fibroblasts, growth factors, and more diverse components of the ECMs specific to the cell type being studied. Incorporating fibroblasts in a co-culturing system with the cell type of interest is a relatively newer approach, which allows for fine-tuning mechanical properties using a hydrogel model while adding biological complexity to the culture through the addition of the secreted matrix from fibroblast cells [156,157]. Lee et al. developed a 3D hydrogel system for co-culturing human liver-cancer cell spheroids with fibroblasts on a micropatterned fibrous scaffold, thereby modeling the three dimensional structure of tumors and the cross-talk between cancer cells and neighboring fibroblasts [156]. Several groups have also developed methods of controlled growth factor release into hydrogel-based cell cultures [158,159,160]. It is crucial to combine the biochemical components of the ECM with the mechanical properties to better mimic the cellular microenvironment in vitro.

2. HYDROGELS AND THEIR APPLICATIONS

Hydrogels are polymer networks that are hydrophilic and have a three-dimensional structure [161]. Hydrogels possess a crucial characteristic of being able to absorb water and expand without undergoing dissolution, thanks to the presence of hydrophilic functional groups on the polymer chains of the hydrogel. Due to their high water content, they possess flexibility and a soft tissue-like appearance, which makes them suitable for utilization in biological research [162]. Hydrogels are a diverse group of polymeric compounds, consisting of natural or manufactured components, that find application in several fields. The structural integrity of hydrogels remains intact even when they undergo swelling due to the presence of chemical and/or physical crosslinks within the polymer network. Modifications in the composition, protein concentration, and crosslinking density result in alterations in the elasticity, polymer density, and rate of biodegradation [163]. Hydrogels include several adjustable characteristics that make them appropriate for a diverse array of investigations into mechanical, chemical, and biological conditions in laboratory settings. Additionally, they are highly favorable for potential usage in therapeutic applications within living organisms, such as drug delivery systems [161,162]. In this article, we will discuss the primary categorizations of hydrogels according to their composition, the overall characteristics of hydrogels, and the present and possible experimental and clinical uses of hydrogels, with a specific emphasis on pre-clinical cancer models.

2.1. Types of Hydrogels

There are several classification methods for hydrogels, such as those based on ionic charge, biodegradability, physical properties, crosslinking, and preparation. They are most commonly classified by their source polymer(s): naturally occurring biomaterial, synthetic bio-mimetic, or a hybrid of these two sources [163]. Synthetic materials are traditionally used extensively due to low biodegradation rate, ease of manipulation, and greater control over biochemical interactions [149,164,165]. Natural biomaterials, however, are preferred due to their ability to biologically mimic the structural and biochemical properties of the cellular niche in vitro, and are responsive to cellular activities in terms of biochemical reactivity and degradability [164]. Numerous studies have presented methods for forming hybrid hydrogels that possess properties of both natural and synthetic biomaterials that are mechanically and biochemically responsive and tunable by the surrounding environment [165,166].

2.1.1. Natural Hydrogels

Natural hydrogels are composed of naturally occurring polysaccharides and proteins such as collagen, Matrigel®, hyaluronan, gelatin, and their derivatives, including alginate and chitosan [167,168]. While most studies cite hydrogel formation from one ECM protein, there are several that utilized hydrogels composed of several proteins [167,168]. Either protein concentration or crosslinking density is altered in the hydrogels in order to change the Young's modulus of the gels [167,168]. However, these are not finely tunable or as well-understood as the mechanical properties of synthetic hydrogels [169]. In contrast, the biocompatibility of natural hydrogels and their responsiveness to cellular degradation add an important dimension to biological studies [170].

Many studies incorporate several proteins into biomaterial hydrogels to improve hydrogel stability and similarity to the in vivo cellular microenvironment. Comparing cellular properties in vitro in composite hydrogels versus single-material hydrogels can give more specific information on how cell mechano-sensitivity is influenced by specific cell-ECM protein interactions. A useful representative investigation of the insight obtained by studies undertaken with composite hydrogels is a study by Rao et al. [171] (Figure 4). They mixed collagens I and III with hyaluronic acid (HA), collagen IV with HA, and used collagen (I and III) as a standalone in hydrogel formation to determine modulus values of the different hydrogels and observe changes in cell morphology, spreading, and migration. Composite hydrogels had much higher elastic modulus values than the collagen hydrogel, and increased HA content correlated to increased modulus values and greater cell spreading and migration [171]. Interestingly, cell morphology differed by the type of collagen used in the hydrogels: cancer cells cultured in collagen IV had a rounder cell shape, while those in collagen I and III were spindle shaped [171].

Another common composite hydrogel used in several cell mechanics studies is Matrigel®, a mixture of ECM proteins secreted from mouse sarcoma cells, which is commonly used for spheroid formation in 3D cultures [172]. Several mechanics studies use Matrigel® for the formation of spheroids before seeding onto other types of hydrogels or for studies of cell invasion. For instance, Grundy et al. reported that primary GBM cells that were insensitive to rigidity were more invasive in spheroid cultures in Matrigel® as opposed to rigidity sensitive cells [173].

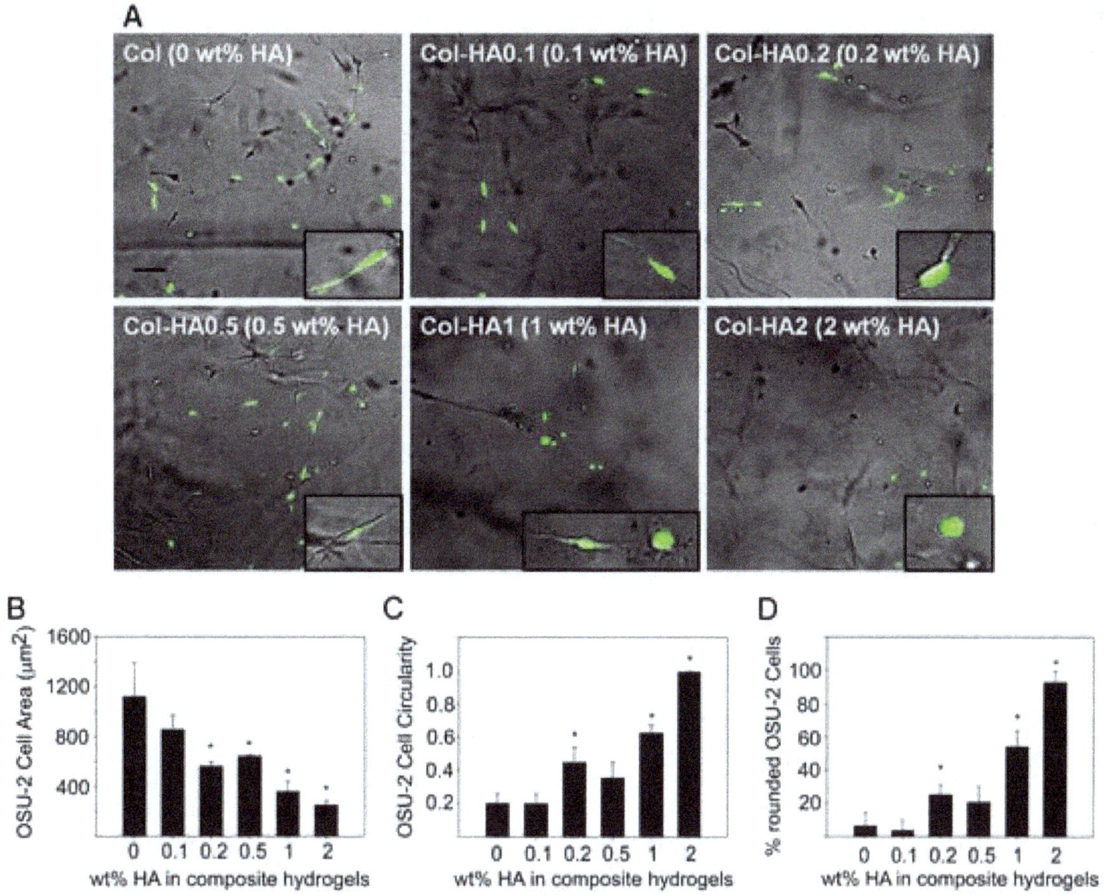

Figure 4. Glioblastoma cell morphologies and migration when cultured on different substrates. OSU-2 glioma cells were cultured on collagen I and III and HA composite hydrogels with different concentrations of HA. Cell morphologies shown in (A) were quantified via cell area (B), circularity (C), and roundness (D). Increased HA content led to lower cell area but increased cell circularity and roundness. The scale bar in (A) indicates 100 μm, * in (B–D) represents a p-value < 0.05 compared to 0% HA condition. Figure reprinted/adapted with permission from Ref. [171]. Copyright 2013, American Chemical Society.

Hydrogels using natural biomaterials are also advantageous, since they allow for the design of an in vitro extracellular environment that can biologically mimic the in vivo ECM and various cellular conditions. Natural hydrogels have been used for tissue engineering and regenerative medicine because they form structures similar to tissue ECM due to the natural proteins and polysaccharides that they are composed of [168]. While the biological similarity is closer to in vivo when natural biomaterials are used in hydrogel preparation, the mechanical properties are not adjustable or robust compared to those of synthetic hydrogels. Additionally, natural hydrogels can lead to great variation from batch to batch due to the components being sourced naturally [168,174]. Recently, however, 'smart' hydrogels, which are robust, have localized mechanical properties, and are responsive to changes in pH, temperature, and light have been developed from natural materials [148]. Smart hydrogels composed of natural materials have many possible applications in tissue engineering, regenerative medicine, and stem cell and cancer research.

2.1.2. Synthetic Hydrogels

Synthetic hydrogels are composed of synthetic materials such as polyethylene glycol (PEG), poly(vinyl alcohol), and poly-2-hydroxy ethyl methacrylate [175,176]. They offer advantages to natural hydrogels in the sense that they offer greater control over gel mechanical properties, have higher capacity for water absorption, do not degrade as rapidly as biomaterial hydrogels, and have great reproducibility [177]. For instance, PEG-based hydrogels are useful in mechanics studies due to stiffness tunability and ability to support long culture periods (several weeks) [178]. Synthetic hydrogels have more utility in studies focused on the effect of isolated mechanical properties on cell behavior without the additional influences of interactions between cells and biochemically active biomaterials [179]. Biomimetic polymers have also been synthesized which may contain a similar component to a natural material, such as a specific amino-acid sequence, which can add biological activity such as degradability or biochemical signaling to the gel in a more controlled environment than if a natural biomaterial were used [180]. Therefore, synthetic hydrogels are extremely customizable and can be synthesized for specific medical or research applications.

Synthetic hydrogels have been used to study the mechanics of cancer, as reported in several previous publications. For instance, 2D "films" of poly-methylphenyl siloxane with increasing stiffness values resulted in increased cell spreading and migration of glioblastoma cells compared to more compliant films [181]. In developing an in vitro drug-screening platform for cancer, synthetic HA derivatives, HA-aldehyde and HA-hydrazide, were crosslinked and formed into a hydrogel where cells were able to form clustered structures similar to tumors, and had greater drug resistance than in 2D cultures [182]. PEG hydrogels are often used for hydrogel preparation due to their tunable properties and high biocompatibility compared to other synthetic materials. They were used in mechanical studies of lung adenocarcinoma where matrix stiffness alterations resulted in changes in cellular morphology [183].

While there are several advantages to using synthetic hydrogels, they are limiting in their ability to mimic the complexity of native ECM. This highlights the importance of efforts to add biochemical reactivity to synthetic hydrogels for in vitro studies that are aiming to mimic cellular microenvironments [184]. For instance, Lutolf et al. engineered synthetic hydrogels that were degradable by matrix metalloproteinases through crosslinking of synthetic substrates into the hydrogels [164]. Smart synthetic hydrogels have also been studied to engineer stimuli-responsive synthetic biomaterials that change when faced with altered temperature, pH, light, and other stimuli [185].

2.1.3. Hybrid Hydrogels

Many hydrogels are composed of synthetic materials that are mixed, conjugated, or coated with biomaterials to provide researchers with insight into the controlled mechanical response and specific cell-protein interactions while maintaining the ability to adjust mechanical properties, and keep a low rate of degradation [186]. In general, a limiting property of hydrogels is their low stiffness and rigidity, which is the opposite of in vivo tissue properties. Hybrid hydrogels have been of interest in studies to improve both the stiffness and rigidity of hydrogels [177,187].

Many types of hybrid hydrogels have been developed for cancer mechanics studies. Fibronectin-coated PA substrates were used to study the invasiveness of different human glioma cell lines with a focus on cell structure, migration, and proliferation [188]. Cells were rounder and had lower migration and proliferation rates in ECM substrates with lower rigidity. GBM migration patterns in the brain white matter tract were mimicked in vitro with electrospun alignment of nanofibers in a mixture of gelatin, poly-ethersulfone, poly-dimethylsiloxane, HA, and collagen in order to study migration patterns of GBM cells, and the addition of HA was seen to have a converse effect on migration [189]. The addition of HA to gelatin and PEG composite hydrogels resulted in dose-dependent glioma malignancy marker expression changes and cell clustering [190]. Another study combined PA hydrogels with HA and either laminin or collagen I and found that collagen and laminin presence was correlated to mechanical response to substrate stiffness [134].

2.2. General Properties of Hydrogels

Hydrogels are defined by their hydrophilicity, which allows them to store water and swell without dissolving [191]. The bio-responsive properties of hydrogels, such as biochemical activity and degradability, allow for culture conditions to be dynamic and receptive to cellular cues. Conversely, since hydrogels are tunable in mechanical properties such as elasticity, compliance, and stiffness, it is possible to study the responses of cells to microenvironmental mechanics. While synthetic hydrogels are easier to adjust mechanically, natural hydrogels are more bioresponsive. In crafting a study, these are important properties to keep in mind, as they introduce variables to whichever system is being studied.

Two important descriptors of hydrogels that determine many physical properties are ionization degree and crosslink density. Crosslinks in a hydrogel are either chemical or physical, and can be introduced to a gel by methods such as irradiation, sulfur vulcanization, or chemical reactions aided by temperature and pressure [144]. Swelling and elastic modulus values are determined by cross-linking degree and charge densities or ionic strength of the polymers in the hydrogel [192]. Greater concentration of cross-linked polymers and the number of ionic groups cause higher elastic modulus values and greater swelling capacity [193]. Similarly, the distribution of proteins or polymers and cross-links in hydrogels, which are generally non-homogenous, are affected by cross-link density and degree of ionization [194,195].

There are numerous properties of hydrogels that can be influenced by internal factors, e.g., composition, protein concentration, and polymer modifications, and by external factors, e.g., UV radiation and temperature [161]. Chemical and physical reproducibility of hydrogels depends on controlled conditions and utility in research can be greatly enhanced by proper knowledge of hydrogel properties and subsequent unique modifications of the scaffold to better mimic whichever system is being studied [196].

2.3. Research Applications of Hydrogels

Research applications of hydrogels are varied in vitro. They are often used as scaffolds in 3D cultures for the study of cellular physiology and characteristics, as they interact with tissue-mimetic hydrogels. While 2D culturing methods have provided valuable information on cellular characteristics, recent studies have undertaken a shift toward preference for 3D-culturing methods for more advanced and sensitive studies that imitate native tissues and cells more closely [126,197]. These 3D platforms provide more physiologically relevant information on cell-environment interactions biochemically and mechanically, including: morphology, cell and environmental stiffness, motility, and signaling [117]. Some 3D platforms that have been used for cell culture include microporous and nanofibrous scaffolds encapsulating cells. However, these either have pore sizes that are too large, which negate the 3D structure and act as a 2D scaffold, or are too weak for mechanical studies [198,199]. Hydrogels are useful as mechanically and biochemically tunable matrices for 3D cultures, which simulate soft tissue structure and have potential for translation and clinical applications [184]. For instance, Jiang et al. found that the formation of a hybrid hydrogel with ultralong hydroxyapatite nanowires and sodium alginate allows for improved mechanical properties of the hydrogels and enhanced biocompatibility for in vitro studies [200].

Drug screening or efficacy assays are prone to showing promising effects in vitro, only to fail or be substantially less effective in animal models and clinical trials [201]. In the search for in vitro cell-culture systems that can provide more accurate and relevant results, 3D scaffolds such as hydrogels have been gaining attention, in part due to their potential to mimic the ECM and inhibit drug delivery [119]. Huber et al. compared the response of 2D cultures of non-small cell lung cancer cultured to 3D microtissues of the same cell line to various drugs and found that drug efficacy was significantly different between the models [127]. Singh et al. developed a hydrogel microarray assay to generate uniform microtumors and subsequently study tumor response to epidermal growth factor (EGF) and cetuximab treatments [202]. Such systems increase the chances of success for translation because they enable studies of treatment response more closely aligned to clinical response.

2.4. Clinical Applications of Hydrogels

A clinical approach to hydrogel scaffolds is found tissue engineering applications, which have gained traction recently as potential solutions to donor shortage problems for tissue or organ transplantations [203,204]. These approaches generally combine cells from a donor with a hydrogel scaffold that is prepared to mimic the extracellular matrix of the tissue or organ being engineered [203]. Hydrogels are suitable tissue-engineering applications due to their ability to uptake water, to encapsulate cells, and to be bio-reactive [8]. Various hydrogels have been used as bio-ink for 3D organ or tissue printing applications in which tissues are built with direct deposition of cells with the bio-ink [205]. Studies have shown applicability of alginate, collagen, and various composite hydrogels for tissue printing methods [206,207,208]. For instance, natural protein and polysaccharide hydrogels have been used for articular cartilage-tissue engineering applications to promote cartilage regeneration [209]. Latifi and colleagues demonstrated the potential of an injectable hybrid hydrogel (collagen I and III) to be applied in soft-tissue engineering, specifically human vocal fold engineering [210]. A long-standing problem of the tissue engineering field is the need for in vitro tissue vascularization to be able to transplant or implant larger portions of tissue into patients. A couple of groups have recently made strides in promoting in vitro vascularization of hydrogel constructs of bone [211] and soft tissue [212].

Injectable hydrogels have been studied for drug delivery and wound healing or dressing applications [213]. Hydrogels have been applied in drug delivery applications due to their ability to give control over the time and/or site of drug delivery for enhanced treatment [214,215]. The biocompatibility, similarity to native tissues, and high-water content all contribute to great applicability of hydrogels for controlled drug release and delivery [216]. Naturally derived injectable hydrogels for the controlled delivery of small molecules to the central nervous system have also been extensively studied. Wang et al. have shown that an injectable hyaluronan-methylcellulose hydrogel enhanced delivery of growth factors [217]. Further, the great tunability of hydrogels has led to development of thermosensitive, pH-sensitive, and temperature-sensitive hydrogels that can be used for drug delivery in distinct biological environments [218,219,220]. Similarly, temperature-sensitive hydrogels have been developed for various wound-healing applications. A PEG-PLGA-PEG composite hydrogel has been developed for delivery of a growth factor linked to tissue repair for diabetic wound healing [221].

Tissue regeneration is another field in which hydrogels are being investigated and have shown to be promising for translation to the clinic due to their biocompatibility and ability to be fine-tuned or adapted for specific applications. For instance, Zheng et al. developed a polyacrylic, acid-alginate-demineralized, bone matrix hybrid double-network hydrogel, which was shown to promote vascular endothelial growth factor (VEGF) synthesis and basic fibroblast growth factor (bFGF) and alkaline phosphatase activity of MG63 osteosarcoma cells to enhance bone regeneration [211]. Stem-cell therapy is a promising solution for injuries that require tissue regeneration, but it is limiting since uncontrolled differentiation can lead to the presence of unnecessary cells at the site of injury and lead to stem cell metastasis and tumorigenesis [43,60,129]. Application of hydrogels can improve stem cell therapy by introducing stem cells and differentiation factors to the injury in a site-specific manner [5]. In a study of spinal cord injury, Mothe et al. investigated a hydrogel-integrated stem cell therapy by encapsulating neural stem cells and differentiation factors in a hyaluronan-based hybrid hydrogel and found that treatment enhanced graft survival, increased oligodendrocytic differentiation, and reduced cavitation in the injury site in rats [222].

Contact lenses are an example of the clinical application of hydrogels [223,224]. While hard contact lenses are composed of hydrophobic materials, soft contact lenses are hydrogel-based [225]. Owing to the wide variety in hydrogel-forming substances, attempts are continually being made to improve the physical and chemical properties of contact lenses [225]. Hydrogels have also been used clinically in dermatology applications such as wound healing and skin regeneration [226,227].

3. CANCER AND THE TUMOR MICROENVIRONMENT

Biomechanical properties of the tumor microenvironment have been shown to be altered compared to the healthy state in many types of cancer to promote processes crucial to tumorigenesis, including cellular proliferation and migration. Since changes in extracellular mechanical properties can induce structural reorganization, morphological changes, and altered signaling, they can cause cancer cells to exhibit mechanical properties differently than healthy cells, e.g., stiffness [228]. This, in turn, can further promote invasive or metastatic phenotypes [229]. For instance, cancer cells also usually have a lower Young's modulus compared to healthy cells of the same type, which can influence deformability and influence migratory ability [230]. Cancer cells also have a more robust ability to respond to ECM conditions, and can alter cytoplasm viscoelasticity in response to increased ECM stiffness and collagen I deposition [231].

Breast cancer is a context in which mechanical properties of tissue and cells have been well described, and many biomechanical contributions to carcinogenesis and metastasis have been identified [96]. Many of the studies involving cell response to mechanical stimuli have been conducted with aggressive breast-cancer cells, especially the MDA-MB-231 cell line, due to their robust response to changes in extracellular mechanics. Clinically, tissue stiffness has served as an indicator of breast tumors and risk of breast cancer [232,233], and the biomolecular consequences of this phenotype has been studied rigorously in many types of 3D hydrogel models in the laboratory, especially in the last decade. Various mechanical stimuli can cause cellular stress and lead to carcinogenesis or increased invasiveness of cancer cells. For instance, several mechanotransduction pathways are linked to carcinogenesis and invasiveness and upregulated in cancer [95,100,234,235]. Application of mechanical load was shown to regulate breast-cancer cell proliferation independent of matrix deformations or stiffness [234]. Mechanical stretch, ECM stiffness, and fluid shear-stress all led to more invasive phenotypes of breast-cancer cells. At the level of response to treatment, ECM stiffness has been characterized to contribute to chemoresistance of breast-cancer cells to doxorubicin [235]. Therefore, a potential therapeutic approach may be to introduce proteinases or drugs that reduce ECM stiffness by degrading certain components to reduce the number of treatment-resistant cells. Lastly, certain microenvironmental properties have been correlated to improved prognosis in breast cancer and can be used to identify diagnostic and prognostic signatures. For instance, the tumor-associated collagen signature consisting of aligned collagen fibers in biopsied tissues from breast-cancer patients has been identified as a prognostic signature for survival [236].

The positive contributions that studying biomechanics has made to our overall understanding of breast cancer highlight the importance of incorporating a biomechanics-based approach to cancer biology in vitro studies. Inspired by the benefits of biomechanics studies on our understanding of breast cancer, we will present the current research in glioblastoma (GBM), which is a disease that has not been well-defined biomechanically. We believe that a better understanding of the biomechanical properties of GBM and its microenvironment can produce translatable results that may contribute to the development of diagnostic and therapeutic approaches to improve the prognosis of this invasive cancer.

3.1. Glioblastoma and the Tumor Microenvironment

Biomechanical and biophysical studies can help to glean valuable insight into a wide variety of diseases, since biomechanics is an integral part of cell proliferative, migration, and survival signaling, all of which are crucial to carcinogenesis and tumorigenesis. GBM is the most common and lethal form of primary brain cancer in adults [237], but there are only two clinically approved chemotherapies targeting it. It is an example of a disease that has not been traditionally studied in terms of biomechanics until approximately the last ten years but is one that may greatly benefit from such studies. Here, we will present an overview of the biomechanical properties known about GBM, the methods and contributions of available 3D culture and mechanotransduction studies of GBM, and the potential translational impact of those studies on the clinic and patient survival.

3.1.1. The Blood-Brain Barrier

The blood-brain barrier (BBB) is also an essential part of the brain microenvironment that is altered in the cancer state. The BBB is formed by vascular endothelial cells lining microvessels in the brain and is essential in regulating brain extracellular conditions to ensure neuronal signaling [238]. The endothelial cells of the BBB limit transcellular and paracellular transit into, and thereby protect, the brain by regulating permeability through tight junctions, adherens junctions, charged moieties, pericytes, etc. [239,240]. In tumor microvessels, however, loss of claudin-1 and claudin-3 and down-regulation of claudin-5 was observed, which correlated to increased permeability [241,242].

3.1.2. Extracellular Matrix of the Brain

The ECM of the brain (Figure 5) is altered when tumorigenesis occurs. Several ECM components, including HA, tenascin-C, and vitronectin, are upregulated in the tumor microenvironment [243]. Studies show that proteins of the tumor niche also tend to be different than healthy brain tissue and that tumor invasion alters ECM composition [244]. Basement membrane components such as laminin, fibronectin, and collagen type IV are more highly secreted by glioma cells and, in turn, alter composition of the local ECM [245,246]. Tumor-associated mesenchymal stem-like cells induce HA synthase 2 activity and lead to greater HA abundance in the tumor niche [247]. In addition to a distinct microenvironment, the GBM tumor has a hypoxic and necrotic core that aids in the cancer cell-induced blood vessel formation by increasing expression of pro-angiogenic VEGF, VEGFR2, and angiopoietin 2, which results in the disorganized network of blood vessels observed in GBM [243,248,249,250].

Many cellular properties are altered in the cancer state. For instance, cell proliferation, migration, and deformability is increased in the GBM state when compared to healthy cells [67,138,251,252,253,254]. The impact of the extracellular niche on these cellular properties and the biology behind the changes have been elucidated by various in vitro and in vivo studies focusing on mechanical cell-ECM interactions [17,101,255,256,257].

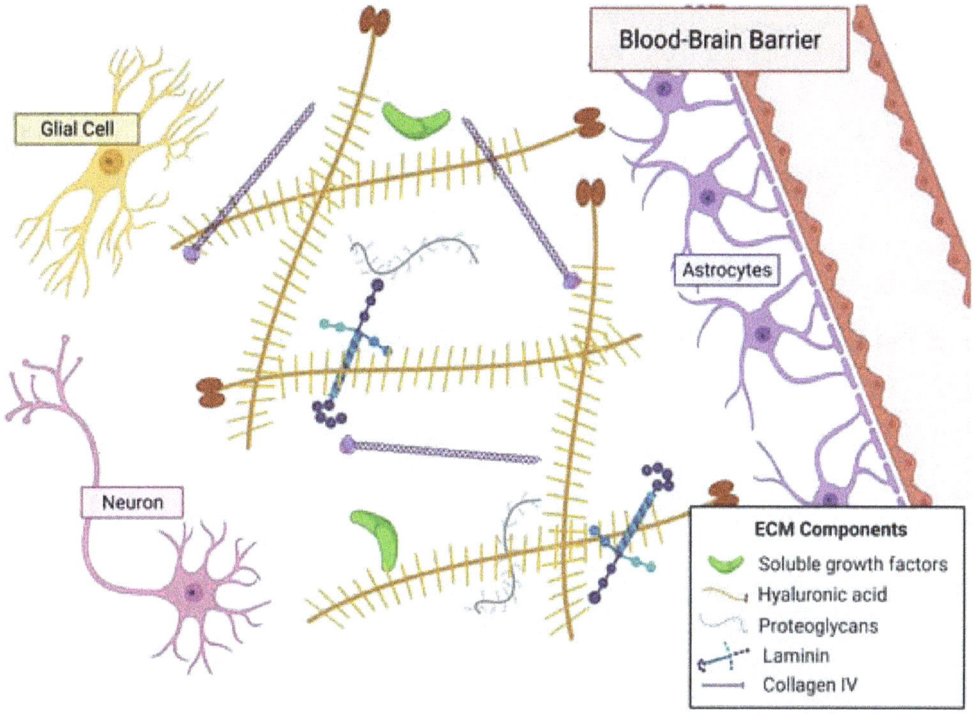

Figure 5. Extracellular matrix (ECM) of the brain. The ECM of the brain has a unique composition, including hyaluronic acid, collagen IV, and other ECM components along with glial cells, neurons, and astrocytes. The blood-brain barrier (BBB) is a neurovascular unit composed of vascular endothelial cells with surface charge modifications, tight junction proteins, pericytes, astrocytes, and other components. The BBB is selectively permeable and can block solutes in the systemic blood from entering the environment of the central nervous system (Created with BioRender.com, accessed on 14 July 2022).

3.1.3. Overview of Microenvironment and Biomechanics of Glioblastoma

A large body of research has been dedicated to studying the differences between the microenvironments, or niche, of tumor and healthy cells. These include studies on overall tissue stiffness, ECM composition, cellular signaling, and the presence/activation of mechanotransducers. There is a general consensus that while cancer cells are less stiff and more deformable than healthy cells [28,105,106,255,256], tumor tissue tends to be stiffer by variable magnitude compared to non-tumor tissues, a trend that has been shown in thyroid, breast, prostate, bladder, and kidney tissues [107,108,109]. The tumor niche is also characterized by altered ECM composition, which may lead to increased invasiveness and metastatic properties of cancer cells [257]. Characterization of the ECM, both biologically and mechanically, in GBM has not reached the depth of understanding that has been achieved for several other types of cancer, such as prostate and breast cancer. However, there are certain differences that have been noted between healthy and cancerous tissue microenvironments, which can lay the groundwork for future studies of GBM microenvironment and mechanics.

Biomechanics of the Glioblastoma Extracellular Matrix

The mechanical properties of the tumor niche in GBM are different than those of the healthy brain ECM due to the altered composition, protein-protein, cell-protein, and cell-cell interactions. Even within the GBM tumor, there are distinct mechanical regions for necrotic and non-necrotic portions [250]. In general, GBM tissue has been found to be stiffer than healthy ECM, with increasing stiffness generally correlating to increased malignancy [109,147]. From a set of human brain biopsies, one study showed that increasing malignancy of tumors gave higher Young's modulus values, with primary GBMs exhibiting stiffness values varying from 70 to 13,500 Pa [258]. Stewart et al. showed that brain tumors had elastic moduli ranging from 170 to 16,060 Pa using a custom-built indenter [259]. Another study correlated increased ECM stiffness to decreased survival of human patients [260]. Altered stiffness results in changed mechanical cues that are relayed to the cell, which then impact cellular gene expression so significantly that the overall behavior of the cell can be drastically changed [9] (Table 1). Several studies have been published that note differences in cellular morphology, deformability, motility, proliferation, and signaling in response to changes in environmental stiffness [104,134].

Modification of the Extracellular Matrix in the Brain by Glioblastoma Cells

The tumor also modifies the local ECM through protein degradation. Some of the most studied cell-secreted proteins to be up-regulated in the cancer state are matrix metalloproteinases (MMPs), which are proteases that remodel the ECM by degrading certain component proteins [261]. The hypoxic core of GBMs has been shown to be a significant contributor to the increased MMP activity [262]. Various MMPs that are up-regulated in GBMs have also been shown to aid in glioma cell invasion [263]. Plasminogen activation by the urokinase pathway, which includes urokinase (uPA), urokinase receptor (uPAR), and plasminogen, is also prominent in GBMs. The urokinase pathway aids in ECM degradation and remodeling by converting plasminogen into active plasmin, which is a serine kinase that degrades certain ECM proteins, and activates MMP-2 and MMP-9 [264]. MMP-2 is a type IV collagenase that has been implicated in invasion and metastasis of GBMs [263]. Lastly, various cathepsins, which are lysosomal cysteine proteases that can be secreted into the ECM, are also up-regulated in GBMs and have been linked to tumorigenesis and invasion [265,266,267].

3.1.4. Glioblastoma Migration, Invasion, and Mechanotransduction

Some hallmarks of GBM are enhanced cellular migration and aggressive invasiveness [268]. These properties are achieved through complex mechano-chemical signaling mechanisms that enable crosstalk between the tumor cells and the tumor microenvironment. The proteins that sense and translate mechanical cues from the ECM or microenvironment and relay them to the cell are called mechanotransducers [54]. Interestingly, one study found that keeping the mechanotransducer RhoA GTPase constitutively active in vitro in U87 cells caused similar toxic responses in 3D environments with varying stiffness, indicating the importance of mechanotransduction in cell response to environmental conditions [133]. Another study found that by altering mechanotransducers in GBM tumor-initiating cells, they were able to alter cell motility and invasion in 3D cultures [67]. Knock-down of CD44, a transmembrane glycoprotein receptor for HA and other ECM components, resulted in decreased microtubule and vimentin expression, hampered migration, and decreased nuclear stiffness compared to control cells [104]. Integrins are well-studied mechanotransducers that are linked to malignancy and are primary communicators in cell-ECM adhesion and signal transduction [134,245,263]. Clinical nuclear medicine studies have evaluated the role of various positron emission tomography (PET) radiotracers (e.g., [18F]Galacto-RGD [269], [68Ga]PRGD2 [270], and [18F]FPPRGD2 [271]) to enable molecular imaging of integrin avb3, a member a class of adhesion molecules that mediate cell–cell and cell-ECM interactions, and which plays an important role in cancer metastasis and angiogenesis [272]. These studies reveal the importance of understanding how mechanotransducers process mechanical cues from the ECM to influence cancer-cell properties [93,94].

Cellular structure and cytoskeletal alterations are under the direct influence of ECM mechanical cues and cellular mechanotransducers [93]. Pathak and Kumar found that culturing cells in narrow versus wide channels of various extracellular stiffnesses led to altered cell morphology, migration, and myosin alignment (Figure 6), underlining the importance of extracellular culture conditions in determining cell behavior. Various integrins serve to form attachments to proteins of the ECM. Once an integrin is bound to an extracellular ligand, focal adhesion clusters form at the surface of the integrin receptor within the cell, which link to the cytoskeleton and function in cell motility by recruiting proteases for ECM degradation and activating signaling pathways that induce cytoskeletal rearrangement [55]. Cytoskeletal rearrangement as orchestrated by cell-ECM interactions are vital to increasing cell deformability and, in turn, enhancing migration.

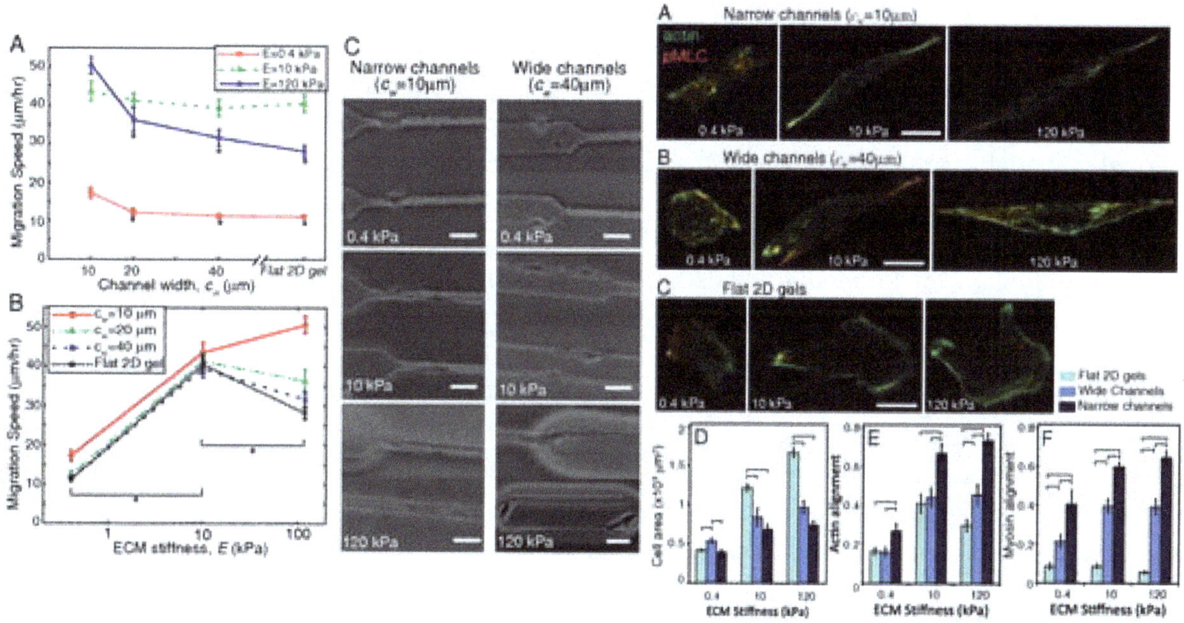

Figure 6. Cell migration, cell area, actin alignment, and myosin alignment when exposed to different channel widths and hydrogel stiffnesses. (A–C, left panel) Migration speed was quantified for cells cultured in narrow or wide channels with varying stiffness. With increased stiffness but decreased channel width, cell migration speed was higher. * $p < 0.05$. (A–D, right panel) Cell area, actin alignment, and myosin alignment were quantified from cells imaged after culture in wide or narrow channels with varying stiffness. Cells cultured in stiffer conditions with narrow channels exhibited increased actin and myosin alignment but lower cell spreading than those in narrow channels. Figure reprinted/ adapted with permission from Ref. [102]. Copyright 2012, National Academy of Sciences.

3.1.5. In Vitro Studies of Mechanotransduction in Glioblastoma

Studies of GBM cell-ECM interactions and mechanics rely on in vitro matrix-mimetics that are composed of either biologically occurring proteins or synthetic materials. In 2D cultures, this is accomplished by coating glass or polystyrene with bio- or synthetic materials and culturing cells on top of the coating. In 3D cultures, hydrogels, defined as crosslinked polymer networks that can retain water, are composed of one or several synthetic or biomaterials to mimic the cell microenvironment. Changes in composition, component concentration, and crosslinking density lead to changes in rheological properties such as elasticity and stiffness [177]. Cell migration, morphology, cytoskeletal structure, invasiveness, and signaling are some of the properties characterized in gels of varying stiffness.

It is common practice in the study of GBM mechanics to use various naturally occurring proteins of the brain ECM to construct hydrogels in which cells are seeded. While most studies cite hydrogel formation from one ECM protein, there are several that utilize hydrogels mixed with several proteins, which are discussed in the next sections. Either protein concentration or crosslinking density is altered in the hydrogels to change Young's modulus of the gels and study the effect of elasticity or stiffness in the microenvironment on the cells. The use of biomaterials in hydrogels offers the additional variable of gel degradation by cells or over time in incubation, which adds an important dimension to understanding tumor-cell function in remodeling the ECM.

3.1.6. Hydrogel Culture Methods

Collagen

Despite collagen type I not being an abundant protein in the GBM and healthy brain ECM, it is often used in mechanics studies of these cells, due to its ability to easily form a gel. Even though collagen IV is upregulated and more abundant in the GBM microenvironment, there is a limited amount of studies in which collagen IV is used for hydrogel formation to investigate GBM mechanics [171]. There are commercially available pre-coated glass slides that make mechanics studies using collagen I simpler. In studies where glioma cells were seeded onto collagen I-coated substrates (glass or polystyrene), there was a trend of cells spreading more extensively in stiffer ECMs [76,273]. One group observed a nearly linear 80 μm extension called an "invadopodia" that stretched from one cell to another, which they hypothesized was a significant factor in cell signaling [273]. The drawback of studies using collagen-coated glass or polystyrene is that it forms a 2D culture, which does not accurately mimic the 3D environment that cells are exposed to in vivo. It was shown that GBM cells exhibited chemoresistance to sunitinib, a kinase inhibitor, in vitro in a 3D collagen-based environment, but not on plastic or collagen-coated surfaces [136]. Additionally, 2D cell migration does not require MMP activity, while 3D culture does, so MMP expression can be studied extensively in relation to mechanics only in 3D culture [274]. For greater insight into in vivo characteristics, it is more useful to employ 3D rather than 2D cultures.

Collagen fiber density or concentration is an important factor to consider in experimental design. A study confirmed that varying the collagen I concentration resulted in varying elastic modulus of hydrogels, but that the difference is not significant with collagen concentrations from 1.5 to 2.5 mg/mL [275]. They also found that a higher concentration of collagen inhibited growth of the GBM in vitro. Decreasing collagen gel stiffness was shown to increase migration distance and velocity of GBM cells [124]. Another study investigated the effects of gelation temperature on collagen I gel pore size and how this impacted glioma invasion. Gelation temperature caused variations in pore size for hydrogels composed of 1 mg/mL collagen and was a significant factor in determining speed of invasion of glioma cells with smaller pore sizes (5-12 μm) hindering cell motility [276]. A study by Hwang et al. showed that actin filamentation of migrating cells in collagen gels is dynamic and undergoes rapid changes, and produces of stress fibers and lamellipodia [277]. This study also drew attention to the migratory patterns of glioma cells, which was shown to mostly be composed of double-nucleated cells that migrated in clusters and had extensive interactions, through actin filament extensions, to collagen in the surrounding gel and nearby cells [277]. Distinct cell shapes, in terms of elongation and roundness, have been observed in different collagen gels [278]. Collagen gels were also used to study the effect of the microenvironment on gene expression and the effects of depletion or overexpression of certain genes on cell ability to retain mechano-responsiveness [278,279,280].

Three-dimensional culturing methods are being modified with novel technologies to model more than just a homogenous tumor microenvironment; they are being used to form models of interacting systems. In a study by Chonan et al., the tumor niche including ECM and blood vessels was mimicked in a microfluidic device using collagen I and human umbilical vein endothelial cells (HUVECs) [281]. Further studies with collagens are required for a more thorough understanding of the effect of these proteins on GBM cells.

Hyaluronan

There is extensive research on the production of hyaluronan (HA)-based gels for studies of GBM mechanics. HA hydrogels have been used clinically as implants for neural regeneration and reduction of scar formation [282,283]. In in vitro studies, HA gels are generally functionalized with peptides or other ECM proteins, such as laminin, RGD (Arg-Gly-Asp peptides), poly-D-lysine, and poly-L-lysine, which allow for cell adhesion to the gel, since HA alone does not attach to the cells [283,284]. In a study with short Arg-Gly-Asp (RGD) peptides incorporated to HA gels, it was noted that glioma cells' actin stress fiber assembly and cell spreading was greater in stiffer gels [147]. Various HA crosslinkers can also be used to increase gel stiffness. Divinyl sulfone, for example, was used to crosslink HA carboxyl groups to varying degrees when different concentrations were used [285]. Chitosan-HA scaffolds, when compared to 2D surfaces, increased invasiveness and chemotherapeutic resistance and were proposed as possible in vitro mimics of the tumor microenvironment for pre-clinical drug effectiveness studies [286].

Other Proteins

Proteins such as fibronectin, laminin, and gelatin on their own are not commonly used for hydrogel formation. Rather, they are more frequently used to coat glass or polystyrene for two-dimensional cell-protein interaction studies. In the formation of hydrogels, they are generally used in combination with synthetic materials. These were also discussed previously in the synthetic hydrogels section of this review.

Methacrylated gelatin (GelMA) has been used on its own as a scaffold for glioma cells in a study of gel biophysical properties and their effect on cell morphology, proliferation, motility, and gene expression. Pedron et al. observed that GelMA biophysical properties could be varied by the methacrylation degree and bulk density. Cell morphology, motility and expression of hypoxia markers (VEGF, MMP-2, MMP-9, and HIF-1) and the ECM protein fibronectin were affected by the environmental properties [287]. Ramamoorthi et al. used an alginate hydrogel with varying stiffness values and observed greater cell sensitivity to toxins in less stiff gels [133].

Composite Biomaterial Hydrogels

There is a paucity of research involving composite hydrogels in the study of GBM mechanics. Researchers developed a composite matrix of HA and collagen oligomers with the addition of Matrigel-coated microfibers with tunable stiffness by varying component protein concentrations [124]. While low stiffness correlated to lower migration velocity and distance, collagen source and concentration was shown to affect these parameters variably. They also reported that in the HA hydrogel, cells exhibited collective migration while in collagen, they relied on single-cell migration [124]. These studies outline some of the different cellular responses to each of the ECM components and serve to emphasize the importance of studying specific cell-ECM interactions when attempting to describe GBM mechanics.

Synthetic Hydrogels

There are several types of synthetic polymers that have been used in the study of GBM mechanics. In 2D culture, "poly-methylphenyl-siloxane film" with greater stiffness values resulted in increased cell spreading and migration compared to more compliant films [181]. Wan et al. developed nanotextured polydimethylsiloxane (PDMS) surfaces with aptamers overexpressing epidermal growth factor receptor (EGFR) to isolate human GBM cells from a mix with fibroblast cells [288]. PA hydrogels are commonly used in GBM mechanics studies. In a novel study of cell mechano-sensitivity to matrix confinement, cells that were seeded onto PA hydrogels in more narrow channels exhibited greater migration speeds when compared to cells seeded on PA hydrogels in wider channels or directly onto 2D surfaces. This effect was abrogated by the inhibition of non-muscle myosin II, implicating this protein as a mechanosensory [102]. In a study investigating patient-derived primary GBM cell sensitivity to gel stiffness or rigidity, cells were seeded onto PA gels with different rigidity measures and the migration rate was correlated to rigidity-sensitivity [173]. Human glioma cell lines U373 and U118 cultured on polyacrylamide gels of normal brain stiffness (1 kPa) and GBM tumor stiffness (12 kPa), had greater proliferation rates in the stiffer substrate [289]. Umesh et al. studied the effect of PA gel stiffness on expression of proteins related to the cell cycle and dependency on EGFR signaling in human GBM cells, and saw that increased stiffness caused increased expression and phosphorylation of EGFR and Akt. Conversely, loss of EGFR, Akt, or phosphoinositide 3-kinase (PI3K) function resulted in decreased stiffness-sensitivity of the cells [252].

4. CHALLENGES AND FUTURE DIRECTIONS

We have endeavored to provide a thorough explanation of the various hydrogel-based methods that are employed for biomechanical studies and an overview of how studying cell-ECM interactions can lead to significant advancements in understanding the pathology of cancers, with a focus on GBM. It is abundantly clear from the state of the field that cellular response to biomechanical cues is a key player in maintaining health and homeostasis. To incorporate biomechanical effects into our current understanding of cell biology, hydrogel-based three-dimensional models of cell culture must become more widely used. The number of publications mentioning hydrogels has increased exponentially over the past 25 years (Figure 7). Furthermore, publications mentioning both hydrogels and cancer are making up a greater proportion of the hydrogel-related manuscripts published each year since 2012, indicating a growing interest in the development and use of hydrogels in the study of cancer. However, a major drawback of these approaches is that any one hydrogel model is not universally applicable due to the unique ECM composition and mechanical properties of the suite of cellular states that may be studied. Additionally, differences in hydrogel composition (and lot-to-lot variability of commercially available hydrogels), cell passage number, cell seeding density, hydrogel crosslinking density, time between cell seeding and microscopy, and other factors make it difficult to obtain consistent results [290,291]. It is worthy of note that ECM composition still has not been characterized based on the abundance of each component and elastic modulus for many tissues and conditions. Protocols unique to the disease model (cell type, microenvironment, perfusion, etc.) need to be established to enable wider adoption of hydrogel-based cell culture studies as an improved biomimetic replacement for conventional two-dimensional cell culture.

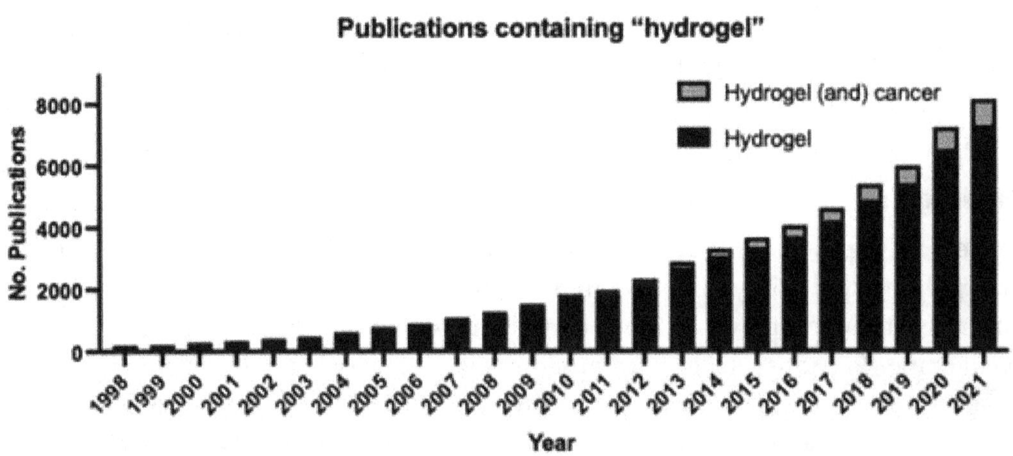

Figure 7. A bar chart showing the number of publications containing the key word "hydrogel" or ("hydrogel" AND "cancer") from the years 1998 to 2021 found in the PubMed database.

This paper examines the utilization of hydrogels as model systems to investigate how cells respond to biomechanical stimuli. The specific focus is on glioblastoma (GBM) as a case study, highlighting the progress made in this field and the remaining gaps in understanding. Breast cancer serves as a prime example of how biomechanical research can significantly influence the methods used for diagnosing, predicting the outcome of, and treating a disease. Hence, we assert that further investigations into the cellular response and adaptation to mechanical stimuli, as well as the potential dysregulation of these responses in different pathologies, are crucial for developing a comprehensive understanding of diseases that have proven challenging to manage and eradicate, such as GBM. Enhanced comprehension of the mechanics of tissues and cells would aid in the advancement of mechanotherapies for regenerative rehabilitation [292,293,294,295]. These research will not only enhance our comprehension of the disease, but they may also offer insights for utilizing mechanical stimuli for therapeutic purposes. Hydrogels are utilized in contact lenses and as carriers for drug administration. They also have the capacity to serve as therapeutic agents in situations where mechanical stimuli, such as stiffness or stress, can impact the resistance to treatment in specific types of cancer. For instance, hydrogels possessing low rigidity can be inserted into patients with invasive tumors following tumor removal to diminish the capacity of cancer cells to move. The utilization of hydrogels in novel ways and the application of insights obtained via biomechanical investigations might enhance our comprehension and management of cancer.

AT A GLANCE

Hydrogels are polymers that are compatible with living organisms and may be adjusted to suit the specific conditions of the study. As a result, they have extensive applications in the fields of medicine, bioprinting, tissue engineering, and biomechanics. Hydrogels are employed to replicate the intricate three-dimensional milieu of tissues, a crucial aspect in comprehending cell-cell interactions and intracellular signaling pathways such as proliferation, apoptosis, growth, and survival. Recent findings indicate that the harmful characteristics of cancer cells are influenced by mechanical signals that result from alterations in their surrounding microenvironment. The mechanobiological signals that influence cancer proliferation and invasion include stiffness, shear stress, and pressure. These hydrogels can be adjusted to replicate these mechanobiological characteristics of tissues. Despite a growing interest and research focus on the biomedical uses of hydrogels over the last 25 years, there is still a significant amount of knowledge to be gained regarding the creation of biomimetic hydrogels and their potential applications in biomedical and clinical environments. This review examines the use of hydrogels in creating pre-clinical cancer models and their potential for being applied to human disease. The focus is on evaluating the effectiveness of these models in investigating the progression of glioblastoma.

REFERENCES

1. Shakiba, D.; Babaei, B.; Saadat, F.; Thomopoulos, S.; Genin, G.M. The Fibrous Cellular Microenvironment, and How Cells Make Sense of a Tangled Web. Proc. Natl. Acad. Sci. USA 2017, 114, 5772–5774.

2. Lodish, H.; Berk, A.; Zipursky, S.L.; Matsudaira, P.; Baltimore, D.; Darnell, J. Noncollagen Components of the Extracellular Matrix. Molecular Cell Biology. 4th Edition. 2000. Available online: https://www.ncbi.nlm.nih.gov/books/NBK21706/ (accessed on 1 March 2022).

3. Frantz, C.; Stewart, K.M.; Weaver, V.M. The Extracellular Matrix at a Glance. J. Cell Sci. 2010, 123, 4195–4200.

4. Chaffey, N. Alberts, B., Johnson, A., Lewis, J., Raff, M., Roberts, K. and Walter, P. Molecular Biology of the Cell. 4th Edn. Ann. Bot. 2003, 91, 401.

5. Akhmanova, M.; Osidak, E.; Domogatsky, S.; Rodin, S.; Domogatskaya, A. Physical, Spatial, and Molecular Aspects of Extracellular Matrix of In Vivo Niches and Artificial Scaffolds Relevant to Stem Cells Research. Available online: https://www.hindawi.com/journals/sci/2015/167025/ (accessed on 25 May 2019).

6. Muiznieks, L.D.; Keeley, F.W. Molecular Assembly and Mechanical Properties of the Extracellular Matrix: A Fibrous Protein Perspective. Biochim. Et Biophys. Acta (BBA)-Mol. Basis Dis. 2013, 1832, 866–875.

7. Yue, B. Biology of the Extracellular Matrix: An Overview. J. Glaucoma 2014, S20–S23.

8. Kular, J.K.; Basu, S.; Sharma, R.I. The Extracellular Matrix: Structure, Composition, Age-Related Differences, Tools for Analysis and Applications for Tissue Engineering. J. Tissue Eng. 2014, 5, 2041731414557112.

9. Nelson, C.M.; Bissell, M.J. Of Extracellular Matrix, Scaffolds, and Signaling: Tissue Architecture Regulates Development, Homeostasis, and Cancer. Annu. Rev. Cell Dev. Biol. 2006, 22, 287–309.

10. Gattazzo, F.; Urciuolo, A.; Bonaldo, P. Extracellular Matrix: A Dynamic Microenvironment for Stem Cell Niche. Biochim. Biophys. Acta 2014, 1840, 2506–2519.

11. Ahmed, M.; ffrench-Constant, C. Extracellular Matrix Regulation of Stem Cell Behavior. Curr. Stem Cell Rep. 2016, 2, 197–206.

12. Kalluri, R. Angiogenesis: Basement Membranes: Structure, Assembly and Role in Tumour Angiogenesis. Nat. Rev. Cancer 2003, 3, 422–433.

13. Malik, R.; Lelkes, P.I.; Cukierman, E. Biomechanical and Biochemical Remodeling of Stromal Extracellular Matrix in Cancer. Trends Biotechnol. 2015, 33, 230–236.

14. Katz, B.-Z.; Zamir, E.; Bershadsky, A.; Kam, Z.; Yamada, K.M.; Geiger, B.; Hynes, R. Physical State of the Extracellular Matrix Regulates the Structure and Molecular Composition of Cell-Matrix Adhesions. Mol. Biol. Cell 2000, 11, 1047–1060.

15. Mui, K.L.; Chen, C.S.; Assoian, R.K. The Mechanical Regulation of Integrin–Cadherin Crosstalk Organizes Cells, Signaling and Forces. J. Cell Sci. 2016, 129, 1093–1100.

16. Parsons, J.T.; Horwitz, A.R.; Schwartz, M.A. Cell Adhesion: Integrating Cytoskeletal Dynamics and Cellular Tension. Nat. Rev. Mol. Cell Biol. 2010, 11, 633–643.

17. Latif, N.; Sarathchandra, P.; Taylor, P.M.; Antoniw, J.; Yacoub, M.H. Molecules Mediating Cell-ECM and Cell-Cell Communication in Human Heart Valves. Cell Biochem. Biophys. 2005, 43, 275–287.

18. Seong, J.; Wang, N.; Wang, Y. Mechanotransduction at Focal Adhesions: From Physiology to Cancer Development. J. Cell Mol. Med. 2013, 17, 597–604.

19. Bernfield, M.; Götte, M.; Park, P.W.; Reizes, O.; Fitzgerald, M.L.; Lincecum, J.; Zako, M. Functions of Cell Surface Heparan Sulfate Proteoglycans. Annu. Rev. Biochem. 1999, 68, 729–777.

20. Salmivirta, M.; Jalkanen, M. Syndecan Family of Cell Surface Proteoglycans: Developmentally Regulated Receptors for Extracellular Effector Molecules. Experientia 1995, 51, 863–872.

21. Mythreye, K.; Blobe, G.C. Proteoglycan Signaling Co-Receptors: Roles in Cell Adhesion, Migration and Invasion. Cell. Signal. 2009, 21, 1548–1558.

22. Lin, X. Functions of Heparan Sulfate Proteoglycans in Cell Signaling during Development. Development 2004, 131, 6009–6021.

23. Kirkbride, K.C.; Ray, B.N.; Blobe, G.C. Cell-Surface Co-Receptors: Emerging Roles in Signaling and Human Disease. Trends Biochem. Sci. 2005, 30, 611–621.

24. Orr, A.W.; Helmke, B.P.; Blackman, B.R.; Schwartz, M.A. Mechanisms of Mechanotransduction. Dev. Cell 2006, 10, 11–20.

25. Wang, N. Review of Cellular Mechanotransduction. J. Phys. D Appl. Phys. 2017, 50, 233002.

26. Bauer, J.; Emon, M.A.B.; Staudacher, J.J.; Thomas, A.L.; Zessner-Spitzenberg, J.; Mancinelli, G.; Krett, N.; Saif, M.T.; Jung, B. Increased Stiffness of the Tumor Microenvironment in Colon Cancer Stimulates Cancer Associated Fibroblast-Mediated Prometastatic Activin A Signaling. Sci. Rep. 2020, 10, 50.

27. Emon, B.; Bauer, J.; Jain, Y.; Jung, B.; Saif, T. Biophysics of Tumor Microenvironment and Cancer Metastasis-A Mini Review. Comput. Struct. Biotechnol. J. 2018, 16, 279–287.

28. Xu, W.; Mezencev, R.; Kim, B.; Wang, L.; McDonald, J.; Sulchek, T. Cell Stiffness Is a Biomarker of the Metastatic Potential of Ovarian Cancer Cells. PLoS ONE 2012, 7, e46609.

29. Hayashi, K.; Iwata, M. Stiffness of Cancer Cells Measured with an AFM Indentation Method. J. Mech. Behav. Biomed. Mater. 2015, 49, 105–111.

30. Raudenska, M.; Kratochvilova, M.; Vicar, T.; Gumulec, J.; Balvan, J.; Polanska, H.; Pribyl, J.; Masarik, M. Cisplatin Enhances Cell Stiffness and Decreases Invasiveness Rate in Prostate Cancer Cells by Actin Accumulation. Sci. Rep. 2019, 9, 1660.

31. Xiao, W.; Sohrabi, A.; Seidlits, S.K. Integrating the Glioblastoma Microenvironment into Engineered Experimental Models. Future Sci. OA 2017, 3, FSO189.

32. Pepin, K.M.; McGee, K.P.; Arani, A.; Lake, D.S.; Glaser, K.J.; Manduca, A.; Parney, I.F.; Ehman, R.L.; Huston, J. MR Elastography Analysis of Glioma Stiffness and IDH1-Mutation Status. Am. J. Neuroradiol. 2017.

33. Streitberger, K.-J.; Reiss-Zimmermann, M.; Freimann, F.B.; Bayerl, S.; Guo, J.; Arlt, F.; Wuerfel, J.; Braun, J.; Hoffmann, K.-T.; Sack, I. High-Resolution Mechanical Imaging of Glioblastoma by Multifrequency Magnetic Resonance Elastography. PLoS ONE 2014, 9, e110588.

34. Madsen, C.D.; Cox, T.R. Relative Stiffness Measurements of Tumour Tissues by Shear Rheology. Bio-Protoc. 2017, 7, e2265.

35. Deng, X.; Xiong, F.; Li, X.; Xiang, B.; Li, Z.; Wu, X.; Guo, C.; Li, X.; Li, Y.; Li, G.; et al. Application of Atomic Force Microscopy in Cancer Research. J. Nanobiotechnology 2018, 16.

36. Song, E.J.; Sohn, Y.-M.; Seo, M. Tumor Stiffness Measured by Quantitative and Qualitative Shear Wave Elastography of Breast Cancer. Br. J. Radiol. 2018, 91.

37. Bondu, V.; Wu, C.; Cao, W.; Simons, P.C.; Gillette, J.; Zhu, J.; Erb, L.; Zhang, X.F.; Buranda, T. Low-Affinity Binding in Cis to P2Y2R Mediates Force-Dependent Integrin Activation during Hantavirus Infection. Mol. Biol. Cell 2017, 28, 2887–2903.

38. Nordenfelt, P.; Elliott, H.L.; Springer, T.A. Coordinated Integrin Activation by Actin-Dependent Force during T-Cell Migration. Nat. Commun. 2016, 7, 13119.

39. Li, J.; Springer, T.A. Integrin Extension Enables Ultrasensitive Regulation by Cytoskeletal Force. Proc. Natl. Acad. Sci. USA 2017, 114, 4685–4690.

40. Shimizu, T.; Osanai, Y.; Tanaka, K.F.; Abe, M.; Natsume, R.; Sakimura, K.; Ikenaka, K. YAP Functions as a Mechanotransducer in Oligodendrocyte Morphogenesis and Maturation. Glia 2017, 65, 360–374.

41. Conway, D.E.; Schwartz, M.A. Mechanotransduction of Shear Stress Occurs through Changes in VE-Cadherin and PECAM-1 Tension: Implications for Cell Migration. Cell Adh. Migr. 2014, 9, 335–339.

42. Kale, G.R.; Yang, X.; Philippe, J.-M.; Mani, M.; Lenne, P.-F.; Lecuit, T. Distinct Contributions of Tensile and Shear Stress on E-Cadherin Levels during Morphogenesis. Nat. Commun. 2018, 9, 5021.

43. Wang, C.-H.; Cherng, W.-J.; Verma, S. Drawbacks to Stem Cell Therapy in Cardiovascular Diseases. Future Cardiol. 2008, 4, 399–408.

44. Snyder, J.L.; McBeath, E.; Thomas, T.N.; Chiu, Y.J.; Clark, R.L.; Fujiwara, K. Mechanotransduction Properties of the Cytoplasmic Tail of PECAM-1. Biol. Cell 2017, 109, 312–321.

45. Gliemann, L.; Rytter, N.; Piil, P.; Nilton, J.; Lind, T.; Nyberg, M.; Cocks, M.; Hellsten, Y. The Endothelial Mechanotransduction Protein Platelet Endothelial Cell Adhesion Molecule-1 Is Influenced by Aging and Exercise Training in Human Skeletal Muscle. Front. Physiol. 2018, 9, 1807.

46. Chachisvilis, M.; Zhang, Y.-L.; Frangos, J.A. G Protein-Coupled Receptors Sense Fluid Shear Stress in Endothelial Cells. Proc. Natl. Acad. Sci. USA 2006, 103, 15463–15468.

47. Beckmann, R.; Houben, A.; Tohidnezhad, M.; Kweider, N.; Fragoulis, A.; Wruck, C.J.; Brandenburg, L.O.; Hermanns-Sachweh, B.; Goldring, M.B.; Pufe, T.; et al. Mechanical Forces Induce Changes in VEGF and VEGFR-1/SFlt-1 Expression in Human Chondrocytes. Int. J. Mol. Sci. 2014, 15, 15456–15474.

48. Ju, L.; Chen, Y.; Xue, L.; Du, X.; Zhu, C. Cooperative Unfolding of Distinctive Mechanoreceptor Domains Transduces Force into Signals. eLife 2016, 5.

49. Wang, N.; Tytell, J.D.; Ingber, D.E. Mechanotransduction at a Distance: Mechanically Coupling the Extracellular Matrix with the Nucleus. Nat. Rev. Mol. Cell Biol. 2009, 10, 75–82.

50. Poh, Y.-C.; Na, S.; Chowdhury, F.; Ouyang, M.; Wang, Y.; Wang, N. Rapid Activation of Rac GTPase in Living Cells by Force Is Independent of Src. PLoS ONE 2009, 4, e7886.

51. Wang, S.; Stoops, E.; CP, U.; Markus, B.; Reuveny, A.; Ordan, E.; Volk, T. Mechanotransduction via the LINC Complex Regulates DNA Replication in Myonuclei. J. Cell Biol. 2018, 217, 2005–2018.

52. del Alamo, J.C.; Norwich, G.N.; Li, Y.J.; Lasheras, J.C.; Chien, S. Anisotropic Rheology and Directional Mechanotransduction in Vascular Endothelial Cells. Proc. Natl. Acad. Sci. USA 2008, 105, 15411–15416.

53. Na, S.; Collin, O.; Chowdhury, F.; Tay, B.; Ouyang, M.; Wang, Y.; Wang, N. Rapid Signal Transduction in Living Cells Is a Unique Feature of Mechanotransduction. Proc. Natl. Acad. Sci. USA 2008, 105, 6626–6631.

54. Dupont, S.; Morsut, L.; Aragona, M.; Enzo, E.; Giulitti, S.; Cordenonsi, M.; Zanconato, F.; Le Digabel, J.; Forcato, M.; Bicciato, S.; et al. Role of YAP/TAZ in Mechanotransduction. Nature 2011, 474, 179–183.

55. Provenzano, P.P.; Keely, P.J. Mechanical Signaling through the Cytoskeleton Regulates Cell Proliferation by Coordinated Focal Adhesion and Rho GTPase Signaling. J. Cell Sci. 2011, 124, 1195–1205.

56. Isenberg, B.C.; DiMilla, P.A.; Walker, M.; Kim, S.; Wong, J.Y. Vascular Smooth Muscle Cell Durotaxis Depends on Substrate Stiffness Gradient Strength. Biophys. J. 2009, 97, 1313–1322.

57. Brouzés, E.; Farge, E. Interplay of Mechanical Deformation and Patterned Gene Expression in Developing Embryos. Curr. Opin. Genet. Dev. 2004, 14, 367–374.

58. Wozniak, M.A.; Chen, C.S. Mechanotransduction in Development: A Growing Role for Contractility. Nat. Rev. Mol. Cell Biol. 2009, 10, 34–43.

59. Mammoto, T.; Ingber, D.E. Mechanical Control of Tissue and Organ Development. Development 2010, 137, 1407–1420.

60. Engler, A.J.; Sen, S.; Sweeney, H.L.; Discher, D.E. Matrix Elasticity Directs Stem Cell Lineage Specification. Cell 2006, 126, 677–689.

61. Matthews, B.D.; Overby, D.R.; Mannix, R.; Ingber, D.E. Cellular Adaptation to Mechanical Stress: Role of Integrins, Rho, Cytoskeletal Tension and Mechanosensitive Ion Channels. J. Cell Sci. 2006, 119, 508–518.

62. Yeh, Y.-C.; Ling, J.-Y.; Chen, W.-C.; Lin, H.-H.; Tang, M.-J. Mechanotransduction of Matrix Stiffness in Regulation of Focal Adhesion Size and Number: Reciprocal Regulation of Caveolin-1 and Β1 Integrin. Sci. Rep. 2017, 7, 15008.

63. Dong, Y.; Xie, X.; Wang, Z.; Hu, C.; Zheng, Q.; Wang, Y.; Chen, R.; Xue, T.; Chen, J.; Gao, D.; et al. Increasing Matrix Stiffness Upregulates Vascular Endothelial Growth Factor Expression in Hepatocellular Carcinoma Cells Mediated by Integrin Β1. Biochem. Biophys. Res. Commun. 2014, 444, 427–432.

64. van der Pijl, R.J.; Granzier, H.; Ottenheijm, C.A.C. Diaphragm Contractile Weakness Due to Altered Mechanical Loading: Role of Titin. Am. J. Physiol. Cell Physiol. 2019.

65. Burridge, K.; Wittchen, E.S. The Tension Mounts: Stress Fibers as Force-Generating Mechanotransducers. J. Cell Biol. 2013, 200, 9–19.

66. Lee, H.T.; Sharek, L.; O'Brien, E.T.; Urbina, F.L.; Gupton, S.L.; Superfine, R.; Burridge, K.; Campbell, S.L. Vinculin and Metavinculin Exhibit Distinct Effects on Focal Adhesion Properties, Cell Migration, and Mechanotransduction. PLoS ONE 2019, 14, e0221962.

67. Wong, S.Y.; Ulrich, T.A.; Deleyrolle, L.P.; MacKay, J.L.; Lin, J.-M.G.; Martuscello, R.T.; Jundi, M.A.; Reynolds, B.A.; Kumar, S. Constitutive Activation of Myosin-Dependent Contractility Sensitizes Glioma Tumor-Initiating Cells to Mechanical Inputs and Reduces Tissue Invasion. Cancer Res. 2015, 75, 1113–1122.

68. Oldenburg, J.; van der Krogt, G.; Twiss, F.; Bongaarts, A.; Habani, Y.; Slotman, J.A.; Houtsmuller, A.; Huveneers, S.; de Rooij, J. VASP, Zyxin and TES Are Tension-Dependent Members of Focal Adherens Junctions Independent of the α-Catenin-Vinculin Module. Sci. Rep. 2015, 5, 17225.

69. Mack, J.J.; Mosqueiro, T.S.; Archer, B.J.; Jones, W.M.; Sunshine, H.; Faas, G.C.; Briot, A.; Aragón, R.L.; Su, T.; Romay, M.C.; et al. NOTCH1 Is a Mechanosensor in Adult Arteries. Nat. Commun. 2017, 8, 1620.

70. Li, J.; Hou, B.; Tumova, S.; Muraki, K.; Bruns, A.; Ludlow, M.J.; Sedo, A.; Hyman, A.J.; McKeown, L.; Young, R.S.; et al. Piezo1 Integration of Vascular Architecture with Physiological Force. Nature 2014, 515, 279–282.

71. Swift, J.; Ivanovska, I.L.; Buxboim, A.; Harada, T.; Dingal, P.C.D.P.; Pinter, J.; Pajerowski, J.D.; Spinler, K.R.; Shin, J.-W.; Tewari, M.; et al. Nuclear Lamin-A Scales with Tissue Stiffness and Enhances Matrix-Directed Differentiation. Science 2013, 341.

72. Lammerding, J.; Schulze, P.C.; Takahashi, T.; Kozlov, S.; Sullivan, T.; Kamm, R.D.; Stewart, C.L.; Lee, R.T. Lamin A/C Deficiency Causes Defective Nuclear Mechanics and Mechanotransduction. J. Clin. Investig. 2004, 113, 370–378.

73. Wozniak, M.A.; Modzelewska, K.; Kwong, L.; Keely, P.J. Focal Adhesion Regulation of Cell Behavior. Biochim. Biophys. Acta 2004, 1692, 103–119.

74. Kubow, K.E.; Vukmirovic, R.; Zhe, L.; Klotzsch, E.; Smith, M.L.; Gourdon, D.; Luna, S.; Vogel, V. Mechanical Forces Regulate the Interactions of Fibronectin and Collagen I in Extracellular Matrix. Nat. Commun. 2015, 6, 8026.

75. Liu, B.; Chen, W.; Evavold, B.D.; Zhu, C. Accumulation of Dynamic Catch Bonds between TCR and Agonist Peptide-MHC Triggers T Cell Signaling. Cell 2014, 157, 357–368.

76. Sen, S.; Ng, W.P.; Kumar, S. Contributions of Talin-1 to Glioma Cell–Matrix Tensional Homeostasis. J. R. Soc. Interface 2012, 9, 1311–1317.

77. Alcaino, C.; Knutson, K.R.; Treichel, A.J.; Yildiz, G.; Strege, P.R.; Linden, D.R.; Li, J.H.; Leiter, A.B.; Szurszewski, J.H.; Farrugia, G.; et al. A Population of Gut Epithelial Enterochromaffin Cells Is Mechanosensitive and Requires Piezo2 to Convert Force into Serotonin Release. Proc. Natl. Acad. Sci. USA 2018, 115, E7632–E7641.

78. Rubashkin, M.G.; Cassereau, L.; Bainer, R.; DuFort, C.C.; Yui, Y.; Ou, G.; Paszek, M.J.; Davidson, M.W.; Chen, Y.-Y.; Weaver, V.M. Force Engages Vinculin and Promotes Tumor Progression by Enhancing PI3K Activation of Phosphatidylinositol (3,4,5)-Triphosphate. Cancer Res. 2014, 74, 4597–4611.

79. Sawada, Y.; Tamada, M.; Dubin-Thaler, B.J.; Cherniavskaya, O.; Sakai, R.; Tanaka, S.; Sheetz, M.P. Force Sensing by Mechanical Extension of the Src Family Kinase Substrate P130Cas. Cell 2006, 127, 1015–1026.

80. Julien, M.A.; Haller, C.A.; Wang, P.; Wen, J.; Chaikof, E.L. Mechanical Strain Induces a Persistent Upregulation of Syndecan-1 Expression in Smooth Muscle Cells. J. Cell. Physiol. 2007, 211, 167–173.

81. Cappelli, H.C.; Kanugula, A.K.; Adapala, R.K.; Amin, V.; Sharma, P.; Midha, P.; Paruchuri, S.; Thodeti, C.K. Mechanosensitive TRPV4 Channels Stabilize VE-Cadherin Junctions to Regulate Tumor Vascular Integrity and Metastasis. Cancer Lett. 2019, 442, 15–20.

82. Martinac, B. Mechanosensitive Ion Channels: Molecules of Mechanotransduction. J. Cell Sci. 2004, 117, 2449–2460.

83. Zhang, W.; Deng, W.; Zhou, L.; Xu, Y.; Yang, W.; Liang, X.; Wang, Y.; Kulman, J.D.; Zhang, X.F.; Li, R. Identification of a Juxtamembrane Mechanosensitive Domain in the Platelet Mechanosensor Glycoprotein Ib-IX Complex. Blood 2015, 125, 562–569.

84. Science, A.A. for the A. of PECAM-1 as Fluid Shear Stress Sensor. Sci. Signal. 2005, 2005, tw333.

85. Xu, J.; Mathur, J.; Vessières, E.; Hammack, S.; Nonomura, K.; Favre, J.; Grimaud, L.; Petrus, M.; Francisco, A.; Li, J.; et al. GPR68 Senses Flow and Is Essential for Vascular Physiology. Cell 2018, 173, 762–775.e16.

86. Cha, B.; Srinivasan, R.S. Mechanosensitive β-Catenin Signaling Regulates Lymphatic Vascular Development. BMB Rep. 2016, 49, 403–404.

87. Peng, Y.; Chen, Y.; Qin, X.; Li, S.; Liu, Y. Unveiling the Mechanotransduction Mechanism of Substrate Stiffness-Modulated Cancer Cell Motility via ROCK1 and ROCK2 Differentially Regulated Manner. FASEB J. 2019, 33, 644.4.

88. Hadden, W.J.; Young, J.L.; Holle, A.W.; McFetridge, M.L.; Kim, D.Y.; Wijesinghe, P.; Taylor-Weiner, H.; Wen, J.H.; Lee, A.R.; Bieback, K.; et al. Stem Cell Migration and Mechanotransduction on Linear Stiffness Gradient Hydrogels. Proc. Natl. Acad. Sci. USA 2017, 114, 5647–5652.

89. Wang, Y.; Xiao, B. The Mechanosensitive Piezo1 Channel: Structural Features and Molecular Bases Underlying Its Ion Permeation and Mechanotransduction. J. Physiol. 2018, 596, 969–978.

90. Zeng, Y.; Shen, Y.; Huang, X.-L.; Liu, X.-J.; Liu, X.-H. Roles of Mechanical Force and CXCR1/CXCR2 in Shear-Stress-Induced Endothelial Cell Migration. Eur. Biophys. J. 2012, 41, 13–25.

91. Shieh, A.C.; Rozansky, H.A.; Hinz, B.; Swartz, M.A. Tumor Cell Invasion Is Promoted by Interstitial Flow-Induced Matrix Priming by Stromal Fibroblasts. Cancer Res. 2011, 71, 790–800.

92. Burridge, K.; Monaghan-Benson, E.; Graham, D.M. Mechanotransduction: From the Cell Surface to the Nucleus via RhoA. Philos. Trans. R. Soc. Lond. B Biol. Sci. 2019, 374.

93. Chin, L.; Xia, Y.; Discher, D.E.; Janmey, P.A. Mechanotransduction in Cancer. Curr. Opin. Chem. Eng. 2016, 11, 77–84.

94. McKenzie, A.J.; Hicks, S.R.; Svec, K.V.; Naughton, H.; Edmunds, Z.L.; Howe, A.K. The Mechanical Microenvironment Regulates Ovarian Cancer Cell Morphology, Migration, and Spheroid Disaggregation. Sci. Rep. 2018, 8, 7228.

95. Thoppil, R.J.; Cappelli, H.C.; Adapala, R.K.; Kanugula, A.K.; Paruchuri, S.; Thodeti, C.K. TRPV4 Channels Regulate Tumor Angiogenesis via Modulation of Rho/Rho Kinase Pathway. Oncotarget 2016, 7, 25849–25861.

96. Broders-Bondon, F.; Ho-Bouldoires, T.H.N.; Fernandez-Sanchez, M.-E.; Farge, E. Mechanotransduction in Tumor Progression: The Dark Side of the Force. J. Cell Biol. 2018, 217, 1571–1587.

97. Wei, S.C.; Fattet, L.; Tsai, J.H.; Guo, Y.; Pai, V.H.; Majeski, H.E.; Chen, A.C.; Sah, R.L.; Taylor, S.S.; Engler, A.J.; et al. Matrix Stiffness Drives Epithelial-Mesenchymal Transition and Tumour Metastasis through a TWIST1-G3BP2 Mechanotransduction Pathway. Nat. Cell Biol. 2015, 17, 678–688.

98. Biggs, M.J.P.; Dalby, M.J. Focal Adhesions in Osteoneogenesis. Proc. Inst. Mech Eng. H 2010, 224, 1441–1453.

99. Dugina, V.; Fontao, L.; Chaponnier, C.; Vasiliev, J.; Gabbiani, G. Focal Adhesion Features during Myofibroblastic Differentiation Are Controlled by Intracellular and Extracellular Factors. J. Cell Sci. 2001, 114, 3285–3296.

100. Liu, Y.-S.; Liu, Y.-A.; Huang, C.-J.; Yen, M.-H.; Tseng, C.-T.; Chien, S.; Lee, O.K. Mechanosensitive TRPM7 Mediates Shear Stress and Modulates Osteogenic Differentiation of Mesenchymal Stromal Cells through Osterix Pathway. Sci. Rep. 2015, 5, 16522.

101. Jaalouk, D.E.; Lammerding, J. Mechanotransduction Gone Awry. Nat. Rev. Mol. Cell Biol. 2009, 10, 63–73.

102. Pathak, A.; Kumar, S. Independent Regulation of Tumor Cell Migration by Matrix Stiffness and Confinement. Proc. Natl. Acad. Sci. USA 2012, 109, 10334–10339.

103. Desgrosellier, J.S.; Cheresh, D.A. Integrins in Cancer: Biological Implications and Therapeutic Opportunities. Nat. Rev. Cancer 2010, 10, 9–22.

104. Maherally, Z.; Smith, J.R.; Ghoneim, M.K.; Dickson, L.; An, Q.; Fillmore, H.L.; Pilkington, G.J. Silencing of CD44 in Glioma Leads to Changes in Cytoskeletal Protein Expression and Cellular Biomechanical Deformation Properties as Measured by AFM Nanoindentation. BioNanoScience 2016, 6, 54–64.

105. Prabhune, M.; Belge, G.; Dotzauer, A.; Bullerdiek, J.; Radmacher, M. Comparison of Mechanical Properties of Normal and Malignant Thyroid Cells. Micron 2012, 43, 1267–1272.

106. Ramos, J.R.; Pabijan, J.; Garcia, R.; Lekka, M. The Softening of Human Bladder Cancer Cells Happens at an Early Stage of the Malignancy Process. Beilstein J. Nanotechnol. 2014, 5, 447–457.

107. Lekka, M.; Pabijan, J. Measuring Elastic Properties of Single Cancer Cells by AFM. Methods Mol. Biol. 2019, 1886, 315–324.

108. Plodinec, M.; Loparic, M.; Monnier, C.A.; Obermann, E.C.; Zanetti-Dallenbach, R.; Oertle, P.; Hyotyla, J.T.; Aebi, U.; Bentires-Alj, M.; Lim, R.Y.H.; et al. The Nanomechanical Signature of Breast Cancer. Nat. Nanotechnol. 2012, 7, 757–765.

109. Paszek, M.J.; Zahir, N.; Johnson, K.R.; Lakins, J.N.; Rozenberg, G.I.; Gefen, A.; Reinhart-King, C.A.; Margulies, S.S.; Dembo, M.; Boettiger, D.; et al. Tensional Homeostasis and the Malignant Phenotype. Cancer Cell 2005, 8, 241–254.

110. Nishida-Aoki, N.; Gujral, T.S. Emerging Approaches to Study Cell–Cell Interactions in Tumor Microenvironment. Oncotarget 2019, 10, 785–797.

111. Zervantonakis, I.K.; Poskus, M.D.; Scott, A.L.; Selfors, L.M.; Lin, J.-R.; Dillon, D.A.; Pathania, S.; Sorger, P.K.; Mills, G.B.; Brugge, J.S. Fibroblast–Tumor Cell Signaling Limits HER2 Kinase Therapy Response via Activation of MTOR and Antiapoptotic Pathways. Proc. Natl. Acad. Sci. USA 2020, 117, 16500–16508.

112. Mueller, M.; Rasoulinejad, S.; Garg, S.; Wegner, S.V. The Importance of Cell–Cell Interaction Dynamics in Bottom-Up Tissue Engineering: Concepts of Colloidal Self-Assembly in the Fabrication of Multicellular Architectures. Nano Lett. 2020, 20, 2257–2263.

113. Mayor, R.; Etienne-Manneville, S. The Front and Rear of Collective Cell Migration. Nat. Rev. Mol. Cell Biol. 2016, 17, 97–109.

114. Chen, C.S.; Tan, J.; Tien, J. Mechanotransduction at Cell-Matrix and Cell-Cell Contacts. Annu. Rev. Biomed. Eng. 2004, 6, 275–302.

115. Giladi, A.; Cohen, M.; Medaglia, C.; Baran, Y.; Li, B.; Zada, M.; Bost, P.; Blecher-Gonen, R.; Salame, T.-M.; Mayer, J.U.; et al. Dissecting Cellular Crosstalk by Sequencing Physically Interacting Cells. Nat. Biotechnol. 2020, 38, 629–637.

116. Melissaridou, S.; Wiechec, E.; Magan, M.; Jain, M.V.; Chung, M.K.; Farnebo, L.; Roberg, K. The Effect of 2D and 3D Cell Cultures on Treatment Response, EMT Profile and Stem Cell Features in Head and Neck Cancer. Cancer Cell Int. 2019, 19, 16.

117. Pampaloni, F.; Reynaud, E.G.; Stelzer, E.H.K. The Third Dimension Bridges the Gap between Cell Culture and Live Tissue. Nat. Rev. Mol. Cell Biol. 2007, 8, 839–845.

118. Lee, J.; Cuddihy, M.J.; Kotov, N.A. Three-Dimensional Cell Culture Matrices: State of the Art. Tissue Eng. Part. B Rev. 2008, 14, 61–86.

119. Edmondson, R.; Broglie, J.J.; Adcock, A.F.; Yang, L. Three-Dimensional Cell Culture Systems and Their Applications in Drug Discovery and Cell-Based Biosensors. Assay Drug Dev. Technol. 2014, 12, 207–218.

120. Hsieh, C.-H.; Chen, Y.-D.; Huang, S.-F.; Wang, H.-M.; Wu, M.-H. The Effect of Primary Cancer Cell Culture Models on the Results of Drug Chemosensitivity Assays: The Application of Perfusion Microbioreactor System as Cell Culture Vessel. Biomed. Res. Int. 2015, 2015, 470283.

121. Bonnier, F.; Keating, M.E.; Wróbel, T.P.; Majzner, K.; Baranska, M.; Garcia-Munoz, A.; Blanco, A.; Byrne, H.J. Cell Viability Assessment Using the Alamar Blue Assay: A Comparison of 2D and 3D Cell Culture Models. Toxicol. Vitr. 2015, 29, 124–131.

122. Duval, K.; Grover, H.; Han, L.-H.; Mou, Y.; Pegoraro, A.F.; Fredberg, J.; Chen, Z. Modeling Physiological Events in 2D vs. 3D Cell Culture. Physiology 2017, 32, 266–277.

123. Petrie, R.J.; Yamada, K.M. At the Leading Edge of Three-Dimensional Cell Migration. J. Cell Sci. 2012, 125, 5917–5926.

124. Herrera-Perez, M.; Voytik-Harbin, S.L.; Rickus, J.L. Extracellular Matrix Properties Regulate the Migratory Response of Glioblastoma Stem Cells in Three-Dimensional Culture. Tissue Eng. Part A 2015, 21, 2572–2582.

125. Petrie, R.J.; Gavara, N.; Chadwick, R.S.; Yamada, K.M. Nonpolarized Signaling Reveals Two Distinct Modes of 3D Cell Migration. J. Cell Biol. 2012, 197, 439–455.

126. Antoni, D.; Burckel, H.; Josset, E.; Noel, G. Three-Dimensional Cell Culture: A Breakthrough in Vivo. Int. J. Mol. Sci. 2015, 16, 5517–5527.

127. Huber, J.M.; Amann, A.; Koeck, S.; Lorenz, E.; Kelm, J.M.; Obexer, P.; Zwierzina, H.; Gamerith, G. Evaluation of Assays for Drug Efficacy in a Three-Dimensional Model of the Lung. J. Cancer Res. Clin. Oncol. 2016, 142, 1955–1966.

128. Das, S.; Sahan, A.Z.; Lim, E.; Suarez, K. The Potential Use of a Three Dimensional Cell Culture of Spheroids in the Study of Various Therapeutic Approaches. MOJ Immunol. 2016, 3, 11–12.

129. Barthes, J.; Özçelik, H.; Hindié, M.; Ndreu-Halili, A.; Hasan, A.; Vrana, N.E. Cell Microenvironment Engineering and Monitoring for Tissue Engineering and Regenerative Medicine: The Recent Advances. Biomed. Res. Int. 2014, 2014.

130. Gentili, C.; Cancedda, R. Cartilage and Bone Extracellular Matrix. Curr. Pharm. Des. 2009, 15, 1334–1348.

131. Liu, M.; Zeng, X.; Ma, C.; Yi, H.; Ali, Z.; Mou, X.; Li, S.; Deng, Y.; He, N. Injectable Hydrogels for Cartilage and Bone Tissue Engineering. Bone Res. 2017, 5, 17014.

132. Charrier, E.E.; Pogoda, K.; Wells, R.G.; Janmey, P.A. Control of Cell Morphology and Differentiation by Substrates with Independently Tunable Elasticity and Viscous Dissipation. Nat. Commun. 2018, 9, 449.

133. Ramamoorthi, K.; Hara, J.; Ito, C.; Asuri, P. Role of Three-Dimensional Matrix Stiffness in Regulating the Response of Human Neural Cells to Toxins. Cell. Mol. Bioeng. 2014, 7, 278–284.

134. Pogoda, K.; Bucki, R.; Byfield, F.J.; Cruz, K.; Lee, T.; Marcinkiewicz, C.; Janmey, P.A. Soft Substrates Containing Hyaluronan Mimic the Effects of Increased Stiffness on Morphology, Motility, and Proliferation of Glioma Cells. Biomacromolecules 2017, 18, 3040–3051.

135. Koch, D.; Rosoff, W.J.; Jiang, J.; Geller, H.M.; Urbach, J.S. Strength in the Periphery: Growth Cone Biomechanics and Substrate Rigidity Response in Peripheral and Central Nervous System Neurons. Biophys. J. 2012, 102, 452–460.

136. Fernandez-Fuente, G.; Mollinedo, P.; Grande, L.; Vazquez-Barquero, A.; Fernandez-Luna, J.L. Culture Dimensionality Influences the Resistance of Glioblastoma Stem-like Cells to Multikinase Inhibitors. Mol. Cancer 2014, 13, 1664–1672.

137. Wong, S.; Guo, W.-H.; Wang, Y.-L. Fibroblasts Probe Substrate Rigidity with Filopodia Extensions before Occupying an Area. Proc. Natl. Acad. Sci. USA 2014, 111, 17176–17181.

138. Rao, S.S.; Bentil, S.; DeJesus, J.; Larison, J.; Hissong, A.; Dupaix, R.; Sarkar, A.; Winter, J.O. Inherent Interfacial Mechanical Gradients in 3D Hydrogels Influence Tumor Cell Behaviors. PLoS ONE 2012, 7, e35852.

139. Stabenfeldt, S.E.; LaPlaca, M.C. Variations in Rigidity and Ligand Density Influence Neuronal Response in Methylcellulose–Laminin Hydrogels. Acta Biomater. 2011, 7, 4102–4108.

140. Pereira, R.F.; Bártolo, P.J. 3D Bioprinting of Photocrosslinkable Hydrogel Constructs. J. Appl. Polym. Sci. 2015, 132.

141. Bessonov, I.V.; Rochev, Y.A.; Arkhipova, A.Y.; Kopitsyna, M.N.; Bagrov, D.V.; Karpushkin, E.A.; Bibikova, T.N.; Moysenovich, A.M.; Soldatenko, A.S.; Nikishin, I.I.; et al. Fabrication of Hydrogel Scaffolds via Photocrosslinking of Methacrylated Silk Fibroin. Biomed. Mater. 2019, 14, 034102.

142. Seiffert, S.; Oppermann, W.; Saalwächter, K. Hydrogel Formation by Photocrosslinking of Dimethylmaleimide Functionalized Polyacrylamide. Polymer 2007, 48, 5599–5611.

143. O'Connell, C.D.; Zhang, B.; Onofrillo, C.; Duchi, S.; Blanchard, R.; Quigley, A.; Bourke, J.; Gambhir, S.; Kapsa, R.; Bella, C.D.; et al. Tailoring the Mechanical Properties of Gelatin Methacryloyl Hydrogels through Manipulation of the Photocrosslinking Conditions. Soft Matter 2018, 14, 2142–2151.

144. Maitra, J.; Shukla, V.K. Cross-Linking in Hydrogels—A Review. Am. J. Polym. Sci. 2014, 4, 25–31.

145. Baday, M.; Ercal, O.; Sahan, A.Z.; Sahan, A.; Ercal, B.; Inan, H.; Demirci, U. Density Based Characterization of Mechanical Cues on Cancer Cells Using Magnetic Levitation. Adv. Healthc. Mater. 2019, 8, 1801517.

146. Akalin, O.B.; Bayraktar, H. Alteration of Cell Motility Dynamics through Collagen Fiber Density in Photopolymerized Polyethylene Glycol Hydrogels. Int. J. Biol. Macromol. 2020, 157, 414–423.

147. Ananthanarayanan, B.; Kim, Y.; Kumar, S. Elucidating the Mechanobiology of Malignant Brain Tumors Using a Brain Matrix-Mimetic Hyaluronic Acid Hydrogel Platform. Biomaterials 2011, 32, 7913–7923.

148. Samal, S.K.; Dash, M.; Dubruel, P.; Van Vlierberghe, S. Chapter 8—Smart Polymer Hydrogels: Properties, Synthesis and Applications. In Smart Polymers and their Applications; Aguilar, M.R., San Román, J., Eds.; Woodhead Publishing: Sawston, UK, 2014; pp. 237–270. ISBN 978-0-85709-695-1.

149. Kopecek, J. Hydrogel Biomaterials: A Smart Future? Biomaterials 2007, 28, 5185–5192.

150. Liu, Z.; Faraj, Y.; Ju, X.-J.; Wang, W.; Xie, R.; Chu, L.-Y. Nanocomposite Smart Hydrogels with Improved Responsiveness and Mechanical Properties: A Mini Review. J. Polym. Sci. Part B Polym. Phys. 2018, 56, 1306–1313.

151. Lim, H.L.; Hwang, Y.; Kar, M.; Varghese, S. Smart Hydrogels as Functional Biomimetic Systems. Biomater. Sci. 2014, 2, 603–618.

152. Dzobo, K.; Motaung, K.S.C.M.; Adesida, A. Recent Trends in Decellularized Extracellular Matrix Bioinks for 3D Printing: An Updated Review. Int. J. Mol. Sci. 2019, 20, 4628.

153. Tam, R.Y.; Smith, L.J.; Shoichet, M.S. Engineering Cellular Microenvironments with Photo- and Enzymatically Responsive Hydrogels: Toward Biomimetic 3D Cell Culture Models. Acc. Chem. Res. 2017, 50, 703–713.

154. DeQuach, J.A.; Mezzano, V.; Miglani, A.; Lange, S.; Keller, G.M.; Sheikh, F.; Christman, K.L. Simple and High Yielding Method for Preparing Tissue Specific Extracellular Matrix Coatings for Cell Culture. PLoS ONE 2010, 5, e13039.

155. Scherzer, M.T.; Waigel, S.; Donninger, H.; Arumugam, V.; Zacharias, W.; Clark, G.; Siskind, L.J.; Soucy, P.; Beverly, L. Fibroblast-Derived Extracellular Matrices: An Alternative Cell Culture System That Increases Metastatic Cellular Properties. PLoS ONE 2015, 10, e0138065.

156. Lee, H.W.; Kook, Y.-M.; Lee, H.J.; Park, H.; Koh, W.-G. A Three-Dimensional Co-Culture of HepG2 Spheroids and Fibroblasts Using Double-Layered Fibrous Scaffolds Incorporated with Hydrogel Micropatterns. RSC Adv. 2014, 4, 61005–61011.

157. Hunt, N.C.; Shelton, R.M.; Grover, L. An Alginate Hydrogel Matrix for the Localised Delivery of a Fibroblast/Keratinocyte Co-Culture. Biotechnol. J. 2009, 4, 730–737.

158. Yamamoto, M.; Ikada, Y.; Tabata, Y. Controlled Release of Growth Factors Based on Biodegradation of Gelatin Hydrogel. J. Biomater. Sci. Polym. Ed. 2001, 12, 77–88.

159. Tabata, Y. Tissue Regeneration Based on Growth Factor Release. Tissue Eng. 2003, 9 (Suppl. S1), S5–S15.

160. Wang, Z.; Wang, Z.; Lu, W.W.; Zhen, W.; Yang, D.; Peng, S. Novel Biomaterial Strategies for Controlled Growth Factor Delivery for Biomedical Applications. NPG Asia Mater. 2017, 9, e435.

161. Zhu, J.; Marchant, R.E. Design Properties of Hydrogel Tissue-Engineering Scaffolds. Expert Rev. Med. Devices 2011, 8, 607–626.

162. Saha, N.; Saarai, A.; Roy, N.; Kitano, T.; Saha, P. Polymeric Biomaterial Based Hydrogels for Biomedical Applications. J. Biomater. Nanobiotechnology 2011, 2, 85.

163. Patel, A.; Mequanint, K. Hydrogel Biomaterials. Biomed. Eng.-Front. Chall. 2011.

164. Lutolf, M.P.; Lauer-Fields, J.L.; Schmoekel, H.G.; Metters, A.T.; Weber, F.E.; Fields, G.B.; Hubbell, J.A. Synthetic Matrix Metalloproteinase-Sensitive Hydrogels for the Conduction of Tissue Regeneration: Engineering Cell-Invasion Characteristics. Proc. Natl. Acad. Sci. USA 2003, 100, 5413–5418.

165. Li, J.; Illeperuma, W.R.K.; Suo, Z.; Vlassak, J.J. Hybrid Hydrogels with Extremely High Stiffness and Toughness. ACS Macro Lett. 2014, 3, 520–523.

166. Dehne, H.; Hecht, F.M.; Bausch, A.R. The Mechanical Properties of Polymer–Colloid Hybrid Hydrogels. Soft Matter 2017, 13, 4786–4790.

167. Singh, M.R.; Patel, S.; Singh, D. Chapter 9—Natural Polymer-Based Hydrogels as Scaffolds for Tissue Engineering. In Nanobiomaterials in Soft Tissue Engineering; Grumezescu, A.M., Ed.; William Andrew Publishing: Norwich, NY, USA, 2016; pp. 231–260. ISBN 978-0-323-42865-1.

168. Vieira, S.; da Silva Morais, A.; Silva-Correia, J.; Oliveira, J.M.; Reis, R.L. Natural-Based Hydrogels: From Processing to Applications. In Encyclopedia of Polymer Science and Technology; American Cancer Society: Atlanta, GA, USA, 2017; pp. 1–27. ISBN 978-0-471-44026-0.

169. Caliari, S.R.; Burdick, J.A. A Practical Guide to Hydrogels for Cell Culture. Nat. Methods 2016, 13, 405–414.

170. Tan, H.; Marra, K.G. Injectable, Biodegradable Hydrogels for Tissue Engineering Applications. Materials 2010, 3, 1746.

171. Rao, S.S.; DeJesus, J.; Short, A.R.; Otero, J.J.; Sarkar, A.; Winter, J.O. Glioblastoma Behaviors in Three-Dimensional Collagen-Hyaluronan Composite Hydrogels. ACS Appl. Mater. Interfaces 2013, 5, 9276–9284.

172. Hughes, C.S.; Postovit, L.M.; Lajoie, G.A. Matrigel: A Complex Protein Mixture Required for Optimal Growth of Cell Culture. Proteomics 2010, 10, 1886–1890.

173. Grundy, T.J.; De Leon, E.; Griffin, K.R.; Stringer, B.W.; Day, B.W.; Fabry, B.; Cooper-White, J.; O'Neill, G.M. Differential Response of Patient-Derived Primary Glioblastoma Cells to Environmental Stiffness. Sci. Rep. 2016, 6, 23353.

174. Cruz-Acuña, R.; Quirós, M.; Farkas, A.E.; Dedhia, P.H.; Huang, S.; Siuda, D.; García-Hernández, V.; Miller, A.J.; Spence, J.R.; Nusrat, A.; et al. Synthetic Hydrogels for Human Intestinal Organoid Generation and Colonic Wound Repair. Nat. Cell Biol. 2017, 19, 1326–1335.

175. Sawhney, A.S.; Pathak, C.P.; Hubbell, J.A. Bioerodible Hydrogels Based on Photopolymerized Poly(Ethylene Glycol)-Co-Poly(.Alpha.-Hydroxy Acid) Diacrylate Macromers. Macromolecules 1993, 26, 581–587.

176. Chirila, T.V.; Constable, I.J.; Crawford, G.J.; Vijayasekaran, S.; Thompson, D.E.; Chen, Y.C.; Fletcher, W.A.; Griffin, B.J. Poly(2-Hydroxyethyl Methacrylate) Sponges as Implant Materials: In Vivo and in Vitro Evaluation of Cellular Invasion. Biomaterials 1993, 14, 26–38.

177. Ahmed, E.M. Hydrogel: Preparation, Characterization, and Applications: A Review. J. Adv. Res. 2015, 6, 105–121.

178. Wang, C.; Tong, X.; Yang, F. Bioengineered 3D Brain Tumor Model to Elucidate the Effects of Matrix Stiffness on Glioblastoma Cell Behavior Using PEG-Based Hydrogels. Mol. Pharm. 2014, 11, 2115–2125.

179. Wu, J.; Li, P.; Dong, C.; Jiang, H.; Xue, B.; Gao, X.; Qin, M.; Wang, W.; Chen, B.; Cao, Y. Rationally Designed Synthetic Protein Hydrogels with Predictable Mechanical Properties. Nat. Commun. 2018, 9, 620.

180. Langer, R.; Tirrell, D.A. Designing Materials for Biology and Medicine. Nature 2004, 428, 487.

181. Thomas, T.W.; DiMilla, P.A. Spreading and Motility of Human Glioblastoma Cells on Sheets of Silicone Rubber Depend on Substratum Compliance. Med. Biol. Eng. Comput. 2000, 38, 360–370.

182. Gurski, L.A.; Jha, A.K.; Zhang, C.; Jia, X.; Farach-Carson, M.C. Hyaluronic Acid-Based Hydrogels as 3D Matrices for in Vitro Evaluation of Chemotherapeutic Drugs Using Poorly Adherent Prostate Cancer Cells. Biomaterials 2009, 30, 6076–6085.

183. Gill, B.J.; Gibbons, D.L.; Roudsari, L.C.; Saik, J.E.; Rizvi, Z.H.; Roybal, J.D.; Kurie, J.M.; West, J.L. A Synthetic Matrix with Independently Tunable Biochemistry and Mechanical Properties to Study

Epithelial Morphogenesis and EMT in a Lung Adenocarcinoma Model. Cancer Res. 2012, 72, 6013–6023.

184. Tibbitt, M.W.; Anseth, K.S. Hydrogels as Extracellular Matrix Mimics for 3D Cell Culture. Biotechnol. Bioeng. 2009, 103, 655–663.

185. Ratner, B.D.; Bryant, S.J. Biomaterials: Where We Have Been and Where We Are Going. Annu. Rev. Biomed. Eng. 2004, 6, 41–75.

186. Ratner, B.D.; Hoffman, A.S. Synthetic Hydrogels for Biomedical Applications. In Hydrogels for Medical and Related Applications; ACS Symposium Series; American Chemical Society: Washington, NY, USA, 1976; Volume 31, pp. 1–36. ISBN 978-0-8412-0338-9.

187. Zheng, S.Y.; Ding, H.; Qian, J.; Yin, J.; Wu, Z.L.; Song, Y.; Zheng, Q. Metal-Coordination Complexes Mediated Physical Hydrogels with High Toughness, Stick–Slip Tearing Behavior, and Good Processability. Macromolecules 2016, 49, 9637–9646.

188. Ulrich, T.A.; de Juan Pardo, E.M.; Kumar, S. The Mechanical Rigidity of the Extracellular Matrix Regulates the Structure, Motility, and Proliferation of Glioma Cells. Cancer Res. 2009, 69, 4167–4174.

189. Rao, S.S.; Nelson, M.T.; Xue, R.; DeJesus, J.K.; Viapiano, M.S.; Lannutti, J.J.; Sarkar, A.; Winter, J.O. Mimicking White Matter Tract Topography Using Core-Shell Electrospun Nanofibers to Examine Migration of Malignant Brain Tumors. Biomaterials 2013, 34, 5181–5190.

190. Pedron, S.; Becka, E.; Harley, B.A.C. Regulation of Glioma Cell Phenotype in 3D Matrices by Hyaluronic Acid. Biomaterials 2013, 34, 7408–7417.

191. Shibayama, M.; Tanaka, T. Volume Phase Transition and Related Phenomena of Polymer Gels. In Responsive Gels: Volume Transitions I; Dušek, K., Ed.; Advances in Polymer Science; Springer: Berlin/Heidelberg, Germany, 1993; pp. 1–62. ISBN 978-3-540-47737-2.

192. Taraban, M.B.; Feng, Y.; Hammouda, B.; Hyland, L.L.; Yu, Y.B. Chirality-Mediated Mechanical and Structural Properties of Oligopeptide Hydrogels. Chem. Mater. 2012, 24, 2299–2310.

193. Hydrogel Sensors and Actuators: Engineering and Technology; Gerlach, G.; Arndt, K.-F. (Eds.) Springer Series on Chemical Sensors and Biosensors; Springer: Berlin/Heidelberg, Germany, 2010; ISBN 978-3-540-75644-6.

194. Kizilay, M.Y.; Okay, O. Effect of Initial Monomer Concentration on Spatial Inhomogeneity in Poly(Acrylamide) Gels. Macromolecules 2003, 36, 6856–6862.

195. Lindemann, B.; Schröder, U.P.; Oppermann, W. Influence of the Cross-Linker Reactivity on the Formation of Inhomogeneities in Hydrogels. Macromolecules 1997, 30, 4073–4077.

196. Oldenziel, W.H.; Beukema, W.; Westerink, B.H.C. Improving the Reproducibility of Hydrogel-Coated Glutamate Microsensors by Using an Automated Dipcoater. J. Neurosci. Methods 2004, 140, 117–126.

197. Petersen, O.W.; Rønnov-Jessen, L.; Howlett, A.R.; Bissell, M.J. Interaction with Basement Membrane Serves to Rapidly Distinguish Growth and Differentiation Pattern of Normal and Malignant Human Breast Epithelial Cells. Proc. Natl. Acad. Sci. USA 1992, 89, 9064–9068.

198. Baino, F.; Novajra, G.; Vitale-Brovarone, C. Bioceramics and Scaffolds: A Winning Combination for Tissue Engineering. Front. Bioeng. Biotechnol. 2015, 3.

199. Viswanathan, P.; Chirasatitsin, S.; Ngamkham, K.; Engler, A.J.; Battaglia, G. Cell Instructive Microporous Scaffolds through Interface Engineering. J. Am. Chem. Soc. 2012, 134, 20103–20109.

200. Jiang, Y.-Y.; Zhu, Y.-J.; Li, H.; Zhang, Y.-G.; Shen, Y.-Q.; Sun, T.-W.; Chen, F. Preparation and Enhanced Mechanical Properties of Hybrid Hydrogels Comprising Ultralong Hydroxyapatite Nanowires and Sodium Alginate. J. Colloid Interface Sci. 2017, 497, 266–275.

201. Suggitt, M.; Bibby, M.C. 50 Years of Preclinical Anticancer Drug Screening: Empirical to Target-Driven Approaches. Clin. Cancer Res. 2005, 11, 971–981.

202. Singh, M.; Close, D.A.; Mukundan, S.; Johnston, P.A.; Sant, S. Production of Uniform 3D Microtumors in Hydrogel Microwell Arrays for Measurement of Viability, Morphology, and Signaling Pathway Activation. Assay Drug Dev. Technol. 2015, 13, 570–583.

203. Lee, K.Y.; Mooney, D.J. Hydrogels for Tissue Engineering. Chem. Rev. 2001, 101, 1869–1879.

204. Putnam, A.J.; Mooney, D.J. Tissue Engineering Using Synthetic Extracellular Matrices. Nat. Med. 1996, 2, 824–826.

205. Wu, Z.; Su, X.; Xu, Y.; Kong, B.; Sun, W.; Mi, S. Bioprinting Three-Dimensional Cell-Laden Tissue Constructs with Controllable Degradation. Sci. Rep. 2016, 6, 24474.

206. Lee, K.Y.; Mooney, D.J. Alginate: Properties and Biomedical Applications. Prog. Polym. Sci. 2012, 37, 106–126.

207. Kim, G.; Ahn, S.; Kim, Y.; Cho, Y.; Chun, W. Coaxial Structured Collagen–Alginate Scaffolds: Fabrication, Physical Properties, and Biomedical Application for Skin Tissue Regeneration. J. Mater. Chem. 2011, 21, 6165–6172.

208. Tan, Z.; Parisi, C.; Silvio, L.D.; Dini, D.; Forte, A.E. Cryogenic 3D Printing of Super Soft Hydrogels. Sci. Rep. 2017, 7, 16293.

209. Zhao, W.; Jin, X.; Cong, Y.; Liu, Y.; Fu, J. Degradable Natural Polymer Hydrogels for Articular Cartilage Tissue Engineering. J. Chem. Technol. Biotechnol. 2013, 88, 327–339.

210. Yu, L.; Ding, J. Injectable Hydrogels as Unique Biomedical Materials. Chem. Soc. Rev. 2008, 37, 1473–1481.

211. Zheng, Y.; Huang, K.; You, X.; Huang, B.; Wu, J.; Gu, Z. Hybrid Hydrogels with High Strength and Biocompatibility for Bone Regeneration. Int. J. Biol. Macromol. 2017, 104, 1143–1149.

212. Wu, D.-Q.; Chu, C.-C. Biodegradable Hydrophobic-Hydrophilic Hybrid Hydrogels: Swelling Behavior and Controlled Drug Release. J. Biomater. Sci. Polym. Ed. 2008, 19, 411–429.

213. Thambi, T.; Li, Y.; Lee, D.S. Injectable Hydrogels for Sustained Release of Therapeutic Agents. J. Control. Release 2017, 267, 57–66.

214. Gil, M.S.; Cho, J.; Thambi, T.; Giang Phan, V.H.; Kwon, I.; Lee, D.S. Bioengineered Robust Hybrid Hydrogels Enrich the Stability and Efficacy of Biological Drugs. J. Control. Release 2017, 267, 119–132.

215. Kang, C.E.; Poon, P.C.; Tator, C.H.; Shoichet, M.S. A New Paradigm for Local and Sustained Release of Therapeutic Molecules to the Injured Spinal Cord for Neuroprotection and Tissue Repair. Tissue Eng. Part A 2008, 15, 595–604.

216. Nguyen, M.K.; Lee, D.S. Injectable Biodegradable Hydrogels. Macromol. Biosci. 2010, 10, 563–579.

217. Wang, Y.; Lapitsky, Y.; Kang, C.E.; Shoichet, M.S. Accelerated Release of a Sparingly Soluble Drug from an Injectable Hyaluronan-Methylcellulose Hydrogel. J. Control. Release 2009, 140, 218–223.

218. Jeong, B.; Bae, Y.H.; Lee, D.S.; Kim, S.W. Biodegradable Block Copolymers as Injectable Drug-Delivery Systems. Nature 1997, 388, 860–862.

219. Qiu, Y.; Park, K. Environment-Sensitive Hydrogels for Drug Delivery. Adv. Drug Deliv. Rev. 2001, 53, 321–339.

220. Yubao, G.; Hecheng, M.; Jianguo, L. Controlled WISP-1 ShRNA Delivery Using Thermosensitive Biodegradable Hydrogel in the Treatment of Osteoarthritis. J. Bionic Eng. 2015, 12, 285–293.

221. Lee, P.Y.; Cobain, E.; Huard, J.; Huang, L. Thermosensitive Hydrogel PEG–PLGA–PEG Enhances Engraftment of Muscle-Derived Stem Cells and Promotes Healing in Diabetic Wound. Mol. Ther. 2007, 15, 1189–1194.

222. Mothe, A.J.; Tam, R.Y.; Zahir, T.; Tator, C.H.; Shoichet, M.S. Repair of the Injured Spinal Cord by Transplantation of Neural Stem Cells in a Hyaluronan-Based Hydrogel. Biomaterials 2013, 34, 3775–3783.

223. Caló, E.; Khutoryanskiy, V.V. Biomedical Applications of Hydrogels: A Review of Patents and Commercial Products. Eur. Polym. J. 2015, 65, 252–267.

224. Agaoglu, S.; Diep, P.; Martini, M.; Kt, S.; Baday, M.; Emre Araci, I. Ultra-Sensitive Microfluidic Wearable Strain Sensor for Intraocular Pressure Monitoring. Lab A Chip 2018, 18, 3471–3483.

225. Lloyd, A.W.; Faragher, R.G.; Denyer, S.P. Ocular Biomaterials and Implants. Biomaterials 2001, 22, 769–785.

226. Salmon, D.; Roussel, L.; Gilbert, E.; Kirilov, P.; Pirot, F. Percutaneous Penetration Enhancers Chemical Methods in Penetration Enhancement; Springer: Berlin, Germany, 2015.

227. Qi, C.; Xu, L.; Deng, Y.; Wang, G.; Wang, Z.; Wang, L. Sericin Hydrogels Promote Skin Wound Healing with Effective Regeneration of Hair Follicles and Sebaceous Glands after Complete Loss of Epidermis and Dermis. Biomater. Sci. 2018, 6, 2859–2870.

228. Baker, E.L.; Lu, J.; Yu, D.; Bonnecaze, R.T.; Zaman, M.H. Cancer Cell Stiffness: Integrated Roles of Three-Dimensional Matrix Stiffness and Transforming Potential. Biophys. J. 2010, 99, 2048–2057.

229. Gkretsi, V.; Stylianopoulos, T. Cell Adhesion and Matrix Stiffness: Coordinating Cancer Cell Invasion and Metastasis. Front. Oncol. 2018, 8, 145.

230. Quan, F.-S.; Kim, K.S. Medical Applications of the Intrinsic Mechanical Properties of Single Cells. Acta Biochim. Biophys. Sin. 2016, 48, 865–871.

231. Wullkopf, L.; West, A.-K.V.; Leijnse, N.; Cox, T.R.; Madsen, C.D.; Oddershede, L.B.; Erler, J.T. Cancer Cells' Ability to Mechanically Adjust to Extracellular Matrix Stiffness Correlates with Their Invasive Potential. Mol. Biol. Cell 2018, 29, 2378–2385.

232. Boyd, N.F.; Li, Q.; Melnichouk, O.; Huszti, E.; Martin, L.J.; Gunasekara, A.; Mawdsley, G.; Yaffe, M.J.; Minkin, S. Evidence That Breast Tissue Stiffness Is Associated with Risk of Breast Cancer. PLoS ONE 2014, 9, e100937.

233. Fenner, J.; Stacer, A.C.; Winterroth, F.; Johnson, T.D.; Luker, K.E.; Luker, G.D. Macroscopic Stiffness of Breast Tumors Predicts Metastasis. Sci. Rep. 2014, 4, 5512.

234. Olcum, M.; Ozcivici, E. Daily Application of Low Magnitude Mechanical Stimulus Inhibits the Growth of MDA-MB-231 Breast Cancer Cells in Vitro. Cancer Cell Int. 2014, 14, 102.

235. Joyce, M.H.; Lu, C.; James, E.R.; Hegab, R.; Allen, S.C.; Suggs, L.J.; Brock, A. Phenotypic Basis for Matrix Stiffness-Dependent Chemoresistance of Breast Cancer Cells to Doxorubicin. Front. Oncol. 2018, 8, 337.

236. Conklin, M.W.; Eickhoff, J.C.; Riching, K.M.; Pehlke, C.A.; Eliceiri, K.W.; Provenzano, P.P.; Friedl, A.; Keely, P.J. Aligned Collagen Is a Prognostic Signature for Survival in Human Breast Carcinoma. Am. J. Pathol 2011, 178, 1221–1232.

237. Wen, P.Y.; Weller, M.; Lee, E.Q.; Alexander, B.M.; Barnholtz-Sloan, J.S.; Barthel, F.P.; Batchelor, T.T.; Bindra, R.S.; Chang, S.M.; Chiocca, E.A.; et al. Glioblastoma in Adults: A Society for Neuro-Oncology (SNO) and European Society of Neuro-Oncology (EANO) Consensus Review on Current Management and Future Directions. Neuro-Oncol. 2020, 22, 1073–1113.

238. Abbott, N.J.; Rönnbäck, L.; Hansson, E. Astrocyte-Endothelial Interactions at the Blood-Brain Barrier. Nat. Rev. Neurosci. 2006, 7, 41–53.

239. Blanchette, M.; Daneman, R. Formation and Maintenance of the BBB. Mech. Dev. 2015, 138, 8–16.

240. Tajes, M.; Ramos-Fernández, E.; Weng-Jiang, X.; Bosch-Morató, M.; Guivernau, B.; Eraso-Pichot, A.; Salvador, B.; Fernàndez-Busquets, X.; Roquer, J.; Muñoz, F.J. The Blood-Brain Barrier: Structure, Function and Therapeutic Approaches to Cross It. Mol. Membr. Biol. 2014, 31, 152–167.

241. Wolburg, H.; Wolburg-Buchholz, K.; Kraus, J.; Rascher-Eggstein, G.; Liebner, S.; Hamm, S.; Duffner, F.; Grote, E.-H.; Risau, W.; Engelhardt, B. Localization of Claudin-3 in Tight Junctions of the Blood-Brain Barrier Is Selectively Lost during Experimental Autoimmune Encephalomyelitis and Human Glioblastoma Multiforme. Acta Neuropathol. 2003, 105, 586–592.

242. Liebner, S.; Fischmann, A.; Rascher, G.; Duffner, F.; Grote, E.H.; Kalbacher, H.; Wolburg, H. Claudin-1 and Claudin-5 Expression and Tight Junction Morphology Are Altered in Blood Vessels of Human Glioblastoma Multiforme. Acta Neuropathol. 2000, 100, 323–331.

243. Bellail, A.C.; Hunter, S.B.; Brat, D.J.; Tan, C.; Van Meir, E.G. Microregional Extracellular Matrix Heterogeneity in Brain Modulates Glioma Cell Invasion. Int. J. Biochem. Cell Biol. 2004, 36, 1046–1069.

244. Demuth, T.; Berens, M.E. Molecular Mechanisms of Glioma Cell Migration and Invasion. J. Neurooncol. 2004, 70, 217–228.

245. Mahesparan, R.; Read, T.-A.; Lund-Johansen, M.; Skaftnesmo, K.O.; Bjerkvig, R.; Engebraaten, O. Expression of Extracellular Matrix Components in a Highly Infiltrative in Vivo Glioma Model. Acta Neuropathol. 2003, 105, 49–57.

246. Payne, L.S.; Huang, P.H. The Pathobiology of Collagens in Glioma. Mol. Cancer Res. 2013, 11, 1129–1140.

247. Lim, E.-J.; Suh, Y.; Yoo, K.-C.; Lee, J.-H.; Kim, I.-G.; Kim, M.-J.; Chang, J.H.; Kang, S.-G.; Lee, S.-J. Tumor-Associated Mesenchymal Stem-like Cells Provide Extracellular Signaling Cue for Invasiveness of Glioblastoma Cells. Oncotarget 2017, 8, 1438–1448.

248. Gilbertson, R.J.; Rich, J.N. Making a Tumour's Bed: Glioblastoma Stem Cells and the Vascular Niche. Nat. Rev. Cancer 2007, 7, 733–736.

249. Heddleston, J.M.; Li, Z.; McLendon, R.E.; Hjelmeland, A.B.; Rich, J.N. The Hypoxic Microenvironment Maintains Glioblastoma Stem Cells and Promotes Reprogramming towards a Cancer Stem Cell Phenotype. Cell Cycle 2009, 8, 3274–3284.

250. Ciasca, G.; Sassun, T.E.; Minelli, E.; Antonelli, M.; Papi, M.; Santoro, A.; Giangaspero, F.; Delfini, R.; Spirito, M.D. Nano-Mechanical Signature of Brain Tumours. Nanoscale 2016, 8, 19629–19643.

251. Diao, W.; Tong, X.; Yang, C.; Zhang, F.; Bao, C.; Chen, H.; Liu, L.; Li, M.; Ye, F.; Fan, Q.; et al. Behaviors of Glioblastoma Cells in in Vitro Microenvironments. Sci. Rep. 2019, 9, 85.

252. Umesh, V.; Rape, A.D.; Ulrich, T.A.; Kumar, S. Microenvironmental Stiffness Enhances Glioma Cell Proliferation by Stimulating Epidermal Growth Factor Receptor Signaling. PLoS ONE 2014, 9, e101771.

253. Memmel, S.; Sisario, D.; Zöller, C.; Fiedler, V.; Katzer, A.; Heiden, R.; Becker, N.; Eing, L.; Ferreira, F.L.R.; Zimmermann, H.; et al. Migration Pattern, Actin Cytoskeleton Organization and Response to PI3K-, MTOR-, and Hsp90-Inhibition of Glioblastoma Cells with Different Invasive Capacities. Oncotarget 2017, 8, 45298–45310.

254. Masoumi, S.; Harisankar, A.; Gracias, A.; Bachinger, F.; Fufa, T.; Chandrasekar, G.; Gaunitz, F.; Walfridsson, J.; Kitambi, S.S. Understanding Cytoskeleton Regulators in Glioblastoma Multiforme for Therapy Design. Drug Des. Devel. 2016, 10, 2881–2897.

255. Efremov, Y.M.; Dokrunova, A.A.; Efremenko, A.V.; Kirpichnikov, M.P.; Shaitan, K.V.; Sokolova, O.S. Distinct Impact of Targeted Actin Cytoskeleton Reorganization on Mechanical Properties of Normal and Malignant Cells. Biochim. Biophys. Acta 2015, 1853, 3117–3125.

256. Li, Q.S.; Lee, G.Y.H.; Ong, C.N.; Lim, C.T. AFM Indentation Study of Breast Cancer Cells. Biochem. Biophys. Res. Commun. 2008, 374, 609–613.

257. Lu, P.; Weaver, V.M.; Werb, Z. The Extracellular Matrix: A Dynamic Niche in Cancer Progression. J. Cell Biol. 2012, 196, 395–406.

258. Miroshnikova, Y.A.; Mouw, J.K.; Barnes, J.M.; Pickup, M.W.; Lakins, J.N.; Kim, Y.; Lobo, K.; Persson, A.I.; Reis, G.F.; McKnight, T.R.; et al. Tissue Mechanics Promote IDH1-Dependent HIF1α-Tenascin C Feedback to Regulate Glioblastoma Aggression. Nat. Cell Biol. 2016, 18, 1336–1345.

259. Stewart, D.C.; Rubiano, A.; Dyson, K.; Simmons, C.S. Mechanical Characterization of Human Brain Tumors from Patients and Comparison to Potential Surgical Phantoms. PLoS ONE 2017, 12, e0177561.

260. Barnes, J.M.; Przybyla, L.; Weaver, V.M. Tissue Mechanics Regulate Brain Development, Homeostasis and Disease. J. Cell Sci. 2017, 130, 71–82.

261. Wiranowska, M.; Rojiani, M.V. Extracellular Matrix Microenvironment in Glioma Progression. Glioma-Explor. Its Biol. Pract. Relev. 2011.

262. Brat, D.J.; Castellano-Sanchez, A.A.; Hunter, S.B.; Pecot, M.; Cohen, C.; Hammond, E.H.; Devi, S.N.; Kaur, B.; Van Meir, E.G. Pseudopalisades in Glioblastoma Are Hypoxic, Express Extracellular Matrix Proteases, and Are Formed by an Actively Migrating Cell Population. Cancer Res. 2004, 64, 920–927.

263. Nakada, M.; Okada, Y.; Yamashita, J. The Role of Matrix Metalloproteinases in Glioma Invasion. Front. Biosci. 2003, 8, e261–e269.

264. Schuler, P.J.; Bendszus, M.; Kuehnel, S.; Wagner, S.; Hoffmann, T.K.; Goldbrunner, R.; Vince, G.H. Urokinase Plasminogen Activator, UPAR, MMP-2, and MMP-9 in the C6-Glioblastoma Rat Model. In Vivo 2012, 26, 571–576.

265. Mikkelsen, T.; Yan, P.-S.; Ho, K.-L.; Sameni, M.; Sloane, B.F.; Rosenblum, M.L. Immunolocalization of Cathepsin B in Human Glioma: Implications for Tumor Invasion and Angiogenesis. J. Neurosurg. 1995, 83, 285–290.

266. Kobayashi, T.; Honke, K.; Gasa, S.; Fujii, T.; Maguchi, S.; Miyazaki, T.; Makita, A. Proteolytic Processing Sites Producing the Mature Form of Human Cathepsin D. Int. J. Biochem. 1992, 24, 1487–1491.

267. Rempel, S.A.; Rosenblum, M.L.; Mikkelsen, T.; Yan, P.S.; Ellis, K.D.; Golembieski, W.A.; Sameni, M.; Rozhin, J.; Ziegler, G.; Sloane, B.F. Cathepsin B Expression and Localization in Glioma Progression and Invasion. Cancer Res. 1994, 54, 6027–6031.

268. Armento, A.; Ehlers, J.; Schötterl, S.; Naumann, U. Molecular Mechanisms of Glioma Cell Motility. In Glioblastoma; De Vleeschouwer, S., Ed.; Codon Publications: Brisbane, Australia, 2017; ISBN 978-0-9944381-2-6.

269. Schnell, O.; Krebs, B.; Carlsen, J.; Miederer, I.; Goetz, C.; Goldbrunner, R.H.; Wester, H.-J.; Haubner, R.; Pöpperl, G.; Holtmannspötter, M.; et al. Imaging of Integrin Alpha(v)Beta(3) Expression in Patients with Malignant Glioma by [18F] Galacto-RGD Positron Emission Tomography. Neuro-Oncol. 2009, 11, 861–870.

270. Li, D.; Zhao, X.; Zhang, L.; Li, F.; Ji, N.; Gao, Z.; Wang, J.; Kang, P.; Liu, Z.; Shi, J.; et al. 68Ga-PRGD2 PET/CT in the Evaluation of Glioma: A Prospective Study. Mol. Pharm. 2014, 11, 3923–3929.

271. Iagaru, A.; Mosci, C.; Mittra, E.; Zaharchuk, G.; Fischbein, N.; Harsh, G.; Li, G.; Nagpal, S.; Recht, L.; Gambhir, S.S. Glioblastoma Multiforme Recurrence: An Exploratory Study of (18)F FPPRGD2 PET/CT. Radiology 2015, 277, 497–506.

272. Ellert-Miklaszewska, A.; Poleszak, K.; Pasierbinska, M.; Kaminska, B. Integrin Signaling in Glioma Pathogenesis: From Biology to Therapy. Int. J. Mol. Sci. 2020, 21, 888.

273. Fillmore, H.L.; Chasiotis, I.; Cho, S.W.; Gillies, G.T. Atomic Force Microscopy Observations of Tumour Cell Invadopodia: Novel Cellular Nanomorphologies on Collagen Substrates. Nanotechnology 2002, 14, 73–76.

274. Kim, H.-D.; Guo, T.W.; Wu, A.P.; Wells, A.; Gertler, F.B.; Lauffenburger, D.A. Epidermal Growth Factor–Induced Enhancement of Glioblastoma Cell Migration in 3D Arises from an Intrinsic Increase in Speed But an Extrinsic Matrix- and Proteolysis-Dependent Increase in Persistence. Mol. Biol. Cell 2008, 19, 4249–4259.

275. Kaufman, L.J.; Brangwynne, C.P.; Kasza, K.E.; Filippidi, E.; Gordon, V.D.; Deisboeck, T.S.; Weitz, D.A. Glioma Expansion in Collagen I Matrices: Analyzing Collagen Concentration-Dependent Growth and Motility Patterns. Biophys. J. 2005, 89, 635–650.

276. Yang, Y.; Motte, S.; Kaufman, L.J. Pore Size Variable Type I Collagen Gels and Their Interaction with Glioma Cells. Biomaterials 2010, 31, 5678–5688.

277. Hwang, Y.-J.; Kolettis, N.; Yang, M.; Gillard, E.R.; Sanchez, E.; Sun, C.-H.; Tromberg, B.J.; Krasieva, T.B.; Lyubovitsky, J.G. Multiphoton Imaging of Actin Filament Formation and Mitochondrial Energetics of Human ACBT Gliomas. Photochem. Photobiol. 2011, 87, 408–417.

278. Zhong, J.; Bach, C.T.; Shum, M.S.Y.; O'Neill, G.M. NEDD9 Regulates 3D Migratory Activity Independent of the Rac1 Morphology Switch in Glioma and Neuroblastoma. Mol. Cancer Res. 2014, 12, 264–273.

279. Ardebili, S.Y.; Zajc, I.; Gole, B.; Campos, B.; Herold-Mende, C.; Drmota, S.; Lah, T.T. CD133/Prominin1 Is Prognostic for GBM Patient's Survival, but Inversely Correlated with Cysteine Cathepsins' Expression in Glioblastoma Derived Spheroids. Radiol. Oncol. 2011, 45, 102–115.

280. Chigurupati, S.; Venkataraman, R.; Barrera, D.; Naganathan, A.; Madan, M.; Paul, L.; Pattisapu, J.V.; Kyriazis, G.A.; Sugaya, K.; Bushnev, S.; et al. Receptor Channel TRPC6 Is a Key Mediator of Notch-Driven Glioblastoma Growth and Invasiveness. Cancer Res. 2010, 70, 418–427.

281. Chonan, Y.; Taki, S.; Sampetrean, O.; Saya, H.; Sudo, R. Endothelium-Induced Three-Dimensional Invasion of Heterogeneous Glioma Initiating Cells in a Microfluidic Coculture Platform. Int. Biol. 2017, 9, 762–773.

282. Burdick, J.A.; Prestwich, G.D. Hyaluronic Acid Hydrogels for Biomedical Applications. Adv. Mater. Weinh. 2011, 23, H41–H56.

283. Ouasti, S.; Donno, R.; Cellesi, F.; Sherratt, M.J.; Terenghi, G.; Tirelli, N. Network Connectivity, Mechanical Properties and Cell Adhesion for Hyaluronic Acid/PEG Hydrogels. Biomaterials 2011, 32, 6456–6470.

284. Wang, X.; He, J.; Wang, Y.; Cui, F.-Z. Hyaluronic Acid-Based Scaffold for Central Neural Tissue Engineering. Interface Focus 2012, 2, 278–291.

285. Borzacchiello, A.; Russo, L.; Malle, B.M.; Schwach-Abdellaoui, K.; Ambrosio, L. Hyaluronic Acid Based Hydrogels for Regenerative Medicine Applications. Available online: https://www.hindawi.com/journals/bmri/2015/871218/ (accessed on 17 June 2020).

286. Florczyk, S.J.; Wang, K.; Jana, S.; Wood, D.L.; Sytsma, S.K.; Sham, J.; Kievit, F.M.; Zhang, M. Porous Chitosan-Hyaluronic Acid Scaffolds as a Mimic of Glioblastoma Microenvironment ECM. Biomaterials 2013, 34, 10143–10150.

287. Pedron, S.; Harley, B.A.C. Impact of the Biophysical Features of a 3D Gelatin Microenvironment on Glioblastoma Malignancy. J. Biomed. Mater. Res. Part A 2013, 101, 3404–3415.

288. Wan, Y.; Mahmood, M.A.I.; Li, N.; Allen, P.B.; Kim, Y.; Bachoo, R.; Ellington, A.D.; Iqbal, S.M. Nano-Textured Substrates with Immobilized Aptamers for Cancer Cell Isolation and Cytology. Cancer 2012, 118, 1145–1154.

289. Niu, C.J.; Fisher, C.; Scheffler, K.; Wan, R.; Maleki, H.; Liu, H.; Sun, Y.; A Simmons, C.; Birngruber, R.; Lilge, L. Polyacrylamide Gel Substrates That Simulate the Mechanical Stiffness of Normal and Malignant Neuronal Tissues Increase Protoporphyin IX Synthesis in Glioma Cells. J. Biomed. Opt. 2015, 20, 098002.

290. Cruz-Acuña, R.; García, A.J. Synthetic Hydrogels Mimicking Basement Membrane Matrices to Promote Cell-Matrix Interactions. Matrix Biol. 2017, 57–58, 324–333.

291. Trujillo, S.; Gonzalez-Garcia, C.; Rico, P.; Reid, A.; Windmill, J.; Dalby, M.J.; Salmeron-Sanchez, M. Engineered 3D Hydrogels with Full-Length Fibronectin That Sequester and Present Growth Factors. Biomaterials 2020, 252, 120104.

292. Huang, C.; Holfeld, J.; Schaden, W.; Orgill, D.; Ogawa, R. Mechanotherapy: Revisiting Physical Therapy and Recruiting Mechanobiology for a New Era in Medicine. Trends Mol. Med. 2013, 19, 555–564.

293. Tadeo, I.; Berbegall, A.P.; Escudero, L.M.; Alvaro, T.; Noguera, R. Biotensegrity of the Extracellular Matrix:Physiology,DynamicMechanicalBalance,andImplicationsinOncologyandMechanotherapy. Front. Oncol. 2014, 4, 39.

294. Vining, K.H.; Mooney, D.J. Mechanical Forces Direct Stem Cell Behaviour in Development and Regeneration. Nat. Rev. Mol. Cell Biol. 2017, 18, 728–742.

295. Sheridan, C. Pancreatic Cancer Provides Testbed for First Mechanotherapeutics. Nat. Biotechnol. 2019, 37, 829–831.

Chapter 17

AN UPDATED COMPREHENSIVE REVIEW BASED ON SOURCE IS PRESENTED ON NOVEL HYDROGELS FOR TOPICAL APPLICATIONS

1. INTRODUCTION: AN OVERVIEW

Typically, active pharmaceutical ingredients (API) or medications are not provided in their pure chemical form (for the purpose of preventing or treating diseases). However, APIs are typically administered in pharmaceutical formulations, which are commonly referred to as dosage forms [1]. The procedure involves combining an active pharmaceutical ingredient (API) with other components, known as excipients or pharmaceutical inactive ingredients, to create a valuable therapeutic product [2]. The substances have undergone conversion into an appropriate dose formulation and have been administered through different methods of administration [1]. The dose forms are often provided in solid, semi-solid, and liquid states. The solid dosage forms encompass tablets, capsules, and powders. The different semi-solid pharmaceutical formulations encompass creams, gels, emulgels, ointments, pastes, and other similar forms. The several categories of liquid pharmaceutical formulations encompass emulsions, injections, lotions, ophthalmic preparations, suspensions, syrups, and more [3].

Topical administration is the favored method for administering therapeutic drugs locally due to its ease and affordability. A topical drug delivery system (TDDS) is primarily designed to treat localized disorders by administering therapeutic substances to superficial areas of the body, such as the skin, eyes, nose, and vaginal canal. Topical application, unlike oral treatment, bypasses the liver's first-pass metabolism, variations in gastric pH in the stomach, and fluctuations in plasma concentrations. Additional advantages of a topical drug delivery system include patient satisfaction and acceptance, simple and convenient application, reduced pain and noninvasive nature, enhanced absorption of the active pharmaceutical ingredient (API), improved physiological and pharmacological effects, as well as minimal toxicity to the body as a whole and limited exposure of the API to non-infected tissues or sites [4].

Semi-solid goods have been manufactured for many years and are commonly utilized in the pharmaceutical, cosmetics, and health supplement industries. Semi-solid dosage form (SSDF) products are usually either topically or inserted into a bodily opening [5]. This dosage form has several advantages, including the direct application of drugs to the affected area and the convenient administration to patients of all age groups. One of the difficulties in SSDFs is the requirement to administer the API through the patient's skin or other physical membranes in order to access the intended system [6]. Throughout history, many types of topical semi-solid dosage forms have been extensively utilized in human culture to treat a range of skin ailments [7]. Topically applied semi-solid dose forms often include creams, gels, ointments, or pastes [8]. The field of dermatology offers a wide range of medications that come in different forms and textures. However, among these options, semi-solid dosage forms are the most widely used [9].

The skin is the primary and biggest organ in the human body. The skin is responsible for a range of essential tasks, namely providing protection against germs and the external environment, and preventing excessive water loss. Additionally, it helps regulate body temperature [10,11]. According to TDDS, the skin is considered to be one of the main and easily reachable organs [9]. The majority of topical semi-solid formulations are specifically formulated to target the skin or the tissues beneath it [7]. The stratum corneum (SC), one of the three layers of the skin along with the epidermis and dermis, has the main function of safeguarding the underlying tissues. The majority of APIs are unable to penetrate the skin easily due to the highly efficient barrier of the stratum corneum [10]. The stratum corneum (SC) limits the entry of most big and water-loving active pharmaceutical ingredient (API) molecules, such as proteins, peptides, nucleotides, and oligonucleotides [12].

Novel semi-solids, although they have water bases that may be washed off, do not have a greasy texture. As a result, they cause less irritation to the skin and have greater efficacy compared to traditional semi-solid forms of medication. Novel semi-solid formulations can be safely administered to pediatric, geriatric, and pregnant patients without the risk of inducing allergic responses. Novel semi-solids are anticipated to regulate the release pattern, as stated in reference [13]. Topical dose types designed to provide localized or systemic effects. Traditional transdermal drug delivery systems (TDDS) have notable disadvantages. Acne, alopecia, and psoriasis are dermatological conditions that originate from within the skin. Conventional transdermal drug delivery systems (TDDS) are insufficient in treating the aforementioned skin illness because of the limited absorption of active pharmaceutical ingredients (APIs) into the skin.

Conventional topical dose formulations seem to be unsuccessful in treating these disorders because the skin does not retain them well. Novel transdermal drug delivery systems (TDDS) have been increasingly popular in recent decades due to their high patient satisfaction, safety, efficacy, practicality, and long shelf life. The research findings on novel transdermal drug delivery systems (TDDS) indicate that transitioning from traditional TDDS to novel TDDS using carriers requires thorough investigation and holds promise for treating a range of disorders [9].

Gels, commonly referred to as jellies, are semi-solid systems composed of either a suspension containing minute inorganic particles or organic macromolecules (mostly polymers) dispersed in a significant amount of liquid, resulting in the formation of an unending stiff network structure, as stated by the USP [14]. Rathod and Mehta [15] and Jeganath and Jeevitha [16] have identified various methods for classifying gels. These methods include categorizing gels based on their colloidal phases (inorganic and organic), the type of solvents used (emulgel, hydrogels, organogels, and xerogels), their composition (flexible and inflexible gels), and their rheological properties (plastic, pseudoplastic, and thixotropic gels). Paul et al. [17] state that the new formulation can be incorporated into several types of gels, including hydrogels, organogels, in situ gels, emulgels, microgels, nanogels, and vesicular gels (such as liposomal gel, niosomal gel, and transferosomal gel).

A semi-solid condition is beneficial for effectively applying the required dose to the skin. Pharmaceutical companies are showing a growing interest in hydrogels because of their ability to control the release of drugs [18]. Initially, numerous research studies were conducted on hydrogels by introducing different medications either in their pure form or dissolved in appropriate solvents into the created hydrogel basis. However, over the past five to ten years, a multitude of research articles on innovative hydrogels have been published. Novel hydrogels consist of innovative compositions such as nanoparticles, nanoemulsions, microemulsions, liposomes, self-nano emulsifying drug delivery systems, cubosomes, and other similar formulations. Novel hydrogels are utilized for many applications.

Only a limited number of review articles on innovative hydrogels have been previously published in different publications. Torres et al. [19] conducted a comprehensive examination of hydrogel classification, focusing on distinctions between physical and chemical interactions, as well as the inclusion of stimuli-sensitive compounds. The study conducted by Voci et al. [20] examined the use of Taxol, a formulation containing Paclitaxel, in combination with biocompatible nanocarriers such as nanoparticles, microparticles, micelles, and liposomes. These nanocarriers were inserted into hydrogels made from natural polymers like hyaluronic acid, gellan gum, alginate, and chitosan. Alven and Aderibigbe [21] conducted a review on a hydrogel made from chitosan and cellulose, which is used for treating wounds. Michalik and Wandzik [22] conducted a review on the use of chitosan-based hydrogel in the field of agriculture to achieve sustained activity. Cai et al. [23] conducted a comprehensive analysis and discussion on the most recent discoveries in advanced hybrid-based hydrogel technologies that include nanostructures and microstructures. They examined the formulation of these technologies and explored their prospective applications, with a particular focus on tissue engineering and anticancer delivery systems. The purpose of this review is to emphasize the significant advancements resulting from the use of several new formulations that incorporate the active pharmaceutical ingredient (API) into polymer-based hydrogels derived from different sources. This review article examines prior research on topical new hydrogels formulated using several sources, such as natural and synthetic gelling agents.

2. METHODOLOGY

To obtain the appropriate literature, relevant keywords like hydrogels, topical hydrogels, novel topical hydrogels, etc. have been used in search engines such as Google Scholar, PubMed, etc. The information was mainly collected from research and review articles published between 2010 to 2022.

3. HYDROGELS

Hydrogel refers to a three-dimensional arrangement made of hydrophilic polymers with a high capacity to interact with and retain a vast amount of water and biological fluids due to several functional grouping [like amino (–NH2), the carboxylic acid (–COOH), a hydroxyl group (–OH), amide (–CONH), sulfo groups (–SO3H)] in the polymer chains [24]. As per Lee, Kwon, and Park, the term "hydrogel" comes from an article published in 1894 and the first cross-linked network substance that showed up in publications [25]. In 1960, Wichterle O from Prague and Lim D from the Czech Republic were the first to discover hydrogels. They used synthetic polymer poly-2-hydroxyethyl methacrylate for preparing contact lenses [4,26]. Due to the high-water holding capacity, elastic in nature, compatibility with living tissue, and adaptability made hydrogels as a broad choice of use in various fields [27]. As part of the second stage, commencement in the 1970s, scientists began developing a more complex hydrogel capable of responding to pH and temperature and producing a precise response. Third-stage hydrogels are made from supramolecular inclusion complexes that are biocompatible and versatile. Hydrogels developed in the third stage of development led to creating "smart hydrogels" [28]. Since the 1980s, hydrogels have been used for various biomedical disciplines like contact lenses, absorbent cotton, suture, and cell engineering [29], and biosensor, drug delivery, cell therapy, and 3D cell culture [27] to incorporate various conventional and novel formulations [18,29].

4. CLASSIFICATIONS OF HYDROGELS

Hydrogels can be classified into multiple ways with various viewpoints based on the literature. Based on phase transition like gel-sol reactions due to physical or chemical, or biochemical stimulation. The physical stimulants comprise temperature, electric fields, magnetic fields, solvent composition, light strength, and stress. In contrast, chemical stimulation includes pH, ionic strength, specific chemical compositions, and biochemical stimulation by enzymes and amino acids [24]. Based on their pore size (nanogels and microgels), based on the polymer structure (homopolymers and copolymers), based on cross-linking (physically cross-linked and chemically cross-linked), based on degradation (biodegradable, non-degradable) and their source (natural, synthetic, or hybrid), and based on physical properties (conventional, and smart) [20], the classification of hydrogel is shown in Figure 1.

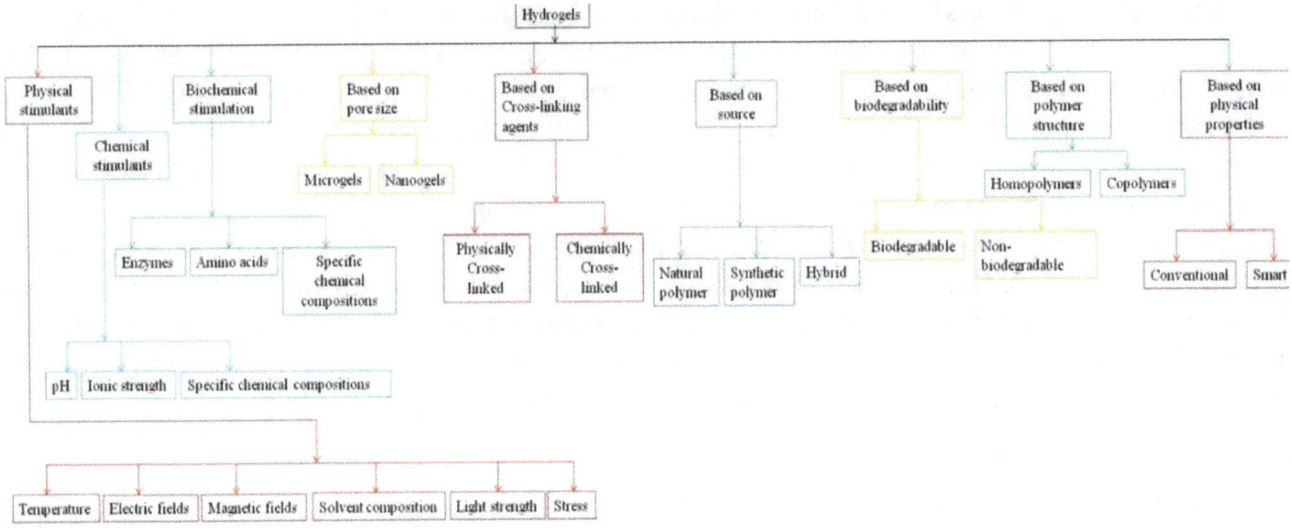

Figure 1. Classification of hydrogel.

Hydrogels provide comfortable drug delivery methods because of their tunable properties, rapid expandable degradation, and attainable preparation. [30]. A hydrogel is an excellent carrier for delivering APIs, especially to the skin. Ex vivo porcine ear skin is a widely used model due to its similarity to the human skin in terms of behavior, composition, and permeability [31].

5. HYDROGELS PREPARED WITH NATURAL POLYMERS

Natural hydrogels were prepared with natural constituents recommended for maximum biocompatibility as they are obtained from natural sources [28]. Novel formulation formulated with natural gelling agents, preparation methods, mixing with the hydrogels, and its use are shown in Table 1. Different novel formulation evaluation studies before incorporating into hydrogel are shown in Table 2. Physical evaluation studies for novel natural hydrogel formulation prepared with natural gelling agent are shown in Table 3. Novel natural hydrogel formulation in vitro, in vivo evaluation studies details are shown in Table 4.

Table 1. Novel formulation formulated with natural gelling agents, preparation methods, mixing with the hydrogels, and its use.

API	Novel Formulation	Novel Formulation Method	Gelling Agent for Making Hydrogels	Novel Formulation and Hydrogel Mixing Method	Use	References
Resveratrol	Liposomes	Film hydration method	Chitosan	Hand stirring method	Vaginal chlamydia infection	[32]
Vitamin C	Self-double-emulsifying drug delivery system	Two-step emulsification method	Xanthan gum	Mixing with a mechanical stirrer	Penetration enhance the skin	[33]
Rosmarinic acid	Nanoemulsions	Spontaneous emulsification method	Hydroxyethyl cellulose	Rosmarinic acid-nanoemulsion added to hydroxyethyl cellulose and stirred for 15 min	New anti-ageing skin products	[34]
Pentyl Gallate	Nanoemulsions	Spontaneous emulsification method	Chitosan	Chitosan added to pentyl gallate nanoemulsions	Increase skin penetration for herpis labialis	[35]
Phenytoin	Nanocapsule	Interfacial deposition	Chitosan	Chitosan was dispersed in the nanoemulsion and nanocapsule	Skin permeation and wound healing activity	[36]
	Nanoemulsion	Spontaneous emulsification method	Chitosan			
Ibuprofen	Microemulsion	NM	Xanthan gum	Mixing	To enhance percutaneous delivery	[37]
Terbinafine hydrochloride	Microemulsion	NM	Chitosan, and Natrosol 250	Gelling agent added to Terbinafine hydrochloride microemulsion	Anti fungal activity	[38]
Curcumin	Microemulsion	NM	Xanthan and galactomannan	Curcumin microemulsion was added to hydrogel preparation	To increase skin penetration and anti-inflammatory activity	[39]
Baicalin	Nanocrystals	Coupling homogenization technology followed by spray-drying technology	Hyaluronic acid	Baicalin nanocrystals was added into hyaluronic acid hydrogel and mixed	To improve skin permeation	[40]
Silver sulfadiazine	Cubosome	Emulsification method	Chitosan	Cubosome incorporated into chitosan	Increasing skin permeation and for treating topical burn	[41]

NM—Not mentioned by the researcher.

Table 2. Different novel formulation evaluation studies before incorporating into hydrogel.

API-Novel Formulation	Droplet/Particle Range	Drug Content/ Entrapment Efficiency (%)	PDI	Zeta Potential (mV)	pH	Viscosity	Shape	Time	DR (%)	References
RVT-LS	100 to 200 nm (Average 158 ± 22 nm)	85 ± 2	0.077	-6.72 ± 2.47	NP	appropriate viscosity (NM)	Spherical	8 h	61	[32]
Vit C-SDEDDS	0.06 to 60.26 μm (Average 17.13 ± 2.50 μm)	NP	NP	NP	NP	NP	NP	NP	NP	[33]
ROS-NE	180.57 ± 1.82 to 224.67 ± 2.31 nm	98.59 ± 2.12 to 107.69 ± 6.28	0.123 ± 0.021 to 0.230 ± 0.036	-39.65 ± 1.53 to 46.17 ± 3.90	3.85 ± 0.07 to 4.73 ± 0.07	1.1 to 1.3 cps	NP	8 h	71.8 ± 1.98	[34]
PG-NE	164.3 ± 7.4 nm	96.2 ± 3.4	0.12 ± 0.03	-48.9 ± 2.1	5.5 ± 0.2	Near to 1 cps	NM	NP	NP	[35]
PENY-NC	161 ± 4 nm	95.2 ± 1.4	0.14 ± 0.01	-15.7 ± 0.3	5.6 ± 0.1	NP	Spherical	3 h 24 h	27 ± 1 67 ± 1	[36]
PENY-NE	125 ± 6 nm	88.7 ± 1.1	0.12 ± 0.01	-10.8 ± 0.4	5.0 ± 0.7	NP	Spherical	3 h 24 h	36 ± 2 77 ± 2	[36]
IBU-ME	14.34 ± 0.98 nm	NP	0.220 ± 0.075	NP	5.23	0.2025 ± 0.003 Pas	NP	12 h	46.78 ± 4.59	[37]
TER-ME	44.98 ± 27.34 nm	NP	NP	NP	NP	77.98 ± 0.75 cp	Spherical	NP	NP	[38]
CUR-ME	231.8 ± 7.6 nm	99.50	NP	NP	NP	NP	NP	NP	NP	[39]
BCN-NCY	189.21 ± 0.36 nm	NP	NP	NP	NP	NP	Spherical Gel TEM showed network structure	4 h	65.3 ± 3.2	[40]
SSD-CUBO	152.3 to 389.6 nm	86.05 ± 3.86 to 94.56 ± 1.40	0.25 ± 0.004 to 0.65 ± 0.45	NP	NP	NP	Cubic	NP	NP	[41]

NP: Not performed by the researchers; NM: Not mentioned by the researchers.

Table 3. Physical evaluation studies for novel hydrogel formulation prepared with natural gelling agent.

API-Novel Formulation	Hydrogel Made of	Hydrogel Concentration	Formulation Concentration	Loading Efficient (%)	Ph	Droplet SIZE (nm)	Zeta Potential	Viscosity	SEM	References
RVT-LS	CHI	2.5% w/w	20% w/w	NP	NP	NP	NP	NP	NP	[32]
Vit-C-SEDDS	XG	2%	5%	NP	5.5 ± 0.1	NP	NP	4.62 ± 0.50	The structure of Vit-C was completely destroyed in freeze-dried hydrogel	[33]
ROS-NE	HEC	1% w/v	0.1% w/v	98.50 ± ٣,٥٩ to ± ١٠٠,٧٩ ١,٩٨	3.83 ± 0.05 to 4.73 ± 0.07	NP	NP	NP	NP	[34]
PG-NE	CHI	2.5% w/w	0.5% w/w	94.4 ± 4.8	5.0 ± 0.3	297.0 ± 8.6	52.6 ± 0.1	NP	NP	[35]
PENY-NCY	CHI	2.75% w/v	0.025% w/v	0.24 ± 0.01 mg/gm	4.8 ± 0.1	NP	NP	24.23 ± 2.70 pasn	NP	[36]
PENY-NE	CHI	2.75% w/v	0.025% w/v	0.25 ± 0.01 mg/gm	4.7 ± 0.2	NP	NP	24.53 ± 3.71 pasn	NP	[36]
IBU-ME	XG	0.25–1%	5%	NP	NP	5.17 ± 0.01	NP	1.12 ± 0.15 to 6.80 ± 0.02	NP	[37]
TER-ME	CHI, NAT, and CAR	CHI-1%, NAT-4%, CAR-1%	1%	NP	3.04 ± 0.02	NP	NP	5044.03 ± 22.43	NP	[38]
CUR-ME	X-GAL	1.25%	NP	103.90	5.3	NP	NP	NP	Network structure	[39]
BCN-NCY	HA	0.5%, 1%, 1.5% and 2%, w/v	1%	NP	NP	NP	NP	NP	Porous structure	[40]
SSD-CUBO	CHI	NP	NP	NP	4	NP	NP	NP	NP	[41]

NP: Not performed by the researchers.

Table 4. Novel natural hydrogel formulation in vitro, in vivo evaluation studies details.

API-Formulation-Hydrogel	In Vitro Release Study		In Vitro Kinetics		In Vivo Skin Studies			Animal Used	References
	Time	Drug Released (%)	Model	Mechanism	Model Skin	Time	DR		
RVT-LS-CHI	8 h	38	NP	NP	NP	NP	NP	NP	[32]
Vit C-SEDDS-XG	6 h	72.33	Weibull model	Fickian diffusion and Case-II transport	Porcine abdominal skin	12 h	12%	NP	[33]
ROS-NE-HEC	8 h	57 ± 0.36	NP	NP	Pig ear skin	8 h	0.65 ± 0.08 μg/cm^2	NP	[34]
PG-NE-CHI	24 h	Not shown	NP	NP	Porcine ear skin	NP	NP	NP	[35]
PENY-NC-CHI	12 h	43 ± 1	NP	NP	Porcine ear skin	12 h	NP	Male Wistar rats	[36]
PENY-NE-CHI	12 h	53 ± 1	NP	NP	Porcine ear skin	12 h	NP	Male Wistar rats	[36]
IBU-ME-XG	12	ME-XG-H1-23%, ME-XG-H2-16%, ME-XG-H3-14%, and ME-XG-4-11%	Zero order	NP	NP	NP	NP	Male Wistar rats	[37]
TER-ME-CHI TER-ME-NAT	7 h	8.70	Zero order	NP	NP	NP	NP	NP	[38]
CUR-ME-X-GAL	10 h	<60	Higuchi	Diffusion controlled	Porcine ear skin	NP	NP	NP	[39]
BCN-NCY-HA	6 h	0.5 and 1% CAR->95% 1.5%CAR-85.4% 2%CAR-72.3%	NP	NP	Mouse abdominal skin	12 h	NM	NP	[40]
SSD-CUBO-CHI	NP	NP	Zero order	Diffusion controlled	NP	NP	NP	NP	[41]

NP: Not performed by the researchers; NM: Not mentioned by the researchers.

5.1. Hydrogels Loaded with Liposomes

Liposomes (LS) are nanocarriers primarily composed of phospholipids and cholesterol [42]. Chlamydia trachomatis is the causative agent of sexually transmitted infections. The current treatment option is oral azithromycin or doxycycline, both of which have potential side effects. To reduce the side effect and effectively treat C. trachomatis, Jraholmen et al. [32] used natural polyphenol Resveratrol (RVT) LS incorporated into a natural CHI hydrogel. To maximize RVT's potential therapeutic activity, they used LS as the primary release medium and CHI hydrogel as a supplemental medium. Since RVT inhibits biofilm formation, LS are preferred nanocarriers for topical therapy since it does not interact with vaginal flora, and chitosan hydrogel effectively obstructs vaginal biofilms. RVT in LS preparation was found to increase RVT solubility, deliver sustained action, and enhance chemical stability, allowing for medical uses. In addition, RVT in LS formulation improves RVT's ability to bind to microbes, producing a more potent antimicrobial action even at a low dosage. The study showed that RVT-LS in CHI hydrogel could inhibit nitric oxide, the chief free radical that causes inflammation. The RVT-LS incorporated into the CHI hydrogel delivery system improved the anti-chlamydial effect of RVT at lower concentrations and highlighted the existence of a delivery system to ensure effective treatment.

5.2. Hydrogels Loaded with Self-Double-Emulsifying Drug Delivery System

Self-double-emulsifying drug delivery system (SDEDDS) is a mix ture of water-soluble surfactants and water-in-oil (w/o) emulsions, which can instantaneously emulsify to water-in-oil-in-water (w/o/w) double emulsions, with water-soluble drugs present in the internal aqueous phase [43]. Vitamin C is one of the most effective antioxidants for preventing skin damage and a whitening agent. Once exposed to oxygen, basic pH, and high temperatures, vitamin C is highly unsteady, rapidly degraded, and discolored. As a result, a promising strategy for resolving these weaknesses is required to enable its clinical implementation. To overcome the above and improve the vitamin C penetration in the skin, Wang Q et al. [33] prepared a vitamin C-loaded SDEDDS, followed by blending it in a xanthan gum (XG) hydrogel. Vitamin C-loaded SDEDDS showed improved physical potency upon incorporation into hydrogels, implying that the shell is more secure to protect vitamin C from degradation, mainly from ionization solution and oxygen exposure. Incorporating vitamin C in SEDDS-based hydrogels increases the vitamin C permeation in the skin due to the bioadhesive property of XG. The oil vesicle of the SDEDDS could even act as a protective coating for vitamin C, allowing for improved vitamin C-controlled release from the SDEDDS formulation. Vitamin C encapsulation in SDEDDS, hydrogels, or even both might greatly increase vitamin C permeability and distribution inside the skin. Overall, vitamin C-incorporated SDEDDS-mixed with XG hydrogels will greatly enhance skin penetration.

5.3. Hydrogels Loaded with Microparticles

Polymeric microparticles (MP) have been extensively researched as a beneficial and novel carrier for the sustained and controlled release of vast drugs. Chronic wound healing treatment usually requires the administration of drugs at regular intervals for a more extended period. Long-term sustained release treatment might decrease the frequency of administration while maintaining drug concentration at the site of a wound. Based on this knowledge, Yasasvini S et al. [44] created simvastatin (SIM) CHI CHI-MP and incorporated them into polyvinyl alcohol (PVA) hydrogels to improve wound healing activity. After seven days, 92% of SIM was released from 5% PVA hydrogel for a dose of 2.5 mg. This SIM release from 5% PVA was correlated with the swelling index. In low dose (2.5 mg), the swelling index value was more when compared to 5 and 10 mg SIM concentration. The in vivo wound healing activity showed SIM released in controlled manner results in continuous wound healing. The combination of APIs in MP formulation incorporated into hydrogels could be best for releasing the APIs in a sustained manner and a successful topical wound healing activity from the above results.

5.4. Hydrogels Loaded with Nanoemulsion

Nanoemulsions (NE) are thermodynamically stable and spherical structures where a thin layer of emulsifying agent stabilizes the oil droplets that contain the drug. NE is getting popular for enhancing the skin permeation of lipophilic drugs [45]. Rosmarinic acid (ROS) is a natural polyphenol. Due to its antioxidant property, it can be used as anti-aging compound. But its use is limited due to poor water solubility, high instability, and poor permeability through biological barriers. To overcome the above drawbacks Marafon et al. [34] created ROS loaded NE for topical use. The pH of the ROS-NE was ranged from 3.80 to 5.80, making the formulation physically stable and suitable for topical application. The permeability studies in porcine ear skin showed that nearly 1.5 fold higher ROS was penetrated when compared to pure ROS. The absence of necrosis in keratinocyte cells indicates the formulation's safety. Their findings showed that ROS as a NE prepared with tween-80 and incorporated in HEC-based gelling agents could be used as an appropriate carrier for skin application in novel anti-aging skincare preparation.

Another example for NE encapsulated hydrogel formulation was prepared with Pentyl Gallate (PG) by Kelmann RG and colleagues [35]. PG-NE was prepared and encapsulated into a hydrogel formulation containing natural polymer CHI and a synthetic polymer Aristoflex AVC (ART). PG-NE-ART's viscosity is less than PG-NE-CHI; this increase in viscosity is due to the degree of deacetylation of CHI. The pH of both the PG-NE-CHI and PG-NE-ART was reduced after 90 days due to the hydrolysis of the triglyceride and phospholipid moieties, which resulted in the release of free fatty acids. The presence of a CHI or an ART did not affect the skin retention of PG from the NE. ART compared to the two gelling agents tested, ART appeared to be more intriguing since it initially did not affect PG-NE's droplet size and zeta potential. The PG-NE-ART also demonstrated improved intrinsic stability, viscosity, and spreadability. According to the authors, PG can be an attractive topical antiherpetic agent.

Phenytoin (PENY) is an antiepileptic drug on prolonged oral administration that results in gingival hyperplasia. Previous studies confirmed that PENY could be used successfully as a wound-healing agent, but none of the studies clearly states the permeation of PENY across the skin. Cardoso et al. [36] prepared PENY-NE and nanocapsule (NC), to which CHI was dispersed and converted into novel hydrogels, and then evaluated PENY skin penetration and wound healing activity. The in vitro release study results showed that PENY diffuses through the NE and NC and then through the hydrogel network to arrive at the release medium in a controlled manner. The higher PENY accumulation in the dermis of injured skin for PENY-NC-CHI and PENY-NE-CHI hydrogels compared to other skin layers suggested that such hydrogels might be helpful to formulations for wound treatment. When PENY-loaded NC and NE hydrogels were compared to the non-encapsulated form, the amount of PENY that reached the dermis and the receptor medium was significantly lower. It indicates that PENY as a wound-healing agent poses a low risk of systemic absorption.

5.5. Hydrogels Loaded with Microemulsion

The formation of microemulsion (ME) is straightforward, comprising an oil phase, a surfactant, a cosurfactant, and aqueous phases, and can easily be prepared through gentle stirring. ME is an encouraging nanocarrier for the topical administration of poorly water-soluble drugs [46]. Ibuprofen (IBU) is a non-steroidal anti-inflammatory poor water-soluble drug. The skin restricted its percutaneous penetration from conventional preparations. Nanocarriers play a vital role in enhancing the IBU permeation in the skin. To increase IBU percutaneous delivery, Djekic L et al. [37] created an o/w ME with IBU and incorporated it into hydrogel made up of XG. ME-XG-hydrogel prepared with the least XG (0.25%) had the most excellent drug release rate and spreadability. Artificially stimulated inflammation in male Wistar rats via intraplantar injection of carrageenan dispersed in saline. The ME-XG-H1 formulation produced approximately 65% antihyperalgesic effect and 74% antiedematous after 3 h, showing a considerable lowering in paw inflammation caused by carrageenan. The hydrogel-thickened ME could offer promise carriers with improved percutaneous delivery to prevent pain and, to a lesser extent, inflammation.

Invasive fungal infections pose a significant and growing risk to human healthiness. Terbinafine hydrochloride (TER) is an anti-fungal agent used in topical and oral forms. Celebi et al. [38] prepared TER-ME in which natural gelling agents such as CHI (1%) and natrosol 250 (4%), as well as synthetic gelling agents such as CAR-974 (1%), are dispersed. The anti-fungal activities of the TER-ME-loaded hydrogels were evaluated against a variety of strains. The anti-fungal activity was determined against three yeast species (Candida albicans, Candida krusei, and Candida parapsilosis), two nondermatophytic fungi (Aspergillus niger and Penicillium), and two dermatophytes (Microsporum spp. and Trichophyton rubrum). TER loaded in CHI hydrogel was the most effective against Candida albicans and Candida krusei. TER loaded in natrosol hydrogel was most effective against Trichophyton rubrum. Based on the findings of their study, they concluded that TER in CHI, CAR-974, natrosol hydrogel, and microemulsion formulations had similar anti-fungal activity as the marketable product (Lamisil).

Curcumin (CUR) administration via percutaneous route could be appropriate for topical and systemic therapeutic applications. Koop et al. [39] created an ME containing the hydrophobic drug CUR and incorporated it into a hydrogel combining xanthan and galactomannan (X-GAL) for topical use. About <60% of CUR was released from CUR-ME-X-GAL. The skin permeation study was carried out on porcine ear skin, and the results revealed that similar amounts of CUR (2.17 to 2.47 μg/mL) were found in the SC, epidermis, and dermis of the skin. In vivo anti-inflammatory study in male Swiss mice showed the inhibition of inflammation of 63.2% for CUR-ME-X-GAL. The researchers hypothesized that combining CUR's bioactivity with X-GAL hydrogels aided tissue restoration.

5.6. Hydrogel Loaded with Nanocrystals

Nanocrystals (NCY) are colloidal carriers of nano-sized particles stabilized in a dispersion medium by least polymeric and/or amphiphilic stabilizing agents. NCY was designed to enhance the absorption and permeation of lipophilic drugs [47]. Baicalin (BCN) cannot be used in topical applications due to its lipophilicity and poor permeation. Wei S et al. [40] attempted to prepare BCN-NCY and incorporate them into a hyaluronic acid (HA) hydrogel to improve BCN topical permeation. The 6-h in vitro release study revealed that more than 95% of BCN was released from BCN-NCY incorporated into 0.5 and 1% HA, while 85.4 and 72.3% of BCN was released from BCN-NCY incorporated into 1.5 and 2% HA, respectively. Over 12 h, the cumulative amount of BCN that permeated rat skin from BCN-NCY-hydrogel with 0.5 and 1% HA was significantly more significant than that of BCN-NCY hydrogel with 1.5 and 2% HA. This is because the viscosity of 0.5 and 1% HA gels was lower than that of 1.5 and 2% HA hydrogels. The 1% HA is relatively advantageous for preparing excellent BCN-NCY hydrogel. According to their study, HA-based NCY-hydrogel can deliver poorly water-soluble drugs topically, safely, and effectively.

5.7. Hydrogel Loaded with Cubosomes

Cubosomes (CUBO) are thermodynamically stable; self-assembled nano-sized liquid crystalline particles prepared by combining a specific ratio of lipids with suitable surfactants, water, and temperature conditions [48,49,50]. Silver sulfadiazine (SSD) helps treat burns, but it has drawbacks such as poor permeation into the burn wound and cytotoxicity. However, permeation can be increased and can reduce

SSD cytotoxicity with the help of nanocarriers. Morsi et al. [41] prepared a CUBO based on silver sulfadiazine (SSD). They incorporated it into a CHI (1.5%), CAR 940 (1%), or a combination of CHI-CAR-based hydrogels to create cubosomal hydrogels or cubogel. The release of SSD from SSD-CUBO formulation is controlled; thereby, the cytotoxic effect of silver is avoided. Formulating SSD-CUBO into cubogels using CHI and CAR-934 showed advantages over the marketed product Dermazin. From the first day of treatment, there was no interference in the healing process as well as it being compatible with a biological fluid. Healing tissues started earlier (day 9) for SSD-cubogel than Dermazin (day 15). SSD cubogel could thus be used very effectively in administering deep second-degree burns, resulting in improved healing outcomes with few adverse effects compared to most formulations on the market.

6. HYDROGELS PREPARED WITH SYNTHETIC POLYMERS

Synthetic hydrogels are prepared with synthetic gelling agents that seem to be more reproducible, have greater flexibility for adjusting their chemical or mechanical qualities, and have more closely controlled structures. However, synthetic hydrogels cannot offer similar biocompatibility to natural hydrogels [28]. Novel formulation formulated with synthetic gelling agents, preparation methods, mixing with the hydrogels, and its use are shown in Table 5. Different novel formulation evaluation studies before incorporating into hydrogel are shown in Table 6. Physical evaluation studies for novel hydrogel formulation prepared with synthetic gelling agent are shown in Table 7. Novel synthetic hydrogel formulation in vitro, in vivo evaluation studies details are shown in Table 8.

Table 5. Novel formulation formulated with synthetic gelling agents, preparation methods, mixing with the hydrogels, and its use.

API	Novel Formulation	Novel Formulation Method	Gelling Agent for Making Hydrogels	Formulation and Hydrogel Mixing Method	Use	References
IR780 iodide and IR792 perchlorate	LP	Thin-film hydration method	Poloxamer 407, and 188	Gelling agents added to novel formulation	Targeted tumor photothermal therapy	[51]
Buparvaquone	SNDDS	NM	Carbopol 940	Carbopol was mixed with novel formulation	Cutaneous leishmaniasis	[52]
Escin and escin β-sitosterol phytosome	PHY	NM	Carbopol 934	Hydrogel added dropwise to the novel formulation	Antihyperalgesic activity	[53]
Pentyl Gallate	NE	Spontaneous emulsification method	Aristoflex AVC	Gelling agent added to novel formulation	Increase skin penetration for herpis labialis	[35]
Simvastatin	MP	Ionic gelation method	Poly vinyl alcohol	Chemical cross linking method	Sustained SIM release and wound healing activity	[44]
5,10,15,20-tetrakis(1-methylpyridinium-4-yl)-porphyrin tetra-iodide	NP	Solvent evaporation method	Carbopol-940	Novel formulation added to CAR-940	Photodynamic applications	[31]
Phenytoin	NLC	Hot homogenization followed by ultrasonication method	Carbomer 934	CAR dispersed in the NLC suspension	Increasing the entrapment efficacy and to sustained release.	[54]
Tenoxicam	ME	NM	Carbopol 940	TEN-ME gelled with CAR-940	Arthritis	[55]
Genistein	NE	Spontaneous emulsification process	Carbopol 940	Hand stirring method	To enhance skin permeation	[56]
Terbinafine hydrochloride	ME	NM	Carbopol 974	Mixing in magnetic stirrer	Anti fungal activity	[38]
Silver sulfadiazine	CUBO	Emulsification method	Carbomer 934	Cubosome incorporated into CAR-934	For improving skin permeation and to treat topical burn	[41]
Resveratrol 3,5,4′-trihydroxy-trans-stilbene	ME	NM	Carbopol 940	CAR dispersed in novel formulation	Treatment of osteoarthritis	[57]
Articaine	NC and SLNP	NM	Aristoflex AVC	NC and SLNP was incorporated into ART hydrogel	In-vitro release studies	[58]
ketoconazole	CUBO	Hot emulsification method	Carbopol 971P	CUBO added to CAR-971P hydrogel and stirred (350 rpm)	In-vitro release and ex vivo penetration studies	[59]
Clobetasol propionate	NS	NM	Carbopol 934	CP-NS incorporated into CAR-934 hydrogel	Anti-psoriatic studies	[60]

NM: Not mentioned by the researchers.

Table 6. Different novel formulation evaluation studies before incorporating into hydrogel.

API-Novel Formulation	Droplet/ Particle Range (nm)	Entrapment Efficiency (%)	PDI	Zeta Potential (mV)	pH	Viscosity	Shape	SEM	DR Time	DR %	Kinetics	References
IR 780-LS	Around 130	NP	0.185	NP	NP	NP	Spherical	Freeze dried formulation- porous sponge-like structures	NP	NP	NP	[51]
IR 792-LS	122	NP	NP	NP	NP	NP	Spherical	NP	NP	NP	NP	
BQ-SEDDS	255 ± 37	NP	0.685 ± 0.085	-13.5 ± 0.2	NP	NP	Spherical	NP	NP	NP	NP	[52]
SIM-MP	between 0.5 µm <10 µm	51 ± 0.7 to 82 ± 0.3	NP	NP	NP	NP	Spherical	Rough	8 h	3.8 ± 1.1 to 9.9 ± 0.4	NP	[44]
TMPTI-NPT	118 ± 5 and 133 ± 2	55.8 ± 1.1 to 92.5 ± 3.5	0.17 ± 0.01 to 0.18 ± 0.03	-21.6 ± 1.0 to 26.7 ± 3.0	NP	NP	Spherical	NP	NP	NP	NP	[31]
PENY-NLC	121.45 ± 2.65 to 258.24 ± 6.59	55.24 ± 1.60 to 88.80 ± 4.13	0.18 ± 0.01 to 0.41 ± 0.02	-15.44 ± 0.87 to -32.26 ± 1.68	5.67 ± 0.02 to 6.49 ± 0.23	NP	Spherical	Smooth surface	48 h	73.47 ± 2.45	Higuchi model	[54]
TEN-ME	106 to 122	99	NP	Near zero	5.5 to 5.7	11,100 to 12,000 cps	Spherical	NP	NP	NP	NP	[55]
GEN-NE	GEN-NE-MCT-240 ± 28	93.00 ± 2.00	<0.25	-37 ± 4	5.8 ± 0.3	1.50 ± 0.10	NP	NP	NP	NP	NP	[56]
	GEN-NE-ODD-247 ± 23	96.00 ± 1.00	<0.25	-36 ± 4	5.9 ± 0.2	1.80 ± 0.07	NP	NP	NP	NP	NP	
SSD-CUBO	150 to 400	74.93 ± 0.903 to 92.10 ± 0.250	0.25 ± 0.004 to 0.65 ± 0.45	-17.61 to 7.41	NP	NP	Cubic	NP	NP	NP	Zero order	[41]
RTTS-ME	17.5 ± 1.4	NP	0.068 ± 0.016	-11.8 ± 0.5	NP	14.2 ± 0.1 mPa s	Spherical	NP	NP	NP	NP	[57]
ART-NC	4455 ± 21	78.10	0068 ± 0005	NP	8.1 ± 1.2	NP	Spherical	Smooth surface	400 min	50	Higuchi model	[58]
ART-SLNP	2499 ± 22	65.70	0113 ± 0008	NP	7.9 ± 0.9	NP	Spherical	Smooth surface	300 min	50	NP	
KETO-CUBO	188.6 ± 5.992 to 381 ± 2.082	15.79 ± 1.23 to 72.22 ± 1.08	0.437 ± 0.032 to 0.918 ± 0.06	NP	NP	NP	Cubic	NP	24 h	67	Korsmeyer–Peppas model	[59]
CP-NS	194.27 ± 49.24 nm	56.33 ± 0.94%	0.498 ± 0.095	-21.83 ± 0.95	NP	NP	Porus and crystalline nature	Freeze dried formulation- porous sponge-like structures	1st h 6th h 24th h	32.39 ± 0.10 55.81 ± 0.60 86.25 ±	Higuchi model	[60]

Table 7. Physical evaluation studies for novel hydrogel formulation prepared with synthetic gelling agent.

API-Novel Formulation	Hydrogel Made of	Hydrogel Concentration	Formulation Concentration	Particle Size	PDI	ZP	Loading Efficient (%)	pH	Viscosity	References
IR 780-LS	Pol-407 and 188	NM	NM	NP	NP	NP	NP	NP	NP	[51]
Escin and escin β-sitosterol PHY	CAR-934	1%	1–5%	NP	NP	NP	NP	4.95 to 6.3	1.0 ± 0.4 to 31.7 ± 0.5 Pas	[53]
BQ-SNDDS	CAR-940	1%	2%	266 ± 99	0.609 ± 0.046	−28.7 ± 1.1	NP	NP	Appropriate for skin application (NM)	[52]
BQ-SNDDS	CAR-940	2%	2%	260 ± 35	0.758 ± 0.072	−34.5 ± 1.2	NP	NP	Appropriate for skin application (NM)	
SIM-MP	PVA	5, 7 and 9% *w/v*	2.5, 5, and 10 mg	NP	NP	NP	NP	NP	NP	[44]
TMPTI-NPT	CAR-940	NM	NM	NP	NP	NP	NP	5.7 to 6.6	NP	[31]
PENY-NLC	CAR-934	1% *w/v*	0.05%	NP	NP	NP	90 to 100	6.88 ± 0.30 and 7.27 ± 0.16	16 to 18 ps	[54]
TEN-ME	CAR-940	NM	NM	NP	NP	NP	NP	NP	NP	[55]
GEN-NE-MCT	CAR-940	0.5%	0.1 (1 mg/gm)	NP	NP	NP	92.00 ± 3.00	7	25–33 cP	[56]
GEN-NE-ODD	CAR-940	0.5%	0.1 (1 mg/gm)	NP	NP	NP	91.00 ± 6.00	7	58–64 cP	
TER-ME	CAR-940	1%	NP	NP	NP	NP	NP	3.04 ± 0.02	5044.03 ± 22.43	[38]
SSD-CUBO	CAR-934	0.5, 1, 1.5 and 2%	NP	NP	NP		76 to 91	8	925 to 982 cps at 10 rpm	[41]
RTTS-ME	CAR-940	1.50%	2%	NP	NP	NP	NP	6.7	171.1 ± 0.3 mPa·s	[57]
ATC-NC	ART	2%	20 mg/gm	463.2 ± 24.7 nm	0.190 ± 0.013	NP	NP	NP	19,554.99 Pa·s^{-1}	[58]
ATC-SLNP	ART	2%	20 mg/gm	315.3 ± 20.1 nm	0.206 ± 0.009	NP	NP	NP	22,090.23 Pa·s^{-1}	
PG-NE	ART	2% *w/w*	0.5% *w/w*	97.3 ± 2.7	5.1 ± 0.2	162.1 ± 1.1	-46.5 ± 1.3	NP	NP	[35]
KETO-CUBO	CAR-971P	1% *w/w*	0.2% *w/w*	NP	NP	NP	96.81 ± 4.50	NP	25,586.67 ± 743.32 at 1.0 rpm	[59]
CP-NS	CAR-934	NP	NP	NP	NP	NP	NP	NP	NP	[60]

NP: Not performed by the researchers; NM: Not mentioned by the researchers.

Table 8. Novel synthetic hydrogel formulation in vitro, in vivo evaluation studies details.

API-Novel Formulation-Hydrogel	In Vitro Release Study				Ex Vivo Skin Studies					Animal Used	References
	Time	Drug Released (%)	Kinetics	Mechanism	Model Skin	Time	DR (%)	Kinetics	Mechanism		
IR 780-LS-POL	12 h	>90	NP	NP	NP	NP	NP	NP	NP	CT-26 cancer bearing mice	[51]
BQ-SNDDS-CAR-940	NP	NP	NP	NP	BALB/c mouse skin	NP	NP	Zero order	Case II drug transport	BALB/c mouse	[52]
Escin and ES-PHY-CAR-934	NP	NP	NP	NP	NP	NP	NP	NP	NP	Wistar rat	[53]
SIM-MP-PVA	7 days	2.5 mg SIM-92% 5 mg SIM-60% 10 mg SIM-36%	NP	NP	NP	NP	NP	NP	NP	Wistar rat	[44]
TMPTI-NPT-CAR-940	4.5 h 24 h	20 40	Korsmeyer–Peppas model	Non-fickian diffusion	Porcine skin	24 h	Not detected	NP	NP	NP	[31]
PENY-NLC-CAR-934	48 h	51.13 ± 1.69	Korsmeyer–Peppas model	Non-fickian diffusion	NP	NP	NP	NP	NP	NP	[54]
TEN-ME-CAR-940	NP	NP	NP	NP	Laca mouse skin	24 h	64–71	NP	NP	Rat Sprague-Dawley	[55]
GEN-NE-MCT-CAR-940	8 h	NP	NP	NP	Porcine skin	In 8 h	100 µg/cm^2	NP	NP	NP	[56]
GEN-NE-ODD-CAR-940	8 h	NP	NP	NP	Porcine skin	In 8 h	150 µg/cm^2	NP	NP	NP	
SSD-CUBO-CAR-934	12 h	76 to 98	NP	NP	NP	NP	NP	NP	NP	Male adult Wister rats	[41]
RTTS-ME-CAR-940	NP	NP	NP	NP	Porcine abdominal skin	NP	NP	NP	NP	Rabbit	[57]
ATC-NC-ART	8 h	NP	Higuchi	Diffusion	NP	NP	NP	NP	NP	NP	[58]
ATC-SLNP-ART	NP	NP	NP	NP	NP	NP	NP	NP	NP	NP	
PG-NE	24	Not shown	NP	NP	Porcine ear skin	NP	NP	NP	NP	NP	[35]
KETO-CUBO-CAR-971P	NP	NP	NP	NP	Goat skin	24 h	92.73	NP	NP	NP	[59]
CP-NS-CAR-934	NP	NP	NP	NP	NP	NP	NP	NP	NP	Male Swiss albino mice	[60]

NP: Not performed by the researchers.

6.1. Hydrogels Loaded with Liposomes

Photothermal therapy transmitted by near-infrared light (NIR) has emerged as an attractive cancer cell treatment method and an alternative to conventional tumor therapy. Chen G et al. [51] prepared IR780 iodide (tumor-targeting photosensitizer) and IR792 perchlorate (tumor non-target photosensitizer) LS loaded in hydrogels for systematically targeted cancer photothermal therapy. Images of photosensitizers against CT-26 tumor behavior in Balb/c mice showed that compared to IR792, IR780 resulted in the most significant tumor accumulation and a strong fluorescence in tumor tissue parts for the patches applied to mice's backs. The IR780-LS in hydrogel could systematically target tumors after topical administration. IR780-LSs applied topically in hydrogel increased lung fluorescence signals, confirming IR780 to be targeted effectively against deep metastases. The anti-cancer effect in CT-26 colon tumor mice showed that IR780-LSs in hydrogel-treated mice showed outstanding anti-cancer activity (which could be attributed to a higher concentration of photosensitizers at the tumor site) are nontoxic under laser irradiation. The hydrogel's biosafety of the IR780-LS was tested in a mouse skin model. The absence of any skin reaction or toxicity during the seven-day course of therapy indicated that IR780/LP in the hydrogel applied topically were secure for the skin. IR780-LS in hydrogel forms a potent combination formulation for topical administration against tumor cells.

6.2. Hydrogels Loaded with Self-Nanoemulsifying Drug Delivery Systems

Self-nanoemulsifying drug delivery systems (SNEDDS) are homogeneous mixtures of API combined with lipids, emulsifiers, and hydrophilic co-solvents/solubilizers, which create nanosize emulsions (generally <50 nm) upon constant mixing in the aqueous phase. SNEDDS have a high capacity to solubilize lipophilic APIs, and they also improve the skin permeation [61]. Cutaneous leishmaniasis (CL) is a significantly ignored tropical skin disease that is the most widespread form of leishmaniasis, evidenced by skin lesions that can lead to ulcers, scars, impairment, and stigma [62]. Liposomal amphotericin B intravenously showed good efficacy, but treating with a topical application is preferred due to it being self-administration and cheaper. For the effective and safe treatment option, CL Lalatsa et al. [52] prepared topical buparvaquone (BQ) SNEDDS-enabled carbopol hydrogels. Treatment with BQ-SNEDDS gel indicates no skin modifications (like inflammatory conditions or skin redness); the epidermis and dermis layers were free of inflammatory cells, with no acanthosis or hyperkeratosis. After seven days, the topical application of BQ-SNEDDS gels reduced the parasite load by 99%, very similar to the intralesional administration of Glucantime. Histology studies confirmed that the Balb/c mice treated with BQ-SNEDDS hydrogels showed a reduction in parasitism and evidence of healing. In conclusion, their findings suggest that nano-enabled BQ hydrogels may offer a safe, non-invasive therapy for CL.

6.3. Hydrogels Loaded with Phytosomes

Phytosomes (PHY) are typically made by combining APIs with phospholipids in a precise molar ratio under controlled conditions. PHY are an advanced lipid-based delivery method with a LS-like structure that can entrap various polyphenolic-based phytoconstituents to improve absorption when administered [63]. Topical escin-containing components have conventionally been used to relieve leg discomfort and heaviness caused by mild vascular circulatory disturbances and reduce bruises. However, it was unclear if escin itself could decrease pain hypersensitivity in the inflammatory process. To confirm the anti-inflammatory effect of escin, Djekic, et al. [53] and colleagues created topical hydrogels containing escin-sitosterol phytosomes (ES) and escin for antihyperalgesic activity. ES and escin-incorporated hydrogels had significant, concentration-dependent antihyperalgesic effects in a rat inflammatory pain model. CAR-934 hydrogels containing 1–2% ES can be recognized as potential topical formulations that are highly significant to hydrogels containing the same concentrations of escin. The skin irritation test on male Wistar rat skin demonstrated that it could use it safely on human skin. The above research results might imply that topical monocomponent hydrogels containing ES and escin could effectively treat inflammation-related pain. ES-incorporated hydrogels were considerably more efficient than those incorporated with escin.

6.4. Hydrogels Loaded with Nanoparticles

Alginate (ALG) hydrogel holds active compounds such as various drugs, signaling molecules, or stem cells with soft flexible gels in ALG rich in M-blocks and firm gels in ALG in rich G-blocks. A few examples of novel approaches based on alginate hydrogels are: Porous 3D hydrogel calcium alginate (Ca-ALG) has great swelling capacity in wounds, providing slow drug release. It is used to entrap cells for tissue regeneration and engineering, as physical support for cells or tissue, or as a hurdle between two media. It protects the cells from the host's immune system until it reaches the targeted area. The encapsulated fibroblasts represent an excellent example of a dual-layered structure made from alginate hydrogel with apical keratinocytes. Hydrogel film based on poly (N-vinyl caprolactam)-calcium alginate (PVCL/PV-Ca-ALG) loaded with thrombin receptor agonist peptide has shown a beneficial effect on wound healing and tissue regeneration. A relatively recent study compared a sodium alginate-acacia gum-based hydrogel loaded with zinc oxide nanoparticles (ZnO-NPs) to only ZnO-NPs by their healing effects and activity against B. cereus and P. aeruginosa [64].

Polymeric nanoparticles (NPT) are colloidal particles that are sub-micron (1 to 1000 nm) in size and contain APIs entrapped inside or adsorbed to the polymer [65]. The 5,10,15,20-tetrakis(1-methylpyridinium-4-yl)-porphyrin tetra-iodide (TMPTI) is a well-known porphyrin that is broadly used for the inactivation of different types of microorganisms. TMPTI, on the other hand, is not widely used due to accumulation in healthy cells. Nanoparticles can rectify it; by keeping this concept, Gonzalez et al. [31], with his research team, synthesized TMPTI-NPT by a solvent evaporation method encapsulated into CAR hydrogel. A skin permeability study in domestic porcine skin showed a better TMPTI permeated deeper skin levels with no surrounding damages. It shows that the TMPTI-NPT encapsulated in hydrogel formulation does not affect the normal cells. These findings indicate that encapsulating topical pharmaceutical carriers like hydrogels may successfully treat topical skin diseases, including skin cancer.

Solid lipid nanoparticles (SLNP) are made up of solid lipid compounds, incorporated with an API in them, and coated with a surfactant to stabilize their structure [66]. Articaine (ATC) is a local anesthetic drug, but continuous usage leads to prolonged and permanent paresthesia. Melo et al. [58] aimed to create NC and SLNP containing ART, which was used topically by incorporating ART hydrogels attained a sustained release of ATC from the NC and SLNP formulation. The encapsulation of ATC in NC and SLNP results in higher cellular viability values. The formulation containing ATC-NC showed the best permeation characteristics compared to SLNP. This study opens up possibilities for the future use of nanostructured delivery systems for ATC local anesthetics.

6.5. Hydrogels Loaded with Nanostructured Lipid Carrier

Nanostructured lipid carrier (NLC) are second-generation lipid nanomaterials [67] composed of a lipid component with solid and liquid lipids distributed in an aqueous emulsifier solution [68]. PENY is an antiepileptic medication; its noticeable stimulatory effect on connective tissue development suggests that it could be used topically as a wound-healing agent. However, it is not widely used due to its low solubility, bioavailability, and ineffective distribution during topical application. Motawea et al. [54] made an effort to sustain the release of PENY by making it into nanostructured lipid carriers (NLCs) and incorporating it into a hydrogel made up of CAR-934. The intermolecular hydrogen bond interaction between the PENY-NLC components evidenced the PENY entrapped inside the lipid matrix. After 48 h, the release of PENY from the hydrogel preparations was slower than their NLC. The obtained reduction in drug release results could be believed to be due to the polymer network of CAR-934. According to their results PENY-NLCs loaded in CAR-934 hydrogels could be novel, cost-efficient, and economically feasible carriers with significant potential in topical administration.

6.6. Hydrogels Loaded with Microemulsion

Tenoxicam (TEN) is an NSAID that is used to treat arthritis. However, long-term oral administration results in peptic ulcers. To overcome this adverse effect, Goindi S et al. [55] formulated a novel TEN-ME and incorporated it into a CAR-940 hydrogel for treating arthritis. The TEN-ME formulations produce higher levels of anti-inflammatory activity better than conventional topical dosage forms with effectiveness comparable to an oral formulation. Their research concluded that TEN-ME formulations could be a viable alternatives to the TEN oral formulations.

Res, 3,5,4′-trihydroxy-trans-stilbene (RTTS) is an effective treatment for osteoarthritis. Long-term oral administration of RTTS causes kidney damage. Because osteoarthritis is a topical disease, topical therapeutic delivery is much preferred. Hu et al. [57] developed RTTS-ME and incorporated it into CAR-940 hydrogel to treat osteoarthritis in a rabbit. Due to the decrease in particle size, the penetration of RTTS was more from ME and hydrogel formulations. It appears that papain-induced osteoarthritis in animals mimicked human osteoarthritis, as papain affects cartilage, but it did not interfere with the cartilage's repair mechanism. Topical delivery of RTTS-ME incorporated in hydrogel noticeably relieved osteoarthritis signs by decreasing pro-inflammatory cytokines and improved macroscopic cartilage restoration. As a result, the hydrogel preparation could be an effective carrier for the topical administration of RTTS-ME to treat osteoarthritis.

6.7. Hydrogels Loaded with Nanoemulsion

Genistein (GEN) is an isoflavone that has recently received a lot of attention because of its effects on avoiding skin carcinoma and dermal aging on exposure to ultraviolet light. Due to its poor aqueous solubility, it cannot incorporate it into a topically applied form. To improve the solubility and permeability, Vargas et al. [56] created topical hydrogels encapsulated in GEN nanoemulsions. CAR-940-based hydrogel containing GEN-NE penetrates the skin in a sustained manner. They concluded that compared to octyldodecanol, there was detected a higher amount of GEN in the skin from the formulation composed of medium chain triglycerides (MCT) as the oily core. It is a promising delivery system for the skin.

6.8. Hydrogel Loaded with Cubosomes

The higher concentration of surfactant affects the size of the particle, entrapment efficiency of API, release from the formulation, and causes an adverse or toxic effect to the body. To overcome the above-mentioned drawbacks, Rapalli et al. [59] used the 'Quality by Design (QbD) method to create a topical hydrogel comprising ketoconazole (KETO)-entrapped cubosomes with lesser surfactant concentrations. They were successful in their research, using the QbD method, by formulating the KETO-CUBO with a lower amount of surfactant than those reported in previous works of literature.

6.9. Hydrogel Loaded with Nanosponge

Nanosponge (NS) is an innovative formulation consisting of a nanoporous carrier composed of hyper cross-linking polymeric materials. This carrier has a sponge-like network structure, forming three-dimensional covalent structures. It is designed to include nanoparticles in a non-collapsible and porous formation [69,70]. Clobetasol propionate (CP) is a strong corticosteroid that is applied topically and has a great ability to treat psoriasis. However, the use of this treatment topically is limited due to typical negative effects, including skin deterioration, steroidal acne, skin pigmentation, and allergic skin reactions. Kumar et al. [60] developed a CP-NS using the information gathered from these observations. The CAR-934 hydrogel was utilized to integrate it, aiming to mitigate the negative consequences and regulate the release of CP. The researchers effectively synthesized cyclodextrin-based CP-NS, resulting in the release of around 86% of CP within 24 hours. The appropriateness of topical administration was boosted by incorporating CP-NS with CAR-934 hydrogel. The efficacy of the developed nanoformulation in treating psoriasis was further confirmed through in vivo experiments utilizing a mouse tail model. The histological and biochemical studies demonstrated significant anti-psoriatic action of the nanogel that was synthesized.

7. CONCISE SUMMARY

The primary objective of the formulations is to efficiently deliver the active pharmaceutical ingredients (APIs) while ensuring they are non-toxic and have a prolonged duration of action. The innovative formulations offer advantages compared to existing formulations, including enhanced solubility, increased bioavailability, protection against toxicity, improved pharmacological action, enhanced stability, better dispersion in tissue macrophages, sustained delivery, and protection against physiochemical deterioration. The integration of various APIs into innovative formulations such as nanoparticles, nanocapsules, liposomes, phytosomes, nanoemulsions, microemulsions, cubosomes, SLNP, NLC, nanosponge, and other novel formulations achieves the aforementioned advantages. Typically, topical treatments are used for common skin illnesses to ensure that patients follow the treatment plan, while inflicting minimal negative effects on the body as a whole. The objective of dermatologists, patients, and pharmaceutical corporations is to treat skin problems using a straightforward semi-solid dosage form. The primary obstacle in accomplishing this objective is obtaining sufficient medication penetration, while minimizing adverse effects. There are many challenges involved in treating the skin topically. Consequently, the process of delivering drugs through the skin is exceedingly intricate. Comprehensive examination of the physical and chemical characteristics, as well as the methods of distribution, is necessary to assess and enhance the properties of topical preparations. For studying drug penetration and evaluating the clinical efficiency and effectiveness of new topical drug delivery systems, in vitro, ex vivo, and in vivo animal investigations are often the most reliable approach. Hydrogels have been widely employed in topical applications due to their exceptional biocompatibility, water solubility, and three-dimensional pore structure that aligns with the extracellular matrix. Research on hydrogels, mostly using natural and synthetic polymers, has recently garnered attention due to their exceptional physicochemical features that could be beneficial in the treatment of many skin ailments. This discovery has the potential to enable the targeted delivery of APIs through topical applications in order to enhance their effectiveness. Research on new hydrogels has significantly increased in the past decade. Novel hydrogels effectively increased the permeation of APIs, reducing the chances of percutaneous absorption. These new hydrogels exhibit strong pharmacological activity against several skin disorders. The integration of innovative formulations with hydrogels is a promising opportunity to address several skin ailments that are currently prevalent. During this analysis, it was observed that numerous researchers developed innovative formulations, while only a few conducted comprehensive studies on the characteristics of these formulations, such as particle size, shape, entrapment efficacy, zeta potential, viscosity, pH, and in-vitro release studies. Subsequently, the new formulation was predominantly combined with hydrogel preparations. The team conducted evaluation studies on the new hydrogel formulation. Animal tests have shown that several new hydrogels have proven to be highly effective against a range of topical ailments. Additionally, stability experiments have confirmed the durability of this particular hydrogel. Novel hydrogels, derived from both natural and synthetic polymers, are likely to undergo clinical studies and become available on the market in the near future.

AT A GLANCE

Active pharmaceutical ingredients (API) or medications are typically not administered in their pure chemical form for the purpose of preventing or treating diseases. APIs are typically delivered in prepared formulations, which are also referred to as dosage forms. The use of topical administration is prevalent for localized delivery of therapeutic substances due to its convenience and cost-effectiveness. Throughout ancient civilizations, a variety of topical semi-solid dose forms have been widely employed in healthcare to effectively treat a range of skin ailments. A topical medication delivery system is primarily meant to treat localized disorders by administering therapeutic substances to superficial areas of the body, such as the skin, eyes, nose, and vaginal cavity. Currently, innovative semi-solid substances can be safely administered to pediatric, geriatric, and pregnant patients without the risk of triggering any allergic reactions. Hydrogels, a novel type of material, are currently being utilized in a diverse array of applications. Initially, multiple hydrogel research studies were conducted by including different APIs in their pure state or dissolved in various solvents into the pre-made hydrogel basis. Several scholarly articles on innovative hydrogels have been published in the past five to ten years. Novel hydrogels are anticipated to have the ability to regulate the release pattern of APIs. Novel hydrogels consist of innovative compositions, including nanoparticles, nanoemulsions, microemulsions, liposomes, self-nano emulsifying drug delivery systems, cubosomes, and similar formulations. This review specifically examines the innovative formulas used in the hydrogel, which is made from both natural and synthetic polymers.

REFERENCES

1. Fakhree, M.; Ahmadian, S. Pharmaceutical dosage forms: Past, present, future. In Drug Delivery; Nova Science Publishers, Inc.: Hauppauge, NY, USA, 2011; pp. 51–89.
2. Afrin, S.; Gupta, V. Pharmaceutical Formulation; StatPearls Publishing: Treasure Island, FL, USA, 2020.
3. Noordin, M.I. Advance delivery system dosage form for analgesic, their rationale, and specialty. In Pain Relief-From Analgesics to Alternative Therapies; IntechOpen: London, UK, 2017.
4. Singh Malik, D.; Mital, N.; Kaur, G. Topical drug delivery systems: A patent review. Expert Opin. Ther. Pat. 2016, 26, 213–228.
5. Sharma, S.; Singh, S. Dermatological preparations, formulation and evaluation of various semi-solid dosage form. Asian J. Pharm. Res. Dev. 2014, 2, 10–25.
6. Panwar, A.; Upadhyay, N.; Bairagi, M.; Gujar, S.; Darwhekar, G.; Jain, D. Emulgel: A review. Asian J. Pharm. Life Sci. 2011, 2231, 4423.
7. Ilić, T.; Pantelić, I.; Savić, S. The implications of regulatory framework for topical semisolid drug products: From critical quality and performance attributes towards establishing bioequivalence. Pharmaceutics 2021, 13, 710.
8. Bora, A.; Deshmukh, S.; Swain, K. Recent advances in semisolid dosage form. Int. J. Pharm. Sci. Res. 2014, 5, 3594–3608.

9. Sharadha, M.; Gowda, D.; Gupta, V.; Akhila, A. An overview on topical drug delivery system–updated review. Int. J. Res. Pharm. Sci 2020, 11, 368–385.

10. Yu, Y.-Q.; Yang, X.; Wu, X.-F.; Fan, Y.-B. Enhancing Permeation of Drug Molecules Across the Skin via Delivery in Nanocarriers: Novel Strategies for Effective Transdermal Applications. Front. Bioeng. Biotechnol. 2021, 9, 646554.

11. Nastiti, C.M.; Ponto, T.; Abd, E.; Grice, J.E.; Benson, H.A.; Roberts, M.S. Topical nano and microemulsions for skin delivery. Pharmaceutics 2017, 9, 37.

12. Choi, H.; Kwon, M.; Choi, H.E.; Hahn, S.K.; Kim, K.S. Non-Invasive Topical Drug-Delivery System Using Hyaluronate Nanogels Crosslinked via Click Chemistry. Materials 2021, 14, 1504.

13. Bharat, P.; Paresh, M.; Sharma, R.; Tekade, B.; Thakre, V.; Patil, V. A review: Novel advances in semisolid dosage forms & patented technology in semisolid dosage forms. Int. J. PharmTech Res. 2011, 3, 420–430.

14. Badola, A.; Goyal, M.; Baluni, S. Gels And Jellies A Recent Technology In Semisolids: A. World J. Pharm. Res. 2021, 10, 461–475.

15. Rathod, H.J.; Mehta, D.P. A review on pharmaceutical gel. Int. J. Pharm. Sci. 2015, 1, 33–47.

16. Jeganath, S.; Jeevitha, E. Pharmaceutical Gels and Recent Trends-A Review. Res. J. Pharm. Technol. 2019, 12, 6181–6186.

17. Paul, S.D.; Sharma, H.; Jeswani, G.; Jha, A.K. Novel gels: Implications for drug delivery. In Nanostructures for Drug Delivery; Elsevier: Amsterdam, The Netherlands, 2017; pp. 379–412.

18. Zielińska, A.; Eder, P.; Rannier, L.; Cardoso, J.C.; Severino, P.; Silva, A.M.; Souto, E.B. Hydrogels for modified-release drug delivery systems. Curr. Pharm. Des. 2021.

19. Bustamante-Torres, M.; Romero-Fierro, D.; Arcentales-Vera, B.; Palomino, K.; Magaña, H.; Bucio, E. Hydrogels Classification According to the Physical or Chemical Interactions and as Stimuli-Sensitive Materials. Gels 2021, 7, 182.

20. Voci, S.; Gagliardi, A.; Molinaro, R.; Fresta, M.; Cosco, D. Recent Advances of Taxol-Loaded Biocompatible Nanocarriers Embedded in Natural Polymer-Based Hydrogels. Gels 2021, 7, 33.

21. Alven, S.; Aderibigbe, B.A. Chitosan and Cellulose-Based Hydrogels for Wound Management. Int. J. Mol. Sci. 2020, 21, 9656.

22. Michalik, R.; Wandzik, I. A mini-review on chitosan-based hydrogels with potential for sustainable agricultural applications. Polymers 2020, 12, 2425.

23. Cai, M.-H.; Chen, X.-Y.; Fu, L.-Q.; Du, W.-L.; Yang, X.; Mou, X.-Z.; Hu, P.-Y. Design and development of hybrid hydrogels for biomedical applications: Recent trends in anticancer drug delivery and tissue engineering. Front. Bioeng. Biotechnol. 2021, 9.

24. Bahram, M.; Mohseni, N.; Moghtader, M. An introduction to hydrogels and some recent applications. In Emerging Concepts in Analysis and Applications of Hydrogels; IntechOpen: London, UK, 2016.

25. Russo, E.; Villa, C. Poloxamer hydrogels for biomedical applications. Pharmaceutics 2019, 11, 671.

26. Patel, K.D.; Silva, L.B.; Park, Y.; Shakouri, T.; Keskin-Erdogan, Z.; Sawadkar, P.; Cho, K.J.; Knowles, J.C.; Chau, D.Y.; Kim, H.-W. Recent advances in drug delivery systems for glaucoma treatment. Mater. Today Nano 2022, 100178.

27. Veloso, S.R.; Andrade, R.G.; Castanheira, E.M. Review on the advancements of magnetic gels: Towards multifunctional magnetic liposome-hydrogel composites for biomedical applications. Adv. Colloid Interface Sci. 2021, 288, 102351.
28. Harrison, I.P.; Spada, F. Hydrogels for atopic dermatitis and wound management: A superior drug delivery vehicle. Pharmaceutics 2018, 10, 71.
29. Singh, G.; Lohani, A.; Bhattacharya, S.S. Hydrogel as a novel drug delivery system: A review. J. Fundam. Pharm. Res 2014, 2, 35–48.
30. Gallo, E.; Diaferia, C.; Rosa, E.; Smaldone, G.; Morelli, G.; Accardo, A. Peptide-Based Hydrogels and Nanogels for Delivery of Doxorubicin. Int. J. Nanomed. 2021, 16, 1617.
31. Gonzalez-Delgado, J.A.; Castro, P.M.; Machado, A.; Araujo, F.; Rodrigues, F.; Korsak, B.; Ferreira, M.; Tome, J.P.; Sarmento, B. Hydrogels containing porphyrin-loaded nanoparticles for topical photodynamic applications. Int. J. Pharm. 2016, 510, 221–231.
32. Jøraholmen, M.W.; Johannessen, M.; Gravningen, K.; Puolakkainen, M.; Acharya, G.; Basnet, P.; Škalko-Basnet, N. Liposomes-In-Hydrogel Delivery System Enhances the Potential of Resveratrol in Combating Vaginal Chlamydia Infection. Pharmaceutics 2020, 12, 1203.
33. Wang, Q.; Zhang, H.; Huang, J.; Xia, N.; Li, T.; Xia, Q. Self-double-emulsifying drug delivery system incorporated in natural hydrogels: A new way for topical application of vitamin C. J. Microencapsul. 2018, 35, 90–101.
34. Marafon, P.; Fachel, F.N.S.; Dal Prá, M.; Bassani, V.L.; Koester, L.S.; Henriques, A.T.; Braganhol, E.; Teixeira, H.F. Development, physico-chemical characterization and in-vitro studies of hydrogels containing rosmarinic acid-loaded nanoemulsion for topical application. J. Pharm. Pharmacol. 2019, 71, 1199–1208.
35. Kelmann, R.G.; Colombo, M.; Nunes, R.J.; Simões, C.M.; Koester, L.S. Nanoemulsion-loaded hydrogels for topical administration of pentyl gallate. Aaps Pharmscitech 2018, 19, 2672–2678.
36. Cardoso, A.M.; de Oliveira, E.G.; Coradini, K.; Bruinsmann, F.A.; Aguirre, T.; Lorenzoni, R.; Barcelos, R.C.S.; Roversi, K.; Rossato, D.R.; Pohlmann, A.R. Chitosan hydrogels containing nanoencapsulated phenytoin for cutaneous use: Skin permeation/penetration and efficacy in wound healing. Mater. Sci. Eng. C 2019, 96, 205–217.
37. Djekic, L.; Martinovic, M.; Stepanović-Petrović, R.; Micov, A.; Tomić, M.; Primorac, M. Formulation of hydrogel-thickened nonionic microemulsions with enhanced percutaneous delivery of ibuprofen assessed in vivo in rats. Eur. J. Pharm. Sci. 2016, 92, 255–265.
38. Çelebi, N.; Ermiş, S.; Özkan, S. Development of topical hydrogels of terbinafine hydrochloride and evaluation of their antifungal activity. Drug Dev. Ind. Pharm. 2015, 41, 631–639.
39. Koop, H.S.; de Freitas, R.A.; de Souza, M.M.; Savi-Jr, R.; Silveira, J.L.M. Topical curcumin-loaded hydrogels obtained using galactomannan from Schizolobium parahybae and xanthan. Carbohydr. Polym. 2015, 116, 229–236.
40. Wei, S.; Xie, J.; Luo, Y.; Ma, Y.; Tang, S.; Yue, P.; Yang, M. Hyaluronic acid based nanocrystals hydrogels for enhanced topical delivery of drug: A case study. Carbohydr. Polym. 2018, 202, 64–71.
41. Morsi, N.M.; Abdelbary, G.A.; Ahmed, M.A. Silver sulfadiazine based cubosome hydrogels for topical treatment of burns: Development and in vitro/in vivo characterization. Eur. J. Pharm. Biopharm. 2014, 86, 178–189.

42. Furlani, F.; Rossi, A.; Grimaudo, M.A.; Bassi, G.; Giusto, E.; Molinari, F.; Lista, F.; Montesi, M.; Panseri, S. Controlled Liposome Delivery from Chitosan-Based Thermosensitive Hydrogel for Regenerative Medicine. Int. J. Mol. Sci. 2022, 23, 894.

43. Bhattacharjee, A.; Verma, S.; Verma, P.R.P.; Singh, S.K.; Chakraborty, A. Fabrication of liquid and solid self-double emulsifying drug delivery system of atenolol by response surface methodology. J. Drug Deliv. Sci. Technol. 2017, 41, 45–57.

44. Yasasvini, S.; Anusa, R.; VedhaHari, B.; Prabhu, P.; RamyaDevi, D. Topical hydrogel matrix loaded with Simvastatin microparticles for enhanced wound healing activity. Mater. Sci. Eng. C 2017, 72, 160–167.

45. Rashid, S.A.; Bashir, S.; Naseem, F.; Farid, A.; Rather, I.A.; Hakeem, K.R. Olive Oil Based Methotrexate Loaded Topical Nanoemulsion Gel for the Treatment of Imiquimod Induced Psoriasis-like Skin Inflammation in an Animal Model. Biology 2021, 10, 1121.

46. Hung, W.-H.; Chen, P.-K.; Fang, C.-W.; Lin, Y.-C.; Wu, P.-C. Preparation and Evaluation of Azelaic Acid Topical Microemulsion Formulation: In Vitro and In Vivo Study. Pharmaceutics 2021, 13, 410.

47. Im, S.H.; Jung, H.T.; Ho, M.J.; Lee, J.E.; Kim, H.T.; Kim, D.Y.; Lee, H.C.; Choi, Y.S.; Kang, M.J. Montelukast nanocrystals for transdermal delivery with improved chemical stability. Pharmaceutics 2020, 12, 18.

48. Khan, S.; Jain, P.; Jain, S.; Jain, R.; Bhargava, S.; Jain, A. Topical delivery of erythromycin through cubosomes for acne. Pharm. Nanotechnol. 2018, 6, 38–47.

49. Sharma, P.; Dhawan, S.; Nanda, S. Cubosome: A Potential Liquid Crystalline Carrier System. Curr. Pharm. Des. 2020, 26, 3300–3316.

50. Kaur, S.D.; Singh, G.; Singh, G.; Singhal, K.; Kant, S.; Bedi, N. Cubosomes as Potential Nanocarrier for Drug Delivery: A Comprehensive Review. J. Pharm. Res. Int. 2021, 33, 118–135.

51. Chen, G.; Ullah, A.; Xu, G.; Xu, Z.; Wang, F.; Liu, T.; Su, Y.; Zhang, T.; Wang, K. Topically applied liposome-in-hydrogels for systematically targeted tumor photothermal therapy. Drug Deliv. 2021, 28, 1923–1931.

52. Lalatsa, A.; Statts, L.; de Jesus, J.A.; Adewusi, O.; Dea-Ayuela, M.A.; Bolas-Fernandez, F.; Laurenti, M.D.; Passero, L.F.D.; Serrano, D.R. Topical buparvaquone nano-enabled hydrogels for cutaneous leishmaniasis. Int. J. Pharm. 2020, 588, 119734.

53. Djekic, L.; Čalija, B.; Micov, A.; Tomić, M.; Stepanović-Petrović, R. Topical hydrogels with escin β-sitosterol phytosome and escin: Formulation development and in vivo assessment of antihyperalgesic activity. Drug Dev. Res. 2019, 80, 921–932.

54. Motawea, A.; Borg, T.; Abd El-Gawad, A.E.-G.H. Topical phenytoin nanostructured lipid carriers: Design and development. Drug Dev. Ind. Pharm. 2018, 44, 144–157.

55. Goindi, S.; Narula, M.; Kalra, A. Microemulsion-based topical hydrogels of tenoxicam for treatment of arthritis. Aaps Pharmscitech 2016, 17, 597–606.

56. De Vargas, B.A.; Bidone, J.; Oliveira, L.K.; Koester, L.S.; Bassani, V.L.; Teixeira, H.F. Development of topical hydrogels containing genistein-loaded nanoemulsions. J. Biomed. Nanotechnol. 2012, 8, 330–336.

57. Hu, X.-B.; Kang, R.-R.; Tang, T.-T.; Li, Y.-J.; Wu, J.-Y.; Wang, J.-M.; Liu, X.-Y.; Xiang, D.-X. Topical delivery of 3, 5, 4′-trimethoxy-trans-stilbene-loaded microemulsion-based hydrogel for the treatment of osteoarthritis in a rabbit model. Drug Deliv. Transl. Res. 2019, 9, 357–365.

58. Melo, N.F.S.d.; Campos, E.V.R.; Franz-Montan, M.; Paula, E.D.; Silva, C.M.G.d.; Maruyama, C.R.; Stigliani, T.P.; Lima, R.D.; Araújo, D.R.d.; Fraceto, L.F. Characterization of articaine-loaded poly (ε-caprolactone) nanocapsules and solid lipid nanoparticles in hydrogels for topical formulations. J. Nanosci. Nanotechnol. 2018, 18, 4428–4438.

59. Rapalli, V.K.; Banerjee, S.; Khan, S.; Jha, P.N.; Gupta, G.; Dua, K.; Hasnain, M.S.; Nayak, A.K.; Dubey, S.K.; Singhvi, G. QbD-driven formulation development and evaluation of topical hydrogel containing ketoconazole loaded cubosomes. Mater. Sci. Eng. C 2021, 119, 111548.

60. Kumar, S.; Prasad, M.; Rao, R. Topical delivery of clobetasol propionate loaded nanosponge hydrogel for effective treatment of psoriasis: Formulation, physicochemical characterization, antipsoriatic potential and biochemical estimation. Mater. Sci. Eng. C 2021, 119, 111605.

61. Ponto, T.; Latter, G.; Luna, G.; Leite-Silva, V.R.; Wright, A.; Benson, H.A. Novel Self-Nano-Emulsifying Drug Delivery Systems Containing Astaxanthin for Topical Skin Delivery. Pharmaceutics 2021, 13, 649.

62. Akuffo, R.; Sanchez, C.; Chicharro, C.; Carrillo, E.; Attram, N.; Mosore, M.-T.; Yeboah, C.; Kotey, N.K.; Boakye, D.; Ruiz-Postigo, J.-A. Detection of cutaneous leishmaniasis in three communities of Oti Region, Ghana. PLoS Negl. Trop. Dis. 2021, 15, e0009416.

63. Alharbi, W.S.; Almughem, F.A.; Almehmady, A.M.; Jarallah, S.J.; Alsharif, W.K.; Alzahrani, N.M.; Alshehri, A.A. Phytosomes as an emerging nanotechnology platform for the topical delivery of bioactive phytochemicals. Pharmaceutics 2021, 13, 1475.

64. Barbu, A.; Neamtu, B.; Zăhan, M.; Iancu, G.M.; Bacila, C.; Mireșan, V. Current trends in advanced alginate-based wound dressings for chronic wounds. J. Pers. Med. 2021, 11, 890.

65. Prabha, A.S.; Dorothy, R.; Jancirani, S.; Rajendran, S.; Singh, G.; Kumaran, S.S. Recent advances in the study of toxicity of polymer-based nanomaterials. In Nanotoxicity; Elsevier: Amsterdam, The Netherlands, 2020; pp. 143–165.

66. Musielak, E.; Feliczak-Guzik, A.; Nowak, I. Synthesis and Potential Applications of Lipid Nanoparticles in Medicine. Materials 2022, 15, 682.

67. Gundogdu, E.; Demir, E.-S.; Ekinci, M.; Ozgenc, E.; Ilem-Ozdemir, D.; Senyigit, Z.; Gonzalez-Alvarez, I.; Bermejo, M. An Innovative Formulation Based on Nanostructured Lipid Carriers for Imatinib Delivery: Pre-Formulation, Cellular Uptake and Cytotoxicity Studies. Nanomaterials 2022, 12, 250.

68. Varrica, C.; Carvalheiro, M.; Faria-Silva, C.; Eleutério, C.; Sandri, G.; Simões, S. Topical Allopurinol-Loaded Nanostructured Lipid Carriers: A Novel Approach for Wound Healing Management. Bioengineering 2021, 8, 192.

69. Ghose, A.; Nabi, B.; Rehman, S.; Md, S.; Alhakamy, N.A.; Ahmad, O.A.; Baboota, S.; Ali, J. Development and Evaluation of Polymeric Nanosponge Hydrogel for Terbinafine Hydrochloride: Statistical Optimization, In Vitro and In Vivo Studies. Polymers 2020, 12, 2903.

70. Mashaqbeh, H.; Obaidat, R.; Al-Shar'i, N. Evaluation and Characterization of Curcumin-β-Cyclodextrin and Cyclodextrin-Based Nanosponge Inclusion Complexation. Polymers 2021, 13, 4073.

INDEX

S

U

V

W

X

Z

www.ingramcontent.com/pod-product-compliance
Lightning Source LLC
LaVergne TN
LVHW080338200526
839322LV00005B/763